Linear Isentropic Oscillations of Stars

Preface

This monograph attempts to provide a systematic and consistent survey of the fundamentals of the theory of free, linear, isentropic oscillations in spherically symmetric, gaseous equilibrium stars, whose structure is affected neither by axial rotation, nor by the tidal action of a companion, nor by a magnetic field.

Three parts can be distinguished. The first part, consisting of Chaps. 1–8, covers the basic concepts and equations, the distinction between spheroidal and toroidal normal modes, the solution of Poisson's differential equation for the perturbation of the gravitational potential, and Hamilton's variational principle. The second part, consisting of Chaps. 9–13, is devoted to the possible existence of waves propagating in the radial direction, the origin and classification of normal modes, the completeness of the normal modes, and the relation between the local stability with respect to convection and the global stability of a star. In the third part, Chaps. 14–18 contain asymptotic representations of normal modes. Chapter 19 deals with slow period changes in rapidly evolving pulsating stars.

The theory is developed within the framework of the Newtonian theory of gravitation and the hydrodynamics of compressible fluids. It is described in its present status, with inclusion of open questions.

We give preference to the use of the adjective "isentropic" above that of the adjective "adiabatic", since, from a thermodynamic point of view, these stellar oscillations are described as reversible adiabatic processes and thus as processes that take place at constant entropy.

The subject requires the use of spherical coordinates, which are a particular type of generalised coordinates. Therefore, we adopt generalised coordinates in several sections and, in association with them, use notions of tensor calculus and Riemannian geometry.

In the chapters devoted to asymptotic representations of oscillation modes of stars, and slow period changes of oscillations in rapidly evolving stars, we apply two-variable expansion procedures and boundary-layer theory. These methods are convenient for perturbation problems involving two highly different length or time scales and thin boundary layers in which terms of differential equations become large because of singularities or vanish because of turning points.

It is our hope that the present comprehensive presentation may fill a gap for doctoral students and researchers who want to get more acquainted with the fundamental aspects of the theory of the linear, isentropic oscillations in spherically symmetric stars. In order to facilitate their task, we give intermediate steps in the mathematical derivations. Furthermore, we refer to original scientific papers and reviews as much as possible. However, it is beyond the scope of this monograph to present an exhaustive list of all papers and reviews that have been published about the subject.

This monograph has grown from lectures given, and research done, for more than thirty years at the Katholieke Universiteit Leuven. I like to express my deep gratitude to the academic authorities for the possibilities they offered me. I also like to express my sincere thanks to the Belgian National Fund for Scientific Research, and the Flemisch Fund for Scientific Research, and especially its former Secretary general, Mr. José Traest, for their year-long support. Without them, this monograph would never have come about. I take the opportunity to thank my former students and collaborators for their interest. From their questions and feedback, I have learnt a lot.

I am heavily indebted to my colleague Dr. Christoffel Waelkens, Head of the Institute of Astronomy, for having allowed me to continue to use the facilities of the institute after my retirement, and for the enthusiasm he has continually manifested for this undertaking.

I am also much indebted to other colleagues. I especially mention my colleague Dr. Johan Quaegebeur of the Department of Mathematics for the precious help he provided in the domain of the functional analysis. My colleagues Dr. Christian Maes of the Department of Physics and Astronomy, Dr. Hans Van Winckel of the Institute of Astronomy, and Dr. Patricia Lampens of the Royal Observatory of Belgium helped me in the redaction and the correction of this and previous versions of the text. I thank them sincerely for their willing assistance.

The text of the monograph, with its mathematical parts, has been typeset, formatted, and composed by the LaTeX document preparation system, and the figures have been drawn by means of the WINDOWS version of the Maple VIII language. For this, I benefitted much from competent technical help of my colleague Dr. Lea Vermeire of the Campus Kortrijk of the university.

For the informatics, I greatly enjoyed the readiness and the competency of Dr. Bart Vandenbussche, Mr. Wim De Meester, and Mr. Bram Vandoren of the Institute of Astronomy.

In drawing up the list of references, I very much appreciated the assiduous help of Christophe Nassen, Head of the Circulation Department in the Campus Library Arenberg of the university, and his collaborators.

Last but not least, my warm and special thanks are due to Dr. Tim Van Hoolst, who has had a very large influence on the final result.

The responsibility for all inaccuracies and errors which may still be found in this monograph rests exclusively with me.

Finally, I thank the copyright holders Astronomy & Astrophysics and the Taylor & Francis Group for kindly granting the permission to reproduce articles and figures from their publications.

Louvain
August 2010

Paul Smeyers

Contents

Introduction

The study of the stellar oscillations is the preeminent way for the investigation of the stability, and the interpretation of the variability of stars, as it has strikingly been stated by Ledoux (1963a):

> Comme nous ne pouvons effectuer des sondages dans une étoile qui nous révéleraient directement sa constitution et sa composition chimique internes, nous en sommes réduits à construire des modèles stellaires dont les caractéristiques macroscopiques, masse M, luminosité L, rayon R ou température effective T_e doivent être comparables à celles des étoiles réelles et satisfaire à diverses relations, relations $M - R$, $L - R$ ou $L - T_e$, révélées par l'observation ...
>
> Le diagramme de Hertzsprung-Russell ... constitue un des moyens les plus heureux de résumer un grand nombre de propriétés d'une population stellaire ...
>
> Non seulement les modèles statiques et la théorie de l'évolution des étoiles doivent-ils restituer les valeurs de M, R, L observées pour des étoiles individuelles et les relations qui existent entre ces grandeurs mais encore, pour la majorité des étoiles, ces modèles doivent être stables. Ainsi, on peut considérer les différents critères de stabilité ... comme des conditions supplémentaires imposées aux modèles et, partant, comme autant de guides dans l'élaboration de ceux-ci.
>
> D'autre part, ... beaucoup ... d'étoiles ... *ne sont pas stables*. Et d'abord toutes les étoiles variables intrinsèques dont les variations doivent trouver leur origine dans quelque forme d'instabilité capable de se manifester pour une petite perturbation qui tendra à grandir jusqu'à ce que des facteurs non linéaires la stabilisent à une amplitude finie ...
>
> Ainsi les critères de stabilité fournissent également une approche naturelle pour l'étude de l'origine tout au moins des variations plus ou moins importantes et plus ou moins régulières dont un grand nombre d'étoiles sont le siège ...
>
> Il existe essentiellement deux approches pour discuter la stabilité d'un système. Nous pouvons lui appliquer une petite perturbation, si petite que, dans les équations qui régissent l'évolution de cette perturbation, nous pouvons nous contenter de ne garder que les termes du premier degré en cette perturbation. Les équations ainsi obtenues sont dites linéarisées et leur solution décrit le comportement de la perturbation.
>
> ... Ceci naturellement laisse échapper des cas d'instabilité vis-à-vis de perturbations finies.
>
> L'autre méthode se rattache au principe de Dirichlet suivant lequel l'état d'équilibre d'un système mécanique conservatif est stable s'il correspond à un minimum de l'énergie potentielle ... Cette dernière méthode a été peu employée dans le problème stellaire.[1]

[1] As we cannot make soundings in a star which would reveal us its internal constitution and chemical composition directly, we are restricted to construct stellar models whose macroscopic

P. Smeyers and T. Van Hoolst, *Linear Isentropic Oscillations of Stars: Theoretical Foundations*, Astrophysics and Space Science Library 371,
DOI 10.1007/978-3-642-13030-4_1, © Springer-Verlag Berlin Heidelberg 2010

About the possible methods, Ledoux (1963b) also stated:

> Two methods are available. In the *energy method*, a generalized potential (including the effects of thermodynamical or other nonmechanical factors) must go through an absolute minimum (i.e. min. along *all* independent variables) in any equilibrium state ...
>
> In the *small perturbation method*, a small disturbance is applied to the equilibrium configuration and its further behaviour under the natural forces of the system is studied. It is assumed that if the initial perturbation is small enough, the solutions of the *linearized* equations will reveal the trend of this motion in a small region around the equilibrium state ...

A study in which the stability of a star was investigated by means of an energy method was that of Tolman (1939) on the stability with respect to radial displacement fields.

The study of the linear, isentropic oscillations of a star has had its very outset in 1863 by a study of Kelvin (Thomson 1863) on the non-radial, linear, isentropic oscillations of a liquid, and thus *incompressible*, spherically symmetric mass in hydrostatic equilibrium of uniform mass density. In astrophysics, non-radial displacement fields in a spherically symmetric mass are displacement fields that involve transverse components, i.e. components perpendicular to the local radial direction, in addition to the radial components. From Kelvin's study, it results that, for an incompressible equilibrium sphere of uniform mass density, with mass M

---

characteristics, mass M, luminosity L, radius R, or effective temperature T_{e}, must be comparable to those of the real stars and satisfy various relations, relations $M - R$, $L - R$ or $L - T_{\mathrm{e}}$, revealed by the observations ...

The diagram of Hertzsprung-Russell ... is one of the most effective means to summarise a great number of properties of a stellar population ...

Not only the static models and the theory of stellar evolution must render the values of M, R, L observed for individual stars and the relations existing between these quantities, but also, for the majority of the stars, these models must be stable. So, one can consider the various criterions ... as additional conditions imposed on the models and, consequently, as as many guides in their elaboration.

On the other hand, ... many ... stars ... *are not stable*. And first all intrinsic variable stars whose variations must have an origin in some form of instability capable of manifesting itself for a small perturbation that will tend to grow until nonlinear factors stabilise it at a finite amplitude ...

In that way the stability criteria also furnish a natural approach to the study of the origin of at least plus or minus important and plus or minus regular variations of which a large number of stars are the seat ...

Essentially two approaches exist for the discussion of the stability of a system. We can apply a small perturbation to it, so small that, in the equations that govern the evolution of this perturbation, we can content ourselves with keeping only the terms of the first degree in this perturbation. The equations so obtained are said to be linearised and their solution describes the behaviour of the perturbation.

... This naturally lets escape cases of instability with respect to finite perturbations.

The other method is related to the principle of Dirichlet according to which the equilibrium state of a conservative mechanical system is stable if it corresponds to a minimum of the potential energy ... This last method has not much been used in the stellar problem.

and radius R, the period of the linear, non-radial, isentropic oscillation that belongs to a spherical harmonic of degree ℓ, with $\ell = 1, 2, 3, \ldots$, is given by

$$\Pi = 2\pi \left[\frac{2\ell + 1}{2\ell(\ell - 1)} \frac{R^3}{GM} \right]^{1/2} . \tag{1}$$

For $\ell = 1$, the oscillation period is infinite, so that the associated displacement fields are time-independent. These displacement fields correspond to uniform translations of the whole mass. With $\ell = 2$, an oscillation is associated by which the configuration passes on from a sphere to a spheroid, and vice versa. Kelvin determined the period of this oscillation for a rigid Earth and obtained 94 min (Dahlen & Tromp 1998). For an equilibrium sphere with a mass and a radius corresponding respectively to the mass and the radius of the Sun, the oscillation period is equal to 3 h 07 min.

The theory of the linear, isentropic oscillations of isolated gaseous stars, and thus of *compressible* spherically symmetric equilibrium configurations, has largely been developed from the point of view of the hypothesis of the physical radial pulsations of stars. The hypothesis was brought to the foreground by Ritter in a series of publications from 1878 to 1883, in particular in a publication of 1879 (Ritter 1878–1883, 1879). About the pioneer work of Ritter, Emden wrote in his book entitled *Gaskugeln* (Emden 1907):

> Besondere Aufmerksamkeit widmet Ritter dem Pulsationsproblem. Im Kap. ... habe ich die Schwingungen von Gaskugeln untersucht, die denjenigen inkompressibler Flüssigkeiten entsprechen ... Ritter untersucht eine andere Klasse von Schwingungen, bei denen die Kompressibilität des Gases zur Geltung kommt; die Teilchen verschieben sich längs des Radius, die Dichtigkeit bleibt in konzentrischen Schichten konstant, und die Kugel behält Kugelgestalt bei. Das Problem wurde gelöst für kleine Schwingungen unter den Annahmen 1. die Kugel ist von konstanter Dichte, 2. die Kugel bleibt während den Pulsationen von konstanter Dichte, 3. die Teilchen folgen während ihrer Bewegung der Poissonschen Gleichung $p\, v^{\gamma} = $ konst.[2]

Assumptions 2 and 3 are compatible with each other in the particular case of the fundamental radial oscillation of the compressible equilibrium sphere of uniform mass density. The oscillation period is then given by

$$\Pi = 2\pi \left(\frac{1}{3\gamma - 4} \frac{R^3}{GM} \right)^{1/2} , \tag{2}$$

[2] Ritter devoted a particular attention to the pulsation problem. In Chap. ... I have investigated the oscillations of gaseous spheres, which correspond to those of incompressible fluids ... Ritter investigates another class of oscillations, those by which the compressibility of the gas appears to full advantage; the particles move along the radius, the density remains constant in concentric layers, and the sphere keeps its spherical form. The problem is solved for small oscillations on the assumptions that 1. the sphere is of constant density, 2. the sphere remains of constant density during the pulsation, 3. the particles follow Poisson's equation $p\, v^{\gamma} =$ const. during their motion.

where γ is the ratio of the specific heat at constant pressure to the specific heat at constant volume. When the equilibrium sphere is composed of a monoatomic gas with $\gamma = 5/3$ and has a mass and a radius equal to the mass and the radius of the Sun, the oscillation period is 2 h 47 min.

The hypothesis of the radial pulsations of stars may also have been suggested by Umoff in 1896 during the defense of a thesis on the discovery of Doppler shifts of spectral lines in the atmosphere of the star δ Cephei (Zhevakin 1963).

In 1914, Shapley felt it still necessary to emphasise that the phenomena observed in Cepheids are most likely to be attributed to pulsations (Shapley 1914):

> It seems a misfortune, perhaps, for the progress of research on the causes of light-variation of the Cepheid type, that the oscillations of the spectral lines in nearly every case can be so readily attributed, by means of the Doppler principle, to elliptical motion in a binary system. The natural conclusion that all Cepheid variables are spectroscopic binaries has been the controlling and fundamental assumption in all the recently attempted interpretations of their light-variability, and the possibility of intrinsic light-fluctuations of a single star has received little attention.
>
> . . . The main conclusion is that the Cepheid and cluster variables are not binary systems, and that the explanation of their light-changes can much more likely be found in a consideration of internal or surface pulsations of isolated stellar bodies.

The theory of the radial stellar pulsations has largely been developed in the first half of the 20th century. Overviews of this development are found in the textbooks of Rosseland (1949) and Ledoux & Walraven (1958). In the development of the theory, a main role was played by Eddington, about whom Chandrasekhar testified in 1988 in the preface to a reedition of Eddington's important book *The Internal Constitution of the Stars* of 1926 (Eddington 1988):

> Eddington's interest in the internal constitution of the stars arose from his efforts to find an explanation for stellar variability and the period-luminosity relation exhibited by the Cepheids . . . Eddington first generalised Ritter's earlier analysis of the adiabatic pulsations of gaseous stars in convective equilibrium to the case of a star in radiative equilibrium built on his standard model . . . Then combining the resulting formula for the period with his mass-luminosity relation, Eddington was able to account, in a general way, for the observed period-luminosity relation of the Cepheids. The pulsation theory of stellar variability thus came to be established.

The question of the pulsation mechanism of the classical Cepheids has however not been elucidated before the fifties and sixties. Representative references are Cox & Whitney (1958), Baker & Kippenhahn (1962), Zhevakin (1963), and Baker & Kippenhahn (1965). During the subsequent decades, the insight in the physical origins of the radial pulsations of stars of different types increased considerably. A survey of these developments is presented in Cox (1974). A more recent account of the development and the acceptance of the pulsation theory was given by Gautschy & Saio (1995). Gautschy (1997) presented an extensive account of the development of the observational and theoretical developments from the discovery of δ Cephei in 1784 by Goodricke up to the paper of Baker and Kippenhahn in 1962.

An extension of the theory to *non-radial stellar oscillations* has first been considered by Rosseland (1932) in a paper in which he stated:

> In an earlier paper an attempt was made to develop a systematic theory of infinitely small radial vibrations of a star ... However, it is desiderable for the sake of generality to generalize the theory to include oscillations depending on latitude and longitude.
>
> A strictly radial expansion and contraction is a rather artificial state of motion, which scarcely can be expected to be realized in nature, since it would imply a complete absence of rotation in the star. This is perhaps a serious limitation, because it may tend to conceal important aspects of the general problem. Thus the obvious instability of a star, in which super-adiabatic temperature gradients occur, does not show up at all in the discussion of purely radial motions. Second, the consideration of deformations of a more general kind forms a step toward the study of rotating stars, and the problem of the genesis of binaries.
>
> The generalization in question leads into a field of complex mathematics, where progress is necessarily slow. We shall therefore approach the problem in successive steps, beginning with the simplest case, that of the *homogeneous liquid globe*. This problem was fully solved by W. Thomson ...

In accordance with Rosseland's expectations, the development of the theory of the non-radial stellar oscillations has progressed slowly. Several reasons played a role, as observed by Gautschy (1996):

> For a long time, the possibility of stars admitting nonradial oscillations was considered as possible by the action of external forces only. The typical scenario considered close binaries or close encounters of single stars. Mostly, however, nonradial oscillations were considered as an exercise in mathematical physics without much concern for the real astronomical world. In isolated, pulsating stars nonradial oscillations were thought to suffer from higher viscous dissipation so that they should not win over radial pulsation modes and hence were not expected to be observable.

The contributions of Pekeris (1938) and Cowling (1941) have been of fundamental importance for the development of the theory. An early extensive presentation of the theory was given by Ledoux & Walraven (1958).

By the time of the symposium of the International Astronomical Union held in 1973 in Canberra on the subject *Stellar Instability and Evolution*, the interest in the theory of non-radial stellar oscillations had suddenly increased, so that Ledoux could state (Ledoux 1974):

> There has been lately quite a renewal of interest in the response of stars to non-radial perturbations aroused either by attempts at interpreting some types of variable stars like the β Canis Majoris stars and the new white dwarf variables, or by phenomena in the external layers of the Sun like the 5-min oscillation discovered by Leighton, or by the hope to add somewhat to our knowledge of convection and its penetration in nearby convectively stable zones, or by the desire to explore some new aspects of stellar stability which may be of great importance for the evolution of the star. On the other hand, one must expect that such non-radial motions should be easily excited in a variety of close double stars with eccentric orbits and it is likely that, with the extraordinary progress in observational techniques, these should become observable and be identified as such pretty soon. Finally, there is direct evidence in novae, perhaps even in planetary nebulae, for the presence of non-radial velocity fields.

Further developments of the theory were described in Cox (1980) and Unno et al. (1989).

It has been a great merit of Ledoux of having foreseen that the progress in observational techniques would make possible the detection of non-radial oscillations on surfaces of isolated stars. At present, non-radial oscillations turn out to be present on the surfaces of many of them. They are currently detected by conventional means in δ Scuti stars (see, e.g., Breger et al. 1999). The slowly pulsating mid-B stars, which were detected by Waelkens (1987, 1991), seem to be the seat of non-radial g^+-oscillations. The members of the more recently discovered class of the variable γ Dor stars are considered to display multiperiodic pulsations of higher-order g^+-modes (Eyer & Aerts 2000). First arguments for the presence of non-radial oscillations in RR Lyrae stars were given by Olech et al. (1999). As a result of an earlier stability analysis, Van Hoolst et al. (1998) had already suggested that low-degree non-radial oscillations could be excited in RR Lyrae stars. See also, e.g., Kolenberg et al. (2008).

Ledoux' expectations have largely been surpassed by the observation of numerous non-radial oscillations on the surfaces of the Sun and white dwarfs.

The first non-radial oscillations detected on the surface of the Sun are the 5 min oscillations whose existence was discovered by Leighton (1961) (see also Leighton et al. 1962). They were identified as global p-oscillations of higher degrees and low radial orders by Deubner (1975) on the basis of theoretical studies by Ulrich (1970), Leibacher & Stein (1971), Wolff (1972), and Ando & Osaki (1975).

Furthermore, around 1975, the presence of 5 min oscillations of low degrees was shown by means of observations of integrated Sun light (Fossat & Ricort 1975, Grec & Fossat 1977). It was confirmed by Claverie et al. (1979) by the use of a potassium line of the global solar disc at the wavelength of 769.9 nm.

Nowadays, it is known that, in the Sun, millions of acoustic oscillation modes interfere with each other, of which many are coherent over the total surface. The oscillation periods range from a few minutes to several hours, and the horizontal wavelengths are longer than thousands of kilometers (Christensen-Dalsgaard et al. 1985). Tables of accurate oscillation mode frequencies of the Sun, as functions of the radial order n and the spherical harmonic degree ℓ, have been presented (Libbrecht et al. 1990).

The first white dwarf for which short periodic oscillations were reported, is HL Tau 76 (Landolt 1968). Several tens of variable white dwarfs are known at present. The displacement fields with periods of a few hundredths of seconds observed on the surfaces of these variables are ascribed to non-radial g^+-modes of low degrees and high radial orders (Chanmugam 1972, Warner & Robinson 1972). The first conclusive proof of this was provided by Robinson et al. (1982).

The multiplicity of the oscillations, radial and non-radial, that are observed on the surfaces of the Sun and the variable white dwarfs has provided additional tools for sounding the internal structure of these celestial bodies. From this, the helio- and the asteroseismology have arisen as new domains of research (Libbrecht 1988, Christensen-Dalsgaard 1988, Shibahashi 1990).

In the introduction to a review paper on the early beginning of the asteroseismology, Brown & Gilliland (1994) explained more fully:

> Asteroseismology is commonly understood to mean the study of normal-mode pulsations in stars that, like the Sun, display a large number of simultaneously excited modes. The idea of learning about a physical system by examining its oscillation modes is of course an old one in physics, but it is only fairly recently that data of sufficient quality have become available to apply this technique to stars.
>
> The Sun is (and will likely remain) the outstanding example of the progress that can be made using seismological methods ...
>
> Three properties of the solar pulsations have combined to make this progress possible. First, the Sun manages to excite a very large number of modes simultaneously: Something like 10^7 modes are thought to have amplitudes large enough for observation. Each mode carries information about the solar interior that is somewhat different from that of any other mode. Thus, one may use mode characteristics to fit for a large number of parameters describing the solar structure. Second, the acoustic modes of the Sun are fairly weakly damped, having lifetimes that are typically several thousand oscillation cycles. This allows one to measure the mode properties (particularly the frequencies) very accurately, permitting delicate tests to distinguish between models. Last, the amplitudes of individual oscillation modes are very small (for the modes with largest amplitude, relative displacements at the surface of the Sun are less than 10^{-7} ...). This property assures that the presence of the modes has only a small influence on the structure of the star, and moreover that linear theory is adequate for most purposes involving the character of the modes themselves.
>
> Many stars besides the Sun may be expected to support pulsations with these same three properties. Stars of roughly solar type should of course behave in ways similar to the Sun, and stars of this sort form a large fraction of the potential targets for asteroseismology. But several other kinds of star (δ Scuti stars, roAp stars, and the pulsating white dwarfs) also have the desired pulsation characteristics.

The etymology of the term asteroseismology was analysed by Gough (1996). Besides, the aims of the helio- and asteroseismology were outlined by Dziembowski (1988):

> There is no well established notion among people on solar and stellar oscillations what is the scope of helio- and asteroseismology. Not surprisingly, we do not find the definition of these words in dictionaries of the English language. The word seismology, in *The Webster's Dictionary of the English Language*, is explained as "... the science or study of earthquakes and their phenomena". Other dictionaries provide similar definitions. It is clear that oscillations observed in stars do not result from starquakes. Excitation of the solar oscillations bears only a distant resemblance to excitation of seismic waves ...
>
> It seems that most of us understand helioseismology as seismic sounding of the Sun's interior, which is a narrower scope than the terrestrial seismology. In other stars, however, we are still very far from the true seismic sounding. Therefore, we should accept a somewhat wider definition of our field and understand it as research in solar and stellar oscillations applied to study their physical properties and to testing all aspects of stellar physics.

The true birth of asteroseismology was proclaimed by Gough (2001) in a comment on one of the first communications about the seismology of the solar-type star β Hydri (Bedding et al. 2001). Oscillations were found in solar-type stars as Procyon (Martic et al. 1999), α Cen A (Bouchy & Carrier 2001), δ Eri (Carrier et al. 2003), and in the giant star ξ Hya (Frandsen et al. 2002). More recently, oscillations that are stochastically excited by turbulent convection have probably been found in the red giants ξ Hya, ε Oph, and η Ser by means of precise radial velocity measurements (De Ridder et al. 2006). With this, the existence of solar-like oscillations in red giants seems to be proved. The presence of radial and non-radial oscillation modes in more than 300 giant stars was reported by De Ridder et al. (2009). For at least some of them, the mode lifetimes are of the order of a month.

While attention was largely paid to stars with lower masses that are, or have been, comparable with the Sun's mass, Aerts et al. (2003) examined the oscillation behaviour of the multiperiodic β Cephei star HD 129929, with a mass of 9.5 $M_{\odot}$. By means of accurate photometric observations extending over a period of 21 years, they were able to distinguish six oscillation periods in the star. Moreover, on the basis of these oscillation periods and stellar models, they inferred that the star does not rotate uniformly and that the convective core gives rise to a mixing of matter up to outside the usual boundary of this core (see also Kawaler 2003).

The efficiency of the helio- and asteroseismology is increased by uninterrupted observations of the Sun and stars. A way to achieve this aim is the creation of networks of observational sites spread over different geographical lengths. Longer uninterrupted periods of observation of the Sun and stars are also realised by the use of a telescope mounted on the south pole. In this way, Grec et al. (1980) made continuous observations of the Sun in a period of 5 days during the southern summer 1979–1980. Another possibility of improving the study of solar and stellar oscillations is provided by observations from space (Ulrich 1984): dedicated space missions are MOST (Microvariability & Oscillations of STars), CoRoT (COnvection, ROtation and planetary Transits), and KEPLER, launched respectively in June 2003, December 2006, and March 2009.

At the European level, the "European Network of Excellence in Astero-Seismology" (ENEAS) has been created, which uses the *Communications in Asteroseismology* as medium for its publications (Aerts 2003). Furthermore, the HELAS IT-platform has been founded as a new tool for the European helio- and asteroseismology community within the framework of the European Commission's Sixth Framework Programme (FP6) (Jimenéz-Reyes et al. 2008).

Helio- and asteroseismology have undeniably opened a gate to a new area in the study of the stellar interior and will give rise to yet unpredictable theoretical questions. Lately, Degroote et al. (2010) reported the detection of numerous gravity modes in the young star HD 50230 with a mass of about seven solar masses. The deviations from the constant period spacing which they determined, allowed them to evaluate the characteristics of the mixing processes in the stellar interior. Many aspects of the promising research domain of asteroseismology have been expounded in the book *Asteroseismology* by Aerts et al. (2010).

In conclusion, a quotation from Christensen-Dalsgaard & Gough (2001) is here apposite:

> Although of a somewhat formal nature, investigations of the mathematical properties of stellar adiabatic oscillations are of substantial importance to our understanding of the nature of the oscillations and their relation to the structure of the star. Such insight is crucial to the use of stellar oscillation frequencies for asteroseismic investigations of stellar interiors.

Chapter 1
Basic Concepts

1.1 The Lagrangian Displacement of a Mass Element

A star is regarded as a compressible fluid composed of a mixture of gas and radiation. The macroscopic motions of the fluid are described by hydrodynamics, in which the existence of particles is left out of consideration, and the fluid is supposed to be continuous. This viewpoint implies that any mass element must be sufficiently large to contain a high number of particles.

The fundamental quantity for the description of a fluid's motion is the Lagrangian displacement of a mass element.

In a Lagrangian description of the motion, the coordinates are coordinates of mass elements, which depend on time t and on three parameters a^1, a^2, a^3 characterising these mass elements.

Suppose that a flow in a fluid, at a given time t, is described by the Cartesian coordinates of the mass elements

$$x^i\left(t, a^1, a^2, a^3\right), \qquad i = 1, 2, 3. \tag{1.1}$$

Cartesian coordinates are orthonormal coordinates everywhere in this monograph.

At the same time, a slightly different flow is considered in the fluid. This so-called perturbed flow is described by the Cartesian coordinates of the mass elements

$$x_{\mathrm{p}}^i\left(t, a^1, a^2, a^3\right), \qquad i = 1, 2, 3. \tag{1.2}$$

The Lagrangian displacement of a mass element at a time t is defined as the difference between the position vector of the mass element in the perturbed flow and that of the mass element in the initial flow, both considered at time t. Its components δx^1, δx^2, δx^3 correspond to the differences between the Cartesian coordinates of the mass element in the perturbed flow and those in the initial flow:

$$\delta x^i\left(t, a^1, a^2, a^3\right) = x_{\mathrm{p}}^i\left(t, a^1, a^2, a^3\right) - x^i\left(t, a^1, a^2, a^3\right), \qquad i = 1, 2, 3. \tag{1.3}$$

P. Smeyers and T. Van Hoolst, *Linear Isentropic Oscillations of Stars: Theoretical Foundations*, Astrophysics and Space Science Library 371,
DOI 10.1007/978-3-642-13030-4_2, © Springer-Verlag Berlin Heidelberg 2010

The Lagrangian displacement is represented as

$$\boldsymbol{\xi}\left(t, a^1, a^2, a^3\right). \tag{1.4}$$

By elimination of the parameters a^1, a^2, a^3 by means of relations (1.1), the Lagrangian displacement can also be expressed as

$$\boldsymbol{\xi}\left[t, x^1\left(t, a^1, a^2, a^3\right), x^2\left(t, a^1, a^2, a^3\right), x^3\left(t, a^1, a^2, a^3\right)\right]. \tag{1.5}$$

In the study of the free oscillations of an isolated, non-rotating, and non-magnetic star with spherical symmetry, it is often appropriate to use spherical coordinates: the radial distance from the origin, r, the colatitude, θ, and the longitude, ϕ. Therefore, it is useful to consider changes from Cartesian coordinates x^1, x^2, x^3 to generalised coordinates q^1, q^2, q^3. From a general point of view, the transformation formulae may contain the time explicitly:

$$x^i\left(t, q^1, q^2, q^3\right), \qquad i = 1, 2, 3. \tag{1.6}$$

As well as Cartesian coordinates, generalised coordinates of mass elements are functions of time t and three parameters a^1, a^2, a^3 characterising the mass elements:

$$q^j\left(t, a^1, a^2, a^3\right), \qquad j = 1, 2, 3. \tag{1.7}$$

When generalised coordinates are used, it is convenient to work with respect to the local coordinate basis or natural basis. In modern differential geometry, the vectors of the local coordinate basis are identified with the directional differential operators

$$\frac{\partial}{\partial q^1}, \ \frac{\partial}{\partial q^2}, \ \frac{\partial}{\partial q^3} \tag{1.8}$$

(Misner 1969, Misner et al. 1973, Hartle 2003). Any arbitrary vector $\mathbf{v}$ at a point can be decomposed with respect to the local coordinate basis as

$$\mathbf{v} = v^i \frac{\partial}{\partial q^i}, \tag{1.9}$$

where the v^i are the contravariant components of the vector. Einstein's summation convention is adopted and currently used from here on.

The vectors of a local coordinate basis are not necessarily orthonormal. The product of a basis vector $\partial/\partial q^k$ and a basis vector $\partial/\partial q^\ell$ is equal to the covariant component $g_{k\ell}$ of the metric tensor:

$$\frac{\partial}{\partial q^k} \frac{\partial}{\partial q^\ell} = g_{k\ell}. \tag{1.10}$$

Two basis vectors $\partial/\partial q^k$ and $\partial/\partial q^\ell$, with $\ell \neq k$, are then mutually orthogonal, when the covariant component $g_{k\ell}$ of the metric tensor is equal to zero. Moreover, the squares of the lengths of the basis vectors are determined by the covariant components g_{11}, g_{22}, g_{33} of the metric tensor. If $g_{11} = 1$, $g_{22} = 1$, and $g_{33} = 1$, the basis vectors are unit vectors. A set of mutually orthogonal basis vectors that are unit vectors forms an orthonormal basis.

A covariant component $g_{k\ell}$ of the metric tensor is derived from the transformation formulae (1.6) as

$$g_{k\ell} = \delta_{ij} \frac{\partial x^i}{\partial q^k} \frac{\partial x^j}{\partial q^\ell}, \qquad k, \ell = 1, 2, 3, \tag{1.11}$$

where the δ_{ij} are Kronecker deltas.

A transformation from basis vectors $\partial/\partial x^1$, $\partial/\partial x^2$, $\partial/\partial x^3$ associated with Cartesian coordinates x^1, x^2, x^3 into basis vectors $\partial/\partial q^1$, $\partial/\partial q^2$, $\partial/\partial q^3$ associated with generalised coordinates q^1, q^2, q^3 changes contravariant vector components linearly. The transformation formulae are obtained as

$$\mathbf{v} = v^i \frac{\partial}{\partial x^i} = \left(v^i \frac{\partial q^j}{\partial x^i} \right) \frac{\partial}{\partial q^j} \equiv v'^j \frac{\partial}{\partial q^j}, \tag{1.12}$$

so that the transformed vector components are given by

$$v'^j = v^i \frac{\partial q^j}{\partial x^i}, \qquad j = 1, 2, 3. \tag{1.13}$$

In accordance with this transformation formula, the contravariant components of the Lagrangian displacement of a mass element with respect to the basis vectors $\partial/\partial q^j$ are given by

$$\delta q^j = \frac{\partial q^j}{\partial x^i} \delta x^i, \qquad j = 1, 2, 3. \tag{1.14}$$

The components δq^j are called the generalised components of the Lagrangian displacement of the mass element.

Notice that a generalised component δq^j of the Lagrangian displacement is *not* defined as the difference between the generalised coordinate q^j_{p} of the mass element in the perturbed flow and the corresponding generalised coordinate q^j of the mass element in the initial flow. Generally, it is not equal to that difference. For illustration, it is instructive to expand the difference

$$\Delta q^j = q^j_{\mathrm{p}} - q^j \tag{1.15}$$

into a Taylor series. Since

$$q^j = q^j\left(t, x^1, x^2, x^3\right), \qquad j = 1, 2, 3, \tag{1.16}$$

and

$$q_{\mathrm{p}}^{j} = q^{j}\left(t, x^{1} + \delta x^{1}, x^{2} + \delta x^{2}, x^{3} + \delta x^{3}\right), \qquad j = 1,2,3, \tag{1.17}$$

the difference is given by

$$\Delta q^{j} = \frac{\partial q^{j}}{\partial x^{i}} \delta x^{i} + \frac{1}{2} \frac{\partial^{2} q^{j}}{\partial x^{i} \partial x^{k}} \delta x^{i} \delta x^{k} + \ldots, \qquad j = 1,2,3. \tag{1.18}$$

The equality

$$\delta q^{j} = \Delta q^{j} \tag{1.19}$$

holds only if transformation formulae (1.6) are linear in the generalised coordinates or if the Taylor series is restricted to the terms that are linear in the components of the Lagrangian displacement. The linear approximation is generally adopted subsequently.

By differentiating definition (1.3) partially with respect to time, while the parameters a^1, a^2, a^3 are kept constant, one derives the relation

$$\left(\frac{\partial}{\partial t} \delta x^{i}\right)_{a^{1},a^{2},a^{3}} = \dot{x}_{\mathrm{p}}^{i}\left(t, a^{1}, a^{2}, a^{3}\right) - \dot{x}^{i}\left(t, a^{1}, a^{2}, a^{3}\right), \qquad i = 1,2,3. \tag{1.20}$$

The equality expresses that the partial derivative with respect to time of a component δx^i of the Lagrangian displacement of a mass element, with the parameters a^1, a^2, a^3 kept constant, is equal to the difference between the velocity component $\dot{x}_{\mathrm{p}}^{i}$ of the mass element in the perturbed flow and the corresponding velocity component $\dot{x}^i$ of the mass element in the initial flow.

When equality (1.19) is valid, a similar relation is derived for generalised coordinates:

$$\left(\frac{\partial}{\partial t} \delta q^{i}\right)_{a^{1},a^{2},a^{3}} = \dot{q}_{\mathrm{p}}^{i}\left(t, a^{1}, a^{2}, a^{3}\right) - \dot{q}^{i}\left(t, a^{1}, a^{2}, a^{3}\right), \qquad i = 1,2,3, \tag{1.21}$$

where $\dot{q}^i$ is the generalised velocity component i.

1.2 Lagrangian and Eulerian Perturbations of Physical Quantities

1.2.1 Definitions

With the Lagrangian displacements of the mass elements in a flow, perturbations of the various physical quantities in the flow are associated. A distinction between Lagrangian and Eulerian perturbations is made (see, e.g., Chandrasekhar 1969, Tassoul 1978, Cox 1980).

Be $Q\left(t, x^1, x^2, x^3\right)$ any physical quantity. It can be a scalar quantity, as the pressure, the mass density, the gravitational potential, or a vector component, as a velocity component of a mass element, or even a tensor component.

In the definition of the Lagrangian perturbation of a physical quantity, the quantity is regarded as being associated with a mass element of the fluid. The Lagrangian perturbation of a physical quantity is the difference between the physical quantity associated with the mass element in the perturbed flow and the physical quantity associated with the same mass element in the initial flow, at the time considered. Hence, if $Q\left(t, x^1, x^2, x^3\right)$ is the physical quantity associated with the mass element at the point with Cartesian coordinates x^1, x^2, x^3 in the initial flow, and $Q_{\mathrm{p}}\left(t, x_{\mathrm{p}}^1, x_{\mathrm{p}}^2, x_{\mathrm{p}}^3\right)$ the physical quantity associated with the mass element at the point with Cartesian coordinates x_{p}^1, x_{p}^2, x_{p}^3 in the perturbed flow, the Lagrangian perturbation of $Q\left(t, x^1, x^2, x^3\right)$ is defined as

$$\delta Q\left(t, x^1, x^2, x^3\right) = Q_{\mathrm{p}}\left(t, x_{\mathrm{p}}^1, x_{\mathrm{p}}^2, x_{\mathrm{p}}^3\right) - Q\left(t, x^1, x^2, x^3\right). \tag{1.22}$$

When generalised coordinates are used, the definition of the Lagrangian perturbation of a physical quantity

$$\delta Q\left(t, q^1, q^2, q^3\right) = Q_{\mathrm{p}}\left(t, q_{\mathrm{p}}^1, q_{\mathrm{p}}^2, q_{\mathrm{p}}^3\right) - Q\left(t, q^1, q^2, q^3\right) \tag{1.23}$$

applies only to *scalar* quantities. For physical quantities that are vector components, the Lagrangian perturbation is determined by application of transformation formula (1.13) for vector components. An important example is that of a generalised velocity component $\dot{q}^j$ of a mass element. One then has

$$\delta \dot{q}^j = \frac{\partial q^j}{\partial x^i} \delta \dot{x}^{\,i}, \qquad j = 1, 2, 3. \tag{1.24}$$

For physical quantities that are tensor components, appropriate extensions of transformation formula (1.13) must be used (see, e.g., Van Hoolst 1992).

With regard to relativistic fluids, Taub (1969) introduced a different definition of the Lagrangian perturbation of a vector and a tensor component. In his definition, the Lagrangian perturbation is a measure for the change of the components of the vector or tensor with respect to the system of coordinates that is embedded in the fluid and moves along with it. The frame of reference is Lagrangian, while in the definition given by Eq. (1.22), the Lagrangian perturbation is considered with respect to an Eulerian frame of reference. Only for scalar quantities, Taub's definition corresponds to the definition of the Lagrangian perturbation adopted here (Friedman & Schutz 1978).

On the other hand, in the definition of the Eulerian perturbation of a physical quantity, the quantity is regarded at a given geometrical point of the fluid. The Eulerian perturbation of a physical quantity at a point is the difference between the physical quantity in the perturbed flow and the same physical quantity in the initial flow, at the time considered. Hence, if $Q(t, x^1, x^2, x^3)$ is the physical quantity at the point with Cartesian coordinates x^1, x^2, x^3 in the initial flow, and $Q_\mathrm{p}(t, x^1, x^2, x^3)$ the physical quantity at the point with Cartesian coordinates x^1, x^2, x^3 in the perturbed flow, the Eulerian perturbation of $Q(t, x^1, x^2, x^3)$ is defined as

$$Q'(t, x^1, x^2, x^3) = Q_\mathrm{p}(t, x^1, x^2, x^3) - Q(t, x^1, x^2, x^3) . \tag{1.25}$$

When generalised coordinates are used, the definition can analogously be formulated as

$$Q'(t, q^1, q^2, q^3) = Q_\mathrm{p}(t, q^1, q^2, q^3) - Q(t, q^1, q^2, q^3) . \tag{1.26}$$

It is valid for vector and tensor components as well as for scalar quantities, since coordinate transformations have the same effect on quantities considered at the same point.

A relation between the Lagrangian perturbation and the Eulerian perturbation of a physical quantity can easily be derived. A Taylor expansion in terms of Cartesian coordinates yields

$$Q_\mathrm{p}(\mathbf{r}_\mathrm{p}) = Q_\mathrm{p}(\mathbf{r}) + \delta x^j \left(\frac{\partial Q_\mathrm{p}}{\partial x^j} \right) (\mathbf{r}) + \ldots \tag{1.27}$$

By subtracting $Q(\mathbf{r})$ from both members and using definitions (1.22) and (1.25) of the Lagrangian and Eulerian perturbation of a physical quantity, one obtains

$$\delta Q(\mathbf{r}) = Q'(\mathbf{r}) + \delta x^j \left(\frac{\partial}{\partial x^j} \left[Q + (Q_\mathrm{p} - Q) \right] \right) (\mathbf{r}) + \ldots \tag{1.28}$$

In the linear approximation, it results that

$$\delta Q(\mathbf{r}) = Q'(\mathbf{r}) + \delta x^j \left(\frac{\partial Q}{\partial x^j} \right) (\mathbf{r}) . \tag{1.29}$$

In terms of generalised coordinates, the relation between the Lagrangian and the Eulerian perturbation of a physical quantity takes the form

$$\delta Q = Q' + \delta q^j \nabla_j Q , \tag{1.30}$$

where the operator ∇_j stands for the operator of partial differentiation with respect to the generalised coordinate q^j when applied to a scalar quantity, and for the operator of covariant differentiation with respect to the same coordinate when applied to a vector or tensor component.

The covariant derivative of a contravariant vector component Q^i is given by

$$\nabla_j Q^i = \frac{\partial Q^i}{\partial q^j} + \Gamma^i_{jk} Q^k. \tag{1.31}$$

Note that the contravariant vector components Q^k are considered with respect to the coordinate basis. The second term stems from the change of the coordinate basis from a point to a neighbouring point. The Γ^i_{jk} are Christoffel three-index symbols of the second kind:

$$\Gamma^i_{jk} = \frac{1}{2} g^{is} \left(\frac{\partial g_{sk}}{\partial q^j} + \frac{\partial g_{js}}{\partial q^k} - \frac{\partial g_{jk}}{\partial q^s} \right), \qquad i, j, k = 1, 2, 3. \tag{1.32}$$

The g^{is}, with $i, s = 1, 2, 3$, are the contravariant components of the metric tensor and obey the relations

$$g^{is} g_{ks} = \delta^i_k, \qquad i, k = 1, 2, 3, \tag{1.33}$$

where the δ^i_k are Kronecker deltas. The covariant components of the metric tensor are defined only if $\det(g_{ij}) \neq 0$. When Cartesian coordinates are used,the Christoffel three-index symbols of the second kind are identically zero, so that the covariant derivative of a contravariant vector component with respect to a coordinate x^j reduces to the partial derivative with respect to this coordinate:

$$\nabla_j Q^i = \frac{\partial Q^i}{\partial x^j}. \tag{1.34}$$

1.2.2 Additional Relations

According to the definition of the Lagrangian perturbation of a physical quantity, the right-hand member of Eq. (1.20) corresponds to the Lagrangian perturbation of the velocity component $\dot{x}^i$ of the mass element, so that

$$\left(\frac{\partial}{\partial t} \delta x^i \right)_{a^1, a^2, a^3} \equiv \frac{d}{dt} \delta x^i = \delta \dot{x}^i. \tag{1.35}$$

Hence, when Cartesian coordinates are used, the total time derivative of a component of the Lagrangian displacement of a mass element is equal to the Lagrangian perturbation of the corresponding velocity component of the mass element. The total time derivative is taken with regard to the initial flow.

By partial differentiation of definition (1.22) with respect to the time in which the parameters a^1, a^2, a^3 are kept constant, the relation results

$$\left(\frac{\partial(\delta Q)}{\partial t}\right)_{a^1,a^2,a^3} = \left[\left(\frac{\partial Q}{\partial t}\right)_{a^1,a^2,a^3}\right]_{\mathrm{p}} - \left(\frac{\partial Q}{\partial t}\right)_{a^1,a^2,a^3} = \delta\left(\frac{\partial Q}{\partial t}\right)_{a^1,a^2,a^3}. \tag{1.36}$$

The relation can also be expressed as

$$\frac{d(\delta Q)}{dt} = \delta\frac{dQ}{dt}. \tag{1.37}$$

Hence, the operations of taking the total time derivative of the Lagrangian perturbation of a physical quantity and taking the Lagrangian perturbation of the total time derivative of that quantity are commutative. When Cartesian coordinates are used, the physical quantity may be a scalar and also a vector or a tensor component. However, when generalised coordinates are used, the relation holds only for scalar quantities because of the use that is made of definition (1.23).

Next, by partially differentiating definition (1.26), successively with respect to the time and with respect to a generalised coordinate q^j, one derives that

$$\frac{\partial Q'}{\partial t} = \left(\frac{\partial Q}{\partial t}\right)_{\mathrm{p}} - \frac{\partial Q}{\partial t} = \left(\frac{\partial Q}{\partial t}\right)', \tag{1.38}$$

$$\frac{\partial Q'}{\partial q^j} = \left(\frac{\partial Q}{\partial q^j}\right)_{\mathrm{p}} - \frac{\partial Q}{\partial q^j} = \left(\frac{\partial Q}{\partial q^j}\right)'. \tag{1.39}$$

Hence, the operations of taking the partial time derivative of the Eulerian perturbation of a physical quantity in which the coordinates of the geometrical point are kept constant, and taking the Eulerian perturbation of the partial time derivative of that quantity are commutative. Similarly, the operations of taking the partial derivative of the Eulerian perturbation of a physical quantity with respect to a coordinate q^j, and taking the Eulerian perturbation of the partial derivative of that quantity with respect to the coordinate q^j are commutative. When the physical quantity is a vector or a tensor component, the operator $\partial/\partial q^j$ for partial differentiation should be replaced by the operator ∇_j for covariant differentiation.

A useful relation is

$$\delta\frac{\partial Q}{\partial t} = \frac{\partial(\delta Q)}{\partial t} - \frac{\partial(\delta q^k)}{\partial t}\nabla_k Q. \tag{1.40}$$

The proof runs as follows. Because of relation (1.30), the Lagrangian perturbation of the partial time derivative $\partial Q/\partial t$ can be expressed as

$$\delta\frac{\partial Q}{\partial t} = \left(\frac{\partial Q}{\partial t}\right)' + \delta q^k \nabla_k \frac{\partial Q}{\partial t}. \tag{1.41}$$

Transformation of the first term in the right-hand member by means of equality (1.38) and relation (1.30) yields

$$\delta\frac{\partial Q}{\partial t} = \frac{\partial(\delta Q)}{\partial t} - \frac{\partial}{\partial t}\left(\delta q^k \nabla_k Q\right) + \delta q^k \nabla_k \frac{\partial Q}{\partial t}, \tag{1.42}$$

so that the result follows.

One analogously derives that

$$\delta\left(\nabla_j Q\right) = \nabla_j(\delta Q) - \left(\nabla_j \delta q^k\right)\nabla_k Q, \qquad j = 1,2,3, \tag{1.43}$$

since, in an Euclidean space,

$$\nabla_k\left(\nabla_j Q\right) = \nabla_j(\nabla_k Q). \tag{1.44}$$

1.3 The Eulerian Perturbation of a Velocity Component

By virtue of relation (1.29), the Eulerian perturbation of a velocity component of a mass element can be expressed as

$$\left(\dot{x}^i\right)' = \delta\dot{x}^i - \delta x^j \frac{\partial \dot{x}^i}{\partial x^j}, \qquad i = 1,2,3. \tag{1.45}$$

The first term in the right-hand member can be transformed by means of equality (1.35), so that

$$\left(\dot{x}^i\right)' = \frac{\partial\left(\delta x^i\right)}{\partial t} + \dot{x}^j \frac{\partial\left(\delta x^i\right)}{\partial x^j} - \delta x^j \frac{\partial \dot{x}^i}{\partial x^j}, \qquad i = 1,2,3 \tag{1.46}$$

(see, e.g., Cox 1980).

When equality (1.19) is valid, the Eulerian perturbation of a generalised velocity component can similarly be expressed as

$$\left(\dot{q}^i\right)' = \frac{\partial\left(\delta q^i\right)}{\partial t} + \dot{q}^j \nabla_j\left(\delta q^i\right) - \delta q^j \nabla_j \dot{q}^i, \qquad i = 1,2,3, \tag{1.47}$$

or, after use of equality (1.31), as

$$\left(\dot{q}^i\right)' = \frac{\partial\left(\delta q^i\right)}{\partial t} + \dot{q}^j \frac{\partial\left(\delta q^i\right)}{\partial q^j} - \delta q^j \frac{\partial \dot{q}^i}{\partial q^j}, \qquad i = 1,2,3. \tag{1.48}$$

In the particular case of an initial flow in which the mass elements have no velocity, it results that

$$\left(\dot{q}^i\right)' = \frac{\partial\left(\delta q^i\right)}{\partial t}, \qquad i = 1,2,3. \tag{1.49}$$

The Eulerian perturbation of a generalised velocity component of a mass element is then equal to the partial time derivative of the corresponding generalised component of the mass element's Lagrangian displacement.

1.4 Perturbations of Mass Density, Gravitational Potential, Pressure, and Temperature

Besides the Lagrangian displacements and the perturbations of the velocity components of mass elements, the perturbations of mass density, gravitational potential, pressure, and temperature play an important role. Therefore, it is useful to derive expressions for the perturbations of these quantities in terms of the Lagrangian displacement.

1.4.1 Perturbations of Mass Density

An expression for the Lagrangian perturbation of the mass density can be derived from the mass conservation: mass elements preserve their mass during their displacements in the perturbed flow (Kato & Unno 1967, Roberts 1967).

Imagine an infinitesimal mass element that is characterised by Lagrangian parameters varying from a^1 to a^1+da^1, from a^2 to a^2+da^2, and from a^3 to a^3+da^3. The mass element occupies the infinitesimal volume $dV(\mathbf{r})$ in the initial flow, and the infinitesimal volume $dV_{\mathrm{p}}(\mathbf{r}_{\mathrm{p}})$ in the perturbed flow. These infinitesimal volumes are given by

$$dV(\mathbf{r}) = dx^1 dx^2 dx^3 = \frac{\partial(x^1, x^2, x^3)}{\partial(a^1, a^2, a^3)} da^1 da^2 da^3 \,, \tag{1.50}$$

$$dV_{\mathrm{p}}(\mathbf{r}_{\mathrm{p}}) = dx_{\mathrm{p}}^1\, dx_{\mathrm{p}}^2\, dx_{\mathrm{p}}^3 = \frac{\partial\left(x_{\mathrm{p}}^1, x_{\mathrm{p}}^2, x_{\mathrm{p}}^3\right)}{\partial(a^1, a^2, a^3)} da^1 da^2 da^3. \tag{1.51}$$

Be $\rho(\mathbf{r})$ the mass density of the mass element in the initial flow, and $\rho_{\mathrm{p}}(\mathbf{r}_{\mathrm{p}})$ that in the perturbed flow. On the ground of the mass conservation, one has that

$$\rho_{\mathrm{p}}(\mathbf{r}_{\mathrm{p}})\, dV_{\mathrm{p}}(\mathbf{r}_{\mathrm{p}}) = \rho(\mathbf{r})\, dV(\mathbf{r}) \tag{1.52}$$

or, by relations (1.50) and (1.51), that

$$\rho_{\mathrm{p}}(\mathbf{r}_{\mathrm{p}})\, X(\mathbf{r}) = \rho(\mathbf{r}) \,, \tag{1.53}$$

where $X(\mathbf{r})$ is the determinant of the displacement gradients defined as

$$X(\mathbf{r}) = \frac{\partial\left(x_{\mathrm{p}}^1, x_{\mathrm{p}}^2, x_{\mathrm{p}}^3\right)}{\partial(x^1, x^2, x^3)} . \tag{1.54}$$

From Eq. (1.53), it results that

$$\frac{\delta \rho}{\rho} = \frac{1}{X} - 1. \tag{1.55}$$

Expansion of the determinant of the displacement gradients up to the second order in the components of the Lagrangian displacement yields

$$X = 1 + \frac{\partial(\delta x^j)}{\partial x^j} + \frac{1}{2}\frac{\partial(\delta x^i)}{\partial x^i}\frac{\partial(\delta x^j)}{\partial x^j} - \frac{1}{2}\frac{\partial(\delta x^i)}{\partial x^j}\frac{\partial(\delta x^j)}{\partial x^i}. \tag{1.56}$$

By passing on to generalised coordinates, one has

$$X = 1 + \nabla_j \delta q^j + \frac{1}{2}(\nabla_i \delta q^i)(\nabla_j \delta q^j) - \frac{1}{2}(\nabla_j \delta q^i)(\nabla_i \delta q^j) \tag{1.57}$$

(Van Hoolst 1992).

At the first order in the components of the Lagrangian displacement, it results that

$$\frac{\delta \rho}{\rho} = -\frac{\partial(\delta x^j)}{\partial x^j} \tag{1.58}$$

and, in terms of generalised coordinates, that

$$\frac{\delta \rho}{\rho} = -\nabla_j \delta q^j. \tag{1.59}$$

By use of relation (1.30), it follows that

$$\rho' = -\nabla_j\left(\rho\, \delta q^j\right). \tag{1.60}$$

With the introduction of the specific volume, relation (1.53) can be rewritten as

$$\tau_{\mathrm{p}}(\mathbf{r}_{\mathrm{p}}) = X(\mathbf{r})\, \tau(\mathbf{r}). \tag{1.61}$$

At the first order in the components of the Lagrangian displacement, one has that

$$\frac{\delta \tau}{\tau} = \nabla_j \delta q^j, \tag{1.62}$$

i.e., the relative Lagrangian change of the specific volume of a mass element is equal to the divergence of the Lagrangian displacement of that element.

1.4.2 Perturbations of Gravitational Potential

The gravitational force **F**, here considered as the force acting on a unit mass, is conservative, in the sense that the work done around a closed orbit is zero, i.e.,

$$\oint \mathbf{F} \cdot d\mathbf{s} = 0. \tag{1.63}$$

By Stokes' theorem, the condition can be written as

$$\nabla \times \mathbf{F} = 0, \tag{1.64}$$

so that the gravitational force is derivable from a potential $\Phi(\mathbf{r})$, called the gravitational potential:

$$\mathbf{F} = -\nabla \Phi \tag{1.65}$$

(see, e.g., Goldstein 1957). Throughout this monograph, the convention with the minus sign in the relation between the gravitational force and the gravitational potential is adopted.

The gravitational potential is determined by the distribution of the mass density by means of Poisson's linear partial differential equation

$$\nabla^2 \Phi = 4\pi G \rho, \tag{1.66}$$

where ∇^2 is the Laplacian, and G the gravitational constant. A general solution of Poisson's differential equation for a three-dimensional region V with a regular boundary surface S is given as follows. If $\Phi(\mathbf{r})$ is twice continuously differentiable in V and continuously differentiable on S, and $\Phi(\mathbf{r}) = O(r^{-1})$ as $r \to \infty$, the solution can be presented in the form

$$\begin{aligned} \Phi(\mathbf{r}) = &-G \int_V \frac{\rho(\mathbf{r}')}{|\mathbf{r}' - \mathbf{r}|} dV(\mathbf{r}') \\ &+ \frac{1}{4\pi} \int_S \left[\frac{1}{|\mathbf{r}' - \mathbf{r}|} \frac{\partial \Phi}{\partial x'^k} - \Phi \frac{\partial}{\partial x'^k} \frac{1}{|\mathbf{r}' - \mathbf{r}|} \right] (\mathbf{r}')\, n^k(\mathbf{r}')\, dS(\mathbf{r}'), \end{aligned} \tag{1.67}$$

where $\mathbf{n}(\mathbf{r}')$ is the outward unit normal on the infinitesimal surface $dS(\mathbf{r}')$ (Korn & Korn 1968). The term with the surface integral is a solution of Laplace's differential equation for the gravitational potential,

$$\nabla^2 \Phi = 0, \tag{1.68}$$

which is the homogeneous equation. The form of the solution results from Green's fundamental formula of potential theory (see Appendix A).

Relative to the presence of a solution of Laplace's equation, Morse & Feshbach (1953) noted:

> If Φ is the integral defined in the equation
>
> $$\Phi(x, y, z) = \int\int\int \frac{q(x', y', z')}{4\pi R} dx' dy' dz'$$
>
> where $R = \sqrt{(x - x')^2 + (y - y')^2 + (z - z')^2}$ is the distance from the point x, y, z to the point x', y', z', then Φ is a solution of Poisson's equation $\nabla^2 \Phi = -q$, when q is a reasonable sort of function going to zero at infinity. It is not the only solution, for we can

add any amount of any solution of Laplace's equation $\nabla^2 \Psi = 0$ to Φ and still have a solution of $\nabla^2 \Phi = -q$ for the same q.

What amount and kind of solution Ψ we add depends on the boundary conditions of the individual problem.

When the mass density is piecewise continuous, the gravitational potential and its gradient are continuous at all points. Therefore, boundary conditions have to be imposed at boundaries at which the density distribution is discontinuous (Kellogg 1929, Dahlen & Tromp 1998).

The solution of Laplace's equation is identically zero for a spherically symmetric equilibrium star and for a linearly perturbed star because of the boundary conditions that are imposed on the star's surface in these cases. This is verified in Sects. 2.3 and 7.2. Anticipating these proofs, we express the gravitational potential at a point with position vector $\mathbf{r}$ in a spherically symmetric equilibrium star as

$$\Phi(\mathbf{r}) = -G \int_V \frac{\rho(\mathbf{r}')}{|\mathbf{r}' - \mathbf{r}|} dV(\mathbf{r}'), \tag{1.69}$$

and the gravitational potential at the point with position vector $\mathbf{r}_\mathrm{p}$ in a perturbed star as

$$\Phi_\mathrm{p}(\mathbf{r}_\mathrm{p}) = -G \int_{V_\mathrm{p}} \frac{\rho_\mathrm{p}(\mathbf{r}'_\mathrm{p})}{|\mathbf{r}'_\mathrm{p} - \mathbf{r}_\mathrm{p}|} dV_\mathrm{p}(\mathbf{r}'_\mathrm{p}). \tag{1.70}$$

The Lagrangian perturbation of the gravitational potential at a point with position vector $\mathbf{r}$ is given by

$$\delta\Phi(\mathbf{r}) = \Phi_\mathrm{p}(\mathbf{r}_\mathrm{p}) - \Phi(\mathbf{r}). \tag{1.71}$$

The right-hand member can be developed as follows. The integral over the volume V_p of the perturbed star is reduced to an integral over the volume V of the equilibrium star by use of the mass conservation expressed by Eq. (1.52). Moreover, the function $|\mathbf{r}'_\mathrm{p} - \mathbf{r}_\mathrm{p}|^{-1}$ can be expanded in a Taylor series around the six Cartesian coordinates $x^1, x^2, x^3, x'^1, x'^2, x'^3$ as

$$|\mathbf{r}'_\mathrm{p} - \mathbf{r}_\mathrm{p}|^{-1} = \left(1 + \delta x^j \frac{\partial}{\partial x^j} + \delta x'^j \frac{\partial}{\partial x'^j}\right) |\mathbf{r}' - \mathbf{r}|^{-1}. \tag{1.72}$$

It results that

$$\begin{aligned} \delta\Phi(\mathbf{r}) = {} & \delta x^j(\mathbf{r}) \left\{ \frac{\partial}{\partial x^j} \left[-G \int_V \frac{\rho(\mathbf{r}')}{|\mathbf{r}' - \mathbf{r}|} dV(\mathbf{r}') \right] \right\} (\mathbf{r}) \\ & - G \int_V \rho(\mathbf{r}')\, \delta x'^j(\mathbf{r}') \left(\frac{\partial}{\partial x'^j} |\mathbf{r}' - \mathbf{r}|^{-1} \right) (\mathbf{r}')\, dV(\mathbf{r}'). \end{aligned} \tag{1.73}$$

One derives an expression for the Eulerian perturbation of the gravitational potential at the point with position vector $\mathbf{r}$ by using relation (1.30) and the solution of Poisson's equation for the gravitational potential in the equilibrium star given by Eq. (1.69). One then has

$$\Phi'(\mathbf{r}) = -G \int_V \rho(\mathbf{r}')\, \delta x'^j(\mathbf{r}') \left(\frac{\partial}{\partial x'^j} |\mathbf{r}' - \mathbf{r}|^{-1} \right)(\mathbf{r}')\, dV(\mathbf{r}'). \tag{1.74}$$

The integral can be transformed by use of Gauss' integral theorem (Appendix A.1) and Eq. (1.60), so that

$$\Phi'(\mathbf{r}) = -G \int_V \frac{\rho'(\mathbf{r}')}{|\mathbf{r}' - \mathbf{r}|} dV(\mathbf{r}') - G \int_S \frac{\rho(\mathbf{r}')\, \delta x'^j(\mathbf{r}')}{|\mathbf{r}' - \mathbf{r}|} n_j(\mathbf{r}')\, dS(\mathbf{r}'). \tag{1.75}$$

The surface integral is identically zero, when the mass density vanishes on the surface of the spherically symmetric equilibrium star.

1.4.3 Perturbations of Pressure

When the pressure P is regarded as a function of the specific volume τ and the specific entropy S, the pressure P_{p} of a mass element in a perturbed star, with specific volume τ_{p} and specific entropy S_{p}, can be expanded around the pressure P of the same mass element in the equilibrium star, with specific volume τ and specific entropy S, as

$$\begin{aligned} P\left[\tau_{\mathrm{p}}(\mathbf{r}_{\mathrm{p}}), S_{\mathrm{p}}(\mathbf{r}_{\mathrm{p}})\right] = P[\tau(\mathbf{r}), S(\mathbf{r})] &+ \left(\frac{\partial P}{\partial \tau}\right)_S \left[\tau_{\mathrm{p}}(\mathbf{r}_{\mathrm{p}}) - \tau(\mathbf{r})\right] \\ &+ \left(\frac{\partial P}{\partial S}\right)_\tau \left[S_{\mathrm{p}}(\mathbf{r}_{\mathrm{p}}) - S(\mathbf{r})\right]. \end{aligned} \tag{1.76}$$

By use of definition (Appendix B.10) of the isentropic coefficient Γ_1, one has

$$\left(\frac{\partial P}{\partial \tau}\right)_S = -\frac{\Gamma_1 P}{\tau}. \tag{1.77}$$

Furthermore, by use of the thermodynamic relations (Appendix B.2) and definition (Appendix B.10) of the isentropic coefficient Γ_3, one has

$$\left(\frac{\partial P}{\partial S}\right)_\tau = -\frac{\partial}{\partial S}\frac{\partial U}{\partial \tau} = -\frac{\partial}{\partial \tau}\frac{\partial U}{\partial S} = -\left(\frac{\partial T}{\partial \tau}\right)_S = (\Gamma_3 - 1)\frac{T}{\tau}. \tag{1.78}$$

It then follows that

$$\delta P = -\frac{\Gamma_1 P}{\tau}\delta\tau + (\Gamma_3 - 1)\frac{T}{\tau}\,\delta S \tag{1.79}$$

or, equivalently,

$$\delta P = \frac{\Gamma_1 P}{\rho}\delta\rho + (\Gamma_3 - 1)\,\rho\, T\, \delta S. \tag{1.80}$$

An expression for the Eulerian perturbation of the pressure is readily derived by means of relation (1.30).

1.4.4 Perturbations of Temperature

When the temperature T is regarded as a function of the pressure P and the entropy S the expansion around the equilibrium values yields

$$\delta T = \left(\frac{\partial T}{\partial P}\right)_S \delta P + \left(\frac{\partial T}{\partial S}\right)_P \delta S. \tag{1.81}$$

From definition (Appendix B.10) of the isentropic coefficient Γ_2, it follows that

$$\left(\frac{\partial T}{\partial P}\right)_S = \frac{\Gamma_2 - 1}{\Gamma_2}\frac{T}{P}. \tag{1.82}$$

Furthermore, from Gibb's equation (Appendix B.1), it follows that

$$\left(\frac{\partial S}{\partial T}\right)_P = \frac{C_P}{T}, \tag{1.83}$$

where C_P is the specific heat at constant pressure per unit mass. Consequently, Eq. (1.81) can be rewritten as

$$\delta T = \frac{\Gamma_2 - 1}{\Gamma_2}\frac{T}{P}\delta P + \frac{T}{C_P}\delta S. \tag{1.84}$$

Chapter 2
The Equations Governing Linear Perturbations in a Quasi-Static Star

2.1 System of Coordinates

Consider an isolated, non-rotating and non-magnetic, spherically symmetric star, with total mass M and total radius R, which evolves so slowly that its structure can be regarded as static.

One generally adopts a frame of reference whose origin coincides with the star's mass centre, and considers the frame of reference as inertial. Be x^1, x^2, x^3 three orthonormal Cartesian coordinates. Because of the star's spherical symmetry, the direction and the orientation of the x^3-axis can be chosen arbitrarily. This axis is referred to as the pulsation axis in a number of asteroseismological studies, since it is used as the axis of symmetry for spherical harmonics.

In the frame of reference, one introduces spherical coordinates r, θ, ϕ by means of the transformation formulae from the orthonormal Cartesian coordinates x^1, x^2, x^3

$$\left.\begin{aligned} x^1 &= r \sin\theta \cos\phi, \\ x^2 &= r \sin\theta \sin\phi, \\ x^3 &= r \cos\theta, \end{aligned}\right\} \tag{2.1}$$

which do not involve the time explicitly.

The vectors of the local coordinate basis are then defined as

$$\frac{\partial}{\partial r}, \qquad \frac{\partial}{\partial \theta}, \qquad \frac{\partial}{\partial \phi}. \tag{2.2}$$

By Eq. (1.11), the covariant components of the metric tensor different from zero are

$$g_{11} = 1, \qquad g_{22} = r^2, \qquad g_{33} = r^2 \sin^2\theta. \tag{2.3}$$

P. Smeyers and T. Van Hoolst, *Linear Isentropic Oscillations of Stars: Theoretical Foundations*, Astrophysics and Space Science Library 371,
DOI 10.1007/978-3-642-13030-4_3, © Springer-Verlag Berlin Heidelberg 2010

Since $g_{ij} = 0$ as $i \neq j$, the local coordinate basis is orthogonal. By resolving Eqs. (1.33), one derives the contravariant components of the metric tensor different from zero

$$g^{11} = 1, \qquad g^{22} = \frac{1}{r^2}, \qquad g^{33} = \frac{1}{r^2 \sin^2 \theta}. \tag{2.4}$$

The Christoffel three-index symbols of the second kind different from zero are

$$\left.\begin{array}{lll} \Gamma^1_{22} = -r, & \Gamma^1_{33} = -r \sin^2 \theta, & \Gamma^2_{12} = \dfrac{1}{r}, \\ \Gamma^2_{33} = -\sin\theta \cos\theta, & \Gamma^3_{13} = \dfrac{1}{r}, & \Gamma^3_{23} = \cot\theta. \end{array}\right\} \tag{2.5}$$

2.2 Equation of Motion

With respect to the inertial frame of reference, a contravariant component k of the equation of motion of hydrodynamics for an ideal fluid takes the form

$$\rho \frac{d\dot{x}^k}{dt} = \rho F^k - \delta^{ki} \frac{\partial P}{\partial x^i}, \qquad k = 1, 2, 3, \ldots, \tag{2.6}$$

in which the δ^{ki} are Kronecker deltas, and F^k is the contravariant component of the resultant of the forces acting at distance upon a unit mass. The only force acting at distance that is considered here is the force of self-gravitation, which is derivable from the gravitational potential Φ as

$$F^k = -\delta^{ki} \frac{\partial \Phi}{\partial x^i}, \qquad k = 1, 2, 3. \tag{2.7}$$

The other acting force is the pressure force, which results from the differences in pressure exerted on the various parts of the surface of the mass element.

The contravariant k-component of the equation of motion can be rewritten as

$$\rho \left(\frac{\partial \dot{x}^k}{\partial t} + \dot{x}^i \frac{\partial \dot{x}^k}{\partial x^i} \right) = -\delta^{ki} \rho \frac{\partial \Phi}{\partial x^i} - \delta^{ki} \frac{\partial P}{\partial x^i}, \quad k = 1, 2, 3. \tag{2.8}$$

Since the star is spherically symmetric, it is appropriate to express the equation of motion in terms of the spherical coordinates r, θ, ϕ, which are a particular case of generalised coordinates. In terms of generalised coordinates q^1, q^2, q^3, a contravariant component k of the equation of motion takes the form

$$\rho \left(\frac{\partial \dot{q}^k}{\partial t} + \dot{q}^i \nabla_i \dot{q}^k \right) = -g^{ki} \rho \frac{\partial \Phi}{\partial q^i} - g^{ki} \frac{\partial P}{\partial q^i}, \qquad k = 1, 2, 3, \tag{2.9}$$

where $\nabla_i \dot{q}^k$ is the covariant derivative of the contravariant generalised velocity component $\dot{q}^{\,k}$ with respect to the coordinate q^i. For this equation, the reader is referred to Appendix C.

2.3 Equilibrium State of a Quasi-Static Star

In a quasi-static star, the macroscopic motions of the mass elements are supposed to be negligibly small, so that

$$\dot{q}^i = 0, \qquad i = 1, 2, 3. \tag{2.10}$$

Moreover, the star is, layer by layer, in *hydrostatic equilibrium*: in the radial direction, the inward gravitational attraction acting upon a mass element is equal in magnitude and opposite in sense to the outward pressure force exerted on the mass element. The hydrostatic equilibrium in the radial direction is expressed as

$$\frac{dP}{dr} = -\rho \frac{d\Phi}{dr}. \tag{2.11}$$

Under these conditions, it follows from Eq. (2.9) that the mass elements in a quasi-static star are not accelerated.

The condition of hydrostatic equilibrium can be expressed in the vectorial form

$$\nabla P = -\rho \nabla \Phi. \tag{2.12}$$

It results that the equipotential surfaces in a quasi-static star are also surfaces of equal pressure (isobaric surfaces). By taking the curl of both members of equality (2.12), one has

$$\nabla \rho \times \nabla \Phi = 0. \tag{2.13}$$

Hence, the equipotential surfaces are in addition surfaces of equal mass density (isopycnic surfaces). An equilibrium configuration in which the equipotential, the isobaric, and the isopycnic surfaces coincide is called a barotrope, according to a notion introduced by Laplace (Tassoul 1978). The property of barotropy implies that, in a quasi-static star, relations exist between the gravitational potential, the pressure, and the mass density in the radial direction.

The variation of the gravitational potential in a quasi-static star is solution of Poisson's differential equation (1.66), which reduces to the ordinary linear, nonhomogeneous, second-order differential equation

$$\frac{d^2\Phi}{dr^2} + \frac{2}{r}\frac{d\Phi}{dr} = 4\pi G\rho. \tag{2.14}$$

It can be solved by means of the method of the variation of the constants. The solution that remains finite at $r = 0$ can be expressed as

$$\Phi(r) = D - 4\pi G \left[r^{-1} \int_0^r \rho\left(r'\right) r'^2 \, dr' - \int_0^r \rho\left(r'\right) r' \, dr' \right], \tag{2.15}$$

where D is an arbitrary constant.

The external gravitational potential is determined by Laplace's equation (1.68), which reduces to

$$\frac{d^2\Phi}{dr^2} + \frac{2}{r}\frac{d\Phi}{dr} = 0. \tag{2.16}$$

The solution that tends to zero as $r \to \infty$ has the form

$$\Phi_e(r) = A\, r^{-1}, \tag{2.17}$$

where A is an arbitrary constant.

From the continuity of the gravitational potential and its first derivative at $r = R$, it results that

$$A = -GM, \qquad D = -4\pi G \int_0^R \rho\left(r'\right) r' dr'. \tag{2.18}$$

Solution (2.15) for the internal gravitational potential can then be rewritten as

$$\Phi(r) = -4\pi G \left[r^{-1} \int_0^r \rho\left(r'\right) r'^2 \, dr' + \int_r^R \rho\left(r'\right) r' dr' \right]. \tag{2.19}$$

The gravity inside the star is given by

$$g \equiv \frac{d\Phi}{dr} = \frac{Gm(r)}{r^2}. \tag{2.20}$$

Here $m(r)$ is the mass contained inside the sphere with radius r, which is determined as

$$m(r) = \int_0^r 4\pi\rho\left(r'\right) r'^2 \, dr'. \tag{2.21}$$

The differential form of this definition is

$$\frac{dm(r)}{dr} = 4\pi\rho\, r^2. \tag{2.22}$$

Solution (2.19) for the internal gravitational potential can also be derived from the general solution (1.67) of Poisson's equation. To this end, one uses the expansion for the inverse of the distance between two points with position vectors $\mathbf{r}$ and $\mathbf{r}'$ and with spherical coordinates r, θ, ϕ and r', θ', ϕ':

$$
\begin{aligned}
\left|\mathbf{r}' - \mathbf{r}\right|^{-1} = \sum_{\lambda=0}^{\infty} f_\lambda\left(r, r'\right) \sum_{\mu=0}^{\lambda} \epsilon_\mu \frac{(\lambda-\mu)!}{(\lambda+\mu)!} \\
P_\lambda^\mu(\cos\theta)\, P_\lambda^\mu\left(\cos\theta'\right)\, \cos\left[\mu\left(\phi-\phi'\right)\right],
\end{aligned} \tag{2.23}
$$

where

$$
\left.
\begin{aligned}
f_\lambda\left(r, r'\right) &= r'^{\lambda}/r^{\lambda+1} \quad \text{if} \quad r' < r, \\
f_\lambda\left(r, r'\right) &= r^{\lambda}/r'^{\lambda+1} \quad \text{if} \quad r' > r,
\end{aligned}
\right\} \tag{2.24}
$$

and ϵ_μ is the Neuman factor defined as

$$
\epsilon_0 = 1, \qquad \epsilon_n = 2 \quad \text{for} \quad n = 1, 2, 3, \ldots \tag{2.25}
$$

(Morse & Feshbach 1953, Chandrasekhar & Lebovitz 1964). The Legendre polynomials $P_\lambda(\cos\theta)$ and the associated polynomials $P_\lambda^\mu(\cos\theta)$ are not normalised (see, e.g., Jahnke & Emde 1945; for a detailed account of the Legendre polynomials, see Robin 1957).

Expansion (2.23) can also be expressed in terms of the complex spherical harmonics

$$
Y_\lambda^\mu(\theta, \phi) = P_\lambda^{|\mu|}(\cos\theta)\, \exp(\mathrm{i}\mu\phi) \tag{2.26}
$$

(see Appendix D) and then takes the form

$$
\left|\mathbf{r}' - \mathbf{r}\right|^{-1} = \sum_{\lambda=0}^{\infty} f_\lambda\left(r, r'\right) \sum_{\mu=-\lambda}^{\lambda} \frac{(\lambda-|\mu|)!}{(\lambda+|\mu|)!} \overline{Y_\lambda^\mu}\left(\theta', \phi'\right)\, Y_\lambda^\mu(\theta, \phi). \tag{2.27}
$$

A horizontal line above a quantity denotes the complex conjugate of that quantity. Substitution of the expansion into the second term of general solution (1.67) of Poisson's equation yields

$$
\begin{aligned}
&\frac{1}{4\pi}\int_S \left[\frac{1}{|\mathbf{r}'-\mathbf{r}|}\frac{\partial\Phi}{\partial x'^k} - \Phi\left(\mathbf{r}'\right)\frac{\partial}{\partial x'^k}\frac{1}{|\mathbf{r}'-\mathbf{r}|}\right]\left(\mathbf{r}'\right) n^k\left(\mathbf{r}'\right) dS\left(\mathbf{r}'\right) \\
&= \sum_{\lambda=0}^{\infty} \frac{r^\lambda}{R^{\lambda+1}} \sum_{\mu=-\lambda}^{\lambda} \frac{(\lambda-|\mu|)!}{(\lambda+|\mu|)!} Y_\lambda^\mu(\theta, \phi) \\
&\quad \frac{1}{4\pi}\int_S \overline{Y_\lambda^\mu}\left(\theta', \phi'\right)\left(\frac{d\Phi}{dr}\right)_R R^2 \sin\theta' d\theta' d\phi'
\end{aligned}
$$

$$+\sum_{\lambda=0}^{\infty}(\lambda+1)\frac{r^{\lambda}}{R^{\lambda+2}}\sum_{\mu=-\lambda}^{\lambda}\frac{(\lambda-|\mu|)!}{(\lambda+|\mu|)!}Y_{\lambda}^{\mu}(\theta,\phi)$$
$$\frac{1}{4\pi}\int_{S}\overline{Y_{\lambda}^{\mu}}\left(\theta',\phi'\right)\,\Phi(R)\,R^{2}\sin\theta' d\theta' d\phi'. \tag{2.28}$$

The surface integrals are equal to zero for $\lambda \neq 0$ because of the orthogonality relation (Appendix D.4) between spherical harmonics. It follows that

$$\frac{1}{4\pi}\int_{S}\left[\frac{1}{|\mathbf{r}'-\mathbf{r}|}\frac{\partial\Phi}{\partial x'^{k}}-\Phi\left(\mathbf{r}'\right)\frac{\partial}{\partial x'^{k}}\frac{1}{|\mathbf{r}'-\mathbf{r}|}\right]\left(r'\right)n^{k}\left(\mathbf{r}'\right)dS\left(\mathbf{r}'\right)$$
$$=R\left[\left(\frac{d\Phi}{dr}\right)_{R}+\frac{1}{R}\Phi(R)\right]. \tag{2.29}$$

On account of the external solution (2.17), the continuity of the gravitational potential and its first derivative at $r = R$ implies that

$$\left(\frac{d\Phi}{dr}\right)_{R}+\frac{1}{R}\Phi(R)=0. \tag{2.30}$$

Hence, because of the boundary conditions at $r = R$, the solution of Laplace's equation in general solution (1.67) of Poisson's equation is identically zero, and the solution reduces to solution (1.69). Then, by substituting expansion (2.27) into it and taking into account the orthogonality relation between spherical harmonics, one recovers solution (2.19) for $\Phi(r)$.

2.4 Eulerian Form of the Equations Governing Linear Perturbations

The free small perturbations of an isolated, non-rotating and non-magnetic, spherically symmetric quasi-static star that is in hydrostatic equilibrium are governed by the linearised forms of the components of the equation of motion given by Eqs. (2.9). The equations are perturbed in an Eulerian way by the use of properties and relations presented in Chap. 1. One should keep in mind that the functions of the coordinates of geometrical points, as the components of the metric tensor and the Christoffel three-index symbols of the second kind, remain unperturbed, and that the unperturbed star is in hydrostatic equilibrium and contains no velocity field. It results that

$$\rho\,\frac{\partial^{2}\left(\delta q^{k}\right)}{\partial t^{2}}=-g^{ki}\left(\rho\,\frac{\partial\Phi'}{\partial q^{i}}+\rho'\frac{\partial\Phi}{\partial q^{i}}+\frac{\partial P'}{\partial q^{i}}\right),\qquad k=1,2,3, \tag{2.31}$$

or, more explicitly in terms of spherical coordinates,

$$\left.\begin{aligned}
\rho\,\frac{\partial^2(\delta r)}{\partial t^2} &= -\rho\,\frac{\partial \Phi'}{\partial r} - \rho'\frac{d\Phi}{dr} - \frac{\partial P'}{\partial r},\\
\rho\,\frac{\partial^2(\delta \theta)}{\partial t^2} &= -\rho\,\frac{1}{r^2}\frac{\partial}{\partial \theta}\left(\frac{P'}{\rho} + \Phi'\right),\\
\rho\,\frac{\partial^2(\delta \phi)}{\partial t^2} &= -\rho\,\frac{1}{r^2\sin^2\theta}\frac{\partial}{\partial \phi}\left(\frac{P'}{\rho} + \Phi'\right).
\end{aligned}\right\} \tag{2.32}$$

The linearised components of the equation of motion contain the three components $\delta r(t,r,\theta,\phi)$, $\delta\theta(t,r,\theta,\phi)$, $\delta\phi(t,r,\theta,\phi)$ of the Lagrangian displacement with respect to the orthogonal coordinate basis $\partial/\partial r$, $\partial/\partial\theta$, $\partial/\partial\phi$, and the Eulerian perturbations of the mass density, $\rho'(t,r,\theta,\phi)$, the pressure, $P'(t,r,\theta,\phi)$, and the gravitational potential, $\Phi'(t,r,\theta,\phi)$. Since the number of unknown functions amounts to six, other equations must be added to the three components of the perturbed equation of motion.

2.4.1 *First Additional Equation*

As first additional equation, Eq. (1.60) is adopted, which results from the mass conservation of the elements in the perturbed flow and can be written as

$$\frac{\rho'}{\rho} + \frac{1}{\rho}\frac{d\rho}{dr}\,\delta r + \alpha = 0. \tag{2.33}$$

Here $\alpha = \nabla_j \delta q^j$ is the divergence of the Lagrangian displacement of the mass element. By use of the general formula

$$\nabla_j \delta q^j = \frac{1}{\sqrt{g_{\rm m}}}\frac{\partial}{\partial q^j}\left(\sqrt{g_{\rm m}}\,\delta q^j\right) \tag{2.34}$$

with $g_{\rm m} = \det\left(g_{ij}\right)$ (see, e.g., McConnell 1957), one has, in terms of the components δr, $\delta\theta$, $\delta\phi$ of the Lagrangian displacement,

$$\alpha = \frac{1}{r^2}\frac{\partial}{\partial r}\left(r^2\,\delta r\right) + \frac{1}{\sin\theta}\frac{\partial}{\partial\theta}(\sin\theta\,\delta\theta) + \frac{\partial(\delta\phi)}{\partial\phi}. \tag{2.35}$$

Equation (2.33) corresponds to the perturbed form of the continuity equation after integration with respect to time.

2.4.2 Second Additional Equation

Equation (1.80) serves as second additional equation. After differentiation with respect to time, it takes the form

$$\frac{\partial}{\partial t}\left(\frac{P'}{P}+\frac{1}{P}\frac{dP}{dr}\,\delta r\right)-\Gamma_1\frac{\partial}{\partial t}\left(\frac{\rho'}{\rho}+\frac{1}{\rho}\frac{d\rho}{dr}\,\delta r\right)=\frac{(\Gamma_3-1)\,\rho\,T}{P}\frac{d}{dt}\,\delta S. \quad (2.36)$$

When the stellar perturbation is reversible, the time derivative of the Lagrangian perturbation of the specific entropy, $d(\delta S)/dt$, can be determined by means of the equality

$$T\frac{dS}{dt}=\frac{dQ}{dt}, \quad (2.37)$$

where dQ/dt is the rate of energy absorption, positive or negative, per unit mass. An equation for dQ/dt is provided by the hydrodynamical equation for the rate of change of the thermal energy. For the sake of simplicity, consider a stellar layer in which energy is liberated by thermonuclear reactions and transferred by radiation, so that the equation takes the form

$$\frac{dQ}{dt}=\frac{1}{\rho}\left(\rho\,\varepsilon_1-\nabla\cdot\mathbf{F}_{\mathrm{Rad}}\right). \quad (2.38)$$

ε_1 is the amount of energy that is liberated in the stellar medium by thermonuclear reactions, per unit time and unit mass, and $\mathbf{F}_{\mathrm{Rad}}$, the flux of radiative energy. The terms contained in dQ/dt render the non-adiabatic effects.

Non-adiabatic effects can create entropy. This can be illustrated as follows. Because of the second law of thermodynamics, one has, for a mass M that occupies a volume V surrounded by a surface S',

$$\frac{d}{dt}\int_M S\,dm \geq \int_M \frac{1}{T}\frac{dQ}{dt}\,dm. \quad (2.39)$$

By use of Eq. (2.38) in the right-hand member and application of Gauss' integral theorem, it follows that

$$\frac{d}{dt}\int_M S\,dm+\int_{S'}\frac{\mathbf{F}_{\mathrm{Rad}}\cdot\mathbf{n}}{T}\,dS' \geq \int_V \frac{1}{T}\left(\rho\,\varepsilon_1-\frac{\mathbf{F}_{\mathrm{Rad}}}{T}\cdot\nabla T\right)dV. \quad (2.40)$$

When the flux of radiative energy is given by an equation of the form

$$\mathbf{F}_{\mathrm{Rad}}=-\lambda\,\nabla T, \quad (2.41)$$

it further follows that

$$\frac{d}{dt}\int_M S\,dm-\int_{S'}\frac{\mathbf{F}_{\mathrm{Rad}}\cdot(-\mathbf{n})}{T}dS' \geq \int_V \frac{1}{T}\left[\rho\,\varepsilon_1+\lambda\frac{(\nabla T)^2}{T}\right]dV. \quad (2.42)$$

The left-hand member can be interpreted as the difference between the rate of change of the amount of entropy in the mass and the amount of entropy that flows into it through its surface per unit time. Since this difference is non-negative, the quantity

$$\Theta \equiv \frac{1}{T}\left[\rho\,\varepsilon_1 + \lambda\frac{(\nabla T)^2}{T}\right] \tag{2.43}$$

is interpreted as the amount of entropy that is created inside the mass due to the non-adiabatic effects per unit time and unit volume.

A thermodynamic process is reversible when it produces no entropy in the medium, i.e., if $\Theta = 0$. When the creation of entropy is sufficiently small, the reversibility is often used as an approximation (Thompson 1972). This approximation is generally adopted in the context of stellar oscillations.

In this approximation, taking the Lagrangian perturbation of Eq. (2.37), one has that

$$\frac{d}{dt}\delta S = \delta\left(\frac{1}{T}\frac{dQ}{dt}\right). \tag{2.44}$$

A quasi-static star is regarded to be, layer by layer, in *thermal equilibrium*, i.e.,

$$\frac{dQ}{dt} = 0. \tag{2.45}$$

According to Eq. (2.38), this equilibrium implies that the difference between the amount of energy that is radiated, per unit time, through the spherical surface with radius $r + dr$ and the amount of energy that is radiated, per unit time, through the spherical surface with radius r is due to the rate of liberation of energy by the thermonuclear reactions in the layer between the two spherical surfaces:

$$\frac{d}{dr}\left[4\pi r^2 F_{\text{Rad}}(r)\right] = 4\pi r^2\rho\,\varepsilon_1. \tag{2.46}$$

Because of the thermal equilibrium, Eq. (2.44) reduces to

$$\frac{d}{dt}\delta S = \frac{1}{T}\delta\frac{dQ}{dt}. \tag{2.47}$$

This equation is added to Eq. (2.36).

2.4.3 Third Additional Equation

The third additional equation is Poisson's perturbed differential equation

$$\nabla^2\Phi' = 4\pi G\rho'. \tag{2.48}$$

The Laplacian of a scalar function Φ' is defined as the divergence of the gradient of this function. The gradient of the function has the covariant components

$$\frac{\partial \Phi'}{\partial q^i}, \qquad i = 1, 2, 3.$$

One passes on to the contravariant components by multiplying by the contravariant components of the metric tensor, g^{ij}:

$$g^{ij} \frac{\partial \Phi'}{\partial q^i}, \qquad j = 1, 2, 3.$$

Taking the divergence, one obtains the Laplacian of Φ':

$$\nabla^2 \Phi' = \frac{1}{\sqrt{g_{\mathrm{m}}}} \frac{\partial}{\partial q^j} \left(\sqrt{g_{\mathrm{m}}}\, g^{ij} \frac{\partial \Phi'}{\partial q^i} \right) \tag{2.49}$$

(see, e.g., Adler et al. 1965). In terms of spherical coordinates, Poisson's perturbed differential equation then takes the form

$$\frac{1}{r^2} \frac{\partial}{\partial r} \left(r^2 \frac{\partial \Phi'}{\partial r} \right) - \frac{1}{r^2} \mathscr{L}^2 \Phi' = 4\pi G \rho', \tag{2.50}$$

where $\mathscr{L}^2$ is the Legendrian defined as

$$\mathscr{L}^2 = -\left[\frac{1}{\sin\theta} \frac{\partial}{\partial \theta} \left(\sin\theta \frac{\partial}{\partial \theta} \right) + \frac{1}{\sin^2\theta} \frac{\partial^2}{\partial \phi^2} \right]. \tag{2.51}$$

For conclusion, Eqs. (2.32), (2.33), (2.36), (2.47), and (2.50) form a system of partial differential equations for the three components of the Lagrangian displacement of a mass element and the Eulerian perturbations of the mass density, the pressure, and the gravitational potential. They are expressed in an Eulerian form, apart from the right-hand member of Eq. (2.36), and Eq. (2.47). The system is at least of the seventh order in time. The term "at least" is here required, since the factor $\delta(dQ/dt)$ in Eq. (2.47) may involve additional time derivatives. The coefficients of the partial differential equations are time-independent for a quasi-static star.

2.5 Lagrangian Form of the Equations Governing Linear Perturbations

By use of relation (1.30) between the Lagrangian and the Eulerian perturbation of a physical quantity, and condition of the hydrostatic equilibrium expressed by Eq. (2.11), the components of the perturbed equation of motion given by Eqs. (2.32) can be transformed into

$$\left.\begin{aligned}\rho \frac{\partial^2(\delta r)}{\partial t^2} &= -\rho \frac{\partial(\delta \Phi)}{\partial r} - \delta\rho \frac{d\Phi}{dr} - \frac{\partial(\delta P)}{\partial r},\\ \rho \frac{\partial^2(\delta \theta)}{\partial t^2} &= -\rho \frac{1}{r^2}\frac{\partial}{\partial \theta}\left(\frac{\delta P}{\rho} + \delta\Phi\right),\\ \rho \frac{\partial^2(\delta \phi)}{\partial t^2} &= -\rho \frac{1}{r^2 \sin^2\theta}\frac{\partial}{\partial \phi}\left(\frac{\delta P}{\rho} + \delta\Phi\right).\end{aligned}\right\} \tag{2.52}$$

To the spherical coordinates r, θ, ϕ of the mass elements in the equilibrium star, the role of Lagrangian parameters a^1, a^2, a^3 characterising the mass elements can be attributed. The time derivatives in the left-hand members are taken while these parameters are kept constant. The equations are therefore expressed in a Lagrangian formalism. As observed by Chandrasekhar & Lebovitz (1964), they have the same *form* as Eqs. (2.32).

The additional Eqs. (2.33), (2.36), and (2.50) can also be expressed in a Lagrangian form:

$$\frac{\delta\rho}{\rho} + \alpha = 0, \tag{2.53}$$

$$\frac{d}{dt}\frac{\delta P}{P} - \Gamma_1 \frac{d}{dt}\frac{\delta\rho}{\rho} = \frac{(\Gamma_3 - 1)\,\rho\, T}{P}\frac{d}{dt}\delta S, \tag{2.54}$$

$$\left[\frac{1}{r^2}\frac{\partial}{\partial r}\left(r^2\frac{\partial}{\partial r}\right) - \frac{1}{r^2}\mathscr{L}^2\right]\left(\delta\Phi - \frac{d\Phi}{dr}\delta r\right) = 4\pi G\rho\left(\frac{\delta\rho}{\rho} - \frac{1}{\rho}\frac{d\rho}{dr}\delta r\right). \tag{2.55}$$

The Eulerian form of the linearly perturbed equations is generally a convenient form for the study of oscillations in a quasi-static star. The Lagrangian form may be more appropriate in specific contexts. Pesnell (1990), on his part, expressed a marked preference for the use of Lagrangian perturbations. In this monograph, a Lagrangian description is adopted mainly in the context of Hamilton's variational principle for the isentropic oscillations of a quasi-static star and the determination of the rate of change of an isentropic oscillation period in a rapidly evolving star.

Chapter 3
Deviations from the Hydrostatic and Thermal Equilibrium in a Quasi-Static Star

3.1 Introduction

Insight into the nature of Lagrangian displacement fields in a quasi-static star can be gained by consideration of local deviations from the hydrostatic and thermal equilibrium. Although circumspection is imperative, local analyses may be useful for the understanding of physical properties of Lagrangian displacement fields and lead to qualitatively correct results (see, e.g., Cox 1980).

3.2 Resolution of the Force Acting upon a Moving Mass Element

In order to interpret the acting forces that are represented by the terms in the right-hand members of the components of the linearised equation of motion, we concentrate on an individual mass element with a unit volume that rises from a layer at the radial distance $r - \delta r$ from the star's mass centre up to the layer at the radial distance r.

In the layer from which it rises, the mass element is subject to the gravitational force with radial component

$$\left(-\rho\,\frac{d\Phi}{dr}\right) - \delta r\,\frac{d}{dr}\left(-\rho\,\frac{d\Phi}{dr}\right). \tag{3.1}$$

In the layer of arrival, the mass element is subject to the gravitational force with radial component

$$\left(-\rho\,\frac{d\Phi}{dr}\right) - \delta r\,\frac{d}{dr}\left(-\rho\,\frac{d\Phi}{dr}\right) + \delta\left(-\rho\,\frac{\partial\Phi}{\partial r}\right). \tag{3.2}$$

P. Smeyers and T. Van Hoolst, *Linear Isentropic Oscillations of Stars: Theoretical Foundations*, Astrophysics and Space Science Library 371,
DOI 10.1007/978-3-642-13030-4_4,

The difference between the gravitational force acting on the mass element and that acting on a mass element in the surrounding layer is

$$\left[\left(-\rho\,\frac{d\Phi}{dr}\right)-\delta r\,\frac{d}{dr}\left(-\rho\,\frac{d\Phi}{dr}\right)+\delta\left(-\rho\,\frac{\partial\Phi}{\partial r}\right)\right]-\left(-\rho\,\frac{d\Phi}{dr}\right)$$
$$=-\delta r\,\frac{d}{dr}\left(-\rho\,\frac{d\Phi}{dr}\right)+\delta\left(-\rho\,\frac{\partial\Phi}{\partial r}\right)=\left(-\rho\,\frac{\partial\Phi}{\partial r}\right)'. \tag{3.3}$$

This differential force can be decomposed into a sum of two forces in the radial direction consisting respectively of the buoyancy force of Archimedes and the perturbed gravitational force:

$$\left(-\rho\,\frac{\partial\Phi}{\partial r}\right)'=-\rho'\,\frac{d\Phi}{dr}-\rho\,\frac{\partial\Phi'}{\partial r}. \tag{3.4}$$

The buoyancy force of Archimedes is oriented in the sense of the increasing or decreasing radial coordinate r depending on whether the Eulerian perturbation of the mass density, $\rho'(r)$, is negative or positive. On the line of reasoning followed here, the Eulerian perturbation of the mass density can be interpreted as the difference between the mass density of the risen mass element and that of the surrounding layer. Hence, a mass element is subject to a positive or a negative buoyancy force of Archimedes depending on whether its mass density is smaller or larger than the mass density of the surrounding layer. In illustration, ballooning in the Earth's atmosphere is possible if the balloon is filled with a gas whose mass density is smaller than that of the surrounding atmospheric layer.

Similarly, the force represented by the term $-(\partial P'/\partial r)$ in the right-hand member of the radial component of the linearised equation of motion can be interpreted as a differential force.

The force acting upon a moving mass element can thus be resolved into the sum of the buoyancy force of Archimedes and the perturbed gravitational force, on one side, and the perturbed pressure force, on the other side. In many cases, the perturbed gravitational force is of secondary importance, so that the resolution is mainly a resolution into the buoyancy force of Archimedes and the perturbed pressure force.

The resolution is meaningful for two reasons.

First, it is meaningful because of the relation existing between the duration of the time during which a mass element moves and the force that mainly acts on it. A mass element that moves slowly during a longer time has the possibility of adapting its pressure to a larger extent to the pressure of its surroundings. Hence, the perturbed pressure force is small, and the buoyancy force of Archimedes is the dominant force. On the contrary, when a mass element moves fast during a shorter time, the adjustment of the mass element's pressure to the pressure of the surroundings is smaller, and the perturbed pressure force is the dominant force.

Secondly, the resolution is meaningful relative to the creation of vorticity, for which the curl of the acting force is a measure. Being represented by the gradient of a scalar function, the perturbed pressure force does not contribute to the creation of vorticity. Only the buoyancy force of Archimedes and the perturbed gravitational force do it.

However, with respect to the curl of the buoyancy force of Archimedes, it may be observed that

$$\nabla \times \left(-\rho' \, \nabla \Phi\right) = \rho' \, \nabla \times (\nabla \Phi) - (\nabla \rho') \times (\nabla \Phi). \tag{3.5}$$

The first term in the right-hand member is identically zero. As $\nabla \Phi$ is a purely radial vector, the second term in the right-hand member is a vector perpendicular to the radial direction. Hence, the buoyancy force of Archimedes can create vorticity only around a direction perpendicular to the local radial direction, i.e., around a direction perpendicular to the local normal to the equipotential surface in the equilibrium star.

An analogous property holds for the perturbed gravitational force.

One arrives at the conclusion that faster motions of mass elements in shorter times are dominated by the perturbed pressure force and generate a smaller vorticity, and that slower motions of mass elements in longer times are dominated by the buoyancy force of Archimedes and generate a larger vorticity. In no case do the acting forces generate vorticity around the local normal to the equipotential surface of the equilibrium star.

3.3 The Dynamic Time-Scale of a Star

With each star, a dynamic time-scale can be associated. For its definition, imagine a mass element that is accelerated at a radial distance r of the star's centre under the influence of the perturbed pressure force. Suppose that the effects of the buoyancy force of Archimedes and the perturbed gravitational force are negligible. The radial component of the linearised equation of motion then reduces to

$$\rho \, \frac{\partial^2 (\delta r)}{\partial t^2} = -\frac{\partial P'}{\partial r}. \tag{3.6}$$

Be the Eulerian perturbation of the pressure force a fraction γ of the equilibrium pressure force:

$$\left(-\frac{\partial P}{\partial r}\right)' = \gamma \left(-\frac{dP}{dr}\right) = \gamma \, \rho \, \frac{G \, m(r)}{r^2}. \tag{3.7}$$

Equation (3.6) then becomes

$$\frac{\partial^2 (\delta r)}{\partial t^2} = \gamma \, \frac{G \, m(r)}{r^2}. \tag{3.8}$$

Depending on whether γ is positive or negative, the mass element is accelerated in the sense of the increasing or decreasing radial coordinate.

Following an approach of Cox & Giuli (1968), we assume that the mass element is subject to a uniformly accelerated motion in the sense of the decreasing radial coordinate and moves over a fraction α of the total radius R in a time interval Δt, so that

$$\alpha R = -\frac{1}{2}\frac{\partial^2(\delta r)}{\partial t^2}(\Delta t)^2. \tag{3.9}$$

It follows that

$$\Delta t = \left[\frac{2\,\alpha R}{-\partial^2(\delta r)/\partial t^2}\right]^{1/2} = \left[\frac{2\,\alpha R}{-\gamma\, G\, m(r)/r^2}\right]^{1/2}. \tag{3.10}$$

For a mass element situated on the star's surface, one has

$$\Delta t = \left(\frac{2\,\alpha}{-\gamma}\right)^{1/2}\left(\frac{R^3}{GM}\right)^{1/2}. \tag{3.11}$$

The time interval has the order of magnitude

$$\tau_{\mathrm{dyn}} = \left(\frac{R^3}{GM}\right)^{1/2} = \left(\frac{3}{4\pi G\bar{\rho}}\right)^{1/2}, \tag{3.12}$$

where $\bar{\rho}$ is the star's mean mass density. τ_{dyn} is known as the star's dynamic time-scale and is defined as the time needed by a mass element for moving, in a uniformly accelerated way, from the star's surface over a significant part of the star's radius, due to a deviation of the pressure force from its equilibrium value.

For example, for $\alpha = 0.1$ and $\gamma = -0.01$, one obtains

$$\left(\frac{2\,\alpha}{-\gamma}\right)^{1/2} = 4.5.$$

In the particular case of the Sun, the dynamic time-scale is

$$\left(\tau_{\mathrm{dyn}}\right)_\odot = \left(\frac{R_\odot^3}{GM_\odot}\right)^{1/2} = 1.59\times 10^3\ \mathrm{sec} = 26.6\ \mathrm{min},$$

so that

$$(\Delta t)_\odot = 4.5\times 26.6\ \mathrm{min} \simeq 2\ \mathrm{h}.$$

The dynamic time-scale of a star can alternatively be defined as the time needed by an acoustic wave for propagating over a part of the star's radius. The time needed by an acoustic wave for propagating through a layer with thickness Δr is given by

$$\tau_{\mathrm{ac}} = \frac{\Delta r}{c}, \tag{3.13}$$

where $c \equiv (\Gamma_1 P/\rho)^{1/2}$ is the isentropic sound velocity (see Appendix J.1). When one introduces the usual dimensionless variables P^* and ρ^* as

$$P = \frac{GM^2}{4\pi R^4} P^*, \quad \rho = \frac{M}{4\pi R^3} \rho^* \tag{3.14}$$

and sets $\Delta r = \alpha R$, where α has a value between 0 and 1, one obtains

$$\tau_{\rm ac} = \alpha \left(\frac{\rho^*}{\Gamma_1 P^*} \right)^{1/2} \left(\frac{R^3}{GM} \right)^{1/2} . \tag{3.15}$$

For the choice of a typical value of ρ^*, we set the mass density ρ equal to mean mass density $\bar{\rho} = 3M/\left(4\pi R^3\right)$, so that $\rho^* = 3$. Next, for the choice of a typical value for P^*, we use condition (2.11) of the hydrostatic equilibrium and Eq. (2.20). After division by $4\pi\rho\, r^2$, one has

$$\frac{dP}{dm} = -\frac{G\, m(r)}{4\pi r^4}.$$

The left-hand member can be approximated by the difference between the surface value of the pressure, P_S, and the central value, $P_{\rm c}$, divided by the total mass:

$$\frac{dP}{dm} \simeq \frac{P_S - P_{\rm c}}{M}.$$

In the right-hand member we set $r = R/2$ and $m(r) = M/2$. It results that

$$P_{\rm c} - P_S = \frac{2\,GM^2}{\pi R^4}.$$

By setting $P = (P_{\rm c} - P_S)/2$, one then finds $P^* = 4$. Furthermore, we set $\Gamma_1 = 5/3$. By adopting $\alpha = 1$, one recovers the dynamic time-scale obtained above, apart from a factor of order unity.

3.4 Energy Exchange Between Moving Mass Elements

Mass elements exchange energy with their surroundings during their motions. For an estimation of this exchange, it is appropriate to start from Eqs. (2.36) and (2.47):

$$\frac{\partial}{\partial t} \left(\frac{\delta P}{P} - \Gamma_1 \frac{\delta\rho}{\rho} \right) = \frac{(\Gamma_3 - 1)\,\rho}{P}\, \delta\, \frac{dQ}{dt} . \tag{3.16}$$

Imagine a mass element that is moving during a multiple β of the star's dynamic time-scale. In order to make the left-hand member of the equation dimensionless, one can adopt a dimensionless time variable t^+ that is related to the multiple of the dynamic time-scale and the time t as

$$t = \beta \left(\frac{R^3}{GM} \right)^{1/2} t^+. \tag{3.17}$$

In the right-hand member, the dimensionless pressure P^* and mass density ρ^* defined by Eqs. (3.14) can be used. A dimensionless form for dQ/dt can be introduced by means of Eq. (2.38), so that

$$\frac{dQ}{dt} = \frac{L}{M} \left(\frac{dQ}{dt} \right)^*, \tag{3.18}$$

where L is the star's luminosity, and $(dQ/dt)^*$ is dimensionless.

Equation (3.16) can then be rewritten as

$$\frac{\partial}{\partial t^+} \left(\frac{\delta P}{P} - \Gamma_1 \frac{\delta \rho}{\rho} \right) = \beta\, C \, \frac{(\Gamma_3 - 1)\, \rho^*}{P^*} \, \delta \left(\frac{dQ}{dt} \right)^*. \tag{3.19}$$

C is the ratio of the star's dynamic time-scale to the star's Helmholtz–Kelvin time-scale,

$$\tau_{\mathrm{HK}} = \frac{GM^2}{RL}, \tag{3.20}$$

which was estimated by Helmholtz and Kelvin as the time-scale for a contracting and cooling star (Chandrasekhar 1939, Kippenhahn & Weigert 1990). The ratio C can be expressed in solar units as

$$C = 1.61 \times 10^{-12} \left(\frac{M_\odot}{M} \right)^{5/2} \left(\frac{R}{R_\odot} \right)^{5/2} \frac{L}{L_\odot} \tag{3.21}$$

and is generally very small. It is larger for stars with a higher luminosity and a larger radius. This is particularly the case for very luminous stellar models with a low mass that are considered to be representative for R Coronae Borealis supergiant stars. With

$$\frac{L}{L_\odot} = 1.15 \times 10^4, \quad \frac{M}{M_\odot} = 1, \quad \log T_{\mathrm{e}} = 3.70 \text{ to } 4,$$

C reaches values from 4.6×10^{-3} to 1.4×10^{-4} (Cox et al. 1980).

The effect of the energy exchange is rendered by the right-hand member of Eq. (3.19). When C is small, and β is not too large, the effect is unimportant, except in the surface layers of a star, where the product $\rho^* \, \delta \, (dQ/dt)^* / P^*$ becomes large as $r \to R$. In the main body of a star, the energy exchange has a small cumulative effect that becomes of order unity only after a time comparable with the time-scale of Helmholtz–Kelvin. Therefore, one often uses the approximation

$$\frac{\partial}{\partial t^+} \left(\frac{\delta P}{P} - \Gamma_1 \frac{\delta \rho}{\rho} \right) = 0. \tag{3.22}$$

When one adopts the initial condition

$$\left(\frac{\delta P}{P} - \Gamma_1 \frac{\delta \rho}{\rho} \right)_{t^+ = t_0^+} = 0, \tag{3.23}$$

the equation has the integrated form

$$\frac{\delta P}{P} - \Gamma_1 \frac{\delta \rho}{\rho} = 0 \tag{3.24}$$

and corresponds to Eq. (1.80) in which δS is set equal to zero. Hence, in the main body of a star, the motions of the mass elements can be regarded as being isentropic. As observed in the Preface, we give the preference to the adjective "isentropic" above the adjective "adiabatic" in the line of thought of Tassoul (1978):

> These motions are sometimes called "adiabatic" (lit. "not passing across") as though the terms adiabatic and isentropic were synonymous. In accordance with the current trend in fluid mechanics, here we shall use the term *isentropic* in its strict etymological sense, i.e. adiabatic *and* reversible. To be specific, an adiabatic motion ... is isentropic only if ... no heat is irreversibly generated by viscous friction in the system. See, e.g., Thompson, P.A., *Compressible-Fluid Dynamics*, pp. 59–60, New York: McGraw-Hill Book Co., 1972.

The isentropic approximation is not valid in the surface layers of a star, since the energy exchanges between the mass elements are very large there. It has been argued that the so-called strongly non-adiabatic region of a star contains such a small fraction of the total mass that it cannot compete with the rest of the mass. Therefore, the isentropic approximation is often adopted as a first approximation for the displacement field in the whole star. A reason for the success of this approximation is that it usually leads to good approximations for the oscillation periods.

In the past, it has been customary to estimate the energy exchange between the mass elements by means of the isentropic approximation. With the so-called quasi-isentropic (quasi-adiabatic) approximation, the transition region near the star's surface was determined whereabouts the isentropic approximation becomes invalid. This procedure was followed by Eddington (1926, 1988) for the radial oscillations of a star:

> We shall now investigate the theory of the pulsation of a gaseous star. The exact theory of the changes of temperature and density, taking into account the flow of heat, involves differential equations of the fourth order which at present seem unmanageable. But the problem is simplified by noticing that owing to the high opacity of stellar material the oscillations through the greater part of the interior are approximately adiabatic. We therefore start by considering adiabatic oscillations of a sphere of gas; we can afterwards calculate the flow of heat which would result, and determine whereabouts in the star it becomes so great as to render the adiabatic approximation invalid.

Jones & Roberts (1979) have drawn the attention to the fact that the perturbation problem of the determination of the non-adiabatic effects on a stellar oscillation, with the use of the isentropic approximation as approximation of order zero, is a *singular* perturbation problem of the boundary-layer type. The mathematical reason is that the order of the system of differential equations in the radial coordinate is reduced in comparison with the order of the full non-adiabatic system. Any approximation by which the order of a system of differential equations is reduced, represents a drastic mathematical change. Here it may have the consequence that the isentropic approximation is not a valid approximation of order zero in the star's non-adiabatic region. A simple perturbation problem of the boundary-layer type that illustrates the question is presented in Appendix E.

The non-adiabatic character of motions of mass elements can locally be estimated by the use of a thermal time-scale that is defined as the time needed by a stellar layer for radiating its heat content with the local luminosity:

$$\tau_{\mathrm{th,loc}} = \frac{4\pi\, \rho\, r^2\, \Delta r\, C_V\, T}{L(r)}. \tag{3.25}$$

A somewhat different definition of the local thermal time-scale has been given by Unno et al. (1989). The isentropic approximation is valid when the ratio $\tau_{\mathrm{th,loc}}/\tau_{\mathrm{ac}}$ is sufficiently large. For many stars, this is the case in the interior, but not in a small region near the surface.

Nowadays, the full non-adiabatic system of equations is regularly integrated by the use of appropriate numerical codes.

The study of the non-adiabatic effects on stellar oscillations is important for the investigation of the *vibrational* stability of stars.

3.5 Criterion for Local Stability with Respect to Convection

In the derivation of the criterion for local stability with respect to convection, a limiting case of deviation from the hydrostatic equilibrium is considered in which the buoyancy force of Archimedes is the only acting force. The derivation of the criterion is based on a stability analysis using the bubble method, about which Smith & Fricke (1975) observed:

> An exact stability analysis of an astrophysically interesting problem is usually too complicated to be attempted, and various approximate schemes have been devised ...
>
> An alternative physical derivation of the stability criteria ... is the bubble method, in which the motion of a perturbed element of fluid is followed. It has normally been assumed that this method is equivalent to a local analysis, since a bubble is an essentially local concept, and the criteria found by a local analysis have been reproduced by a bubble method ...

Perdang (1976) tried to link the intuitive concept of "local stellar stability" to the family of continuous spectra with the aim to give it a precise mathematical meaning.

The criterion for local stability with respect to convection is derived on a twofold supposition, which preserves a balance between two extremes. On one side, one supposes that the bubble moves sufficiently slowly over a long time such that it adapts its pressure to that of its surroundings and that the buoyancy force of Archimedes is the only acting force. On the other side, one supposes that the bubble does not move too slowly over a too long time for the isentropic approximation to become invalid.

When the buoyancy force of Archimedes is the only acting force, the radial component of the linearised equation of motion, given by the first Eq. (2.32), reduces to

$$\rho \frac{\partial^2(\delta r)}{\partial t^2} = -\rho' g, \tag{3.26}$$

so that it can be written as

$$\rho \frac{\partial^2(\delta r)}{\partial t^2} = -g \left(\delta\rho - \frac{d\rho}{dr}\, \delta r \right). \tag{3.27}$$

The Lagrangian perturbation of the mass density can be related to the radial displacement of the bubble as follows. First, in the isentropic approximation, the Lagrangian perturbation of the mass density is related to that of the pressure by

$$\delta\rho = \frac{1}{c^2}\, \delta P, \tag{3.28}$$

where $c^2 \equiv \Gamma_1 P/\rho$ is the square of the isentropic sound velocity, as shown in Appendix J.1.

Next, the supposition that the bubble adapts its pressure to that of its surroundings implies that, in the Lagrangian perturbation of the pressure, the part due to the Eulerian perturbation is negligible in comparison with the change in equilibrium pressure the bubble experiences by its radial displacement. The Eulerian perturbation of the pressure is regarded here as the difference between the bubble's pressure and that of the surroundings. One then has that

$$\delta P = \frac{dP}{dr}\, \delta r. \tag{3.29}$$

Taking into account condition (2.11) of the hydrostatic equilibrium, one obtains

$$\delta\rho = -\rho \frac{g}{c^2}\, \delta r. \tag{3.30}$$

Substitution of the expression for $\delta\rho$ into Eq. (3.27) yields the second-order differential equation for the radial displacement of the bubble

$$\frac{\partial^2(\delta r)}{\partial t^2} + N^2\, \delta r = 0 \tag{3.31}$$

with

$$N^2 = -g \left(\frac{g}{c^2} + \frac{d \ln \rho}{dr} \right) \tag{3.32}$$

(see also Dahlen & Tromp 1998). In geophysical fluid dynamics, N is known as the frequency of Brunt-Väisälä (Väisälä 1925, Brunt 1927).

When $N^2 > 0$, the bubble oscillates around its equilibrium position with the frequency N: a risen bubble has a mass density somewhat larger than that of its surroundings and is pushed downwards by the negative buoyancy force of Archimedes; a bubble that has moved down, has a mass density somewhat smaller than that of its surroundings and is lifted upwards by the positive buoyancy force of Archimedes. The layer is then said to be *stable against convection*. However, when $N^2 < 0$, the amplitude of the radial component of the bubble's Lagrangian displacement increases or decreases exponentially with time. Because of the first possibility, the layer is said to be *unstable against convection*. Consequently, the sign of N^2 can be used as a criterion for local stability with respect to convection in the linear approximation.

When $N^2 = 0$, the layer is said to be in *isentropic equilibrium*, since the density gradient is equal to the isentropic density gradient:

$$\frac{d\rho}{dr} = \frac{\rho}{\Gamma_1 P}\frac{dP}{dr} \equiv \left(\frac{d\rho}{dr}\right)_S . \tag{3.33}$$

In this case, the layer does not react upon isentropic displacements of mass elements.

Note that Väisälä himself derived the frequency in terms of the gradients of temperature:

> Nun führen wir mit einfachen Annahmen eine Rechnung durch, die die Schwingungsperiode der Luft als Funktion des Temperaturgradienten gibt. Wir nehmen an, ein Quantum Luft sei durch Hebung oder Senkung aus seiner Gleichgewichtslage gebracht, wobei seine Temperatur sich adiabatisch ändere. Die Kraft, die das Luftquantum in die Gleichgewichtslage zurücktreibt, sei proportional der Differenz der jeweiligen Dichte des Luftquantums und der umgebenden Luft. Wir bedienen uns folgender Bezeichnungen:
>
> z: Höhe des Luftquantums über seinem Gleichgewichtsniveau,
> θ_0: die absolute Temperatur der Luft im Gleichgewichtsniveau,
> ρ', θ': Dichte und Temperatur des Luftquantums in der Höhe z,
> ρ, θ: Dichte und Temp. der umgebenden Luft in der Höhe z,
> γ: Temperaturgradient der Luft, als konstant angenommen,
> γ_0: der adiabatische Temperaturgradient,
> g: die Schwerebeschleunigung.
>
> Es ist
>
> $$\theta = \theta_0 - \gamma z, \quad \theta' = \theta_0 - \gamma_0 z.$$
>
> Weil der Druck im verrückten Luftquantum als gleich dem Drucke der umgebenden Luft in gleicher Höhe anzunehmen ist, haben wir $\rho\,\theta = \rho'\,\theta'$, woraus
>
> $$\rho' - \rho = \rho' \frac{(\gamma_0 - \gamma)\,z}{\theta_0},$$
>
> wo die höheren Potenzen von $\dfrac{\gamma z}{\theta_0}$ nicht berüksichtigt sind. Die auf das Luftquantum wirkende schwingende Kraft ist also für die Masseneinheit
>
> $$-\frac{g\,(\gamma_0 - \gamma)}{\theta_0} z = \frac{d^2 z}{dt^2}.$$

Diese Gleichung gibt integriert

$$z = \frac{A}{a} \sin a\, t,$$

wo

$$a^2 = \frac{g\,(\gamma_0 - \gamma)}{\theta_0}.$$

Stillschweigend haben wir angenommen, dass $\gamma < \gamma_0$ sei.[1]

As it appears from the quotation, the stability criterion has initially been expressed in terms of local gradients of temperature. This form of the criterion rests on a comparison between the temperature of the bubble and that of the surrounding layer. A wording of the criterion based on temperature differences of mass elements

---

[1] Now we perform a calculation with simple assumptions, which gives the oscillation period of the air as a function of the gradients of temperature. We assume that a quantum of air is removed from its equilibrium layer by a rise or a descent, whereby its temperature changes adiabatically. The force, which drives the quantum of air back in the equilibrium layer, be proportional to the difference of the concerned density of the quantum of air, and of the surrounding layer. We use the following denominations:

z: height of the quantum of air above its equilibrium level,
θ_0: the absolute temperature of the air at the equilibrium level,
ρ', θ': density and temperature of the quantum of air at the height z,
ρ, θ: density and temperature of the surrounding air at the height z,
γ: gradient of temperature of the air, assumed as constant,
γ_0: the adiabatic gradient of temperature,
g: the gravitational acceleration.

It is

$$\theta = \theta_0 - \gamma\, z\,, \quad \theta' = \theta_0 - \gamma_0\, z.$$

Since the pressure in the particular quantum of air is assumed to be equal to the pressure of the surrounding air at the same height, we have $\rho\,\theta = \rho'\,\theta'$, from which

$$\rho' - \rho = \rho' \,\frac{(\gamma_0 - \gamma)\, z}{\theta_0},$$

where the higher powers of $\frac{\gamma\, z}{\theta_0}$ are not taken into consideration. The oscillating force acting on the quantum of air is also for the mass unit

$$-\frac{g\,(\gamma_0 - \gamma)}{\theta_0}\, z = \frac{d^2 z}{dt^2}.$$

The equation yields after integration

$$z = \frac{A}{a} \sin a\, t,$$

where

$$a^2 = \frac{g\,(\gamma_0 - \gamma)}{\theta_0}.$$

Tacitly we have assumed that $\gamma < \gamma_0$.

that move in the Earth atmosphere is already found in a work of Reye (1872). In the second chapter, entitled *Ursachen und Entstehung der Wettersäulen*, one reads:

> Versetzen wir ... eine beliebige Luftmasse ohne äusserliche Zuführung von Wärme in eine höhere Schicht der Atmosphäre, so dehnt sie sich aus wegen Verminderung des äusseren Druckes, und ihre Temperatur sinkt gleichzeitig. Ist diese, dem Poisson'schen Spannungsgesetze entsprechende Temperaturabnahme grösser als die atmosphärische, welche der durchlaufenen Höhe entspricht, ist also unser Luftquantum bis unter die Temperatur seiner neuen Umgebung erkaltet, so muss dasselbe, wenn es sich selbst überlassen wird, wieder zu seiner früheren Lage hinabsinken. Das Gleichgewicht der Luft ist dann ein stabiles oder beständiges. Dagegen wird die Luftmasse noch höher steigen, wenn ihre Temperatur-Abnahme kleiner ist als die atmosphärische, und wenn sie deshalb wärmer bleibt als die umgebende Luftschicht; das Gleichgewicht ist in diesem Falle ein labiles oder schwankendes.[2]

The reader is also referred to Brunt (1934).

For the transformation of the stability criterion established up here into a criterion that is expressed in terms of gradients of temperature, it is necessary to eliminate the gradient of mass density. In this elimination, possible gradients of mean molecular weight must be taken into account. Such a gradient can be generated in the surface layers of a star, where the degrees of ionisation of the abundant elements hydrogen and helium vary from layer to layer. It can also be built up in the central layers of a star by the evolution, since the thermonuclear reactions modify the chemical compositions of these layers at different rates. Here one may especially think of the conversion of hydrogen nuclei into helium nuclei in the central layers of main sequence stars. The necessity to take into account the existence of gradients of mean molecular weight in central layers of stars was brought to the attention by Ledoux (1947).

By use of the equation of state in which the pressure P is considered as a function of the mass density ρ, the temperature T, and the mean molecular weight $\bar{\mu}$, the gradient of pressure can be expressed as

$$\frac{dP}{dr} = \chi_\rho \frac{P}{\rho}\frac{d\rho}{dr} + \chi_T \frac{P}{T}\frac{dT}{dr} + \chi_{\bar{\mu}} \frac{P}{\bar{\mu}}\frac{d\bar{\mu}}{dr}, \qquad (3.34)$$

where the notations are used

$$\chi_\rho = \left(\frac{\partial \ln P}{\partial \ln \rho}\right)_{T,\bar{\mu}}, \qquad \chi_T = \left(\frac{\partial \ln P}{\partial \ln T}\right)_{\rho,\bar{\mu}}, \qquad \chi_{\bar{\mu}} = \left(\frac{\partial \ln P}{\partial \ln \bar{\mu}}\right)_{\rho,T}. \qquad (3.35)$$

---

[2] When we displace ... an arbitrary mass of air without any external supply of heat in a higher layer of the atmosphere, it expands because of the decrease of the external pressure, and its temperature sinks at the same time. If the decrease of the temperature resulting from Poisson's tension theorem is larger than the atmospheric one, which is related to the height reached, and if our quantum of air is cooled down below the temperature of its new surroundings, then that same quantum of air, when it is left to itself, must sink again downwards to its earlier layer. The equilibrium of the air is then a stable or lasting equilibrium. On the contrary, the mass of air will rise still higher, when its decrease of temperature is smaller than that of the atmosphere, and when it therefore remains warmer than the surrounding layer of air; the equilibrium is in this case an unstable or vacillating equilibrium.

Elimination of $d\rho/dr$ from definition (3.32) of N^2 yields

$$N^2 = -\frac{g}{\chi_\rho P}\left[\left(1-\frac{\chi_\rho}{\Gamma_1}\right)\frac{dP}{dr} - \chi_T\frac{P}{T}\frac{dT}{dr} - \chi_{\bar{\mu}}\frac{P}{\bar{\mu}}\frac{d\bar{\mu}}{dr}\right]. \tag{3.36}$$

Transformation of the coefficient $\left(1-\chi_\rho/\Gamma_1\right)$ by use of the equality $\chi_\rho = -\chi_\tau$ and the thermodynamic relations (Appendix B.11) and (Appendix B.34) leads to

$$N^2 = \frac{\chi_T}{\chi_\rho}\frac{g}{P}\frac{dP}{dr}\left(\frac{d\ln T}{d\ln P} - \frac{\Gamma_2-1}{\Gamma_2} + \frac{\chi_{\bar{\mu}}}{\chi_T}\frac{d\ln\bar{\mu}}{d\ln P}\right) \tag{3.37}$$

(see also Brassard et al. 1991).

In the denominator, χ_ρ is positive. When one also considers χ_T to be positive and takes into account that dP/dr is negative, the local criterion for stability against convection can be expressed as

$$\frac{d\ln T}{d\ln P} \leq \frac{\Gamma_2-1}{\Gamma_2} - \frac{\chi_{\bar{\mu}}}{\chi_T}\frac{d\ln\bar{\mu}}{d\ln P}. \tag{3.38}$$

This form of the criterion is known as Ledoux' criterion, which involves the effect of a possible gradient of mean molecular weight.

In the absence of any gradient of mean molecular weight, one recovers the usual Schwarzschild criterion by multiplying both members of the inequality by dP/dr and taking the absolute values:

$$\left|\frac{dT}{dr}\right| \leq \left|\frac{\Gamma_2-1}{\Gamma_2}\frac{T}{P}\frac{dP}{dr}\right| \equiv \left|\left(\frac{dT}{dr}\right)_{\mathrm{S}}\right|. \tag{3.39}$$

For local stability with respect to convection, the gradient of temperature must thus be smaller in absolute value than the isentropic gradient of temperature. In the astrophysical literature, the stability criterion is ascribed to K. Schwarzschild, who published an important paper on the radiative equilibrium of the atmosphere of the Sun in 1906 (Schwarzschild 1906).

The derivation of the criterion for local stability with respect to convection is far from being strict. The attention is concentrated on a particular bubble, which is supposed to adapt its pressure to that of the surroundings so that the Eulerian perturbation of the pressure is negligible. In this connection, it is worth noting that, in Chaps. 16 and 17, the Eulerian perturbation of the pressure is seen to be identically zero in the first asymptotic approximation for higher-order g^+-modes in stars containing a convective core or a convective envelope, except for a boundary layer near the star's surface.

Moreover, the relation between a violation of the local criterion in a stellar layer and the global reaction of the star to this violation is an open question at this stage. This question is treated in Chap. 13.

3.6 Deviations from the Thermal Equilibrium

When the time-scale of the motions of the mass elements is of the order of the Helmholtz–Kelvin time-scale, the product $\beta\, C$ becomes of order unity in Eq. (3.19). In this case, $\delta\rho$ can no more be eliminated from Eq. (3.27) by means of the isentropic approximation. One must then return to Eq. (3.16) and take into account the non-adiabatic terms. In view of this, it is appropriate to differentiate Eq. (3.27) with respect to time. After elimination of $(\partial/\partial t)\,(\delta\rho/\rho)$, one has

$$\frac{\partial^3(\delta r)}{\partial t^3} = g\,\frac{\partial}{\partial t}\left[-\frac{1}{\Gamma_1}\,\frac{\delta P}{P} + \frac{1}{\rho}\,\frac{d\rho}{dr}\,\delta r\right] + g\,\frac{(\Gamma_3 - 1)\,\rho}{\Gamma_1 P}\,\delta\,\frac{dQ}{dt}. \tag{3.40}$$

By use of equality (3.29) and the condition of hydrostatic equilibrium, the equation is transformed into

$$\frac{\partial^3(\delta r)}{\partial t^3} = -N^2\,\frac{\partial(\delta r)}{\partial t} + g\,\frac{(\Gamma_3 - 1)\,\rho}{\Gamma_1 P}\,\delta\,\frac{dQ}{dt}. \tag{3.41}$$

Considering displacement fields with negligible dynamic effects, so that the stellar layers remain in hydrostatic equilibrium, and the second partial derivative of δr with respect to time can be neglected, one reduces the equation to

$$\frac{\partial(\delta r)}{\partial t} = \frac{g}{N^2}\,\frac{(\Gamma_3 - 1)\,\rho}{\Gamma_1 P}\,\delta\,\frac{dQ}{dt}. \tag{3.42}$$

The displacement field is now determined by the deviations from the thermal equilibrium. This type of displacement fields is considered for the investigation of the *secular* stability of a star (see, e.g., Cox 1980). The study of the secular stability of stars throws light on questions of stellar structure and evolution (Hansen 1978, Pesnell & Buchler 1986, Gautschy & Glatzel 1991).

Chapter 4
Eigenvalue Problem of the Linear, Isentropic Normal Modes in a Quasi-Static Star

4.1 Time-Dependent Equations and Boundary Conditions Governing Linear, Isentropic Oscillations

The linear, isentropic oscillations in a quasi-static star are governed by the system of Eqs. (2.32), (2.33), and (2.50), in terms of Eulerian perturbations of physical quantities, or the system of Eqs. (2.52), (2.53), and (2.55), in terms of Lagrangian perturbations of the same physical quantities. To these systems, Eq. (3.24), which expresses the isentropic character of the oscillatory motions, must be added:

$$\frac{P'}{P} + \frac{1}{P}\frac{dP}{dr}\,\delta r - \Gamma_1\left(\frac{\rho'}{\rho} + \frac{1}{\rho}\frac{d\rho}{dr}\,\delta r\right) = 0. \tag{4.1}$$

A system of equations formulated in a Lagrangian formalism was given earlier by Ishizuka (1970).

The solutions must satisfy conditions at the boundary points $r = 0$ and $r = R$.

At $r = 0$, the Lagrangian displacement must remain finite. For radial oscillations, the Lagrangian displacement must be equal to zero, since otherwise a vacuum would be created while nature has a *horror vacui*.

At $r = R$, the Lagrangian perturbation of the pressure must vanish:

$$(\delta P)_R = 0. \tag{4.2}$$

This condition results from the principle relative to the rate of change of momentum known in mechanics of continuous media, as is shown in Appendix F in the twofold supposition that a star's surface is a contact surface and that the pressure outside the star is identically zero. Because of Eqs. (2.53) and (3.24), the condition can also be expressed as

$$(\Gamma_1 P\alpha)_R = 0. \tag{4.3}$$

Since $P(R) = 0$, the condition leads to the regularity condition that the divergence of the Lagrangian displacement, α, must remain finite at $r = R$.

P. Smeyers and T. Van Hoolst, *Linear Isentropic Oscillations of Stars: Theoretical Foundations*, Astrophysics and Space Science Library 371,
DOI 10.1007/978-3-642-13030-4_5,

Moreover, on the surface of the perturbed star, the gravitational potential and its gradient must be continuous. These conditions can be expressed as

$$(\Phi_i)_R + (\delta\Phi_i)_R = (\Phi_e)_R + (\delta\Phi_e)_R\,, \tag{4.4}$$

$$\left(\frac{d\Phi_i}{dr}\right)_R + \left(\delta\,\frac{\partial\Phi_i}{\partial r}\right)_R = \left(\frac{d\Phi_e}{dr}\right)_R + \left(\delta\,\frac{\partial\Phi_e}{\partial r}\right)_R\,, \tag{4.5}$$

$$\left(\delta\,\frac{\partial\Phi_i}{\partial\theta}\right)_R = \left(\delta\,\frac{\partial\Phi_e}{\partial\theta}\right)_R\,, \tag{4.6}$$

$$\left(\delta\,\frac{\partial\Phi_i}{\partial\phi}\right)_R = \left(\delta\,\frac{\partial\Phi_e}{\partial\phi}\right)_R\,, \tag{4.7}$$

where Φ_i is the internal, and Φ_e the external gravitational potential. The conditions can be developed by the transition from Lagrangian to Eulerian perturbations. Furthermore, one takes into account that the gravitational potential and its first derivative with respect to the radial coordinate are continuous on the surface of the equilibrium star and eliminates the second derivative of the internal and external gravitational potential on that surface, respectively, by means of Poisson's and Laplace's differential equation. The conditions then become

$$(\Phi_i')_R = (\Phi_e')_R\,, \tag{4.8}$$

$$\left(\frac{\partial\Phi_i'}{\partial r}\right)_R + (4\pi G\rho\,\delta r)_R = \left(\frac{\partial\Phi_e'}{\partial r}\right)_R\,, \tag{4.9}$$

$$\left(\frac{\partial\Phi_i'}{\partial\theta}\right)_R = \left(\frac{\partial\Phi_e'}{\partial\theta}\right)_R\,, \tag{4.10}$$

$$\left(\frac{\partial\Phi_i'}{\partial\phi}\right)_R = \left(\frac{\partial\Phi_e'}{\partial\phi}\right)_R\,. \tag{4.11}$$

4.2 Vectorial Wave Equation with Tensorial Operator U

Application of Hilbert space theory of operators gives insight into mathematical properties of the problem of the linear, isentropic oscillations of a quasi-static star. For this purpose, it is appropriate to express the Eulerian perturbations of the mass density and the pressure in terms of the components of the Lagrangian displacement and their derivatives by means of Eqs. (2.33) and (4.1). Furthermore, the Eulerian perturbation of the gravitational potential is expressed as a volume integral in terms of the components of the Lagrangian displacement by means of solution (1.74) of Poisson's perturbed differential equation.

After multiplication of Eq. (2.31) by the covariant component g_{sk} of the metric tensor, and division by ρ, one obtains the covariant components of the vectorial wave equation for linear, isentropic displacement fields

$$g_{ij}\,\frac{\partial^2\left(\delta q^j\right)}{\partial t^2} + U_{ij}\,\delta q^j = 0, \quad i = 1, 2, 3, \tag{4.12}$$

where the U_{ij} are defined as integro-differential operators by the equations

$$U_{ij}\,\delta q^j = \nabla_i \Phi' - \frac{\rho'}{\rho^2}\,\nabla_i P + \frac{1}{\rho}\,\nabla_i P', \quad i = 1, 2, 3, \tag{4.13}$$

and are the covariant components of a tensorial operator U.

Equivalent expressions for the operators are derived from the components (2.52) of the linearised equation of motion, which involve the Lagrangian perturbations of the pressure, the mass density, and the gravitational potential. In this case, it is appropriate to express $\delta\rho$ and δP in terms of the components of the Lagrangian displacement by means of Eqs. (2.53) and (3.24), and to derive an expression for $\delta\Phi$ by means of relation (1.30) and solution (1.74) of Poisson's perturbed differential equation. The integro-differential operators U_{ij} are then defined by

$$U_{ij}\,\delta q^j = \nabla_i(\delta\Phi) - \frac{\delta\rho}{\rho^2}\,\nabla_i P + \frac{1}{\rho}\,\nabla_i(\delta P), \quad i = 1, 2, 3. \tag{4.14}$$

Wave equation (4.12) can be expressed in the vectorial form

$$\frac{\partial^2 \boldsymbol{\xi}}{\partial t^2} + \mathsf{U}\,\boldsymbol{\xi} = 0, \tag{4.15}$$

where $\boldsymbol{\xi}$ is the Lagrangian displacement of a mass element. A similar form of the wave equation was adopted by Kaniel & Kovetz (1967) and Dyson & Schutz (1979).

4.3 Separation of Time

In the equations and boundary conditions applying to linear, isentropic oscillations of a quasi-static star, the coefficients are related to the equilibrium structure of the star and are therefore time-independent. Solutions for the components of the Lagrangian displacement field and the associated perturbations of the physical quantities can then be sought that are exponential functions of the time of the form

$$f(t, r, \theta, \phi) = \exp(\mathrm{i}\sigma t)\,f(r, \theta, \phi), \tag{4.16}$$

where σ is the generally complex angular frequency. Solutions with this type of time-dependency are called normal *solutions*. Normal *modes* are normal solutions for which both the displacement field and the velocity field satisfy some initial conditions. Very often no attention is paid to the distinction between normal modes and normal solutions, and normal solutions are called normal modes (Dyson & Schutz 1979; see also Chap. 12, below Eq. (12.107)). A similar procedure is adopted here, so that the term "normal mode" currently stands for the term "normal solution".

By the use of equality (4.16) for the components of the Lagrangian displacement and the perturbations of the physical quantities, the time is separated from the equations and boundary conditions. When one also takes into account condition (2.11) for the hydrostatic equilibrium, Eqs. (2.32), (2.33), (4.1), (2.50) become

$$\sigma^2\,\delta r = \frac{\partial \Phi'}{\partial r} - \frac{\rho'}{\rho^2}\frac{dP}{dr} + \frac{1}{\rho}\frac{\partial P'}{\partial r}, \tag{4.17}$$

$$\sigma^2\, r\,\delta\theta = \frac{1}{r}\frac{\partial}{\partial\theta}\left(\Phi' + \frac{P'}{\rho}\right), \tag{4.18}$$

$$\sigma^2\, r\,\sin\theta\,\delta\phi = \frac{1}{r\,\sin\theta}\frac{\partial}{\partial\phi}\left(\Phi' + \frac{P'}{\rho}\right), \tag{4.19}$$

$$\frac{\rho'}{\rho} + \frac{1}{\rho}\frac{d\rho}{dr}\,\delta r + \alpha = 0, \tag{4.20}$$

$$\frac{P'}{P} + \frac{1}{P}\frac{dP}{dr}\,\delta r - \Gamma_1\left(\frac{\rho'}{\rho} + \frac{1}{\rho}\frac{d\rho}{dr}\,\delta r\right) = 0, \tag{4.21}$$

$$\nabla^2\,\Phi' \equiv \frac{1}{r^2}\frac{\partial}{\partial r}\left(r^2\,\frac{\partial\Phi'}{\partial r}\right) - \frac{1}{r^2}\mathscr{L}^2\Phi' = 4\pi G\rho'. \tag{4.22}$$

The time can be separated in a similar way from the equations expressed in terms of Lagrangian perturbations and from the boundary conditions.

Equations (4.17)–(4.22) and the associated boundary conditions form an eigenvalue problem in which σ^2 is the eigenvalue parameter. The eigenvalue problem is of the third degree in the eigenvalue parameter.

After the separation of the time, components (4.12) of the wave equation for the Lagrangian displacement take the form

$$\sigma^2\, g_{ij}\,\delta q^j - U_{ij}\,\delta q^j = 0, \quad i = 1, 2, 3, \tag{4.23}$$

and the vectorial wave equation (4.15) becomes

$$\sigma^2\,\boldsymbol{\xi} - \mathsf{U}\,\boldsymbol{\xi} = 0. \tag{4.24}$$

4.4 Inner Product of Linear, Isentropic Oscillations

The linear, isentropic oscillations of a quasi-static star are complex vector fields over the star considered as a differentiable manifold and form a complex vector space. In this space, the inner product of a pair of displacement fields $\boldsymbol{\xi}_1$ and $\boldsymbol{\xi}_2$ is the complex number that is assigned to the pair, with respect to the non-negative mass density as weighting function or, equivalently, with respect to the integration measure $\rho\, dV$, as

$$(\boldsymbol{\xi}_1, \boldsymbol{\xi}_2) = \int_V \rho\, g_{ij}\, \overline{\delta q_1^i}\, \delta q_2^j\, dV. \tag{4.25}$$

From the definition of the inner product, it follows that

$$(\boldsymbol{\xi}_2, \boldsymbol{\xi}_1) = \overline{(\boldsymbol{\xi}_1, \boldsymbol{\xi}_2)}. \tag{4.26}$$

Another property of the inner product is that the quadratic form $(\boldsymbol{\xi}, \boldsymbol{\xi})$ is positive definite. This property means that

$$(\boldsymbol{\xi}, \boldsymbol{\xi}) \geq 0 \tag{4.27}$$

and that

$$(\boldsymbol{\xi}, \boldsymbol{\xi}) = 0 \quad \text{implies} \quad \boldsymbol{\xi} = 0. \tag{4.28}$$

With the definition of the inner product, the notion of the norm of a displacement field $\boldsymbol{\xi}$ is associated:

$$\|\boldsymbol{\xi}\| = (\boldsymbol{\xi}, \boldsymbol{\xi})^{1/2}. \tag{4.29}$$

The distance between two displacement fields $\boldsymbol{\xi}_1$ and $\boldsymbol{\xi}_2$ is defined as

$$d(\boldsymbol{\xi}_1, \boldsymbol{\xi}_2) = \|\boldsymbol{\xi}_1 - \boldsymbol{\xi}_2\|. \tag{4.30}$$

With the notion of distance, the convergence of a sequence of displacement fields can be considered. A sequence of displacement fields is a Cauchy sequence, if

$$\|\boldsymbol{\xi}_\mu - \boldsymbol{\xi}_\nu\| \to 0 \quad \text{as} \quad \mu, \nu \to \infty. \tag{4.31}$$

An inner-product-space is also called a pre-Hilbert space, since each inner-product-space can be extended to a Hilbert space or, equivalently, a complete space in which it is dense. A space is called a Hilbert space if each Cauchy sequence of vectors has a limit vector. A subset R' of a normed space R is said to be dense in the space, if, for each vector in R, a vector in R' exists such that the norm of the difference between both vectors is arbitrarily small (Friedrichs 1973).

4.5 Symmetry of the Tensorial Operator U

The tensorial operator U applying to linear, isentropic oscillations of a quasi-static star is symmetric with regard to the inner product weighted with the mass density, in the sense that for all displacement fields in the (not nearer specified) domain of the operator the equality holds

$$\int_V \rho\, \delta q_2^i \left(U_{ij}\, \overline{\delta q_1^j} \right) dV = \int_V \rho\, \overline{\delta q_1^i} \left(U_{ij}\, \delta q_2^j \right) dV. \tag{4.32}$$

Beyer & Schmidt (1995) observed that the property is valid for displacement fields that belong to C^2 or that are twice continuously differentiable.

Instead of the term "symmetric", the term "formally self-adjoint" is also used (see, e.g., Friedrichs 1973). The adverb "formally" serves to distinguish the property of being "formally self-adjoint" from the more restrictive property of being "strictly self-adjoint".

The symmetry of the tensorial operator U has been demonstrated by Kaniel & Kovetz (1967) on the basis of definition (4.13) for the components of the operator, and by Lynden-Bell & Ostriker (1967) on the basis of definition (4.14).

4.5.1 *Proof of Kaniel and Kovetz*

After multiplication of definition (4.13) by ρ, one can consider the resolution

$$\rho\, U_{ij}\, \delta q^j = R_{ij}\, \delta q^j + W_{ij}\, \delta q^j, \quad i = 1, 2, 3, \tag{4.33}$$

where

$$R_{ij}\, \delta q^j = -\frac{\rho'}{\rho} \nabla_i P + \nabla_i P', \tag{4.34}$$

$$W_{ij}\, \delta q^j = \rho\, \nabla_i \Phi'. \tag{4.35}$$

We first concentrate on the term $R_{ij}\, \delta q^j$. By elimination of ρ' by means of Eq. (1.60), it follows from Eq. (4.1) that

$$P' = \left(c^2\, \nabla_j \rho - \nabla_j P \right) \delta q^j - c^2\, \nabla_j \left(\rho\, \delta q^j \right). \tag{4.36}$$

Since an equilibrium star is a barotrope, one has

$$\nabla_j P = \frac{\partial P}{\partial \rho} \nabla_j \rho, \quad j = 1, 2, 3. \tag{4.37}$$

The equation can then be rewritten as

$$P' = \left(c^2 - \frac{\partial P}{\partial \rho} \right) \delta q^j \, \nabla_j \rho - c^2 \, \nabla_j \left(\rho \, \delta q^j \right). \tag{4.38}$$

After elimination of ρ' and P', definition (4.34) is applied to the complex conjugate of a first displacement field denoted by the subscript 1. Multiplication of $R_{ij} \, \overline{\delta q_1^j}$ by δq_2^i, where the subscript 2 denotes a second displacement field, and integration over the volume of the equilibrium star lead to

$$\int_V \delta q_2^i \left(R_{ij} \, \overline{\delta q_1^j} \right) dV = \int_V \left\{ \frac{1}{\rho} \left(\delta q_2^i \, \nabla_i P \right) \nabla_j \left(\rho \, \overline{\delta q_1^j} \right) \right.$$
$$\left. + \, \delta q_2^i \, \nabla_i \left[\left(c^2 - \frac{\partial P}{\partial \rho} \right) \overline{\delta q_1^j} \, \nabla_j \rho - c^2 \, \nabla_j \left(\rho \, \overline{\delta q_1^j} \right) \right] \right\} dV. \tag{4.39}$$

By partial integration of the second volume integral in the right-hand member, and elimination of $\nabla_i \, \delta q_2^i$ by means of the equality

$$\nabla_i \, \delta q_2^i = \frac{1}{\rho} \left[\nabla_i \left(\rho \, \delta q_2^i \right) - \left(\delta q_2^i \, \nabla_i \rho \right) \right], \tag{4.40}$$

one obtains

$$\int_V \delta q_2^i \left(R_{ij} \, \overline{\delta q_1^j} \right) dV = \int_V \left\{ -\nabla_i \left[\left(\frac{\partial P}{\partial \rho} \, \overline{\delta q_1^j} \, \nabla_j \rho + \rho \, c^2 \, \nabla_j \, \overline{\delta q_1^j} \right) \delta q_2^i \right] \right.$$
$$- \frac{1}{\rho} \left[\left(c^2 - \frac{\partial P}{\partial \rho} \right) \overline{\delta q_1^j} \, \nabla_j \rho - c^2 \, \nabla_j \left(\rho \, \overline{\delta q_1^j} \right) \right]$$
$$\left[\nabla_i \left(\rho \, \delta q_2^i \right) - \left(\delta q_2^i \, \nabla_i \rho \right) \right]$$
$$\left. + \frac{1}{\rho} \left(\delta q_2^i \, \nabla_i P \right) \nabla_j \left(\rho \, \overline{\delta q_1^j} \right) \right\} dV. \tag{4.41}$$

Application of Gauss' integral theorem, use of the condition that the equilibrium pressure vanishes on the star's surface, and rearrangement of terms yield

$$\int_V \delta q_2^i \left(R_{ij} \, \overline{\delta q_1^j} \right) dV = - \int_S \frac{\partial P}{\partial n} \left(\overline{\delta q_1^j} \, n_j \right) \left(\delta q_2^i \, n_i \right) dS$$
$$+ \int_V \left\{ \frac{c^2}{\rho} \left[\nabla_j \left(\rho \, \overline{\delta q_1^j} \right) \right] \left[\nabla_i \left(\rho \, \delta q_2^i \right) \right] \right.$$
$$- \frac{1}{\rho} \left(c^2 - \frac{\partial P}{\partial \rho} \right) \left(\overline{\delta q_1^j} \, \nabla_j \rho \right) \nabla_i \left(\rho \, \delta q_2^i \right)$$

$$-\frac{1}{\rho}\left(c^2 - \frac{\partial P}{\partial \rho}\right)\left(\delta q_2^i\,\nabla_i \rho\right)\nabla_j\left(\rho\,\overline{\delta q_1^j}\right)$$

$$+\frac{1}{\rho}\left(c^2 - \frac{\partial P}{\partial \rho}\right)\left(\overline{\delta q_1^j}\,\nabla_j \rho\right)\left(\delta q_2^i\,\nabla_i \rho\right)\Bigg\}\,dV. \tag{4.42}$$

From the symmetry in the two displacement fields, it follows that

$$\int_V \delta q_2^i\left(R_{ij}\,\overline{\delta q_1^j}\right)dV = \int_V \overline{\delta q_1^i}\left(R_{ij}\,\delta q_2^j\right)dV. \tag{4.43}$$

Next, we turn to the term $W_{ij}\,\delta q^j$ in the right-hand member of equality (4.33). By means of solution (1.74) for Φ', one has

$$\int_V \delta q_2^i\left(W_{ij}\,\overline{\delta q_1^j}\right)dV = -G\int_V dV\left(\mathbf{r}\right)\int_V dV\left(\mathbf{r}'\right)$$

$$\rho\left(\mathbf{r}\right)\rho\left(\mathbf{r}'\right)\delta q_2^i\left(\mathbf{r}\right)\overline{\delta q_1^j}\left(\mathbf{r}'\right)\nabla_i\,\nabla_{,j}\left|\mathbf{r}'-\mathbf{r}\right|^{-1}. \tag{4.44}$$

From the symmetry in the two displacement fields, it follows that

$$\int_V \delta q_2^i\left(W_{ij}\,\overline{\delta q_1^j}\right)dV = \int_V \overline{\delta q_1^i}\left(W_{ij}\,\delta q_2^j\right)dV. \tag{4.45}$$

From equalities (4.43) and (4.45), the property of symmetry of the operator follows.

4.5.2 *Proof of Lynden-Bell and Ostriker*

After multiplication of definition (4.14) by ρ, rearrangement of terms, ad use of the condition of the hydrostatic equilibrium, one obtains the resolution

$$\rho\,U_{ij}\,\delta q^j = P_{ij}\,\delta q^j + V_{ij}\,\delta q^j, \quad i = 1,2,3, \tag{4.46}$$

where

$$P_{ij}\,\delta q^j = -\frac{\delta\rho}{\rho}\,\nabla_i\,P + \nabla_i\left(\delta P - \delta q^j\,\nabla_j\,P\right) + \delta q^j\,\nabla_j\nabla_i\,P, \tag{4.47}$$

$$V_{ij}\,\delta q^j = \rho\left(\nabla_i\Phi' + \delta q^j\,\nabla_j\,\nabla_i\,\Phi\right). \tag{4.48}$$

From the first term in the resolution, $\delta\rho$ and δP can be eliminated by means of Eqs. (2.53) and (3.24), so that

$$P_{ij}\,\delta q^j = -\nabla_i\left[(\Gamma_1-1)\,P\,\nabla_j\,\delta q^j\right] - P\,\nabla_i\left(\nabla_j\,\delta q^j\right) - \left(\nabla_j P\right)\left(\nabla_i\,\delta q^j\right). \tag{4.49}$$

It results that

$$\begin{aligned}\int_V \delta q_2^i\left(P_{ij}\,\overline{\delta q_1^j}\right)dV = &-\int_V \nabla_i\left[(\Gamma_1-1)\,P\left(\nabla_j\,\overline{\delta q_1^j}\right)\delta q_2^i\right]dV\\ &+\int_V (\Gamma_1-1)\,P\left(\nabla_j\,\overline{\delta q_1^j}\right)\left(\nabla_i\,\delta q_2^i\right)dV\\ &-\int_V\left[P\,\delta q_2^i\,\nabla_i\left(\nabla_j\,\overline{\delta q_1^j}\right)+\delta q_2^i\left(\nabla_j P\right)\left(\nabla_i\,\overline{\delta q_1^j}\right)\right]dV.\end{aligned} \tag{4.50}$$

The first term in the right-hand member is equal to zero since the equilibrium pressure vanishes on the star's surface. The second term is symmetric in the two displacement fields. The third term can be transformed as

$$\begin{aligned}&-\int_V\left[P\,\delta q_2^i\,\nabla_i\left(\nabla_j\,\overline{\delta q_1^j}\right)+\delta q_2^i\left(\nabla_j P\right)\left(\nabla_i\,\overline{\delta q_1^j}\right)\right]dV\\ &= -\int_V \nabla_j\left[P\,\delta q_2^i\,\nabla_i\,\overline{\delta q_1^j}\,\right]dV + \int_V P\left(\nabla_j\,\delta q_2^i\right)\left(\nabla_i\,\overline{\delta q_1^j}\right)dV.\end{aligned} \tag{4.51}$$

To the first term in the right-hand member, Gauss' integral theorem can be applied, so that this term is equal to zero because of the vanishing equilibrium pressure on the star's surface. The second term is symmetric in the two displacement fields.

It follows that

$$\int_V \delta q_2^i\left(P_{ij}\,\overline{\delta q_1^j}\right)dV = \int_V \overline{\delta q_1^i}\left(P_{ij}\,\delta q_2^j\right)dV. \tag{4.52}$$

Next, we turn to the term $\int_V \delta q_2^i\left(V_{ij}\,\overline{\delta q_1^j}\right)dV$. By use of solution (1.75) for Φ' and Eq. (1.60), one has

$$\begin{aligned}&\int_V \delta q_2^i\left(V_{ij}\,\overline{\delta q_1^j}\right)dV\\ &= -G\int_V \rho\,(\mathbf{r})\,\delta q_2^i\,(\mathbf{r})\left[\nabla_i\int_S \frac{\rho\,(\mathbf{r}')\,\overline{\delta q_1^j}\,(\mathbf{r}')}{|\mathbf{r}'-\mathbf{r}|}\,n_j\,(\mathbf{r}')\,dS\,(\mathbf{r}')\right]dV\,(\mathbf{r})\end{aligned}$$

$$+ G\int_V \rho(\mathbf{r})\,\delta q_2^i(\mathbf{r}) \left\{\nabla_i \int_V \frac{1}{|\mathbf{r}'-\mathbf{r}|}\,\nabla_{\prime j}\left[\rho(\mathbf{r}')\,\overline{\delta q_1^j}(\mathbf{r}')\right] dV(\mathbf{r}')\right\} dV(\mathbf{r})$$

$$+ \int_V \rho(\mathbf{r})\,\delta q_2^i(\mathbf{r})\,\overline{\delta q_1^j}(\mathbf{r})\left(\nabla_j \nabla_i \Phi\right)(\mathbf{r})\,dV(\mathbf{r}). \tag{4.53}$$

Partial integration of the first two terms in the right-hand member yields

$$\int_V \delta q_2^i\left(V_{ij}\,\overline{\delta q_1^j}\right) dV$$

$$= -G\int_V \left\{\nabla_i\left[\rho(\mathbf{r})\,\delta q_2^i(\mathbf{r})\int_S \frac{\rho(\mathbf{r}')\,\overline{\delta q_1^j}(\mathbf{r}')}{|\mathbf{r}'-\mathbf{r}|}\,n_j(\mathbf{r}')\,dS(\mathbf{r}')\right]\right\} dV(\mathbf{r})$$

$$+G\int_V \left[\int_S \frac{\rho(\mathbf{r}')\,\overline{\delta q_1^j}(\mathbf{r}')}{|\mathbf{r}'-\mathbf{r}|}\,n_j(\mathbf{r}')\,dS(\mathbf{r}')\right]\left\{\nabla_i\left[\rho(\mathbf{r})\,\delta q_2^i(\mathbf{r})\right]\right\} dV(\mathbf{r})$$

$$+G\int_V \left\{\nabla_i\left\{\rho(\mathbf{r})\,\delta q_2^i(\mathbf{r})\int_V \frac{1}{|\mathbf{r}'-\mathbf{r}|}\left\{\nabla_{\prime j}\left[\rho(\mathbf{r}')\,\overline{\delta q_1^j}(\mathbf{r}')\right]\right\} dV(\mathbf{r}')\right\}\right\} dV(\mathbf{r})$$

$$-G\int_V \left\{\int_V \frac{1}{|\mathbf{r}'-\mathbf{r}|}\,\nabla_{\prime j}\left[\rho(\mathbf{r}')\,\overline{\delta q_1^j}(\mathbf{r}')\right] dV(\mathbf{r}')\right\}\left\{\nabla_i\left[\rho(\mathbf{r})\,\delta q_2^i(\mathbf{r})\right]\right\} dV(\mathbf{r})$$

$$+\int_V \rho(\mathbf{r})\,\delta q_2^i(\mathbf{r})\,\overline{\delta q_1^j}(\mathbf{r})\left(\nabla_j \nabla_i \Phi\right)(\mathbf{r})\,dV(\mathbf{r}). \tag{4.54}$$

Application of Gauss' integral theorem to the first and third term in the right-hand member yields

$$\int_V \delta q_2^i\left(V_{ij}\,\overline{\delta q_1^j}\right) dV$$

$$= -G\int_S\int_S \frac{\left[\rho(\mathbf{r})\,\delta q_2^i(\mathbf{r})\,n_i(\mathbf{r})\right]\left[\rho(\mathbf{r}')\,\overline{\delta q_1^j}(\mathbf{r}')\,n_j(\mathbf{r}')\right]}{|\mathbf{r}'-\mathbf{r}|}\,dS(\mathbf{r}')\,dS(\mathbf{r})$$

$$+ G\int_V \left[\int_S \frac{\rho(\mathbf{r}')\,\overline{\delta q_1^j}(\mathbf{r}')}{|\mathbf{r}'-\mathbf{r}|}\,n_j(\mathbf{r}')\,dS(\mathbf{r}')\right]\left\{\nabla_i\left[\rho(\mathbf{r})\,\delta q_2^i(\mathbf{r})\right]\right\}(\mathbf{r})\,dV(\mathbf{r})$$

$$+ G\int_V \left[\int_S \frac{\rho(\mathbf{r})\,\delta q_2^i(\mathbf{r})}{|\mathbf{r}'-\mathbf{r}|}\,n_i(\mathbf{r})\,dS(\mathbf{r})\right]\left\{\nabla_{\prime j}\left[\rho(\mathbf{r}')\,\overline{\delta q_1^j}(\mathbf{r}')\right]\right\}(\mathbf{r}')\,dV(\mathbf{r}')$$

$$-G\int_V\int_V \frac{\left\{\nabla_{\prime j}\left[\rho\left(\mathbf{r}'\right)\overline{\delta q_1^j}\left(\mathbf{r}'\right)\right]\right\}\left(\mathbf{r}'\right)\left\{\nabla_i\left[\rho\left(\mathbf{r}\right)\delta q_2^i\left(\mathbf{r}\right)\right]\right\}\left(\mathbf{r}\right)}{\left|\mathbf{r}'-\mathbf{r}\right|}\,dV\left(\mathbf{r}'\right)dV\left(\mathbf{r}\right)$$

$$+\int_V \rho\left(\mathbf{r}\right)\delta q_2^i\left(\mathbf{r}\right)\overline{\delta q_1^j}\left(\mathbf{r}\right)\left(\nabla_j\nabla_i\Phi\right)\left(\mathbf{r}\right)dV\left(\mathbf{r}\right). \tag{4.55}$$

From the symmetry in the two displacement fields, it follows that

$$\int_V \delta q_2^i\left(V_{ij}\,\overline{\delta q_1^j}\right)dV = \int_V \overline{\delta q_1^i}\left(V_{ij}\,\delta q_2^j\right)dV. \tag{4.56}$$

In conclusion, equalities (4.52) and (4.56) show that symmetry property (4.32) is satisfied for all pairs of displacement fields in the domain of the tensorial operator U.

4.6 Orthogonality of the Linear, Isentropic Normal Modes

The eigenvalues of a symmetric operator are real. The validity of this property can readily be verified as follows. By virtue of wave equation (4.23) and the symmetry of the tensorial operator U, one has, for any pair of modes 1 and 2,

$$\left(\overline{\sigma_1^2}-\sigma_2^2\right)\int_V \rho\,g_{ij}\,\overline{\delta q_1^i}\,\delta q_2^j\,dV = 0. \tag{4.57}$$

When the displacement fields 1 and 2 are identical, the equality implies that the eigenvalue σ^2 of a linear, isentropic normal mode with a norm different from zero is equal to its complex conjugate, so that it is *real*.

The real character of the eigenvalues leads to the following possibilities for the time behavior of a linear, isentropic normal mode: when $\sigma^2 > 0$, the normal mode represents a global oscillation of the star; when $\sigma^2 < 0$, the normal mode represents a perturbation of the star that either increases or decreases exponentially with time; when $\sigma^2 = 0$, the normal mode is time-independent, or neutral in the sense that it does not awake any reaction inside the star. Therefore, the study of the linear, isentropic normal modes of a star is important for the investigation of the *dynamic stability* of the star.

Furthermore, the real character of the eigenvalues implies that equality (4.57) can be rewritten as

$$\left(\sigma_1^2-\sigma_2^2\right)\int_V \rho\,g_{ij}\,\overline{\delta q_1^i}\,\delta q_2^j\,dV = 0. \tag{4.58}$$

Two possibilities can then be distinguished.

The first possibility is that the normal modes are associated with different eigenvalues σ_1^2 and σ_2^2. It then follows that

$$\int_V \rho\, g_{ij}\, \overline{\delta q_1^i}\, \delta q_2^j\, dV = 0. \tag{4.59}$$

Hence, linear, isentropic normal modes associated with different angular frequencies are orthogonal with respect to each other. The orthogonality relation can be expressed as

$$\int_V \rho\, \overline{\boldsymbol{\xi}_1} \cdot \boldsymbol{\xi}_2\, dV = 0 \quad \text{if} \quad \sigma_1 \neq \sigma_2. \tag{4.60}$$

The second possibility is that the two eigenvalues σ_1^2 and σ_2^2 are equal to each other:

$$\sigma_1^2 = \sigma_2^2. \tag{4.61}$$

Hence, eigenvalues σ^2 may have a multiplicity two. Gabriel (1980) showed that the multiplicity of the eigenvalues is at most two. A case of an eigenvalue with multiplicity two is considered at the end of Sect. 10.4.1.

4.7 Global Translations of a Quasi-Static Star as Normal Linear, Isentropic Modes

A global translation of a star, represented by a time-independent uniform Lagrangian displacement field, belongs to the linear, isentropic normal modes of the star.

A Lagrangian displacement field of a star that is constant at all points,

$$\boldsymbol{\xi} = \mathbf{C}, \tag{4.62}$$

is divergence-free:

$$\alpha = 0. \tag{4.63}$$

It follows from Eqs. (2.53) and (3.24) that

$$\delta\rho = 0, \quad \delta P = 0. \tag{4.64}$$

Since the displacement field is time-independent, it furthermore follows from Eqs. (2.52) that

$$\delta\Phi = A, \tag{4.65}$$

where A is an arbitrary constant.

Poisson's perturbed differential equation (2.55) can then be transformed into the partial differential equation for the radial component of the Lagrangian displacement

$$\left[\frac{1}{r^2}\frac{\partial}{\partial r}\left(r^2\frac{\partial}{\partial r}\right)-\frac{1}{r^2}\mathscr{L}^2\right]\left(\frac{d\Phi}{dr}\,\delta r\right)=4\pi G\,\frac{d\rho}{dr}\,\delta r. \qquad (4.66)$$

The radial component δr is related to the constant Cartesian components δx, δy, δz of the Lagrangian displacement by means of transformation formula (1.14) for contravariant vector components, so that

$$\delta r = \sin\theta\,\cos\phi\,\delta x + \sin\theta\,\sin\phi\,\delta y + \cos\theta\,\delta z. \qquad (4.67)$$

It follows that δr is independent of the radial coordinate r, and that

$$\mathscr{L}^2\delta r = 2\,\delta r. \qquad (4.68)$$

By use of Poisson's differential equation (2.14), which applies to an unperturbed star, one verifies that Eq. (4.66) is identically satisfied.

The time-independent uniform Lagrangian displacement field must also satisfy the boundary conditions. First, condition (4.2) is automatically satisfied. Next, previous to imposing conditions (4.8)–(4.11), one can observe that the internal solution for the Eulerian perturbation of the gravitational potential is given by

$$\Phi_i'(r,\theta,\phi) = A - \frac{d\Phi}{dr}\,(\sin\theta\,\cos\phi\,\delta x + \sin\theta\,\sin\phi\,\delta y + \cos\theta\,\delta z), \qquad (4.69)$$

and that the external solution can be expanded as

$$\Phi_e'(r,\theta,\phi) = \sum_{\ell=0}^{\infty} r^{-(\ell+1)} \sum_{m=0}^{\ell} P_\ell^m(\cos\theta)\,\big[D_{\ell,m}\,\cos(m\phi) + E_{\ell,m}\,\sin(m\phi)\big], \qquad (4.70)$$

where the $D_{\ell,m}$ and $E_{\ell,m}$ are arbitrary constants. By substitution of these solutions into boundary conditions (4.8), (4.10), and (4.11), it follows that

$$\left.\begin{aligned} &D_{0,0} = R\,A, \quad D_{1,0} = -R^2\left(\frac{d\Phi}{dr}\right)_R \delta z\,, \\ &D_{1,1} = -R^2\left(\frac{d\Phi}{dr}\right)_R \delta x, \quad E_{1,1} = -R^2\left(\frac{d\Phi}{dr}\right)_R \delta y\,, \end{aligned}\right\} \qquad (4.71)$$

and that all other constants in the solution for $\Phi_e'(r, \theta, \phi)$ are equal to zero. The solution for $\Phi_e'(r, \theta, \phi)$ then reduces to

$$\Phi_e'(r, \theta, \phi) = R\,A\,r^{-1} - R^2 \left(\frac{d\Phi}{dr}\right)_R r^{-2}\,\delta r, \tag{4.72}$$

so that

$$\left(\frac{\partial \Phi_e'}{\partial r}\right)_R = -\frac{A}{R} + \frac{2}{R}\left(\frac{d\Phi}{dr}\right)_R \delta r. \tag{4.73}$$

Since

$$\left(\frac{\partial \Phi_i'}{\partial r}\right)_R = \left[\frac{2}{R}\left(\frac{d\Phi}{dr}\right)_R - 4\pi G\rho(R)\right]\delta r, \tag{4.74}$$

boundary condition (4.9) is satisfied when

$$A = 0. \tag{4.75}$$

From Eq. (4.65), it then follows that

$$\delta\Phi = 0. \tag{4.76}$$

This requirement is compatible with the Lagrangian displacement field considered. The conclusion then follows.

With the three constant Cartesian components δx, δy, δz of the Lagrangian displacement, three linearly independent translations of a star can be associated.

4.8 Immovability of the Star's Mass Centre

A particular consequence of orthogonality relation (4.60) is that all linear, isentropic normal modes $\boldsymbol{\xi}$ associated with an angular frequency σ different from zero are mutually orthogonal to the normal modes with an angular frequency equal to zero that represent uniform translations of the star. Hence, for each linear, isentropic normal mode with an angular frequency different from zero, the property holds

$$\int_V \rho\,\boldsymbol{\xi}\,dV = 0. \tag{4.77}$$

Since the displacement of a star's mass centre associated with a linear, isentropic normal mode is given by

$$\delta\mathbf{R} = \frac{1}{M}\int_V \rho\,\boldsymbol{\xi}\,dV, \tag{4.78}$$

it results that linear, isentropic normal modes with an angular frequency different from zero do not bring about a displacement of the star's mass centre (Smeyers 1966b). This conclusion is natural, since no external forces are supposed to act on the star.

Chapter 5
Spheroidal and Toroidal Normal Modes

5.1 Introduction

In this chapter, the eigenvalue problem of the linear, isentropic normal modes of a quasi-static star, which is of the third degree in the eigenvalue parameter σ^2, is split up into two partial eigenvalue problems on the ground of the radial component of the vorticity equation. Moreover, the Lagrangian displacement field is resolved into a longitudinal and a transverse field, which itself is resolved into a toroidal and a poloidal field. The scalar functions involved in these three fields are expanded in terms of spherical harmonics of the colatitude and the azimuthal angle. Two types of uncoupled normal modes associated with a single spherical harmonic can then be distinguished: spheroidal normal modes, which are solutions of the first partial eigenvalue problem and consist of a longitudinal and a poloidal field, and time-independent toroidal normal modes, which are solutions of the second partial eigenvalue problem. The first eigenvalue problem is quadratic in the eigenvalue parameter, the second one, linear in the eigenvalue parameter.

5.2 Radial Component of the Vorticity Equation

The vorticity equation is obtained as the curl of the perturbed equation of motion. Before deriving it, we transform the perturbed equation of motion whose components are given by Eqs. (4.17)–(4.19).

Equations (4.18) and (4.19) contain the partial derivative of the sum of functions $\Phi' + P'/\rho$, successively, with respect to the colatitude and with respect to the azimuthal angle. Therefore, it is appropriate to let appear the partial derivative of the sum of functions $\Phi' + P'/\rho$ with respect to the radial coordinate in Eq. (4.17). This equation then becomes

$$\sigma^2 \, \delta r = \frac{\partial}{\partial r}\left(\Phi' + \frac{P'}{\rho}\right) - \frac{\rho'}{\rho^2}\frac{dP}{dr} - P'\frac{d}{dr}\frac{1}{\rho}. \tag{5.1}$$

P. Smeyers and T. Van Hoolst, *Linear Isentropic Oscillations of Stars: Theoretical Foundations*, Astrophysics and Space Science Library 371,
DOI 10.1007/978-3-642-13030-4_6, © Springer-Verlag Berlin Heidelberg 2010

Elimination of P' from the last term in the right-hand member by means of Eq. (4.21) yields

$$\sigma^2\,\delta r = \frac{\partial}{\partial r}\left(\Phi' + \frac{P'}{\rho}\right) + \left(c^2\,\frac{1}{\rho}\,\frac{d\rho}{dr} - \frac{1}{\rho}\,\frac{dP}{dr}\right)\left(\frac{\rho'}{\rho} + \frac{1}{\rho}\,\frac{d\rho}{dr}\,\delta r\right). \tag{5.2}$$

By use of definition (3.32) of N^2, and Eq. (4.20), it results that

$$\sigma^2\,\delta r = \frac{\partial}{\partial r}\left(\Phi' + \frac{P'}{\rho}\right) + \frac{N^2}{g}\,c^2\,\alpha. \tag{5.3}$$

The perturbed equation of motion can then be put into the vectorial form

$$\sigma^2\,\boldsymbol{\xi} = \nabla\left(\Phi' + \frac{P'}{\rho}\right) + \frac{N^2}{g}\,c^2\,\alpha\,\mathbf{1}_r, \tag{5.4}$$

where $\mathbf{1}_r$ is the unit vector in the radial direction. This equation was derived by Ledoux & Walraven (1958). Hence, the sum of the forces that act on a unit mass in a perturbed quasi-static star can be represented as the sum of the gradient of a scalar function, and a purely radial vector.

By taking the curl of both members of Eq. (5.4), one obtains the vorticity equation

$$\sigma^2\,(\nabla \times \boldsymbol{\xi}) = \nabla \times \left(\frac{N^2}{g}\,c^2\,\alpha\,\mathbf{1}_r\right). \tag{5.5}$$

Since the curl of a purely radial vector has no radial component, it follows, for the radial component of the vorticity equation,

$$\sigma^2\,(\nabla \times \boldsymbol{\xi})_r = 0. \tag{5.6}$$

On the ground of this equation, the eigenvalue problem of the linear, isentropic normal modes of a quasi-static star is split up into two partial eigenvalue problems: the eigenvalue problem of the normal modes for which the condition

$$(\nabla \times \boldsymbol{\xi})_r = 0 \tag{5.7}$$

is fulfilled *at all points* in the equilibrium star, and the eigenvalue problem of the normal modes for which this condition is not fulfilled.

To the normal modes of the first eigenvalue problem, the *time-dependent* normal modes belong. The underlying physical reason why, for a time-dependent normal mode, the radial component of the curl of the Lagrangian displacement is identically zero at all points of the equilibrium star is that none of the acting forces is capable of creating vorticity around the normal of the local equipotential surface.

It should be noted that even time-independent normal modes may belong to the first eigenvalue problem.

For the normal modes of the second eigenvalue problem, the property holds that

$$\sigma^2 = 0, \tag{5.8}$$

so that these modes are all *time-independent*.

Hereafter it is shown that the first eigenvalue problem yields spheroidal normal modes, and the second one, toroidal normal modes. Before the introduction of these denominations, a more convenient form of the governing equations is presented, and admissible resolutions of the Lagrangian displacement field are considered.

5.3 Convenient Form of the Governing Equations

The radial component of the vorticity equation given by Eq. (5.6) is equivalent to a combination of the θ- and ϕ-component of the perturbed equation of motion. Another useful combination of the θ- and ϕ-component is given by the equation for the divergence of the horizontal component of the Lagrangian displacement

$$\sigma^2 \left[\frac{1}{\sin\theta} \frac{\partial}{\partial\theta} (\sin\theta\, \xi_\theta) + \frac{1}{\sin\theta} \frac{\partial \xi_\phi}{\partial \phi} \right] = -\frac{1}{r} \mathscr{L}^2 \left(\frac{P'}{\rho} + \Phi' \right). \tag{5.9}$$

In this equation, ξ_r, ξ_θ, ξ_ϕ denote the components of the Lagrangian displacement with respect to the local orthonormal coordinate basis $\partial/\partial r$, $\partial/(r\,\partial\theta)$, $\partial/(r\sin\theta\,\partial\phi)$.

It is appropriate to replace the horizontal components of the perturbed equation of motion by the radial component of the vorticity equation and the equation for the divergence of the horizontal component of the Lagrangian displacement. As before, we also use Eq. (4.20), which expresses the mass conservation of the perturbed elements, Eq. (4.21), which expresses that the mass motions are isentropic, and Eq. (4.22), which is the perturbed differential equation of Poisson.

The system of governing equations from which the time is separated by the factor $\exp(i\sigma t)$ then takes the form

$$\sigma^2\, (\nabla \times \boldsymbol{\xi})_r = 0, \tag{5.10}$$

$$\sigma^2\, \xi_r = \frac{\partial \Phi'}{\partial r} - \frac{\rho'}{\rho}\frac{dP}{dr} + \frac{1}{\rho}\frac{\partial P'}{\partial r}, \tag{5.11}$$

$$\sigma^2 \left[\frac{1}{\sin\theta} \frac{\partial}{\partial\theta} (\sin\theta\, \xi_\theta) + \frac{1}{\sin\theta} \frac{\partial \xi_\phi}{\partial \phi} \right] = -\frac{1}{r} \mathscr{L}^2 \left(\frac{P'}{\rho} + \Phi' \right), \tag{5.12}$$

$$\frac{\rho'}{\rho} + \frac{1}{\rho}\frac{d\rho}{dr}\,\xi_r + \alpha = 0, \tag{5.13}$$

$$\frac{P'}{P} + \frac{1}{P}\frac{dP}{dr}\,\xi_r - \Gamma_1 \left(\frac{\rho'}{\rho} + \frac{1}{\rho}\frac{d\rho}{dr}\,\xi_r \right) = 0, \tag{5.14}$$

$$\nabla^2 \Phi' \equiv \frac{1}{r^2}\frac{\partial}{\partial r}\left(r^2 \frac{\partial \Phi'}{\partial r} \right) - \frac{1}{r^2}\mathscr{L}^2 \Phi' = 4\pi G \rho', \tag{5.15}$$

where the divergence of the Lagrangian displacement can be expressed as

$$\alpha = \frac{1}{r^2}\frac{\partial}{\partial r}\left(r^2\,\xi_r\right) + \frac{1}{r\sin\theta}\frac{\partial}{\partial\theta}\left(\sin\theta\,\xi_\theta\right) + \frac{1}{r\sin\theta}\frac{\partial\xi_\phi}{\partial\phi}. \tag{5.16}$$

5.4 Helmholtz's Resolution Theorem for Vector Fields

In various fields of theoretical physics, it is customary to represent vector fields by means of vector and scalar potentials on which vector operators act. A similar way of representation can usefully be adopted for linear, isentropic displacement fields in stars.

According to a fundamental theorem of vector analysis, known as Helmholtz's resolution theorem, a continuous vector field $\boldsymbol{\xi}$ that is defined everywhere in the infinite space and tends to zero towards infinity together with its first derivatives, can be represented as the sum of a curl-free vector field $\boldsymbol{\xi}_1$ and a divergence-free vector field $\boldsymbol{\xi}_2$:

$$\boldsymbol{\xi} = \boldsymbol{\xi}_1 + \boldsymbol{\xi}_2, \tag{5.17}$$

where

$$\nabla\times\boldsymbol{\xi}_1 = 0, \quad \nabla\cdot\boldsymbol{\xi}_2 = 0, \tag{5.18}$$

and

$$\nabla\cdot\boldsymbol{\xi}_1 = \nabla\cdot\boldsymbol{\xi}, \quad \nabla\times\boldsymbol{\xi}_2 = \nabla\times\boldsymbol{\xi}. \tag{5.19}$$

The representation of the vector field is unique except for a constant vector (Morse & Feshbach 1953).

The uniqueness of the representation remains valid for a vector field, as the Lagrangian displacement field, that is confined to the domain of a spherical equilibrium star, if one imposes adequate conditions on the surface of the star (Sommerfeld 1964a). This can be verified as follows.

Following a procedure of Sommerfeld, we observe that the curl-free part $\boldsymbol{\xi}_1$ of the Lagrangian displacement field $\boldsymbol{\xi}$ can be derived from a scalar potential Ψ as

$$\boldsymbol{\xi}_1 = \nabla\Psi + \mathbf{C}_1, \tag{5.20}$$

where $\mathbf{C}_1$ is a constant vector field. By taking the divergence of both members of the equation, one has

$$\nabla^2\Psi = \nabla\cdot\boldsymbol{\xi}. \tag{5.21}$$

The solution of this non-homogeneous partial differential equation for the scalar potential Ψ has the same form as solution (1.67) of Poisson's differential equation:

$$\Psi(\mathbf{r}) = -\frac{1}{4\pi}\int_V \frac{(\nabla\cdot\xi)(\mathbf{r}')}{|\mathbf{r}'-\mathbf{r}|}\,dV(\mathbf{r}') + \frac{1}{4\pi}\int_S \frac{1}{|\mathbf{r}'-\mathbf{r}|}\left(\frac{\partial\Psi}{\partial n}\right)(\mathbf{r}')\,dS(\mathbf{r}')$$
$$-\frac{1}{4\pi}\int_S \Psi(\mathbf{r}')\left(\frac{\partial}{\partial n}\frac{1}{|\mathbf{r}'-\mathbf{r}|}\right)(\mathbf{r}')\,dS(\mathbf{r}'). \tag{5.22}$$

The difference between the two surface integrals can be developed by the use of expansion (2.27) for $|\mathbf{r}'-\mathbf{r}|^{-1}$ and the expansion of the scalar potential $\Psi(\mathbf{r}')$ in terms of spherical harmonics

$$\Psi(\mathbf{r}') = \sum_{\ell=0}^{\infty}\sum_{m=-\ell}^{\ell} \Psi_{\ell,m}(r')\; Y_\ell^m(\theta',\phi'). \tag{5.23}$$

By taking into account the orthogonality relation (Appendix D.4) for spherical harmonics, one obtains

$$\frac{1}{4\pi}\int_S \frac{1}{|\mathbf{r}'-\mathbf{r}|}\left(\frac{\partial\Psi}{\partial n}\right)(\mathbf{r}')\,dS(\mathbf{r}') - \frac{1}{4\pi}\int_S \Psi(\mathbf{r}')\left(\frac{\partial}{\partial n}\frac{1}{|\mathbf{r}'-\mathbf{r}|}\right)(\mathbf{r}')\,dS(\mathbf{r}')$$
$$= \sum_{\ell=0}^{\infty}\frac{1}{2\ell+1}\frac{1}{R^{\ell-1}}\sum_{m=-\ell}^{\ell}\left[\left(\frac{d\Psi_{\ell,m}}{dr}\right)_R + \frac{\ell+1}{R}\Psi_{\ell,m}(R)\right] r^\ell\, Y_\ell^m(\theta,\phi). \tag{5.24}$$

The difference between the two surface integrals is identically zero when

$$\left(\frac{d\Psi_{\ell,m}}{dr}\right)_R + \frac{\ell+1}{R}\Psi_{\ell,m}(R) = 0, \quad \ell = 0,1,2,\ldots, \quad m = -\ell,\ldots,\ell. \tag{5.25}$$

Under these conditions, the scalar potential Ψ is determined by a volume integral over the bounded domain of a spherical equilibrium star, in which the Lagrangian displacement field is defined:

$$\Psi(\mathbf{r}) = -\frac{1}{4\pi}\int_V \frac{(\nabla\cdot\xi)(\mathbf{r}')}{|\mathbf{r}'-\mathbf{r}|}\,dV(\mathbf{r}'). \tag{5.26}$$

Once the scalar potential Ψ is determined, one derives the curl-free part ξ_1 of the Lagrangian displacement field by means of Eq. (5.20), except for a constant vector field.

Still following Sommerfeld's procedure, we observe that the divergence-free part $\boldsymbol{\xi}_2$ of the Lagrangian displacement field $\boldsymbol{\xi}$ can be derived from a vector potential $\mathbf{A}$ as

$$\boldsymbol{\xi}_2 = \nabla\times\mathbf{A} + \mathbf{C}_2, \tag{5.27}$$

where $\mathbf{C}_2$ is a constant vector field. On the vector potential $\mathbf{A}$, the condition

$$\nabla\cdot\mathbf{A} = 0 \tag{5.28}$$

can be imposed. Indeed, if necessary, one can make the vector potential divergence-free by adding a curl-free term ∇f to it, where f is a scalar function. The curl-free term has no effect on the determination of $\boldsymbol{\xi}_2$ by Eq. (5.27).

By taking the curl of both members of Eq. (5.27), one obtains the non-homogeneous partial differential equation for the vector potential $\mathbf{A}$

$$\nabla \times \nabla \times \mathbf{A} = \nabla \times \boldsymbol{\xi}. \tag{5.29}$$

For Cartesian vector components, the rule of the vector analysis

$$\nabla^2 \mathbf{A} = \nabla (\nabla \cdot \mathbf{A}) - \nabla \times \nabla \times \mathbf{A} \tag{5.30}$$

can be applied. Because of condition (5.28), the first term in the right-hand member is equal to zero, so that differential equation (5.29) reduces to

$$\nabla^2 \mathbf{A} = -\nabla \times \boldsymbol{\xi}. \tag{5.31}$$

The Cartesian components of the vector potential $\mathbf{A}$ satisfy a non-homogeneous partial differential equation of the same form as the non-homogeneous partial differential equation (5.21). Proceeding as above, one sees that the Cartesian components are given by

$$\mathbf{A} = -\frac{1}{4\pi} \int_V \frac{(\nabla \times \boldsymbol{\xi})(\mathbf{r}')}{|\mathbf{r}' - \mathbf{r}|} \, dV\left(\mathbf{r}'\right). \tag{5.32}$$

Once the scalar potential $\mathbf{A}$ is determined, one derives the divergence-free part $\boldsymbol{\xi}_2$ of the Lagrangian displacement field by means of Eq. (5.27), except for a constant vector.

Thus, by imposing conditions on the surface of the equilibrium star comparable with conditions (5.25), one obtains the scalar potential Ψ and the Cartesian components of the vector potential $\mathbf{A}$ in a way similar to that used for a vector field defined at all points of the infinite space. The uniqueness of the resolution of the Lagrangian displacement field in a star, apart from a constant vector field, can then be verified in the same way as for the resolution of a vector field defined at all points of the infinite space.

Suppose that, besides the solution $\boldsymbol{\xi}_1$ and $\boldsymbol{\xi}_2$, another solution $\boldsymbol{\xi}_1^*$ and $\boldsymbol{\xi}_2^*$ exists. The curl-free vector $\boldsymbol{\xi}_1^* - \boldsymbol{\xi}_1$ is then also divergence-free by virtue of the first equality (5.19):

$$\nabla \cdot \left(\boldsymbol{\xi}_1^* - \boldsymbol{\xi}_1\right) = 0. \tag{5.33}$$

Similarly, the divergence-free vector $\boldsymbol{\xi}_2^* - \boldsymbol{\xi}_2$ is curl-free by virtue of the second equality (5.19):

$$\nabla \times \left(\boldsymbol{\xi}_2^* - \boldsymbol{\xi}_2\right) = 0. \tag{5.34}$$

From the determinations of the scalar potential Ψ and the vector potential $\mathbf{A}$ for the divergence- and curl-free vectors $\boldsymbol{\xi}_1^* - \boldsymbol{\xi}_1$ and $\boldsymbol{\xi}_2^* - \boldsymbol{\xi}_2$, it appears that $\Psi = 0$ and $\mathbf{A} = 0$. Because of Eqs. (5.20) and (5.27), it then follows

$$\boldsymbol{\xi}_1^* - \boldsymbol{\xi}_1 = \mathbf{C}_3, \quad \boldsymbol{\xi}_2^* - \boldsymbol{\xi}_2 = \mathbf{C}_4, \tag{5.35}$$

where $\mathbf{C}_3$ and $\mathbf{C}_4$ are constant vectors. The resolution is thus unique, apart from a constant vector.

By resolving the Lagrangian displacement field in a quasi-static star as

$$\boldsymbol{\xi} = \nabla\,\Psi + \nabla \times \mathbf{A}, \tag{5.36}$$

where the scalar potential Ψ and the Cartesian components of the divergence-free vector potential $\mathbf{A}$ are determined respectively by Eqs. (5.26) and (5.32) and satisfy boundary conditions comparable with conditions (5.25), one expresses the three independent components of the Lagrangian displacement in terms of three other independent quantities. The vector field $\nabla\,\Psi$ is called the *longitudinal* part of the vector field $\boldsymbol{\xi}$, and the vector field $\nabla \times \mathbf{A}$, the *transverse* part. The introduction of these terms goes back to a passage in Lamb's publication entitled *On the Vibrations of an Elastic Sphere* (Lamb 1882):

> ... any arbitrary disturbance, originated in any part of an elastic solid, breaks up in general into two distinct waves which travel with different velocities. We have, in the first place, a wave of pure rotation, unaccompanied by dilation, ...; and superposed on this there is a wave of pure dilation, without rotation, ... In *plane* waves of these two kinds, the vibrations are respectively parallel and perpendicular to the wave front, and a similar statement holds good approximately in the general case. Hence the two kinds of waves are conveniently characterized as waves of *transverse* and of *longitudinal* vibration respectively.

5.5 Resolution of the Vector Field $\rho\,\xi$

For the vector field $\rho\,\boldsymbol{\xi}$, a resolution in terms of a curl-free and a divergence-free vector field can be derived from Eqs. (5.13) and (5.15).

In the case of purely radial normal modes, Eqs. (5.13) and (5.15) reduce to

$$\rho' + \frac{1}{r^2}\frac{d}{dr}\left(\rho\,r^2\,\xi_r\right) = 0, \tag{5.37}$$

$$\frac{1}{r^2}\frac{d}{dr}\left(r^2\,\frac{d\Phi'}{dr}\right) = 4\pi G\rho'. \tag{5.38}$$

By elimination of ρ' and integration of the resulting equation, one obtains

$$4\pi G\rho\,r^2\,\xi_r = -r^2\,\frac{d\Phi'}{dr}. \tag{5.39}$$

The integration constant is equal to zero, since both members of the equation vanish at $r = 0$. One then has the equation

$$4\pi G\rho\,\xi_r = -\frac{d\Phi'}{dr}. \tag{5.40}$$

This equation was derived by Ledoux & Walraven (1958) and is a first integral for the radial linear, isentropic oscillations of stars (Takata 2006a). The vector field $\rho\,\boldsymbol{\xi}$ is thus described by the gradient of a scalar potential, which is the opposite of the Eulerian perturbation of the gravitational potential, apart from the constant factor $1/(4\pi G)$. It results that the vector field $\rho\,\boldsymbol{\xi}$ is curl-free.

Woltjer (1958) and Aizenman & Smeyers (1977) have extended the procedure of Ledoux & Walraven (1958) to non-radial normal modes of stars. By elimination of ρ' and integration, it results that

$$4\pi G\rho\,\boldsymbol{\xi} = -\nabla\left(\Phi' + L\right) + \nabla\times\mathbf{B}, \tag{5.41}$$

where $L(r,\theta,\phi)$ is a harmonic function. The opposite of the function $\Phi' + L$ and the vector $\mathbf{B}$ play the role of scalar potential and vector potential, respectively, for the vector field $\rho\,\boldsymbol{\xi}$, except for a factor $1/(4\pi G)$. As has been done above for the vector potential $\mathbf{A}$, one can set

$$\nabla\cdot\mathbf{B} = 0. \tag{5.42}$$

The vector field $\rho\,\boldsymbol{\xi}$ then depends on the scalar potential $\Phi' + L$ and two independent components of the divergence-free vector potential $\mathbf{B}$.

When one takes the curl of both members of Eq. (5.41), one has

$$4\pi G\rho\,\nabla\times\boldsymbol{\xi} + \nabla\rho\times 4\pi G\boldsymbol{\xi} = \nabla\times\nabla\times\mathbf{B}. \tag{5.43}$$

If the equilibrium star is spherically symmetric, the radial component of this equation reduces to

$$\left(\nabla\times\nabla\times\mathbf{B}\right)_r = 4\pi G\rho\left(\nabla\times\boldsymbol{\xi}\right)_r. \tag{5.44}$$

For normal modes for which Eq. (5.7) is satisfied at all points of the equilibrium star, this implies

$$-\left(1-\mu^2\right)^{1/2}\frac{\partial}{\partial\mu}\left\{\left(1-\mu^2\right)^{1/2}\left[\frac{\partial}{\partial r}\left(r\,B_\theta\right) + \left(1-\mu^2\right)^{1/2}\frac{\partial B_r}{\partial\mu}\right]\right\}$$

$$-\frac{1}{\left(1-\mu^2\right)^{1/2}}\frac{\partial^2 B_r}{\partial\phi^2} + \frac{\partial}{\partial r}\left(r\,\frac{\partial B_\phi}{\partial\phi}\right) = 0, \tag{5.45}$$

where $\mu = \cos\theta$. On the other hand, from Eq. (5.42), it follows that

$$r\,\frac{\partial B_\phi}{\partial\phi} = -\left(1-\mu^2\right)^{1/2}\,\frac{\partial}{\partial r}\left(r^2\,B_r\right) + r\,\left(1-\mu^2\right)^{1/2}\,\frac{\partial}{\partial\mu}\left[\left(1-\mu^2\right)^{1/2}\,B_\theta\right]. \tag{5.46}$$

Elimination of $r\,\left(\partial B_\phi/\partial\phi\right)$ leads to the partial differential equation for B_r

$$\frac{\partial^2}{\partial r^2}\left(r^2\,B_r\right) + \frac{\partial}{\partial\mu}\left[\left(1-\mu^2\right)\frac{\partial B_r}{\partial\mu}\right] + \frac{1}{1-\mu^2}\,\frac{\partial^2 B_r}{\partial\phi^2} = 0. \tag{5.47}$$

The equation admits of the solution

$$B_r(r,\theta,\phi) = \sum_{\ell=0}^{\infty}\sum_{m=-\ell}^{\ell}\frac{dR_\ell(r)}{dr}\,P_\ell^{\,m}(\cos\theta)\,\exp(im\phi), \tag{5.48}$$

where $R_\ell(r)$ is solution of the radial Laplace equation

$$\frac{1}{r^2}\,\frac{d}{dr}\left(r^2\,\frac{dR_\ell}{dr}\right) - \frac{\ell(\ell+1)}{r^2}\,R_\ell = 0. \tag{5.49}$$

As concluded by Aizenman & Smeyers (1977), for all normal modes for which Eq. (5.7) is satisfied, the component B_r of the divergence-free vector potential **B** is a well-determined function, and the components B_θ and B_ϕ are linked by means of Eq. (5.46).

The harmonic function $L(r,\theta,\phi)$ in the right-hand member of Eq. (5.41) was omitted by Aizenman & Smeyers (1977). The necessity of its inclusion was stressed by Sobouti (1981) for the reason that the scalar potential and the Eulerian perturbation of the gravitational potential may obey different conditions on a star's surface. In this context, it is useful to consider the non-homogeneous partial differential equation

$$\nabla^2\left(\Phi' + L\right) = -4\pi G\,\nabla\cdot(\rho\,\xi\,). \tag{5.50}$$

Its solution has the same form as solution (5.22) of the non-homogeneous partial differential equation (5.21) for the scalar potential Ψ. The functions Φ' and L can be developed in terms of spherical harmonics as

$$\left.\begin{aligned} \Phi'(r,\theta,\phi) &= \sum_{\ell=0}^{\infty}\sum_{m=-\ell}^{\ell}\Phi'_{\ell,m}(r)\,Y_\ell^m(\theta,\phi),\\ L(r,\theta,\phi) &= \sum_{\ell=0}^{\infty}\sum_{m=-\ell}^{\ell}L_{\ell,m}(r)\,Y_\ell^m(\theta,\phi), \end{aligned}\right\} \tag{5.51}$$

where the functions $L_{\ell,m}(r)$ are solutions of Laplace's radial equation

$$\frac{1}{r^2}\frac{d}{dr}\left(r^2\frac{dL_{\ell,m}}{dr}\right)-\frac{\ell(\ell+1)}{r^2}L_{\ell,m}=0. \tag{5.52}$$

For displacement fields that remain finite at $r=0$, the admissible solutions are

$$L_{\ell,m}(r)=C_{\ell,m}\,r^{\ell}, \tag{5.53}$$

where the coefficients $C_{\ell,m}$ are arbitrary constants.

The solution of Eq. (5.50) then takes the form

$$\Phi'(\mathbf{r})+L(\mathbf{r})=G\int_V\frac{[\nabla\cdot(\rho\,\xi)](\mathbf{r}')}{|\mathbf{r}'-\mathbf{r}|}\,dV(\mathbf{r}')+\sum_{\ell=0}^{\infty}\frac{1}{2\ell+1}\frac{1}{R^{\ell-1}}$$

$$\sum_{m=-\ell}^{\ell}\left\{\left[\frac{d\left(\Phi'_{\ell,m}+L_{\ell,m}\right)}{dr}\right]_R+\frac{\ell+1}{R}\left(\Phi'_{\ell,m}+L_{\ell,m}\right)_R\right\}r^{\ell}\,Y_\ell^m(\theta,\phi). \tag{5.54}$$

The infinite sum of terms is identically zero, if

$$\left[\frac{d\left(\Phi'_{\ell,m}+L_{\ell,m}\right)}{dr}\right]_R+\frac{\ell+1}{R}\left(\Phi'_{\ell,m}+L_{\ell,m}\right)_R=0,$$

$$\ell=0,1,2,\ldots,\quad m=-\ell,\ldots,\ell. \tag{5.55}$$

By using solution (5.53) for the harmonic function $L_{\ell,m}(r)$ and, in anticipation, condition (5.97) for the Eulerian perturbation of the gravitational potential on the star's surface, one obtains

$$C_{\ell,m}=\frac{4\pi G}{2\ell+1}\frac{\left(\rho\,\xi_{\ell,m}\right)_R}{R^{\ell-1}},\quad \ell=0,1,2,\ldots,\quad m=-\ell,\ldots,\ell. \tag{5.56}$$

Hence, the harmonic function $L(r,\theta,\phi)$ is identically zero when the mass density vanishes on the surface of the equilibrium star. The Eulerian perturbation of the gravitational potential is thus the scalar potential for the vector field $\rho\,\xi$ in a broad class of stellar models. The vector field can then be described as

$$\rho\,\xi=\frac{1}{4\pi G}\left(-\nabla\Phi'+\nabla\times\mathbf{B}\right). \tag{5.57}$$

For stars with a vanishing surface density, Takata (2005) determined a divergence-free vector potential $\mathbf{B}$ in the case of axisymmetric displacement fields associated with the spherical harmonic $Y_1(\theta)$, i.e.,

$$\left.\begin{aligned} \xi_r(r,\theta) &= \xi(r)\,\cos\theta, \quad \Phi'(r,\theta) = \Phi'(r)\,\cos\theta, \\ \rho'(r,\theta) &= \rho'(r)\,\cos\theta, \quad P'(r,\theta) = P'(r)\,\cos\theta. \end{aligned}\right\} \tag{5.58}$$

Indeed, for any axisymmetric displacement field, Eq. (5.57) yields

$$\left.\begin{aligned} 4\pi G\rho\,\xi_r &= -\frac{\partial\Phi'}{\partial r} + \frac{1}{r\,\sin\theta}\,\frac{\partial}{\partial\theta}\left(\sin\theta\,B_\phi\right), \\ 4\pi G\rho\,\xi_\theta &= -\frac{1}{r}\,\frac{\partial\Phi'}{\partial\theta} - \frac{1}{r}\,\frac{\partial}{\partial r}\left(r\,B_\phi\right), \\ 4\pi G\rho\,\xi_\phi &= \frac{1}{r}\,\frac{\partial}{\partial r}\left(r\,B_\theta\right) - \frac{1}{r}\,\frac{\partial B_r}{\partial\theta} = 0. \end{aligned}\right\} \tag{5.59}$$

Elimination of ξ_r from Eq. (5.11), use of equalities (5.58), and integration with respect to θ lead to

$$B_\phi = 4\pi G\frac{r}{2}\left[\frac{1}{\sigma^2}\left(\rho\,\frac{d\Phi'}{dr} + \rho'\,\frac{d\Phi}{dr} + \frac{dP'}{dr}\right) + \frac{1}{4\pi G}\,\frac{d\Phi'}{dr}\right]\sin\theta. \tag{5.60}$$

Since Eq. (5.46) and the third Eq. (5.59) can be satisfied by the solutions

$$B_r = 0, \quad B_\theta = 0, \tag{5.61}$$

a vector potential is given by

$$\mathbf{B} = B_\phi\,\mathbf{1}_\phi. \tag{5.62}$$

In their study, Aizenman & Smeyers (1977) resolved the divergence-free part of the vector field $\rho\,\boldsymbol{\xi}$ into a poloidal and a toroidal field, so that the representation of the vector field involves three scalar functions. In addition, they showed that the expansion of the scalar functions in terms of spherical harmonics leads to a resolution of the vector field into spheroidal and toroidal fields.

Although, for stellar models with a vanishing surface density, the scalar potential for the vector field $\rho\,\boldsymbol{\xi}$ is simply related to the Eulerian perturbation of the gravitational potential, it seems preferable to use resolution (5.36) of the Lagrangian displacement field $\boldsymbol{\xi}$ above resolution (5.41) of the vector field $\rho\,\boldsymbol{\xi}$.

5.6 Resolution of the Displacement Field into a Radial and a Horizontal Field

In resolution (5.36) of the Lagrangian displacement field, the divergence-free part $\nabla\times\mathbf{A}$ can be represented as the sum of a toroidal and a poloidal vector field (see, e.g., Morse & Feshbach 1953, Roberts 1967, Rädler 1974). The toroidal vector field has only a horizontal component, while the poloidal vector field has both a radial and a horizontal component, in accordance with equalities (Appendix G.15) and (Appendix G.16).

The resolution of the Lagrangian displacement field then takes the form

$$\boldsymbol{\xi} = \nabla\Psi + \nabla \times (T\,\mathbf{1}_r) + \nabla \times \nabla \times (S\,\mathbf{1}_r), \tag{5.63}$$

where $\Psi(r,\theta,\phi)$, $S(r,\theta,\phi)$, $T(r,\theta,\phi)$ are scalar functions of the coordinates r, θ, ϕ. Hence, at any point in a star, one derives the three components of the Lagrangian displacement by applying vector operations to the scalar functions Ψ, T, S. The components ξ_r, ξ_θ, ξ_ϕ with respect to the local orthonormal basis are given by

$$\left.\begin{aligned}
\xi_r &= \frac{\partial\Psi}{\partial r} + \frac{1}{r^2}\mathscr{L}^2 S,\\
\xi_\theta &= \frac{1}{r}\frac{\partial}{\partial\theta}\left(\Psi + \frac{\partial S}{\partial r}\right) + \frac{1}{r\,\sin\theta}\frac{\partial T}{\partial\phi},\\
\xi_\phi &= \frac{1}{r\,\sin\theta}\frac{\partial}{\partial\phi}\left(\Psi + \frac{\partial S}{\partial r}\right) - \frac{1}{r}\frac{\partial T}{\partial\theta}.
\end{aligned}\right\} \tag{5.64}$$

Moreover, the radial component of the curl of the Lagrangian displacement is given by

$$(\nabla \times \boldsymbol{\xi})_r = \frac{1}{r^2}\mathscr{L}^2 T\,. \tag{5.65}$$

This component is determined exclusively by the toroidal function $T(r,\theta,\phi)$.

In the right-hand members of equalities (5.64), one passes on from the two functions $\Psi(r,\theta,\phi)$ and $S(r,\theta,\phi)$ to two new functions $\xi(r,\theta,\phi)$ and $\eta(r,\theta,\phi)$ by setting

$$\left.\begin{aligned}
\xi(r,\theta,\phi) &= \frac{\partial\Psi}{\partial r} + \frac{1}{r^2}\mathscr{L}^2 S,\\
\eta(r,\theta,\phi) &= \Psi + \frac{\partial S}{\partial r}.
\end{aligned}\right\} \tag{5.66}$$

Furthermore, by the introduction of the horizontal gradient of the function η

$$\nabla_H\,\eta = \left(0\,,\frac{1}{r}\frac{\partial\eta}{\partial\theta}\,,\frac{1}{r\,\sin\theta}\frac{\partial\eta}{\partial\phi}\right), \tag{5.67}$$

resolution (5.63) can be rewritten as

$$\boldsymbol{\xi} = \xi\,\mathbf{1}_r + (\nabla_H\eta)\,\mathbf{1}_h + \nabla \times (T\,\mathbf{1}_r), \tag{5.68}$$

where $\mathbf{1}_h$ is the unit vector in the horizontal direction. The Lagrangian displacement field is now decomposed into a purely *radial* displacement field and a purely *horizontal* displacement field that consists of the sum of the horizontal gradient of a scalar function, and a toroidal vector.

An alternative derivation of the resolution of the horizontal displacement field was presented by Beyer & Schmidt (1995). In the tangent plane in which it is situated, the horizontal displacement $\boldsymbol{\xi}_h$ can be decomposed with respect to the local orthonormal basis vectors $\mathbf{1}_\theta$ and $\mathbf{1}_\phi$:

$$\boldsymbol{\xi}_h = \left(\xi_\theta, \xi_\phi\right). \tag{5.69}$$

The divergence of the horizontal displacement field $\boldsymbol{\xi}_h$ is given by

$$\nabla \cdot \boldsymbol{\xi}_h = \frac{1}{r\,\sin\theta}\frac{\partial}{\partial\theta}(\sin\theta\,\xi_\theta) + \frac{1}{r\,\sin\theta}\frac{\partial\xi_\phi}{\partial\phi}. \tag{5.70}$$

The curl of the horizontal displacement field $\boldsymbol{\xi}_h$ has only a component perpendicular to the tangent plane, which corresponds to the radial component of the curl of the three-dimensional displacement field $\boldsymbol{\xi}$:

$$(\nabla \times \boldsymbol{\xi}_h)_r = \frac{1}{r\,\sin\theta}\frac{\partial}{\partial\theta}\left(\sin\theta\,\xi_\phi\right) - \frac{1}{r\,\sin\theta}\frac{\partial\xi_\theta}{\partial\phi}. \tag{5.71}$$

In their derivation, Beyer & Schmidt introduced the Hodge dual of a horizontal vector field $\boldsymbol{\xi}_h$ as

$$\boldsymbol{\xi}_h^* = \left(-\xi_\phi, \xi_\theta\right). \tag{5.72}$$

The Hodge dual of the Hodge dual of a horizontal displacement field $\boldsymbol{\xi}_h$ is the opposite of the field:

$$\left(\boldsymbol{\xi}_h^*\right)^* = -\boldsymbol{\xi}_h. \tag{5.73}$$

Moreover, the two following properties hold for the Hodge dual of a horizontal vector field $\boldsymbol{\xi}_h$:

$$\left.\begin{aligned} \nabla \cdot \boldsymbol{\xi}_h^* &= -\,(\nabla \times \boldsymbol{\xi}_h)_r, \\ \left(\nabla \times \boldsymbol{\xi}_h^*\right)_r &= \nabla \cdot \boldsymbol{\xi}_h. \end{aligned}\right\} \tag{5.74}$$

Be $\nabla_H f$ the curl-free part of the horizontal displacement field $\boldsymbol{\xi}_h$, where f is a scalar function. Apart from a constant, the function f is determined uniquely as solution of the non-homogeneous partial differential equation

$$\nabla \cdot \nabla_H f = \nabla \cdot \boldsymbol{\xi}_h, \tag{5.75}$$

so that

$$\nabla \cdot (\nabla_H f - \xi_h) = 0. \tag{5.76}$$

By virtue of the second equality (5.74), it follows that

$$\left[\nabla \times (\nabla_H f - \xi_h)^*\right]_r = 0. \tag{5.77}$$

Consequently, the horizontal vector $(\nabla_H f - \xi_h)^*$ can be represented as the horizontal gradient of a scalar function g:

$$(\nabla_H f - \xi_h)^* = \nabla_H g. \tag{5.78}$$

By considering the Hodge dual of both members and taking into account equality (5.73), one obtains

$$\xi_h = \nabla_H f + (\nabla_H g)^* . \tag{5.79}$$

The horizontal displacement is here represented as the sum of the horizontal gradient of a scalar function and the Hodge dual of the horizontal gradient of another scalar function. Since

$$(\nabla_H g)^* = \left(-\frac{1}{r \sin\theta}\frac{\partial g}{\partial \phi}, \frac{1}{r}\frac{\partial g}{\partial \theta}\right), \tag{5.80}$$

the representation is equivalent to that in resolution (5.68).

5.7 Expansion of the Displacement Field in Terms of Spherical Harmonics

Resolution (5.68) of the Lagrangian displacement field is particularly convenient for the representation of the components by means of spherical harmonics. Expansion of the scalar potentials $\xi(r,\theta,\phi)$, $\eta(r,\theta,\phi)$, $T(r,\theta,\phi)$ in terms of spherical harmonics as

$$f(r,\theta,\phi) = \sum_{\lambda=0}^{\infty} \sum_{\mu=-\lambda}^{\lambda} f_{\lambda,\mu}(r)\, Y_\lambda^\mu(\theta,\phi), \tag{5.81}$$

substitution of these expansions into the components of the resolution, and use of the eigenvalue equation (Appendix D.1) for the spherical harmonics yield

$$\left.\begin{aligned}
\xi_r(r,\theta,\phi) &= \sum_{\lambda=0}^{\infty}\sum_{\mu=-\lambda}^{\lambda} \xi_{\lambda,\mu}(r)\, Y_\lambda^\mu(\theta,\phi),\\
\xi_\theta(r,\theta,\phi) &= \frac{1}{r}\sum_{\lambda=0}^{\infty}\sum_{\mu=-\lambda}^{\lambda} \eta_{\lambda,\mu}(r)\,\frac{\partial Y_\lambda^\mu}{\partial\theta}\\
&\quad+\frac{1}{r\sin\theta}\sum_{\lambda=0}^{\infty}\sum_{\mu=-\lambda}^{\lambda} T_{\lambda,\mu}(r)\,\frac{\partial Y_\lambda^\mu}{\partial\phi},\\
\xi_\phi(r,\theta,\phi) &= \frac{1}{r\sin\theta}\sum_{\lambda=0}^{\infty}\sum_{\mu=-\lambda}^{\lambda} \eta_{\lambda,\mu}(r)\,\frac{\partial Y_\lambda^\mu}{\partial\phi}\\
&\quad-\frac{1}{r}\sum_{\lambda=0}^{\infty}\sum_{\mu=-\lambda}^{\lambda} T_{\lambda,\mu}(r)\,\frac{\partial Y_\lambda^\mu}{\partial\theta}.
\end{aligned}\right\} \tag{5.82}$$

Similarly, it follows from equality (5.65) that

$$(\nabla\times\boldsymbol{\xi})_r = \sum_{\lambda=0}^{\infty}\frac{\lambda(\lambda+1)}{r^2}\sum_{\mu=-\lambda}^{\lambda} T_{\lambda,\mu}(r)\, Y_\lambda^\mu(\theta,\phi). \tag{5.83}$$

On these expansions is based the subsequent distinction between spheroidal and toroidal normal modes. The distinction was made by Lamb as early as 1882 in his publication on a perfectly elastic, spherically symmetric Earth (Lamb 1882).

5.8 Spheroidal Normal Modes

5.8.1 Definition

In this section, we consider normal modes in a quasi-static star for which the radial component of the vorticity equation, given by Eq. (5.10), is satisfied by the condition that, at *all* points of the star,

$$(\nabla\times\boldsymbol{\xi}\,)_r = 0. \tag{5.84}$$

Because of expansion (5.83), it then follows that

$$\sum_{\lambda=0}^{\infty}\frac{\lambda(\lambda+1)}{r^2}\sum_{\mu=-\lambda}^{\lambda} T_{\lambda,\mu}(r)\, Y_\lambda^\mu(\theta,\phi) = 0. \tag{5.85}$$

By multiplying by $\overline{Y_\ell^m}\,(\theta,\phi)$, integrating over a spherical surface, and taking into account orthogonality relation (Appendix D.4) of the spherical harmonics, one has

$$T_{\lambda,\mu}(r) = 0 \quad \text{for} \quad \text{all} \quad \lambda, \mu \neq 0. \tag{5.86}$$

The radial function $T_{0,0}(r)$ remains undetermined, but it does not appear in expansions (5.82) for the horizontal components of the Lagrangian displacement, since the spherical harmonic Y_0 is constant, and its first derivatives with respect to the angular variables vanish. Hence, the horizontal components of a Lagrangian displacement field for which condition (5.84) is satisfied at all points of the equilibrium star do not contain any toroidal part. Resolution (5.68) of the Lagrangian displacement field then reduces to

$$\boldsymbol{\xi} = \xi \, \mathbf{1}_r + (\nabla_H \, \eta) \, \mathbf{1}_h. \tag{5.87}$$

Normal modes of this type are called *spheroidal* modes in accordance with a term which is currently used in geophysics relative to normal modes of the Earth (see, e.g., Bolt & Derr 1969, Dahlen & Tromp 1998).

5.8.2 Eigenvalue Problem of the Spheroidal Normal Modes

The equations that govern spheroidal normal modes result from Eqs. (5.11)–(5.16). Into these equations, one substitutes expansions (5.82) for the components of the Lagrangian displacement, in which equalities (5.86) are taken into account. Furthermore, one substitutes expansions for $\rho'(r,\theta,\phi)$, $P'(r,\theta,\phi)$, $\Phi'(r,\theta,\phi)$ in terms of spherical harmonics of the form of expansion (5.81). After multiplication by $\overline{Y_\ell^m}(\theta,\phi)$, integration over a spherical surface, and use of orthogonality relation (Appendix D.4) of the spherical harmonics, one obtains the following system of linear, homogeneous equations for the functions $\xi_{\ell,m}(r)$, $\eta_{\ell,m}(r)$, $\rho'_{\ell,m}(r)$, $P'_{\ell,m}(r)$, $\Phi'_{\ell,m}(r)$ of the radial coordinate:

$$\sigma^2 \, \xi_{\ell,m} = \frac{d\Phi'_{\ell,m}}{dr} - \frac{\rho'_{\ell,m}}{\rho^2} \frac{dP}{dr} + \frac{1}{\rho} \frac{dP'_{\ell,m}}{dr}, \tag{5.88}$$

$$\ell(\ell+1) \, \sigma^2 \, \eta_{\ell,m} = \ell(\ell+1) \left(\Phi'_{\ell,m} + \frac{P'_{\ell,m}}{\rho} \right), \tag{5.89}$$

$$\frac{\rho'_{\ell,m}}{\rho} + \frac{1}{\rho} \frac{d\rho}{dr} \, \xi_{\ell,m} + \alpha_{\ell,m} = 0, \tag{5.90}$$

$$\frac{P'_{\ell,m}}{P} + \frac{1}{P} \frac{dP}{dr} \, \xi_{\ell,m} - \Gamma_1 \left(\frac{\rho'_{\ell,m}}{\rho} + \frac{1}{\rho} \frac{d\rho}{dr} \, \xi_{\ell,m} \right) = 0, \tag{5.91}$$

$$\frac{1}{r^2} \frac{d}{dr} \left(r^2 \, \frac{d\Phi'_{\ell,m}}{dr} \right) - \frac{\ell(\ell+1)}{r^2} \, \Phi'_{\ell,m} = 4\pi G \rho'_{\ell,m}, \tag{5.92}$$

where

$$\alpha_{\ell,m} = \frac{1}{r^2} \frac{d}{dr} \left(r^2 \, \xi_{\ell,m} \right) - \frac{\ell(\ell+1)}{r^2} \eta_{\ell,m} . \tag{5.93}$$

Equation (5.89) is identically satisfied for $\ell = 0$. For $\ell > 0$, both members of the equation can be divided by $\ell(\ell + 1)$.

The system of Eqs. (5.88)–(5.92) is quadratic in σ^2. All functions are associated with a *single* spherical harmonic $Y_\ell^m(\theta, \phi)$, so that the systems of equations associated with different spherical harmonics are *uncoupled* from each other. Therefore, it is possible to determine basic modes by solving the system of Eqs. (5.88)–(5.92) that is associated with a given spherical harmonic and setting identically zero the radial functions that appear in the systems of equations associated with all other spherical harmonics.

A spheroidal normal mode associated with a spherical harmonic $Y_\ell^m(\theta, \phi)$ must satisfy the boundary conditions given in Sect. 4.1. The radial component of the Lagrangian displacement, $\xi_{\ell,m}(r)$, must be finite at $r = 0$. Furthermore, the Lagrangian perturbation of the pressure must vanish at $r = R$:

$$(\delta P)_{\ell,m}(R) = 0. \tag{5.94}$$

By virtue of Eqs. (5.90) and (5.91), the boundary condition can be rewritten as

$$\left(\Gamma_1 P \, \alpha_{\ell,m} \right) (R) = 0. \tag{5.95}$$

Since $P(R) = 0$, the boundary condition implies that $\alpha_{\ell,m}(r)$ must remain finite at $r = R$.

Boundary conditions (4.8) and (4.9) can also be transformed into boundary conditions that are related to a single spherical harmonic $Y_\ell^m(\theta, \phi)$. The external solution for the Eulerian perturbation of the gravitational potential is a solution of Laplace's perturbed differential equation. The solution related to the spherical harmonic $Y_\ell^m(\theta, \phi)$ that remains finite as $r \to \infty$ takes the form

$$\Phi_e'(r, \theta, \phi) = C \, r^{-(\ell+1)} \, Y_\ell^m(\theta, \phi), \tag{5.96}$$

where C is an arbitrary constant. By use of boundary conditions (4.8) and (4.9), and elimination of the constant C, one obtains the boundary condition for the Eulerian perturbation of the gravitational potential at $r = R$

$$\left(\frac{d\Phi_{\ell,m}'}{dr} \right)_R + \frac{\ell + 1}{R} \, \Phi_{\ell,m}'(R) = -4\pi G \rho(R) \, \xi_{\ell,m}(R). \tag{5.97}$$

For radial normal modes, the condition reduces to

$$\Phi'_{0,0}(R) = 0 \tag{5.98}$$

because of Eq. (5.40).

By virtue of the continuity of the Eulerian perturbation of the gravitational potential at $r = R$, boundary conditions (4.10) and (4.11) are also satisfied.

The azimuthal number m is involved neither in the system of Eqs. (5.88)–(5.92) nor in the boundary conditions, so that the eigenvalue problem of the spheroidal normal modes is *degenerate* with respect to that number. A basic normal mode associated with a spherical harmonic $Y_\ell^m(\theta, \phi)$ can then be represented as

$$\boldsymbol{\xi} = \left[\xi_\ell(r)\,\mathbf{1}_r + \frac{\eta_\ell(r)}{r}\left(\mathbf{1}_\theta\,\frac{\partial}{\partial\theta} + \mathbf{1}_\phi\,\frac{1}{\sin\theta}\,\frac{\partial}{\partial\phi}\right)\right] Y_\ell^m(\theta, \phi), \tag{5.99}$$

where $\mathbf{1}_\theta$ is the unit vector in the local direction of the colatitude, and $\mathbf{1}_\phi$, the unit vector in the local direction of the azimuthal angle.

In the particular case of the spherical harmonic of degree zero, the spheroidal normal modes are purely radial normal modes.

The radial and horizontal components of a spheroidal displacement field generally vanish on different internal spherical surfaces. The horizontal component ξ_θ vanishes on the parallels of latitude determined by

$$\frac{\partial}{\partial\theta}\, Y_\ell^m(\theta, \phi) = 0, \tag{5.100}$$

and the horizontal component ξ_ϕ, on the meridians determined by

$$\frac{\partial}{\partial\phi}\, Y_\ell^m(\theta, \phi) = 0 \tag{5.101}$$

(Bolt & Derr 1969).

In illustration, the horizontal spheroidal displacement fields associated with the spherical harmonics $P_\ell^m(\cos\theta)\,\cos(m\phi)$, for $\ell = 1, 2$ and the admissible values of m, are represented in Figs. 5.1–5.5. The azimuthal angle ϕ is plotted in abscissa, and the latitude $b = \pi/2 - \theta$ above or below the xy-plane, in ordinate. Both angles are expressed in radians. The horizontal spheroidal displacement fields display dilations and contractions.

A basic normal mode associated with an eigenvalue σ^2 and an azimuthal number m contains the exponential factor

$$\exp[\mathrm{i}(\sigma t + m\phi)].$$

When σ and m are different from zero, and σ is real, the factor represents a wave running in the longitudinal direction. The wave propagates in the sense of the increasing values of the azimuthal angle if σ and m have opposite signs, and in the sense of the decreasing values of the azimuthal angle, if σ and m have the same

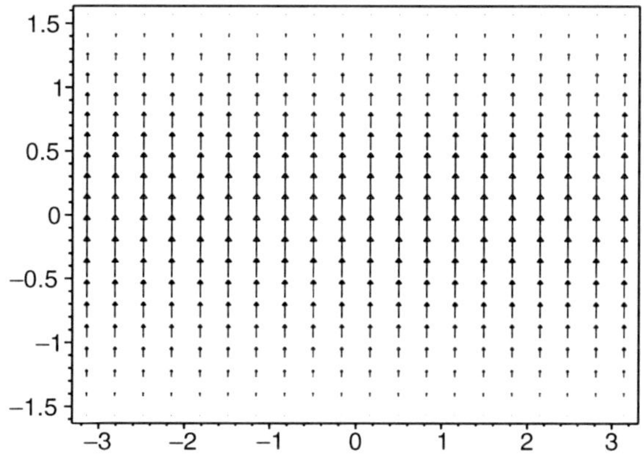

Fig. 5.1 The horizontal spheroidal displacement field associated with the spherical harmonic $P_1(\cos\theta)$, for $\phi = -\pi, \ldots, \pi$ in abscissa, and $b = -\pi/2, \ldots, \pi/2$ in ordinate

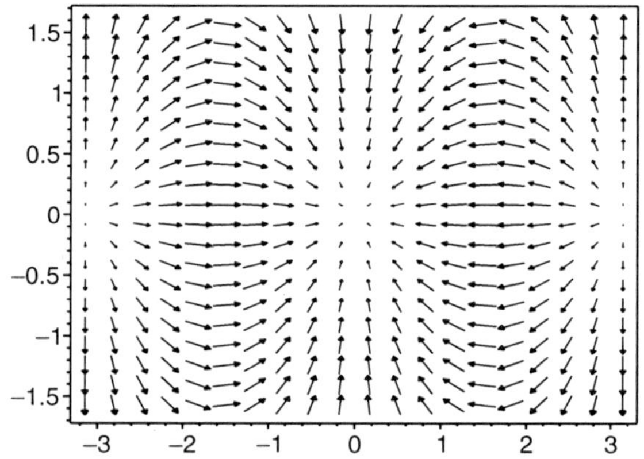

Fig. 5.2 The horizontal spheroidal displacement field associated with the spherical harmonic $P_1^1(\cos\theta)\,\cos\phi$, for $\phi = -\pi, \ldots, \pi$ in abscissa, and $b = -\pi/2, \ldots, \pi/2$ in ordinate

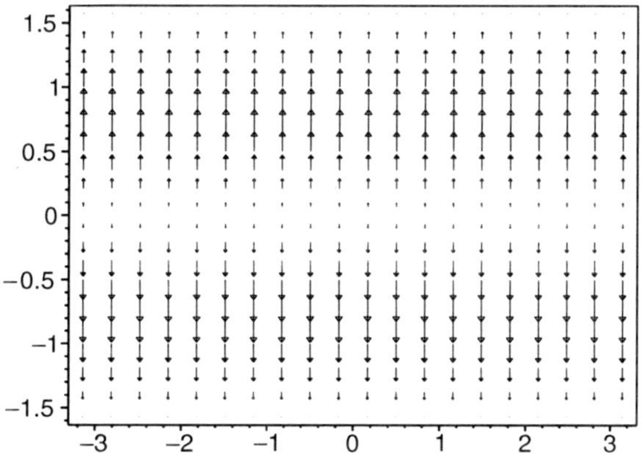

Fig. 5.3 The horizontal spheroidal displacement field associated with the spherical harmonic $P_2(\cos\theta)$, for $\phi = -\pi, \ldots, \pi$ in abscissa, and $b = -\pi/2, \ldots, \pi/2$ in ordinate

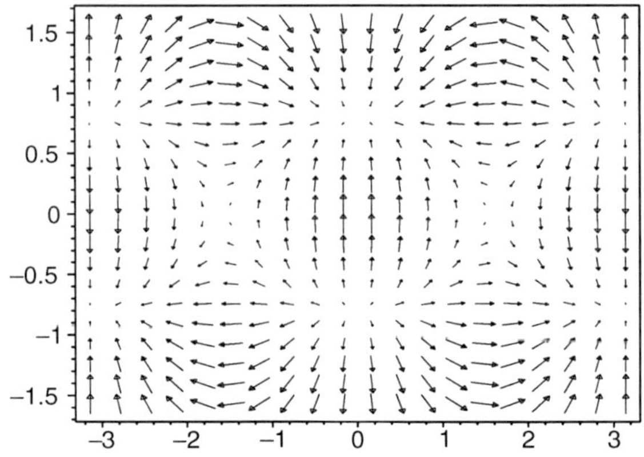

Fig. 5.4 The horizontal spheroidal displacement field associated with the spherical harmonic $P_2^1(\cos\theta)\cos\phi$, for $\phi = -\pi, \ldots, \pi$ in abscissa, and $b = -\pi/2, \ldots, \pi/2$ in ordinate

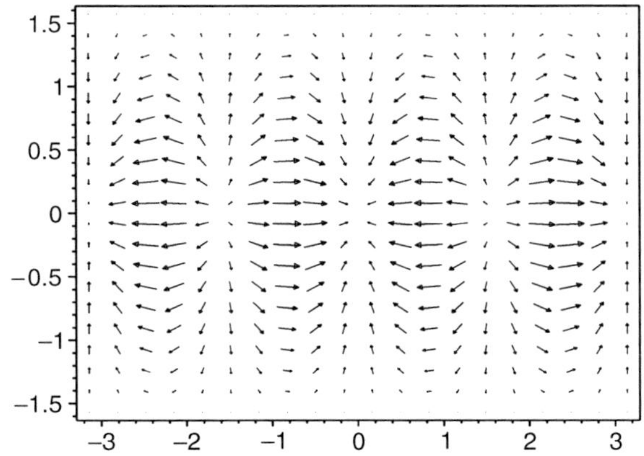

Fig. 5.5 The horizontal spheroidal displacement field associated with the spherical harmonic $P_2^2(\cos\theta)\cos(2\phi)$, for $\phi = -\pi, \ldots, \pi$ in abscissa, and $b = -\pi/2, \ldots, \pi/2$ in ordinate

sign. *Standing* waves are obtained by a superposition of running waves with equal amplitudes that propagate in opposite azimuthal senses:

$$\begin{aligned}\exp[\mathrm{i}(\sigma t + m\phi)] + \exp[-\mathrm{i}(\sigma t + m\phi)] + \exp[\mathrm{i}(\sigma t - m\phi)] + \exp[-\mathrm{i}(\sigma t - m\phi)]\\ = 4\cos(\sigma t)\cos(m\phi)\end{aligned} \tag{5.102}$$

(see, e.g., Unno et al. 1989).

5.8.3 Divergence-Free Spheroidal Normal Modes

The question can be posed whether spheroidal normal modes exist that are divergence-free.

For divergence-free normal modes, Eq. (5.4) reduces to

$$\sigma^2\,\boldsymbol{\xi} = \nabla\left(\Phi' + \frac{P'}{\rho}\right). \tag{5.103}$$

By taking the divergence of both members, one obtains

$$\nabla^2 \left(\Phi' + \frac{P'}{\rho} \right) = 0. \tag{5.104}$$

Substraction of Poisson's perturbed differential equation (2.50) yields

$$\nabla^2 \frac{P'}{\rho} = -4\pi G \rho'. \tag{5.105}$$

On the other hand, from Eqs. (5.13) and (5.14), it follows

$$\rho' = -\frac{d\rho}{dr}\, \xi_r, \quad P' = -\frac{dP}{dr}\, \xi_r, \tag{5.106}$$

so that ρ' and P' can be eliminated from Eq. (5.105). By considering a solution for $\xi_r(r, \theta, \phi)$ that is associated with a spherical harmonic $Y_\ell^m(\theta, \phi)$ as

$$\xi_r(r, \theta, \phi) = \xi(r)\, Y_\ell^m(\theta, \phi), \tag{5.107}$$

and using Poisson's equations (2.14) and (2.20) for the gravity, one derives the second-order differential equation for the function $\xi(r)$

$$\frac{d^2\xi}{dr^2} + 2\left[\frac{1}{m(r)}\,\frac{dm(r)}{dr} - \frac{1}{r}\right]\frac{d\xi}{dr} - \frac{\ell(\ell+1)-2}{r^2}\,\xi = 0. \tag{5.108}$$

This equation was derived by Robe (1965) for divergence-free spheroidal normal modes and by Lebovitz (1965b), in a self-adjoint form, for so-called trivial displacement fields in layers in which $N^2 \neq 0$. It is equivalent to the equations of Clairaut and Radau known in the theory of equilibrium tides that applies to components of close binary stars (see, e.g., Kopal 1959).

Boundary condition (5.97) can also be expressed in terms of the function $\xi(r)$. For $\ell \neq 0$, one derives an equation for $\Phi'(r)$ by means of Eq. (5.89). The functions $\eta(r)$ and $P'(r)/\rho$ can be related to the function $\xi(r)$ by means of Eq. (5.93) and the second Eq. (5.106), so that

$$\Phi' = \frac{\sigma^2}{\ell(\ell+1)}\left(r^2\,\frac{d\xi}{dr} + 2\,r\,\xi\right) - g\,\xi. \tag{5.109}$$

Furthermore, one derives an equation for $d\Phi'/dr$ from the radial component of Eq. (5.103), so that

$$\frac{d\Phi'}{dr} = -g\,\frac{d\xi}{dr} + \left(\sigma^2 + 2\,\frac{g}{r} - 4\pi G\rho\right)\xi. \tag{5.110}$$

Boundary condition (5.97) then becomes

$$\sigma^2\left[\frac{1}{\ell}\left(\frac{d\xi}{dr}\right)_R+\frac{\ell+2}{\ell}\frac{\xi(R)}{R}\right]=\frac{GM}{R^3}\left[\left(\frac{d\xi}{dr}\right)_R+(\ell-1)\frac{\xi(R)}{R}\right]. \quad (5.111)$$

Equation (5.108) admits of a particular solution that remains finite at $r = 0$ for $\ell > 0$ and behaves as $A\,r^{\ell-1}$, where A is an arbitrary constant. For *any* stellar model, this solution is an exact solution that is valid in the whole interval $0 \le r \le R$, when $\ell = 1$. Then the function ξ is independent of the radial coordinate r, i.e.,

$$\xi = A, \quad (5.112)$$

and the function $\eta(r)$ is given by

$$\eta = A\,r. \quad (5.113)$$

From boundary condition (5.111), it follows that $\sigma = 0$.

Thus, for any stellar model, three divergence-free spheroidal normal modes exist that are time-independent and associated with the degree $\ell = 1$. According to Eq. (5.99), these modes can be represented as

$$\left.\begin{aligned}\boldsymbol{\xi}_1 &= A\left(\mathbf{1}_r+\mathbf{1}_\theta\,\frac{\partial}{\partial\theta}\right)Y_1(\theta),\\ \boldsymbol{\xi}_2 &= B\left(\mathbf{1}_r+\mathbf{1}_\theta\,\frac{\partial}{\partial\theta}+\mathbf{1}_\phi\,\frac{1}{\sin\theta}\,\frac{\partial}{\partial\phi}\right)Y_1^1(\theta,\phi),\\ \boldsymbol{\xi}_3 &= C\left(\mathbf{1}_r+\mathbf{1}_\theta\,\frac{\partial}{\partial\theta}+\mathbf{1}_\phi\,\frac{1}{\sin\theta}\,\frac{\partial}{\partial\phi}\right)Y_1^{-1}(\theta,\phi),\end{aligned}\right\} \quad (5.114)$$

where A, B, C are arbitrary constants. They correspond to uniform translations of the star as a whole, which are eigenmodes, as shown in Sect. 4.7. With respect to the orthonormal basis vectors $\partial/\partial x$, $\partial/\partial y$, $\partial/\partial z$, the vectors $\boldsymbol{\xi}_1$, $\boldsymbol{\xi}_2$, $\boldsymbol{\xi}_3$, have the components $(0, 0, A)$, $(B, \mathrm{i}B, 0)$, $(C, -\mathrm{i}C, 0)$, respectively.

Generally, no other divergence-free spheroidal modes exist in an arbitrary stellar model. Indeed, according to vorticity equation (5.5), a divergence-free solution must be time-independent, unless it is also curl-free. But for a time-independent, divergence-free spheroidal mode, boundary condition (5.111) reduces to a relation between the function $\xi(r)$ and its first derivative at $r = R$:

$$\left(\frac{d\xi}{dr}\right)_R+(\ell-1)\frac{\xi(R)}{R}=0. \quad (5.115)$$

This condition will usually not be satisfied, unless $\xi(r)$ is identically zero.

In the particular case of the equilibrium sphere of uniform mass density, differential equation (5.108) admits of divergence-free solutions of the simple form

$$\xi(r) = A\,r^{\ell-1} \quad (5.116)$$

for *any* value of the degree ℓ different from zero. Since these solutions are in addition curl-free, vorticity equation (5.5) can be satisfied even for eigenvalues different from zero. By imposing boundary condition (5.111), one derives the equation for the eigenvalues

$$\sigma^2 = \frac{GM}{R^3} \frac{2\ell(\ell-1)}{2\ell+1}. \tag{5.117}$$

These are the eigenvalues of the modes determined by Kelvin (Thomson 1863) for an *incompressible* equilibrium sphere of uniform mass density [see Eq. (0.1), see also Ledoux & Walraven 1958]. The Kelvin modes are also eigenmodes of the *compressible* equilibrium sphere of uniform mass density, as was shown by Chandrasekhar (1964) and Chandrasekhar & Lebovitz (1964) by means of the variational principle presented in Chap. 8.

5.9 Toroidal Normal Modes

Returning to Eq. (5.6), we now focus our attention on normal modes for which $(\nabla \times \boldsymbol{\xi})_r$ is not equal to zero at all points of the equilibrium star, so that the normal modes are time-independent. In expansion (5.83) for the radial component of the curl of the Lagrangian displacement, at least one function $T_{\lambda,\mu}(r)$ must be different from zero. Be $T_{\ell,m}(r)$ this function.

Moreover, for the toroidal normal modes, Eq. (5.9) implies that

$$\mathscr{L}^2 \left(\frac{P'}{\rho} + \Phi' \right) = 0. \tag{5.118}$$

After expansion of P' and Φ' in terms of spherical harmonics, it results, for $\lambda \neq 0$ and the admissible values of μ, that

$$\frac{P'_{\lambda,\mu}(r)}{\rho} + \Phi'_{\lambda,\mu}(r) = 0. \tag{5.119}$$

Suppose that $N^2 \neq 0$ in the considered region of the star. By using the radial component of vorticity equation (5.4) and expanding the divergence of the Lagrangian displacement in terms of spherical harmonics, one sees that, for $\lambda \neq 0$ and the admissible values of μ,

$$\alpha_{\lambda,\mu}(r) = 0. \tag{5.120}$$

Still for $\lambda \neq 0$ and the admissible values of μ, it follows, from Eqs. (5.13) and (5.14),

$$\frac{\rho'_{\lambda,\mu}(r)}{\rho} + \frac{1}{\rho} \frac{d\rho}{dr} \xi_{\lambda,\mu}(r) = 0, \tag{5.121}$$

$$\frac{P'_{\lambda,\mu}(r)}{P} + \frac{1}{P}\frac{dP}{dr}\xi_{\lambda,\mu}(r) = 0. \tag{5.122}$$

By elimination of $P'_{\lambda,\mu}(r)$ from Eqs. (5.119) and (5.122), it results

$$\Phi'_{\lambda,\mu}(r) = -g\,\xi_{\lambda,\mu}(r). \tag{5.123}$$

Furthermore, elimination of $\rho'_{\lambda,\mu}(r)$ and $\Phi'_{\lambda,\mu}(r)$ from Eq. (5.15) by means of Eqs. (5.121) and (5.123) leads to a second-order differential equation for $\xi_{\lambda,\mu}(r)$ similar to Eq. (5.108), so that generally

$$\xi_{\lambda,\mu}(r) = 0, \tag{5.124}$$

and

$$\Phi'_{\lambda,\mu}(r) = 0, \quad P'_{\lambda,\mu}(r) = 0, \quad \rho'_{\lambda,\mu}(r) = 0. \tag{5.125}$$

From Eq. (5.93), it also follows that

$$\eta_{\lambda,\mu}(r) = 0. \tag{5.126}$$

Next, in the particular case $\lambda = 0$ and $\mu = 0$, it follows, from the radial component of vorticity equation (5.4), that

$$\frac{d}{dr}\left[\Phi'_{0,0}(r) + \frac{P'_{0,0}(r)}{\rho}\right] + \frac{N^2}{g}c^2\,\alpha_{0,0}(r) = 0, \tag{5.127}$$

and, from Eqs. (5.13) and (5.14), that

$$\frac{\rho'_{0,0}(r)}{\rho} + \frac{1}{\rho}\frac{d\rho}{dr}\xi_{0,0}(r) + \alpha_{0,0}(r) = 0, \tag{5.128}$$

$$\frac{P'_{0,0}(r)}{P} + \frac{1}{P}\frac{dP}{dr}\xi_{0,0}(r) + \Gamma_1\,\alpha_{0,0}(r) = 0. \tag{5.129}$$

By elimination of $P'_{0,0}(r)/\rho$ from Eqs. (5.127) and (5.129), one derives

$$\begin{aligned}\frac{d\Phi'_{0,0}}{dr} &= c^2\frac{d\alpha_{0,0}}{dr} + \left(\frac{dc^2}{dr} - \frac{N^2}{g}c^2\right)\alpha_{0,0} \\ &\quad -g\frac{d\xi_{0,0}}{dr} + \left(\frac{2}{r}g - 4\pi G\rho\right)\xi_{0,0}.\end{aligned} \tag{5.130}$$

When one moreover eliminates $d\Phi'_{0,0}/dr$ by means of Eq. (5.40) and relates $\alpha_{0,0}(r)$ to $\xi_{0,0}(r)$ by means of Eq. (5.93), one derives the following linear, homogeneous, second-order differential equation for $\xi_{0,0}(r)$, whose boundary points $r = 0$ and $r = R$ are singular points:

$$\frac{d^2\xi_{0,0}}{dr^2} + \left(\frac{1}{c^2}\frac{dc^2}{dr} + \frac{1}{\rho}\frac{d\rho}{dr} + \frac{2}{r}\right)\frac{d\xi_{0,0}}{dr} + \left(\frac{1}{c^2}\frac{dc^2}{dr} - \frac{N^2}{g} + \frac{g}{c^2} - \frac{1}{r}\right)\frac{2}{r}\,\xi_{0,0} = 0. \tag{5.131}$$

Generally, the only admissible solution is

$$\xi_{0,0}(r) = 0, \tag{5.132}$$

so that

$$\alpha_{0,0}(r) = 0, \quad \Phi'_{0,0}(r) = 0, \quad P'_{0,0}(r) = 0, \quad \rho'_{0,0}(r) = 0. \tag{5.133}$$

Thus, besides the spheroidal modes, the set of the linear, isentropic normal modes of a quasi-static star contains, for $\ell > 0$, time-independent, divergence-free modes with a purely horizontal displacement field depending on a single arbitrary scalar function $T_{\ell,m}(r)$ as

$$\boldsymbol{\xi} = \left[\frac{T_{\ell,m}(r)}{r}\left(\mathbf{1}_\theta\,\frac{1}{\sin\theta}\frac{\partial}{\partial\phi} - \mathbf{1}_\phi\,\frac{\partial}{\partial\theta}\right)\right]Y_\ell^m(\theta,\phi). \tag{5.134}$$

The scalar function $T_{\ell,m}(r)$ must satisfy the condition that the horizontal components of the Lagrangian displacement are finite at all points of the star. Both the Lagrangian and the Eulerian perturbations of the mass density, the pressure, and the gravitational potential are equal to zero at all points, since the purely horizontal displacement field does not perturb the star's hydrostatic equilibrium.

The curl of the Lagrangian displacement field has the radial component

$$(\nabla \times \boldsymbol{\xi})_r = \ell(\ell+1)\,\frac{T_{\ell,m}(r)}{r^2}\,Y_\ell^m(\theta,\phi), \tag{5.135}$$

so that toroidal modes produce vorticity around the local normal to the equipotential surface. Therefore, the toroidal modes are complementary to the spheroidal modes.

The eigenvalue problem of the toroidal modes is of the first order in σ^2.

On the existence of the class of the toroidal normal modes, the attention was first drawn by Perdang (1968) in a group-theoretical study of the non-radial oscillations of stars. The toroidal normal modes are also known in geophysics and are important for Earth seismology (see, e.g., Bolt & Derr 1969, Dahlen & Tromp 1998).

In illustration, the toroidal displacement fields associated with the surface harmonics $P_\ell^m(\cos\theta)\cos(m\phi)$ are represented for $\ell = 1, 2$ and the admissible values of m in Figs. 5.6–5.10. As for the representation of the horizontal spheroidal

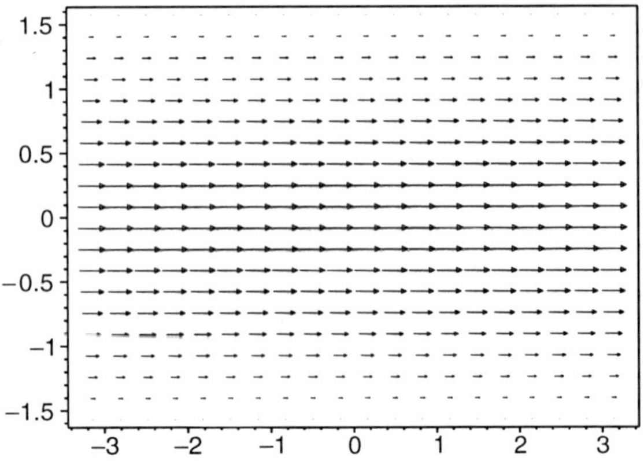

Fig. 5.6 The toroidal displacement field associated with the spherical harmonic $P_1(\cos\theta)$, for $\phi = -\pi, \ldots, \pi$ in abscissa, and $b = -\pi/2, \ldots, \pi/2$ in ordinate

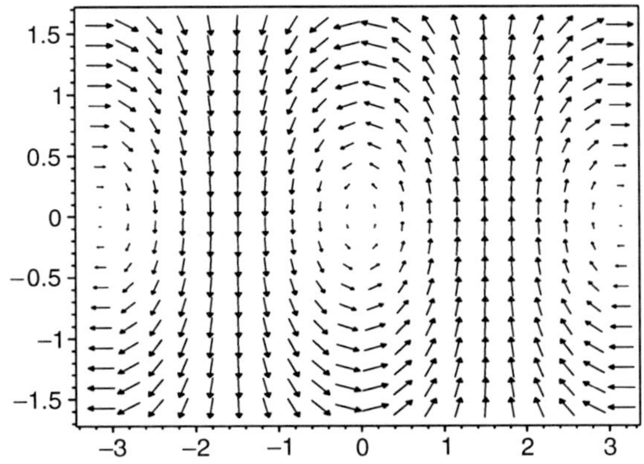

Fig. 5.7 The toroidal displacement field associated with the spherical harmonic $P_1^1(\cos\theta)\cos\phi$, for $\phi = -\pi, \ldots, \pi$ in abscissa, and $b = -\pi/2, \ldots, \pi/2$ in ordinate

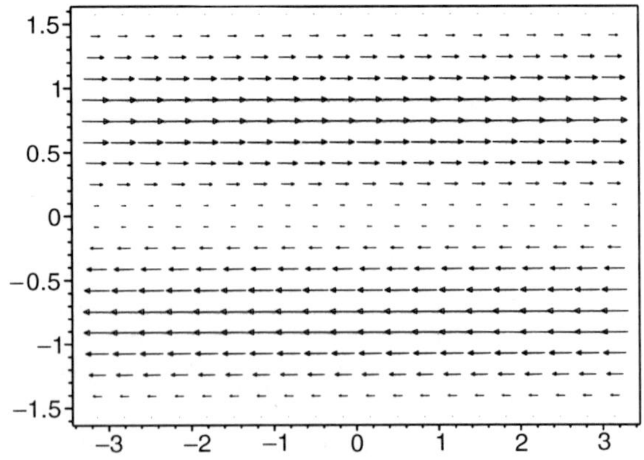

Fig. 5.8 The toroidal displacement field associated with the spherical harmonic $P_2(\cos\theta)$, for $\phi = -\pi, \ldots, \pi$ in abscissa, and $b = -\pi/2, \ldots, \pi/2$ in ordinate

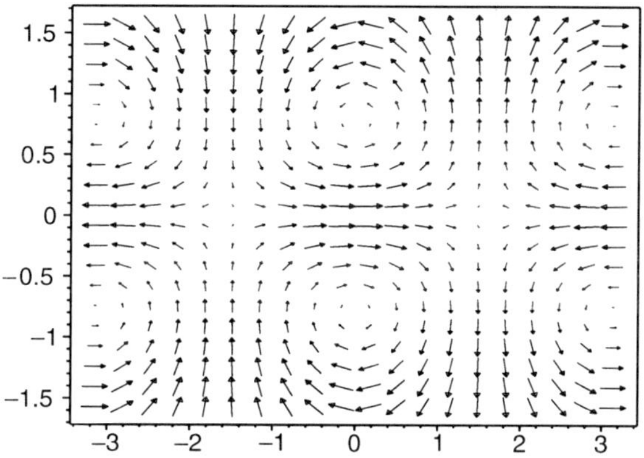

Fig. 5.9 The toroidal displacement field associated with the spherical harmonic $P_2^1(\cos\theta)\cos\phi$, for $\phi = -\pi, \ldots, \pi$ in abscissa, and $b = -\pi/2, \ldots, \pi/2$ in ordinate

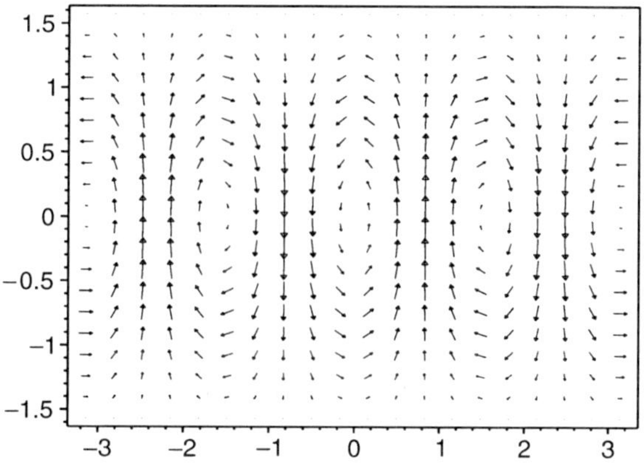

Fig. 5.10 The toroidal displacement field associated with the spherical harmonic $P_2^2(\cos\theta)\cos(2\phi)$, for $\phi = -\pi, \ldots, \pi$ in abscissa, and $b = -\pi/2, \ldots, \pi/2$ in ordinate

displacement fields, the azimuthal angle ϕ is plotted in abscissa, and the latitude $b = \pi/2 - \theta$ above or below the xy-plane, in ordinate. Both angles are still expressed in radians. The toroidal displacement fields display rotations around radial axes.

The component ξ_θ of a toroidal displacement field vanishes on the meridians determined by

$$\frac{\partial}{\partial\phi} Y_\ell^m(\theta,\phi) = 0, \tag{5.136}$$

and the component ξ_ϕ, on the parallels of latitude determined by

$$\frac{\partial}{\partial\theta} Y_\ell^m(\theta,\phi) = 0. \tag{5.137}$$

Small differential rotations of equipotential surfaces around the star's centre given by

$$\boldsymbol{\xi} = \delta \boldsymbol{\Omega}(r) \times \mathbf{r}, \tag{5.138}$$

where $\delta \boldsymbol{\Omega}(r)$ is the rotation vector, correspond to toroidal normal modes associated with spherical harmonics of the degree $\ell = 1$. The property is manifest for a rotation vector whose direction corresponds to that of a coordinate axis. It also holds in the general case in which the rotation vector has an arbitrary direction, since such a rotation vector can be represented as a linear combination of three rotation vectors that each have a direction corresponding to that of a coordinate axis.

When the rotation vector is independent of the radial coordinate r, the star rotates as a rigid body. Consequently, rigid-body rotations of a star belong to the toroidal normal modes associated with the spherical harmonics of the degree $\ell = 1$, just as global translations of the star belong to the spheroidal normal modes associated with these spherical harmonics.

5.10 Inner Products of Normal Modes

In this section, detailed expressions for the inner products of two normal modes are derived from definition (4.25).

5.10.1 Inner Product of Two Spheroidal Modes

After use of properties (Appendix D.4) and (Appendix D.6) of the spherical harmonics, the inner product of a spheroidal normal mode n associated with a spherical harmonic $Y_\ell^m(\theta,\phi)$ and a spheroidal normal mode ν associated with a spherical harmonic $Y_\lambda^\mu(\theta,\phi)$ can be expressed as

$$\int_V \rho\, g_{ij}\, \overline{\delta q^i_{\lambda,\mu,\nu}}\, \delta q^j_{\ell,m,n}\, dV = \delta_{\lambda,\ell}\, \delta_{\mu,m}\, N_{\ell,m}$$
$$\int_0^R \left[\xi_{\lambda,\nu}(r)\, \xi_{\ell,n}(r) + \frac{\ell(\ell+1)}{r^2}\, \eta_{\lambda,\nu}(r)\, \eta_{\ell,n}(r) \right] \rho(r)\, r^2\, dr. \tag{5.139}$$

The inner product is equal to zero not only when $\lambda \neq \ell$ or $\mu \neq m$, but, by virtue of orthogonality relation (4.60), also when the spheroidal normal modes are associated with different eigenvalues σ^2.

Accordingly, the square of the norm of a spheroidal normal mode associated with a spherical harmonic $Y_\ell^m(\theta,\phi)$ is given by

$$\int_V \rho\, g_{ij}\, \overline{\delta q^i}\, \delta q^j\, dV = N_{\ell,m} \int_0^R \left[\xi^2(r) + \frac{\ell(\ell+1)}{r^2}\, \eta^2(r) \right] \rho(r)\, r^2\, dr. \tag{5.140}$$

5.10.2 Inner Product of Two Toroidal Modes

After use of property (Appendix D.6) of the spherical harmonics, the inner product of a toroidal normal mode associated with a spherical harmonic $Y_\ell^m(\theta,\phi)$ and a toroidal normal mode associated with a spherical harmonic $Y_\lambda^\mu(\theta,\phi)$ can be expressed as

$$\int_V \rho\, g_{ij}\, \overline{\delta q^i_{\lambda,\mu}}\, \delta q^j_{\ell,m}\, dV = \delta_{\lambda,\ell}\, \delta_{\mu,m}\, \ell(\ell+1)\, N_{\ell,m} \int_0^R T_{\ell,m}(r)\, T_{\lambda,\mu}(r)\, \rho(r)\, dr. \tag{5.141}$$

The inner product is equal to zero when $\lambda \neq \ell$ or $\mu \neq m$.

Accordingly, the square of the norm of a toroidal mode associated with a spherical harmonic $Y_\ell^m(\theta,\phi)$ is given by

$$\int_V \rho\, g_{ij}\, \overline{\delta q^i}\, \delta q^j\, dV = \ell(\ell+1)\, N_{\ell,m} \int_0^R T^2_{\ell,m}(r)\, \rho(r)\, dr. \tag{5.142}$$

5.10.3 Inner Product of a Spheroidal and a Toroidal Mode

The inner product of a spheroidal normal mode associated with a spherical harmonic $Y_\ell^m(\theta,\phi)$ and a toroidal normal mode associated with a spherical harmonic $Y_\lambda^\mu(\theta,\phi)$ is given by

$$\int_V \rho\, g_{ij}\, \overline{\delta q^i_{\lambda,\mu}}\, \delta q^j_{\ell,m}\, dV = -\mathrm{i} \int_0^R \eta_\ell(r)\, T_{\lambda,\mu}(r)\, \rho\, dr$$

$$\int_0^{2\pi}\int_0^{\pi} \frac{1}{\sin\theta}\left[\mu\, \frac{dP_\ell^{|m|}(\cos\theta)}{d\theta}\, P_\lambda^{|\mu|}(\cos\theta) + m\, P_\ell^{|m|}(\cos\theta)\, \frac{dP_\lambda^{|\mu|}(\cos\theta)}{d\theta}\right]$$

$$\exp\left[\mathrm{i}(m-\mu)\phi\right]\, \sin\theta\, d\theta\, d\phi. \tag{5.143}$$

The integration over the azimuthal angle ϕ leads to

$$\int_V \rho\, g_{ij}\, \overline{\delta q^i_{\lambda,\mu}}\, \delta q^j_{\ell,m}\, dV = -\delta_{\mu,m}\, 2\pi\mathrm{i} \int_0^R \eta_\ell(r)\, T_{\lambda,\mu}(r)\, \rho dr$$

$$\int_0^{\pi} \frac{1}{\sin\theta}\left[\mu\, \frac{dP_\ell^{|m|}(\cos\theta)}{d\theta}\, P_\lambda^{|\mu|}(\cos\theta) + m\, P_\ell^{|m|}(\cos\theta)\, \frac{dP_\lambda^{|\mu|}(\cos\theta)}{d\theta}\right] \sin\theta\, d\theta. \tag{5.144}$$

Next, for the integration over the colatitude θ, the integral can be rewritten as

$$\int_V \rho\, g_{ij}\, \overline{\delta q^i_{\lambda,\mu}}\, \delta q^j_{\ell,m}\, dV = 2\pi \mathrm{i}\, m \int_0^R \eta_\ell(r)\, T_{\lambda,m}(r)\, \rho\, dr$$
$$\int_{-1}^{1} \left[\frac{dP_\ell^{|m|}(\cos\theta)}{d(\cos\theta)}\, P_\lambda^{|m|}(\cos\theta) + P_\ell^{|m|}(\cos\theta)\, \frac{dP_\lambda^{|m|}(\cos\theta)}{d(\cos\theta)} \right] d(\cos\theta), \quad (5.145)$$

so that, by partial integration,

$$\int_V \rho\, g_{ij}\, \overline{\delta q^i_{\lambda,\mu}}\, \delta q^j_{\ell,m}\, dV$$
$$= 2\pi \mathrm{i}\, m \int_0^R \eta_\ell(r)\, T_{\lambda,\mu}(r)\, \rho\, dr \left[P_\ell^{|m|}(\cos\theta)\, P_\lambda^{|m|}(\cos\theta) \right]_{-1}^{1} = 0. \quad (5.146)$$

Consequently, spheroidal and toroidal normal modes are mutually orthogonal.

Chapter 6
Determination of Spheroidal Normal Modes: Mathematical Aspects

6.1 Introduction

For the determination of spheroidal normal modes of a quasi-static star, one might think of deriving a single ordinary differential equation for one of the radial functions that appear in the governing equations after the separations of the time and the angular variables. A fourth-order differential equation was derived by Pekeris (1938) for the function $\alpha(r)$ and by Ledoux & Walraven (1958) for the function $\chi(r) = \Phi'(r) + P'(r)/\rho(r)$. However, these differential equations are of such a great complexity that it is not easy to deal with them. The use of fourth-order systems of differential equations for several radial functions turns out to be simpler. Therefore, in the next section, convenient fourth-order systems of differential equations are presented. The subsequent sections are devoted to a description of the mathematical foundation of the method that is applied, first, for the determination of radial normal modes[1] and, next, for that of spheroidal non-radial normal modes.

6.2 Convenient Fourth-Order Systems of Differential Equations in the Radial Coordinate

6.2.1 *Pekeris' System of Equations*

By taking the divergence of Eq. (5.4), one has

$$\sigma^2 \alpha = \nabla^2 \chi + \frac{1}{r^2} \frac{\partial}{\partial r} \left(r^2 \, \frac{N^2}{g} \, c^2 \, \alpha \right). \tag{6.1}$$

[1] For radial normal modes, the adjective "spheroidal" is dropped, since all radial normal modes are spheroidal.

P. Smeyers and T. Van Hoolst, *Linear Isentropic Oscillations of Stars: Theoretical Foundations*, Astrophysics and Space Science Library 371,
DOI 10.1007/978-3-642-13030-4_7,

On the other hand, from Eqs. (5.13) and (5.14), it follows that

$$\frac{P'}{\rho} = g\,\xi_r - c^2\,\alpha, \tag{6.2}$$

and, from Eqs. (5.13) and (5.15), that

$$\nabla^2\Phi' = -4\pi G\rho\left(\alpha + \frac{1}{\rho}\frac{d\rho}{dr}\,\xi_r\right). \tag{6.3}$$

Elimination of $\nabla^2\Phi'$ and P'/ρ from Eq. (6.1) yields

$$\nabla^2\left(c^2\,\alpha - g\,\xi_r\right) = \frac{1}{r^2}\frac{\partial}{\partial r}\left(r^2\,\frac{N^2}{g}\,c^2\,\alpha\right) - \left(\sigma^2 + 4\pi G\rho\right)\alpha - 4\pi G\frac{d\rho}{dr}\,\xi_r. \tag{6.4}$$

The latter equation and Eqs. (5.3) and (5.16) form a system of three equations for the function $\chi(r,\theta,\phi)$, and the radial component, $\xi_r(r,\theta,\phi)$, and the divergence, $\alpha(r,\theta,\phi)$, of the Lagrangian displacement. Separation of the angular variables θ and ϕ by means of expansions in terms of spherical harmonics and use of the orthogonality relation between them leads to a system of three equations for the radial functions $\xi_{\ell,m}(r)$, $\alpha_{\ell,m}(r)$, $\chi_{\ell,m}(r)$ associated with the spherical harmonic $Y_\ell^m(\theta,\phi)$. When one drops the subscripts ℓ and m for the sake of simplification of the notations, the equations take the form

$$\sigma^2\,\xi = \frac{d\chi}{dr} + \frac{N^2}{g}\,c^2\,\alpha, \tag{6.5}$$

$$\begin{aligned}\frac{1}{r^2}\frac{d}{dr}&\left[r^2\,\frac{d}{dr}\left(c^2\alpha - g\,\xi\right)\right] - \frac{\ell(\ell+1)}{r^2}\left(c^2\alpha - g\,\xi\right)\\ &= \frac{1}{r^2}\frac{d}{dr}\left(r^2\,\frac{N^2}{g}\,c^2\alpha\right) - \left(\sigma^2 + 4\pi G\rho\right)\alpha - 4\pi G\frac{d\rho}{dr}\,\xi,\end{aligned} \tag{6.6}$$

$$\sigma^2\,\alpha = \sigma^2\,\frac{1}{r^2}\frac{d}{dr}\left(r^2\,\xi\right) - \frac{\ell(\ell+1)}{r^2}\,\chi. \tag{6.7}$$

If $\sigma^2 \neq 0$, the elimination of the function $\chi(r)$ from Eqs. (6.5) and (6.7) yields

$$\frac{d^2\xi}{dr^2} + \frac{4}{r}\frac{d\xi}{dr} - \frac{\ell(\ell+1)-2}{r^2}\,\xi = \frac{d\alpha}{dr} - \left[\frac{c^2}{g}\,\frac{K_1(r)}{\sigma^2} - \frac{2}{r}\right]\alpha, \tag{6.8}$$

where the coefficient $K_1(r)$ is defined as

$$K_1(r) = \ell(\ell+1)\,\frac{N^2}{r^2}. \tag{6.9}$$

Next, after elimination of $d^2\xi/dr^2$ by means of Eq. (6.8), Eq. (6.6) can be transformed into

$$\frac{d^2\alpha}{dr^2} + K_2(r)\,\frac{d\alpha}{dr} + \left[\frac{\sigma^2}{c^2} + K_3(r) + \frac{K_1(r)}{\sigma^2}\right]\alpha = -K_4(r)\,\frac{d\xi}{dr}, \tag{6.10}$$

where the coefficients $K_2(r)$, $K_3(r)$, $K_4(r)$ are given by

$$K_2(r) = \frac{2}{r} + \frac{2}{\rho\, c^2}\,\frac{d\left(\rho\, c^2\right)}{dr} - \frac{1}{\rho}\,\frac{d\rho}{dr}, \tag{6.11}$$

$$K_3(r) = -\frac{\ell(\ell+1)}{r^2} + \frac{2g}{c^2}\left(\frac{1}{g}\,\frac{dg}{dr} + \frac{1}{r}\right)$$
$$+\frac{1}{\rho\, c^2}\,\frac{d\left(\rho\, c^2\right)}{dr}\left(\frac{2}{r} - \frac{1}{\rho}\,\frac{d\rho}{dr}\right) + \frac{1}{\rho\, c^2}\,\frac{d^2\left(\rho\, c^2\right)}{dr^2}, \tag{6.12}$$

$$K_4(r) = -\frac{2g}{c^2}\left(\frac{1}{g}\,\frac{dg}{dr} - \frac{1}{r}\right). \tag{6.13}$$

Equations (6.8) and (6.10) form a fourth-order system of differential equations for the radial parts of the divergence, $\alpha(r)$, and the radial component, $\xi(r)$, of the Lagrangian displacement, which is valid for any degree ℓ. This system was first derived by Pekeris (1938). In the particular case of the equilibrium sphere of uniform mass density, $K_4(r) = 0$, so that Eq. (6.10) reduces to a homogeneous second-order differential equation for the function $\alpha(r)$. Pekeris derived the eigenfrequency equation and showed that the eigenfunctions are analytical functions. Various eigenfunctions of lower-degree modes were afterwards determined by Sauvenier-Goffin (1951) (see also Ledoux & Walraven 1958).

No further attention was paid to the system of Eqs. (6.8) and (6.10) during several decades, until Tassoul (1990) adopted it as the basic system for an asymptotic representation of low-degree, non-radial spheroidal modes with large eigenfrequencies, in which the Eulerian perturbation of the gravitational potential was taken into consideration in contrast with previous studies.

6.2.2 Ledoux' System of Equations

From Eqs. (5.88)–(5.92), other systems of differential equations can be derived by elimination of the function $\rho'(r)$, whose first derivative does not appear in the equations. This elimination can be carried out by means of Eq. (5.91).

When $\ell \neq 0$, two possibilities can be considered: either the function $\eta(r)$ or the function $P'(r)/\rho(r)$ is eliminated.

First, when the function $\eta(r)$ is eliminated, and the new dependent variables

$$u(r) = r^2\,\xi(r), \quad y(r) = \frac{P'(r)}{\rho(r)} \tag{6.14}$$

are introduced, one obtains the system of differential equations

$$\frac{du}{dr} = \frac{g}{c^2}\,u + \left[\frac{\ell(\ell+1)}{\sigma^2} - \frac{r^2}{c^2}\right] y + \frac{\ell(\ell+1)}{\sigma^2}\,\Phi', \tag{6.15}$$

$$\frac{dy}{dr} = \left(\sigma^2 - N^2\right)\frac{u}{r^2} + \frac{N^2}{g}\,y - \frac{d\Phi'}{dr}, \tag{6.16}$$

$$\frac{1}{r^2}\frac{d}{dr}\left(r^2\,\frac{d\Phi'}{dr}\right) - \frac{\ell(\ell+1)}{r^2}\,\Phi' = 4\pi G\rho\left(\frac{N^2}{g}\,\frac{u}{r^2} + \frac{1}{c^2}\,y\right). \tag{6.17}$$

This system was derived by Ledoux & Walraven (1958).

Secondly, the elimination of the function $y(r)$ by means of Eq. (5.89) leads to the system of differential equations

$$\frac{du}{dr} = \frac{g}{c^2}\,u + \left[\ell(\ell+1) - \sigma^2\,\frac{r^2}{c^2}\right]\eta + \frac{r^2}{c^2}\,\Phi', \tag{6.18}$$

$$\frac{d\eta}{dr} = \left(1 - \frac{N^2}{\sigma^2}\right)\frac{u}{r^2} + \frac{N^2}{g}\,\eta - \frac{1}{\sigma^2}\,\frac{N^2}{g}\,\Phi', \tag{6.19}$$

$$\frac{1}{r^2}\frac{d}{dr}\left(r^2\,\frac{d\Phi'}{dr}\right) - \left[\frac{\ell(\ell+1)}{r^2} - \frac{4\pi G\rho}{c^2}\right]\Phi' = 4\pi G\rho\left(\frac{N^2}{g}\,\frac{u}{r^2} + \frac{\sigma^2}{c^2}\,\eta\right), \tag{6.20}$$

in which the horizontal component $\eta(r)$ of the Lagrangian displacement appears besides the radial component $\xi(r)$ (Eisenfeld 1969, Smeyers 1984).

In the case $\ell = 0$, the system of differential equations (6.15)–(6.17) can be written as

$$\frac{du}{dr} = \frac{g}{c^2}\,u - \frac{r^2}{c^2}\,y\,, \tag{6.21}$$

$$\frac{dy}{dr} = \left(\sigma^2 - N^2\right)\frac{u}{r^2} + \frac{N^2}{g}\,y - \Psi, \tag{6.22}$$

$$\frac{d\Phi'}{dr} = \Psi, \tag{6.23}$$

$$\frac{d\Psi}{dr} = 4\pi G\rho\left(\frac{N^2}{g}\,\frac{u}{r^2} + \frac{1}{c^2}\,y\right) - \frac{2}{r}\,\Psi. \tag{6.24}$$

The first integral given by Eq. (5.40),

$$\Psi = -4\pi G\rho\,\frac{u}{r^2}, \tag{6.25}$$

is equivalent to an integral of Eq. (6.24). By means of it, the system of differential equations is reduced to a third-order system consisting of Eqs. (6.21)–(6.23) for the three unknown functions $u(r)$, $y(r)$, $\Phi'(r)$. In this system, Eqs. (6.21) and (6.22) form a closed second-order system for the functions $u(r)$ and $y(r)$. The elimination of the function $y(r)$ leads to the second-order differential equation for the function $u(r)$

$$\frac{d}{dr}\left(\frac{\Gamma_1\, P}{r^2}\,\frac{du}{dr}\right)+\frac{\rho}{r^2}\left(\sigma^2+\frac{4g}{r}\right)u=0. \tag{6.26}$$

This equation was given by Rosseland (1949). It can be transformed into an equation for the relative radial displacement $\zeta(r)=\xi(r)/r=u(r)/r^3$ as

$$\frac{d}{dr}\left(r^4\Gamma_1 P\,\frac{d\zeta}{dr}\right)+\left\{\sigma^2\rho\, r^4+r^3\,\frac{d}{dr}\,[(3\Gamma_1-4)\,P]\right\}\zeta=0, \tag{6.27}$$

or, equivalently, as

$$\sigma^2\,\zeta+\frac{1}{r^4\rho}\,\frac{d}{dr}\left(r^4\Gamma_1 P\,\frac{d\zeta}{dr}\right)+\frac{1}{r\,\rho}\left\{\frac{d}{dr}\,[(3\Gamma_1-4)\,P]\right\}\zeta=0. \tag{6.28}$$

The equation was originally established by Eddington (1926) for the case in which the isentropic coefficient Γ_1 is constant. Equation (6.27) corresponds to the equation given by Ledoux & Walraven (1958) for the more general case in which the isentropic coefficient is a function of the radial coordinate.

In the case $\ell=1$, Takata (2005) showed that, in stellar models with a vanishing surface density, the conservation of the mode momentum of each mass $m(r)$ inside the star provides, for axisymmetric modes with an eigenfrequency different from zero, the specific integral

$$P'+\frac{g}{4\pi G}\left(\frac{d\Phi'}{dr}+\frac{2}{r}\,\Phi'\right)=\sigma^2 r\left[\rho\,\xi+\frac{1}{4\pi G}\left(\frac{d\Phi'}{dr}-\frac{\Phi'}{r}\right)\right]. \tag{6.29}$$

In his proof of the existence of the integral, he considered the z-component of the mode momentum of the mass $m(r)$, contained in the volume $V(r)$,

$$P_z(r)=\int_{V(r)}\rho\,\frac{\partial\left(\delta q^i\right)}{\partial t}\,(\mathbf{e}_i\cdot\mathbf{1}_z)\,dV,$$

and the z-component of the displacement of the mass centre of $m(r)$,

$$\delta z_g(r)=\frac{1}{m(r)}\,D_z(r)$$

with

$$D_z(r)=\int_{V(r)}\rho\,\delta q^i\,(\mathbf{e}_i\cdot\mathbf{1}_z)\,dV\,,$$

and observed that

$$\frac{\partial P_z}{\partial t}=\frac{\partial^2 D_z}{\partial t^2}=-\sigma^2 D_z.$$

It may be noted that $D_z(R) = 0$, in accordance with the property mentioned in Sect. 4.8 that the mass centre of the whole star is not subject to any displacement for modes with eigenfrequencies different from zero.

Equation (6.29) is also valid for non-axisymmetric modes in models with a vanishing surface density, since the radial parts of the eigenfunctions are independent of the azimuthal number m.

As well as in the case $\ell = 0$, the third-order system of differential equations which is obtained by use of the integral, can be decomposed into two parts: a closed second-order system of differential equations, from which the eigenfrequencies can be determined, and a first-order differential equation, which can be integrated by quadrature once the solutions of the second-order system are known. Takata (2006a) observed that such a decomposition is possible only when the dependent variables are chosen appropriately. Since the general theory of ordinary differential equations does not always guarantee the existence of such convenient dependent variables, he conjectured that some mathematical character of the problem is reflected in the fact that a closed second-order system of differential equations can be found in the cases $\ell = 0$ and $\ell = 1$. He identified the mathematical character with the property of Hamiltonian systems for which differential equations can always be derived with an order reduced by two in comparison with that of the original equation, with the help of only one integral (Takata 2009). The formulation of the system of equations that govern linear, isentropic oscillations of a quasi-static star as a Hamiltonian system of canonical equations is considered in Sect. 8.8.

The second-order system of differential equations derived by Takata (2005) for the first-degree modes by use of the first integral may display a singularity, so that any additional analysis of the differential equations becomes difficult. Therefore, Takata (2006b) derived an alternative second-order system without singularity, except at the boundaries.

For $\ell > 1$, the mode momentum of the mass $m(r)$ is identically zero for all modes. The fact that this momentum is different from zero for modes belonging to $\ell = 1$ and identically zero for modes belonging to higher degrees is related to the property shown below (Sect. 6.3.1) that the matter in the star's geometrical centre is subject to a purely radial displacement for $\ell = 1$ and is at rest for $\ell > 1$.

6.2.3 Dziembowski's System of Equations

The following four dimensionless variables were introduced by Dziembowski (1971):

$$y_1(r) = \frac{\xi(r)}{r}, \quad y_2(r) = \frac{1}{g(r)\,r}\left(\frac{P'(r)}{\rho(r)} + \Phi'(r)\right),$$
$$y_3(r) = \frac{\Phi'(r)}{g(r)\,r}, \quad y_4(r) = \frac{1}{g(r)}\frac{d\Phi'}{dr} \tag{6.30}$$

(see also Hansen & Kawaler 1994). In terms of these dependent variables, the system of differential equations (6.15)–(6.17) is transformed into

$$\frac{dy_1}{d\ln r} = \left(\frac{V}{\Gamma_1} - 3\right) y_1 + \left[\frac{\ell(\ell+1)}{C\,\omega^2} - \frac{V}{\Gamma_1}\right] y_2 + \frac{V}{\Gamma_1}\, y_3, \tag{6.31}$$

$$\frac{dy_2}{d\ln r} = \left(C\,\omega^2 + A\,r\right) y_1 + (1 - U - A\,r)\, y_2 + A\,r\, y_3, \tag{6.32}$$

$$\frac{dy_3}{d\ln r} = (1 - U)\, y_3 + y_4, \tag{6.33}$$

$$\frac{dy_4}{d\ln r} = -U A\, r y_1 + \frac{U V}{\Gamma_1}\, y_2 + \left[\ell(\ell+1) - \frac{U V}{\Gamma_1}\right] y_3 - U\, y_4, \tag{6.34}$$

where

$$\left.\begin{aligned} U &\equiv \frac{d\ln m(r)}{d\ln r} = \frac{4\pi\rho r^3}{m(r)}, \quad V \equiv -\frac{d\ln P}{d\ln r} = \frac{\rho g r}{P}, \\ C &\equiv \left(\frac{r}{R}\right)^3 \frac{M}{m(r)} = \frac{\bar{\rho}}{\bar{\rho}(r)}, \quad A = -N^2/g, \\ \omega^2 &= \frac{R^3}{GM}\,\sigma^2. \end{aligned}\right\} \tag{6.35}$$

This system of equations is largely used for computational purposes.

6.3 Determination of Radial Normal Modes

Normal radial modes of a star are determined by the resolution of an eigenvalue problem with two end points. The standard procedure is based on the application of initial-value techniques, by which trial solutions are constructed from each end point (Ledoux & Walraven 1958). The second-order differential equation (6.27) can be brought into the standard form

$$\frac{d^2\zeta}{dr^2} + \left[\frac{1}{\rho\, c^2}\frac{d\left(\rho\, c^2\right)}{dr} + \frac{4}{r}\right]\frac{d\zeta}{dr} + \left[\frac{\sigma^2}{c^2} + \frac{3}{r}\frac{1}{\rho\, c^2}\frac{d\left(\rho\, c^2\right)}{dr} + \frac{4}{r}\frac{g}{c^2}\right]\zeta = 0. \tag{6.36}$$

The boundary points $r = 0$ and $r = R$ are isolated singular points of the differential equation and are supposed to be *regular* singular points. A point $x = a$ is a regular singular point of a differential equation

$$f''(x) + P(x)\, f'(x) + Q(x)\, f(x) = 0$$

if $P(x)$ and $Q(x)$ are analytic in some neighbourhood of the point $x = a$, except possibly for a simple pole of $P(x)$ and a pole of $Q(x)$ that is at most of the second order, at that point (see, e.g., Richtmyer 1978). Furthermore, a function is said to be analytic at $x = a$ when it can be expanded in a convergent power series in the neighbourhood of that point.

Admissible solutions are constructed from each regular singular boundary point by the use of the method of Frobenius (1873) for the integration of linear differential equations by means of series expansions. Next, the equation for the eigenvalues is derived by the imposition of conditions of continuity on the admissible solutions and their first derivatives at a point of their common domain of validity. The method is illustrated by its application to the well-known eigenvalue problem of the vibrating string in Appendix H.

6.3.1 Admissible Solutions from the Boundary Point $r = 0$

The mass density is assumed to be analytic at the boundary point $r = 0$. Be ρ_c the central mass density. Because of the spherical symmetry of the equilibrium star, the power series of the mass density about $r = 0$ is of the form

$$\rho(r) = \rho_c \left[1 + \frac{\rho_2}{\rho_c} r^2 + O\left(r^4\right)\right]. \tag{6.37}$$

From definitions (2.20) and (2.21), one derives

$$\begin{aligned} g(r) &= \frac{4\pi G \rho_c}{3} r \left[1 + \frac{3}{5} \frac{\rho_2}{\rho_c} r^2 + O\left(r^4\right)\right] \\ &\equiv g_c\, r \left[1 + \frac{g_2}{g_c} r^2 + O\left(r^4\right)\right]. \end{aligned} \tag{6.38}$$

By again using definition (2.20) and integrating the condition of hydrostatic equilibrium, one obtains

$$\begin{aligned} P(r) &= P_c \left[1 - \frac{2\pi G \rho_c^2}{3P_c} r^2 + O\left(r^4\right)\right] \\ &\equiv P_c \left[1 + \frac{P_2}{P_c} r^2 + O\left(r^4\right)\right], \end{aligned} \tag{6.39}$$

where P_c is the central pressure. In addition, the power series of the following form hold:

$$\left.\begin{aligned} \Gamma_1(r) &= \Gamma_{1,c}\left[1 + \frac{\Gamma_{1,2}}{\Gamma_{1,c}}\, r^2 + O(r^4)\right], \\ c(r) &= c_c\left[1 + \frac{c_2}{c_c}\, r^2 + O(r^4)\right], \\ N^2(r) &= N_c^2\, r^2\left[1 + \frac{N_2^2}{N_c^2}\, r^2 + O(r^4)\right]. \end{aligned}\right\} \tag{6.40}$$

From the introduction of power series (6.37), (6.38), (6.40) into differential equation (6.36), it results that the coefficient of the first derivative $d\zeta/dr$ has a pole of order one at the boundary point $r = 0$. In accordance with the method of Frobenius, a power series of the form

$$\zeta(r) = r^{\nu} \sum_{n=0}^{\infty} a_n\, r^{2n}, \quad \text{with } a_0 \neq 0, \tag{6.41}$$

is adopted as solution. By substituting this series into the differential equation and equating the coefficients of the various powers of r^2 to zero, one derives the indicial equation

$$\nu(\nu + 3) = 0. \tag{6.42}$$

The roots are $\nu_1 = 0$ and $\nu_2 = -3$. Since the difference $\nu_1 - \nu_2$ is an integer, the particular solution associated with the root ν_2 may contain a logarithmic term.

The power series associated with the root ν_1 converges in some neighbourhood of the boundary point $r = 0$ and defines a solution of differential equation (6.36), represented by $\zeta^{(c,1)}(r)$. As $r \to 0$, the solution behaves as r^0, so that the associated radial displacement, $\xi^{(c,1)}(r)$, behaves as r.

A second particular solution can be sought by substitution of the function

$$\zeta^{(c,2)}(r) = \zeta^{(c,1)}(r)\, y^{(c)}(r) \tag{6.43}$$

into the differential equation, where $y^{(c)}(r)$ is a yet unknown function. By setting

$$w^{(c)}(r) = \frac{dy^{(c)}}{dr}, \tag{6.44}$$

one derives the first-order differential equation

$$\frac{1}{w^{(c)}}\frac{dw^{(c)}}{dr} = -\left[\frac{1}{\rho c^2}\frac{d(\rho c^2)}{dr} + \frac{4}{r} + \frac{2}{\zeta^{(c,1)}}\frac{d\zeta^{(c,1)}}{dr}\right], \tag{6.45}$$

since $\zeta^{(c,1)}(r)$ is a particular solution. Integration of this equation and use of definition (6.44) lead to

$$y^{(\mathrm{c})}(r) = \int \left[\rho\, c^2 r^4 \left(\zeta^{(\mathrm{c},1)} \right)^2 \right]^{-1} dr. \tag{6.46}$$

As $r \to 0$, the integrand behaves as

$$\left[\rho\, c^2 r^4 \left(\zeta^{(\mathrm{c},1)} \right)^2 \right]^{-1} = \left(\rho_\mathrm{c}\, c_\mathrm{c}^2\, a_0\, r^4 \right)^{-1} \left[1 + O\left(r^2\right) + O\left(r^4\right) \right], \tag{6.47}$$

so that the function $y^{(\mathrm{c})}(r)$ does not contain a logarithmic term and behaves as r^{-3}. The same holds true for the particular solution $\zeta^{(\mathrm{c},2)}(r)$. The associated radial displacement $\xi^{(\mathrm{c},2)}(r)$ then behaves as r^{-2}. Since it does not remain finite at the boundary point $r = 0$, the second particular solution does not satisfy the regularity condition at the boundary point.

The solution for $\zeta(r)$ that is admissible from the boundary point $r = 0$ thus consists only of the particular solution $\zeta^{(\mathrm{c},1)}(r)$ and involves a single arbitrary constant. For this solution, the matter at the star's centre remains at rest, and the property holds that

$$\left(\frac{d\zeta}{dr} \right)_{r=0} = 0. \tag{6.48}$$

6.3.2 *Admissible Solutions from the Boundary Point $r = R$*

For the study of the behaviour of the particular solutions of differential equation (6.36) in the neighbourhood of the boundary point $r = R$, it is convenient to pass on to the independent variable

$$z = R - r, \tag{6.49}$$

which becomes equal to zero at that point. In terms of this independent variable, the equation becomes

$$\frac{d^2\zeta}{dz^2} - \left[\frac{4}{R-z} - \frac{1}{\rho\, c^2} \frac{d(\rho\, c^2)}{dz} \right] \frac{d\zeta}{dz} + \left[\frac{\sigma^2}{c^2} - \frac{3}{R-z} \frac{1}{\rho\, c^2} \frac{d(\rho\, c^2)}{dz} + \frac{4}{R-z} \frac{g}{c^2} \right] \zeta = 0. \tag{6.50}$$

Suppose that the mass density is also analytic at the boundary point $r = R$ and has a power series of the form

$$\rho(r) = \rho_\mathrm{s}\, z^{n_\mathrm{e}} \left[1 + \frac{\rho_1}{\rho_\mathrm{s}} z + O(z^2) \right], \tag{6.51}$$

where n_e is a positive integer.

A power series for the pressure can be derived by means of condition (2.11) of the hydrostatic equilibrium and equality (2.20) for the gravity. In the supposition that the surface layers contribute so little to the star's total mass that the mass $m(r)$ contained in the sphere with radius r may there be set equal to the total mass M, one has

$$\frac{dP}{dz} = \frac{GM}{R^2} \frac{\rho}{(1 - z/R)^2}. \tag{6.52}$$

Substitution of power series (6.51) for $\rho(r)$ and integration lead to the power series

$$\begin{aligned} P(r) &= \frac{GM}{R^2} \frac{\rho_s}{n_e + 1} z^{n_e+1} \left[1 + \frac{n_e + 1}{n_e + 2} \left(\frac{\rho_1}{\rho_s} + \frac{2}{R} \right) z + O(z^2) \right] \\ &\equiv P_s z^{n_e+1} \left[1 + \frac{P_1}{P_s} z + O(z^2) \right]. \end{aligned} \tag{6.53}$$

From this power series and power series (6.51), it results that, in the lowest-order approximation, the relation between the pressure and the mass density corresponds to a polytropic relation with an effective polytropic index n_e.

For the particular case of the equilibrium sphere of uniform mass density, the reader is referred to Sect. 10.4.1.

From Eq. (6.52), it follows that

$$\begin{aligned} g(r) &= \frac{GM}{R^2} \left[1 + \frac{2}{R} z + O(z^2) \right] \\ &\equiv g_s \left[1 + \frac{2}{R} z + O(z^2) \right]. \end{aligned} \tag{6.54}$$

Furthermore, by use of the power series

$$\Gamma_1(r) = \Gamma_{1,R} \left[1 + \frac{\Gamma_{1,1}}{\Gamma_{1,R}} z + O(z^2) \right], \tag{6.55}$$

one derives the power series

$$\begin{aligned} c(r) &= \left(\frac{GM}{R^2} \right)^{1/2} \left(\frac{\Gamma_{1,R}}{n_e + 1} \right)^{1/2} z^{1/2} \left[1 + \frac{1}{2} \left(\frac{\Gamma_{1,1}}{\Gamma_{1,R}} + \frac{P_1}{P_s} - \frac{\rho_1}{\rho_s} \right) z + O(z^2) \right] \\ &\equiv c_s z^{1/2} \left[1 + \frac{c_1}{c_s} z + O(z^2) \right], \end{aligned} \tag{6.56}$$

$$\begin{aligned} N^2(r) &= -\frac{GM}{R^2} \left(\frac{n_e + 1}{\Gamma_{1,R}} - n_e \right) \frac{1}{z} [1 + O(z)] \\ &\equiv N_s^2 \frac{1}{z} [1 + O(z)]. \end{aligned} \tag{6.57}$$

By virtue of power series (6.51) and (6.53)–(6.56), the coefficients of the first derivative $d\zeta/dz$ and the function $\zeta(z)$ in differential equation (6.50) display a pole of order one at the boundary point $z = 0$. By substituting a power series of the form

$$\zeta(r) = z^{\nu'} \sum_{n=0}^{\infty} b_n z^n, \quad \text{with } b_0 \neq 0, \tag{6.58}$$

into the equation and equating the coefficients of the various powers of z to zero, one derives the indicial equation

$$\nu'\left(\nu' + n_\mathrm{e}\right) = 0 \tag{6.59}$$

with the roots $\nu'_1 = 0$ and $\nu'_2 = -n_\mathrm{e}$. The difference $\nu'_1 - \nu'_2$ is an integer, since n_e is supposed to be a positive integer.

The power series $\zeta^{(\mathrm{s},1)}(r)$ associated with the root ν'_1 converges in some neighbourhood of the boundary point $r = R$ and is a particular solution of differential equation (6.50).

One constructs a second particular solution by proceeding in the same way as for the construction of a second particular solution of differential equation (6.36). By substituting the function

$$\zeta^{(\mathrm{s},2)}(r) = \zeta^{(\mathrm{s},1)}(r)\, y^{(\mathrm{s})}(r) \tag{6.60}$$

into differential equation (6.50) and setting

$$w^{(\mathrm{s})}(r) = \frac{dy^{(\mathrm{s})}}{dz}, \tag{6.61}$$

one derives the first-order differential equation

$$\frac{1}{w^{(\mathrm{s})}} \frac{dw^{(\mathrm{s})}}{dz} = \frac{4}{R - z} - \frac{1}{\rho c^2} \frac{d\left(\rho c^2\right)}{dz} - \frac{2}{\zeta^{(\mathrm{s},1)}} \frac{d\zeta^{(\mathrm{s},1)}}{dz}. \tag{6.62}$$

Integration yields

$$w^{(\mathrm{s})}(r) = \left[\rho c^2 (R - z)^4 \left(\zeta^{(\mathrm{s},1)}\right)^2\right]^{-1}. \tag{6.63}$$

In the neigbhourhood of the boundary point $z = 0$, the solution for $y^{(\mathrm{s})}(r)$ has the form

$$y^{(\mathrm{s})}(r) = \int \left(\rho_\mathrm{s}\, c_\mathrm{s}^2\, R^4\, b_0^2\, z^{n_\mathrm{e}+1}\right)^{-1} [1 + O(z)]\, dz. \tag{6.64}$$

The function $y^{(\mathrm{s})}(r)$ generally contains a logarithmic term and behaves as z^{-n_e} as $z \to 0$. The same holds true for the particular solution $\zeta^{(\mathrm{s},2)}(r)$. Consequently, the second particular solution does not satisfy the regularity condition at the boundary point $r = R$.

The solution for $\zeta(r)$ that is admissible from the boundary point $r = R$ is thus given by the particular solution $\zeta^{(\mathrm{s},1)}(r)$ and involves a single arbitrary constant. At the boundary point itself, the relation holds

$$\left(\frac{d\zeta}{dr}\right)_R - \frac{1}{\Gamma_{1,R}\,R}\left[\frac{\sigma^2}{GM/R^3} + (4 - 3\Gamma_{1,R})\right]\zeta(R) = 0. \qquad (6.65)$$

6.3.3 *Eigenvalue Equation*

Since the radial displacement and its first derivative must be continuous in the interval $[0, R\,]$, one imposes the admissible solutions constructed from the boundary points, and their first derivatives to be continuous at an arbitrary point $r = r_\mathrm{f}$ of their common domain of validity:

$$\left.\begin{aligned} \left(\zeta^{(\mathrm{c},1)}\right)_{r_\mathrm{f}} &= \left(\zeta^{(\mathrm{s},1)}\right)_{r_\mathrm{f}}, \\ \left(\frac{d\zeta^{(\mathrm{c},1)}}{dr}\right)_{r_\mathrm{f}} &= \left(\frac{d\zeta^{(\mathrm{s},1)}}{dr}\right)_{r_\mathrm{f}}. \end{aligned}\right\} \qquad (6.66)$$

It follows that the logarithmic derivatives of the solutions must be continuous at the junction point $r = r_\mathrm{f}$. This condition yields the equation for the eigenvalues σ^2:

$$\left(\frac{1}{\zeta^{(\mathrm{c},1)}}\,\frac{d\zeta^{(\mathrm{c},1)}}{dr}\right)_{r_\mathrm{f}} = \left(\frac{1}{\zeta^{(\mathrm{s},1)}}\,\frac{d\zeta^{(\mathrm{s},1)}}{dr}\right)_{r_\mathrm{f}}. \qquad (6.67)$$

For each eigenvalue, the relation between the two arbitrary constants involved in the admissible solutions that are constructed from the boundary points can be fixed by means of one of the continuity conditions (6.66). The radial displacement, $\xi(r)$, is then determined apart from a factor. Next, the other eigenfunctions can be derived.

6.4 Determination of Non-Radial Spheroidal Normal Modes

As well as the radial normal modes, the non-radial spheroidal normal modes are determined by the resolution of an eigenvalue problem with two end points. The governing equations now consist of a fourth-order linear, homogeneous differential system. The standard procedure of resolution again rests on an application of initial-value techniques by which trial solutions are constructed from each singular end point (Lebovitz 1965a, Hurley et al. 1966, Smeyers 1967, Eisenfeld 1969). Another, less known, procedure has been based on Chebyshev polynomials (Wright 1964, Hurley et al. 1966).

We describe the standard procedure by using the system of Eqs. (6.18)–(6.20), for which the boundary points $r = 0$ and $r = R$ are isolated singular points. Admissible solutions are constructed from each singular boundary point by application of an extension of Frobenius' method to linear, homogeneous systems of first-order differential equations that display an isolated singularity of the first kind. A linear, homogeneous differential system

$$\frac{dw}{dx} = A(x)\, w \tag{6.68}$$

displays a singularity of the first kind at a point $x = x_0$, if the matrix of the coefficients, $A(x)$, has a pole of at most order one at that point (Coddington & Levinson 1955).

6.4.1 Admissible Solutions from the Boundary Point $r = 0$

The system of ordinary differential equations (6.18)–(6.20) can be brought into the form of system (6.68), with a singularity of the first kind at the boundary point $r = 0$, by the introduction of the dependent variables $w_1(r)$, $w_2(r)$, $w_3(r)$, $w_4(r)$ as

$$\left.\begin{aligned} w_1(r) &= \frac{u(r)}{r}, \quad w_2(r) = \eta(r), \\ w_3(r) &= \Phi'(r), \quad w_4(r) = r\,\frac{d\Phi'(r)}{dr}. \end{aligned}\right\} \tag{6.69}$$

The system then becomes

$$\left.\begin{aligned} \frac{dw_1}{dr} &= \left(\frac{g}{c^2} - \frac{1}{r}\right) w_1 + \left[\frac{\ell(\ell+1)}{r^2} - \frac{\sigma^2}{c^2}\right] r\, w_2 + \frac{r}{c^2}\, w_3, \\ \frac{dw_2}{dr} &= \left(1 - \frac{N^2}{\sigma^2}\right) \frac{1}{r}\, w_1 + \frac{N^2}{g}\, w_2 - \frac{1}{\sigma^2}\frac{N^2}{g}\, w_3, \\ \frac{dw_3}{dr} &= \frac{1}{r}\, w_4, \\ \frac{dw_4}{dr} &= 4\pi G\rho\, \frac{N^2}{g}\, w_1 + \sigma^2\, \frac{4\pi G\rho}{c^2}\, r\, w_2 \\ &\quad + \left[\frac{\ell(\ell+1)}{r^2} - \frac{4\pi G\rho}{c^2}\right] r\, w_3 - \frac{1}{r}\, w_4. \end{aligned}\right\} \tag{6.70}$$

The (4×4)-matrix of the coefficients, $A(r)$, can be decomposed in terms of powers of the eigenvalue parameter σ^2 as

$$A(r) = \sigma^{-2} A_1^*(r) + A_2^*(r) + \sigma^2 A_3^*(r). \tag{6.71}$$

Hence, the eigenvalue problem of the non-radial spheroidal normal modes is quadratic, and therefore non-linear, in the eigenvalue parameter.

When one expands the matrix $A(r)$ of the coefficients in a power series of r^2 around the boundary point $r = 0$ by means of power series (6.37)–(6.40), the matrix of the coefficients of the terms that behave as r^{-1} as $r \to 0$ is given by

$$R_c = \begin{pmatrix} -1 & \ell(\ell+1) & 0 & 0 \\ 1 & 0 & 0 & 0 \\ 0 & 0 & 0 & 1 \\ 0 & 0 & \ell(\ell+1) & -1 \end{pmatrix} \tag{6.72}$$

and has the twofold eigenvalues ℓ and $-(\ell+1)$, which differ by an integer. For the application of Frobenius' method, as described by Coddington & Levinson (1955), it is therefore required to transform the system of differential equations $(2\ell+1)$ times, so that both eigenvalues become equal to $-(\ell+1)$. However, these transformations are cumbersome, and their number depends on the degree ℓ of the spherical harmonic considered. One avoids the transformations by passing on from the independent variable r to the independent variable $x = r^2$. By this change, Eqs. (6.70) are transformed into

$$\left.\begin{aligned}
\frac{dw_1}{dx} &= \frac{1}{2}\left(\frac{g}{c^2\sqrt{x}} - \frac{1}{x}\right) w_1 + \frac{1}{2}\left[\frac{\ell(\ell+1)}{x} - \frac{\sigma^2}{c^2}\right] w_2 + \frac{1}{2c^2} w_3, \\
\frac{dw_2}{dx} &= \frac{1}{2}\left(1 - \frac{N^2}{\sigma^2}\right)\frac{1}{x} w_1 + \frac{1}{2}\frac{N^2}{g\sqrt{x}} w_2 - \frac{1}{2}\frac{1}{\sigma^2}\frac{N^2}{g\sqrt{x}} w_3, \\
\frac{dw_3}{dx} &= \frac{1}{2x} w_4, \\
\frac{dw_4}{dx} &= 2\pi G\rho\frac{N^2}{g\sqrt{x}} w_1 + 2\pi G\rho\frac{\sigma^2}{c^2} w_2 \\
&\quad + \frac{1}{2}\left[\frac{\ell(\ell+1)}{x} - \frac{4\pi G\rho}{c^2}\right] w_3 - \frac{1}{2x} w_4.
\end{aligned}\right\} \tag{6.73}$$

Now the matrix of the coefficients of the terms that behave as x^{-1} as $x \to 0$ is given by

$$R_c^* = \begin{pmatrix} -\frac{1}{2} & \frac{1}{2}\,\ell(\ell+1) & 0 & 0 \\ \frac{1}{2} & 0 & 0 & 0 \\ 0 & 0 & 0 & \frac{1}{2} \\ 0 & 0 & \frac{1}{2}\,\ell(\ell+1) & -\frac{1}{2} \end{pmatrix}, \tag{6.74}$$

and has the twofold eigenvalues $\ell/2$ and $-(\ell+1)/2$, which differ no more by an integer. The associated eigenvectors are

$$[\ell, 1, 0, 0], \quad [0, 0, 1, \ell], \quad [-(\ell+1), 1, 0, 0], \quad [0, 0, 1, -(\ell+1)].$$

The fundamental solution matrix is then composed of four linearly independent solution vectors whose components behave as follows as $r \to 0$:

$$w^{(c,1)}(r) = \begin{pmatrix} \ell\, r^{\ell} \\ r^{\ell} \\ \dfrac{4\pi G\rho_c}{2(2\ell+3)}\left(\dfrac{\sigma^2}{c_c^2} + \ell\,\dfrac{N_c^2}{g_c}\right) r^{\ell+2} \\ 4\pi G\rho_c\,\dfrac{\ell+2}{2(2\ell+3)}\left(\dfrac{\sigma^2}{c_c^2} + \ell\,\dfrac{N_c^2}{g_c}\right) r^{\ell+2} \end{pmatrix}, \tag{6.75}$$

$$w^{(c,2)}(r) = \begin{pmatrix} \dfrac{1}{2(2\ell+3)}\left[\dfrac{\ell+2}{c_c^2} - \dfrac{\ell(\ell+1)}{\sigma^2}\,\dfrac{N_c^2}{g_c}\right] r^{\ell+2} \\ \dfrac{1}{2(2\ell+3)}\left(\dfrac{1}{c_c^2} - \dfrac{\ell+3}{\sigma^2}\,\dfrac{N_c^2}{g_c}\right) r^{\ell+2} \\ r^{\ell} \\ \ell\, r^{\ell} \end{pmatrix}, \tag{6.76}$$

$$w^{(c,3)}(r) = \begin{pmatrix} -(\ell+1)\, r^{-(\ell+1)} \\ \\ r^{-(\ell+1)} \\ \\ -\dfrac{4\pi G\rho_c}{2(2\ell-1)} \left[\dfrac{\sigma^2}{c_c^2} - (\ell+1)\, \dfrac{N_c^2}{g_c} \right] r^{-(\ell-1)} \\ \\ 4\pi G\rho_c \, \dfrac{\ell-1}{2(2\ell-1)} \left[\dfrac{\sigma^2}{c_c^2} - (\ell+1)\, \dfrac{N_c^2}{g_c} \right] r^{-(\ell-1)} \end{pmatrix}, \tag{6.77}$$

$$w^{(c,4)}(r) = \begin{pmatrix} \dfrac{1}{2(2\ell-1)} \left[\dfrac{\ell-1}{c_c^2} + \dfrac{\ell(\ell+1)}{\sigma^2} \dfrac{N_c^2}{g_c} \right] r^{-(\ell-1)} \\ \\ -\dfrac{1}{2(2\ell-1)} \left[\dfrac{1}{c_c^2} + \dfrac{\ell-2}{\sigma^2} \dfrac{N_c^2}{g_c} \right] r^{-(\ell-1)} \\ \\ r^{-(\ell+1)} \\ \\ -(\ell+1)\, r^{-(\ell+1)} \end{pmatrix}. \tag{6.78}$$

The solution vectors $w^{(c,3)}(r)$ and $w^{(c,4)}(r)$ contain no logarithmic terms.

Only the solution vectors $w^{(c,1)}(r)$ and $w^{(c,2)}(r)$ yield a radial component of the Lagrangian displacement, $\xi(r)$, that remains finite as $r \to 0$. The admissible solution from the boundary point $r = 0$ thus consists of a linear combination of the two solution vectors $w^{(c,1)}(r)$ and $w^{(c,2)}(r)$. This linear combination can be represented as

$$A_1\, w^{(c,1)}(r) + A_2\, w^{(c,2)}(r), \tag{6.79}$$

where A_1 and A_2 are arbitrary constants. It results that, as $r \to 0$ and $\ell > 0$, the radial component of the Lagrangian displacement, $\xi(r)$, behaves as $r^{\ell-1}$, and the horizontal component, $\eta(r)$, as r^{ℓ}.

For all displacement fields associated with spherical harmonics of degrees $\ell > 1$, the matter at the star's geometrical centre $r = 0$ remains at rest. However, for the displacement fields associated with spherical harmonics of the degree $\ell = 1$, the matter at the star's geometrical centre is subject to a purely radial displacement. Note that the displacement of the matter at the star's geometrical centre does not imply a displacement of the star's mass centre [see equality (4.77)].

6.4.2 Admissible Solutions from the Boundary Point $r = R$

For the study of the solutions of the system of Eqs. (6.70) in the neigbourhood of the boundary point $r = R$, it is convenient to pass on to the independent variable z introduced by definition (6.49). The system then becomes

$$\left.\begin{aligned}
\frac{dw_1}{dz} &= -\left(\frac{g}{c^2} - \frac{1}{R-z}\right) w_1 \\
&\quad - \left[\frac{\ell(\ell+1)}{(R-z)^2} - \frac{\sigma^2}{c^2}\right](R-z)\, w_2 - \frac{R-z}{c^2}\, w_3, \\
\frac{dw_2}{dz} &= -\left(1 - \frac{N^2}{\sigma^2}\right)\frac{1}{R-z}\, w_1 - \frac{N^2}{g}\, w_2 + \frac{1}{\sigma^2}\frac{N^2}{g}\, w_3, \\
\frac{dw_3}{dz} &= -\frac{1}{R-z}\, w_4, \\
\frac{dw_4}{dz} &= -4\pi G\rho\, \frac{N^2}{g}\, w_1 - \sigma^2\, \frac{4\pi G\rho}{c^2}\, (R-z)\, w_2 \\
&\quad - \left[\frac{\ell(\ell+1)}{(R-z)^2} - \frac{4\pi G\rho}{c^2}\right](R-z)\, w_3 + \frac{1}{R-z}\, w_4.
\end{aligned}\right\} \tag{6.80}$$

It is of the form of system (6.68) and displays a singularity of the first kind at the boundary point $r = R$.

The matrix of the coefficients, which is equal to $-A(r)$, can be expanded into a power series of z around $z = 0$ by use of power series (6.51) and (6.53)–(6.57). The exponent n_e is again supposed to be a positive integer. The matrix of the coefficients of the terms that behave as z^{-1} is given by

$$R_s = \begin{pmatrix}
-\dfrac{GM}{R^2}\dfrac{1}{c_s^2} & \sigma^2\dfrac{R}{c_s^2} & -\dfrac{R}{c_s^2} & 0 \\[2ex]
\dfrac{N_s^2}{\sigma^2 R} & -\dfrac{R^2}{GM}N_s^2 & \dfrac{1}{\sigma^2}\dfrac{R^2}{GM}N_s^2 & 0 \\[2ex]
0 & 0 & 0 & 0 \\[2ex]
0 & 0 & 0 & 0
\end{pmatrix} \tag{6.81}$$

and has the threefold eigenvalue 0 and the singular eigenvalue $-n_e$, which differ by an integer.

With the threefold eigenvalue 0, the three eigenvectors can be associated

$$\left[\sigma^2 R^3/(GM), 1, 0, 0\right], \quad \left[-R^3/(GM), 0, 1, 0\right], \quad [0, 0, 0, 1],$$

which lead to three linearly independent vector solutions. For example, when $n_e = 3$, the components of these vector solutions behave as follows as $z \to 0$:

$$w^{(s,1)}(r) = \begin{pmatrix} \sigma^2 \dfrac{R^3}{GM} \\ 1 \\ \pi G\rho_s \sigma^2 \dfrac{R^2}{GM} z^4 \\ -4\pi G\rho_s \sigma^2 \dfrac{R^3}{GM} z^3 \end{pmatrix}, \tag{6.82}$$

$$w^{(s,2)}(r) = \begin{pmatrix} -\dfrac{R^3}{GM} \\ \dfrac{1}{4} \dfrac{R^2}{GM} \left[1 + \dfrac{4}{\Gamma_{1,R}} + \dfrac{4}{\sigma^2} \dfrac{GM}{R^3} \left(\dfrac{4}{\Gamma_{1,R}} - 3\right)\right] z \\ 1 \\ -\dfrac{\ell(\ell+1)}{R} z \end{pmatrix}, \tag{6.83}$$

$$w^{(s,3)}(r) = \begin{pmatrix} \dfrac{R^2}{GM} \dfrac{1}{\Gamma_{1,R}} z \\ \dfrac{1}{4} \dfrac{1}{\sigma^2} \dfrac{1}{R} \left(\dfrac{4}{\Gamma_{1,R}} - 3\right) z \\ -\dfrac{1}{R} z \\ 1 \end{pmatrix}. \tag{6.84}$$

For the construction of a fourth linearly independent solution vector, it is appropriate to apply a method by which the order of the linear homogeneous system is reduced by three, so that only a first-order differential equation needs to be integrated (Coddington & Levinson 1955). Be $V(r)$ the (4×4)-matrix defined as

$$V(r) = \begin{pmatrix} w_1^{(s,1)}(r) \; w_1^{(s,2)}(r) \; w_1^{(s,3)}(r) \; 0 \\ w_2^{(s,1)}(r) \; w_2^{(s,2)}(r) \; w_2^{(s,3)}(r) \; 0 \\ w_3^{(s,1)}(r) \; w_3^{(s,2)}(r) \; w_3^{(s,3)}(r) \; 0 \\ w_4^{(s,1)}(r) \; w_4^{(s,2)}(r) \; w_4^{(s,3)}(r) \; 1 \end{pmatrix}, \tag{6.85}$$

and $y^{(s)}(r)$ the vector of four yet undetermined functions $y_1(r), y_2(r), y_3(r), y_4(r)$. After substitution of the vector

$$w^{(s)}(r) = V(r)\, y^{(s)}(r) \tag{6.86}$$

into system (6.80), one obtains the system of four equations

$$\left.\begin{aligned} w_1^{(s,1)} \frac{dy_1}{dz} + w_1^{(s,2)} \frac{dy_2}{dz} + w_1^{(s,3)} \frac{dy_3}{dz} &= 0, \\ w_2^{(s,1)} \frac{dy_1}{dz} + w_2^{(s,2)} \frac{dy_2}{dz} + w_2^{(s,3)} \frac{dy_3}{dz} &= 0, \\ w_3^{(s,1)} \frac{dy_1}{dz} + w_3^{(s,2)} \frac{dy_2}{dz} + w_3^{(s,3)} \frac{dy_3}{dz} &= -\frac{1}{R-z} y_4, \\ w_4^{(s,1)} \frac{dy_1}{dz} + w_4^{(s,2)} \frac{dy_2}{dz} + w_4^{(s,3)} \frac{dy_3}{dz} + \frac{dy_4}{dz} &= \frac{1}{R-z} y_4. \end{aligned}\right\} \tag{6.87}$$

From the first three equations, the first derivatives $dy_1/dz, dy_2/dz, dy_3/dz$ are solved in terms of the unknown function $y_4(r)$. Still when $n_e = 3$, they behave as

$$\left.\begin{aligned} \frac{dy_1}{dz} &= \left[-\frac{1}{\sigma^2}\frac{1}{R}\left(\frac{4}{\Gamma_{1,R}} - 3\right) + O(z)\right] y_4, \\ \frac{dy_2}{dz} &= \left[\frac{3}{R} + O(1)\right] y_4, \\ \frac{dy_3}{dz} &= \left[\frac{4}{z} + O(z)\right] y_4, \end{aligned}\right\} \tag{6.88}$$

as $z \to 0$. Substitution of the three first derivatives into the fourth equation leads to a first-order differential equation for the function $y_4(r)$ of the form

$$\frac{1}{y_4}\frac{dy_4}{dz} = -\frac{4}{z} + O(1). \tag{6.89}$$

Integration yields

$$y_4(r) = z^{-4}\,[1 + O(z)] \equiv z^{-4}\, f(z), \tag{6.90}$$

where $f(z)$ is an analytic function tending to 1 as $z \to 0$. The solutions for $y_1(r)$, $y_2(r)$, $y_3(r)$ generally contain a logarithmic term, so that they are of the form

$$\left.\begin{aligned} y_1(r) &= \frac{1}{3}\frac{1}{\sigma^2}\frac{1}{R}\left(\frac{4}{\Gamma_{1,R}} - 3\right) z^{-3}\,[1 + O(z)] + C_1 \ln z, \\ y_2(r) &= -\frac{1}{R} z^{-3}\,[1 + O(z)] + C_2 \ln z, \\ y_3(r) &= -z^{-4}\,[1 + O(z)] + C_3 \ln z, \end{aligned}\right\} \tag{6.91}$$

where C_1, C_2, C_3 are arbitrary constants.

A fundamental solution matrix of system (6.70) is then given by

$$\tau_s(r) = \begin{pmatrix} w_1^{(s,1)}(r) & w_1^{(s,2)}(r) & w_1^{(s,3)}(r) & 0 \\ w_2^{(s,1)}(r) & w_2^{(s,2)}(r) & w_2^{(s,3)}(r) & 0 \\ w_3^{(s,1)}(r) & w_3^{(s,2)}(r) & w_3^{(s,3)}(r) & 0 \\ w_4^{(s,1)}(r) & w_4^{(s,2)}(r) & w_4^{(s,3)}(r) & 1 \end{pmatrix} \begin{pmatrix} 1 & 0 & 0 & y_1(r) \\ 0 & 1 & 0 & y_2(r) \\ 0 & 0 & 1 & y_3(r) \\ 0 & 0 & 0 & y_4(r) \end{pmatrix}. \tag{6.92}$$

As $z \to 0$, the components of the fourth solution vector behave as

$$\left.\begin{aligned} w_1^{(s,4)}(r) &= \frac{1}{3}\frac{R^2}{GM}\frac{1}{\Gamma_{1,R}} z^{-3} + O\left(z^{-2}\right) + C_4 \ln z \\ &= w_{4,1}\, z^{-3} + O\left(z^{-2}\right) + C_4 \ln z, \\ w_2^{(s,4)}(r) &= \frac{1}{12}\frac{1}{\sigma^2}\frac{1}{R}\left(\frac{4}{\Gamma_{1,R}} - 3\right) z^{-3} + O\left(z^{-2}\right) + C_1 \ln z \\ &= w_{4,2}\, z^{-3} + O\left(z^{-2}\right) + C_1 \ln z, \\ w_3^{(s,4)}(r) &= w_{4,3}\, z^{-2} + O\left(z^{-1}\right) + C_2 \ln z, \\ w_4^{(s,4)}(r) &= w_{4,4}\, z^{-3} + O\left(z^{-2}\right) + C_3 \ln z, \end{aligned}\right\} \tag{6.93}$$

where C_4 is another arbitrary constant.

Since it does not lead to displacement fields remaining finite at $r = R$, the fourth solution vector must be discarded. The admissible solution from the boundary point $r = R$ thus consists of a linear combination of the first three solution vectors,

$$B_1 \, w^{(s,1)}(r) + B_2 \, w^{(s,2)}(r) + B_3 \, w^{(s,3)}(r), \tag{6.94}$$

where B_1, B_2, B_3 are arbitrary constants. The linear combination must satisfy boundary condition (5.97) on the Eulerian perturbation of the gravitational potential. This requirement leads to the relation between the three constants B_1, B_2, B_3

$$B_3 = -(\ell + 1) \, B_2. \tag{6.95}$$

The admissible solution from the boundary point $r = R$ can then be represented as

$$B_1 \, w^{(s,1)}(r) + B_2 \, w^{(s,2')}(r), \tag{6.96}$$

where the solution vector $w^{(s,2')}(r)$ is defined as

$$w^{(s,2')}(r) = w^{(s,2)}(r) - (\ell + 1) \, w^{(s,3)}(r). \tag{6.97}$$

It involves the two arbitrary constants B_1 and B_2.

6.4.3 Eigenvalue Equation

From the foregoing analysis, it results that the admissible solutions of system (6.70) from the boundary point $r = 0$ consist of the linear combination (6.79) of the two solution vectors $w^{(c,1)}(r)$ and $w^{(c,2)}(r)$ and that the admissible solutions from the boundary point $r = R$ consist of the linear combination (6.96) of the two solution vectors $w^{(s,1)}(r)$ and $w^{(s,2')}(r)$. The two admissible solutions must be continuous at any point $r = r_f$ of their common domain of validity. The conditions of continuity lead to a linear, homogeneous system of four algebraic equations for the four constants A_1, A_2, B_1, B_2. A necessary and sufficient condition for the existence of a non-trivial solution is that the determinant of the matrix of the coefficients, which depend on the eigenvalue parameter σ^2, vanishes at the fitting point $r = r_f$ considered. The resulting condition takes the form

$$\Delta\left(\sigma^2\right) \equiv \begin{vmatrix} w_1^{(c,1)}(r_f) & w_1^{(c,2)}(r_f) & w_1^{(s,1)}(r_f) & w_1^{(s,2')}(r_f) \\ w_2^{(c,1)}(r_f) & w_2^{(c,2)}(r_f) & w_2^{(s,1)}(r_f) & w_2^{(s,2')}(r_f) \\ w_3^{(c,1)}(r_f) & w_3^{(c,2)}(r_f) & w_3^{(s,1)}(r_f) & w_3^{(s,2')}(r_f) \\ w_4^{(c,1)}(r_f) & w_4^{(c,2)}(r_f) & w_4^{(s,1)}(r_f) & w_4^{(s,2')}(r_f) \end{vmatrix} = 0. \tag{6.98}$$

The eigenvalues σ^2 are thus obtained as the zeros of the function $\Delta\left(\sigma^2\right)$.

It may be noted that, when σ^2 differs from all eigenvalues, the matrix of the coefficients whose determinant is $\Delta\left(\sigma^2\right)$, is a fundamental solution matrix of system (6.70) at distances sufficiently large from the two singular boundary points $r = 0$ and $r = R$ (Eisenfeld 1969).

Chapter 7
The Eulerian Perturbation of the Gravitational Potential

7.1 As Solution of Poisson's Perturbed Differential Equation

A solution for the Eulerian perturbation of the gravitational potential, $\Phi'(r)$, at a point of a perturbed star can be derived by integration of Poisson's perturbed non-homogeneous, second-order differential equation (5.92). The integration is performed by means of the method of the variation of the constants in a way similar to the integration of Eq. (2.14) for the gravitational potential at a point in an equilibrium star. The solution that remains finite at $r = 0$ can be written as

$$\Phi'(r) = D r^{\ell} + \frac{4\pi G}{2\ell+1}\left[r^{\ell} \int_0^r \rho'\left(r'\right) r'^{-(\ell-1)}\, dr' \right. \\ \left. - r^{-(\ell+1)} \int_0^r \rho'\left(r'\right) r'^{(\ell+2)}\, dr' \right], \tag{7.1}$$

where D is an arbitrary constant. The constant is fixed by the condition that the solution must satisfy boundary condition (5.97). One then obtains

$$\Phi'(r) = -\frac{4\pi G}{2\ell+1}\left[r^{-(\ell+1)} \int_0^r \rho'\left(r'\right) r'^{(\ell+2)}\, dr' \right. \\ \left. + r^{\ell} \int_r^R \rho'\left(r'\right) r'^{-(\ell-1)}\, dr' + \frac{\rho(R)\,\xi(R)}{R^{\ell-1}}\, r^{\ell} \right]. \tag{7.2}$$

This solution is expressed in terms of the distribution of the Eulerian perturbation of the mass density, $\rho'(r)$, inside the perturbed star. When the mass density vanishes on the surface of the equilibrium star, the solution corresponds to the solutions given by Cowling (1941) and Ledoux & Walraven (1958).

Differentiation of solution (7.2) yields

$$\frac{d\Phi'(r)}{dr} = -\frac{4\pi G}{2\ell+1}\left[-(\ell+1)\, r^{-(\ell+2)} \int_0^r \rho'\left(r'\right) r'^{(\ell+2)}\, dr' \right.$$

P. Smeyers and T. Van Hoolst, *Linear Isentropic Oscillations of Stars: Theoretical Foundations*, Astrophysics and Space Science Library 371,
DOI 10.1007/978-3-642-13030-4_8, © Springer-Verlag Berlin Heidelberg 2010

$$+\,\ell\, r^{\ell-1} \int_r^R \rho'\left(r'\right) r'^{-(\ell-1)}\, dr' + \ell\, \frac{\rho(R)\, \xi(R)}{R^{\ell-1}}\, r^{\ell-1} \Bigg]. \qquad (7.3)$$

The Eulerian perturbation of the gravitational potential can also be expressed in terms of the distribution of the components of the Lagrangian displacement, $\xi(r)$ and $\eta(r)$. By elimination of $\rho'\left(r'\right)$ by means of Eq. (5.90), use of Eq. (5.93), and partial integration, one obtains

$$\Phi'(r) = -\frac{4\pi G}{2\ell+1} \Bigg\{ \ell\, r^{-(\ell+1)} \int_0^r \rho\left(r'\right) r'^{\ell} \left[r'\, \xi\left(r'\right) + (\ell+1)\, \eta\left(r'\right)\right] dr'$$

$$-(\ell+1)\, r^{\ell} \int_r^R \rho\left(r'\right) r'^{-(\ell+1)} \left[r'\, \xi\left(r'\right) - \ell\, \eta\left(r'\right)\right] dr' \Bigg\}. \qquad (7.4)$$

Differentiation yields

$$\frac{d\Phi'(r)}{dr} = -4\pi G \rho\, \xi$$

$$+\, 4\pi G\, \frac{\ell(\ell+1)}{2\ell+1} \Bigg\{ r^{-(\ell+2)} \int_0^r \rho\left(r'\right) r'^{\ell} \left[r'\, \xi\left(r'\right) + (\ell+1)\, \eta\left(r'\right)\right] dr'$$

$$+\, r^{\ell-1} \int_r^R \rho\left(r'\right) r'^{-(\ell+1)} \left[r'\, \xi\left(r'\right) - \ell\, \eta\left(r'\right)\right] dr' \Bigg\}. \qquad (7.5)$$

For radial normal modes, it follows from solution (7.4) that $\Phi'(R) = 0$, in agreement with condition (5.98). Moreover, solution (7.5) reduces to its first term, in agreement with Eq. (5.40).

7.2 Derivation from the General Integral Solution of Poisson's Equation

Solution (7.2) can also be derived from the general integral solution (1.67) of Poisson's equation. We verify that, for a perturbed star, the sum of the terms with the surface integrals in the solution, which corresponds to a solution of Laplace's equation, is identically zero because of boundary condition (5.97).

To this end, it is useful to derive a mathematical representation of the star's perturbed surface. An appropriate way for doing this is to follow the mass elements located on the surface of the equilibrium star in their Lagrangian displacement to the surface of the perturbed star. Consider a mass element at a point with spherical coordinates R, θ, ϕ on the surface of the equilibrium star that moves to the point with spherical coordinates R_{p}, θ_{p}, ϕ_{p} on the surface of the perturbed star. By virtue of transformation formulae (2.1), the Cartesian coordinates of the mass element on the surface of the equilibrium star can be expressed in terms of the spherical coordinates R, θ, ϕ as

$$\left.\begin{aligned} x &= R \sin\theta \cos\phi, \\ y &= R \sin\theta \sin\phi, \\ z &= R \cos\theta. \end{aligned}\right\} \tag{7.6}$$

Similarly, the Cartesian coordinates of the mass element on the surface of the perturbed star can be expressed in terms of the spherical coordinates $R_{\rm p}, \theta_{\rm p}, \phi_{\rm p}$ as

$$\left.\begin{aligned} x_{\rm p} &= R_{\rm p} \sin\theta_{\rm p} \cos\phi_{\rm p}, \\ y_{\rm p} &= R_{\rm p} \sin\theta_{\rm p} \sin\phi_{\rm p}, \\ z_{\rm p} &= R_{\rm p} \cos\theta_{\rm p}. \end{aligned}\right\} \tag{7.7}$$

Since

$$R_{\rm p} = R + \delta r(R,\theta,\phi), \quad \theta_{\rm p} = \theta + \delta\theta(R,\theta,\phi), \quad \phi_{\rm p} = \phi + \delta\phi(R,\theta,\phi),$$

the Cartesian coordinates $x_{\rm p}, y_{\rm p}, z_{\rm p}$ can be expressed as

$$\left.\begin{aligned} x_{\rm p} &= [R + \delta r(R,\theta,\phi)] \sin[\theta + \delta\theta(R,\theta,\phi)] \cos[\phi + \delta\phi(R,\theta,\phi)], \\ y_{\rm p} &= [R + \delta r(R,\theta,\phi)] \sin[\theta + \delta\theta(R,\theta,\phi)] \sin[\phi + \delta\phi(R,\theta,\phi)], \\ z_{\rm p} &= [R + \delta r(R,\theta,\phi)] \cos[\theta + \delta\theta(R,\theta,\phi)], \end{aligned}\right\} \tag{7.8}$$

so that, in the linear approximation,

$$\left.\begin{aligned} x_{\rm p} &= [R + \delta r(R,\theta,\phi)] \sin\theta \cos\phi \\ &\quad + R \cos\theta \cos\phi \, \delta\theta(R,\theta,\phi) - R \sin\theta \sin\phi \, \delta\phi(R,\theta,\phi), \\ y_{\rm p} &= [R + \delta r(R,\theta,\phi)] \sin\theta \sin\phi \\ &\quad + R \cos\theta \sin\phi \, \delta\theta(R,\theta,\phi) + R \sin\theta \cos\phi \, \delta\phi(R,\theta,\phi), \\ z_{\rm p} &= [R + \delta r(R,\theta,\phi)] \cos\theta - R \sin\theta \, \delta\theta(R,\theta,\phi). \end{aligned}\right\} \tag{7.9}$$

These equations give a parametric representation of the surface of the perturbed star in terms of the angular variables θ and ϕ of the mass elements on the surface of the equilibrium star (Heynderickx et al. 1994).

On the surface of the perturbed star, the products of the components of the outward unit normal $\mathbf{n}_\mathrm{p}$ with respect to the orthonormal basis vectors $\partial/\partial x$, $\partial/\partial y$, $\partial/\partial z$ and the surface element dS_p are determined by the equations

$$\left.\begin{aligned} n_\mathrm{p}^x dS_\mathrm{p} &= \frac{\partial\left(y_\mathrm{p}, z_\mathrm{p}\right)}{\partial\left(\theta_\mathrm{p}, \phi_\mathrm{p}\right)}\, d\theta_\mathrm{p}\, d\phi_\mathrm{p}, \\ n_\mathrm{p}^y dS_\mathrm{p} &= \frac{\partial\left(z_\mathrm{p}, x_\mathrm{p}\right)}{\partial\left(\theta_\mathrm{p}, \phi_\mathrm{p}\right)}\, d\theta_\mathrm{p}\, d\phi_\mathrm{p}, \\ n_\mathrm{p}^z dS_\mathrm{p} &= \frac{\partial\left(x_\mathrm{p}, y_\mathrm{p}\right)}{\partial\left(\theta_\mathrm{p}, \phi_\mathrm{p}\right)}\, d\theta_\mathrm{p}\, d\phi_\mathrm{p} \end{aligned}\right\} \tag{7.10}$$

(Korn & Korn 1968). Since

$$d\theta_\mathrm{p}\, d\phi_\mathrm{p} = \frac{\partial\left(\theta_\mathrm{p}, \phi_\mathrm{p}\right)}{\partial\left(\theta, \phi\right)}\, d\theta d\phi, \tag{7.11}$$

it follows

$$\left.\begin{aligned} n_\mathrm{p}^x dS_\mathrm{p} &= \frac{\partial\left(y_\mathrm{p}, z_\mathrm{p}\right)}{\partial(\theta, \phi)}\, d\theta d\phi, \\ n_\mathrm{p}^y dS_\mathrm{p} &= \frac{\partial\left(z_\mathrm{p}, x_\mathrm{p}\right)}{\partial(\theta, \phi)}\, d\theta d\phi, \\ n_\mathrm{p}^z dS_\mathrm{p} &= \frac{\partial\left(x_\mathrm{p}, y_\mathrm{p}\right)}{\partial(\theta, \phi)}\, d\theta d\phi, \end{aligned}\right\} \tag{7.12}$$

or, more explicitly,

$$\begin{aligned} n_\mathrm{p}^x dS_\mathrm{p} = \Bigg[& \sin\theta\cos\phi + 2\sin\theta\cos\phi\,\frac{\delta r}{r} + 2\cos\theta\cos\phi\,\delta\theta - \sin\theta\sin\phi\,\delta\phi \\ & - \frac{1}{r}\cos\theta\cos\phi\,\frac{\partial(\delta r)}{\partial\theta} + \frac{\sin\phi}{r\sin\theta}\,\frac{\partial(\delta r)}{\partial\phi} \\ & + \sin\theta\cos\phi\,\frac{\partial(\delta\theta)}{\partial\theta} + \sin\theta\cos\phi\,\frac{\partial(\delta\phi)}{\partial\phi} \Bigg]_\mathbf{R} dS\left(\mathbf{R}\right), \end{aligned} \tag{7.13}$$

$$\begin{aligned} n_\mathrm{p}^y dS_\mathrm{p} = \Bigg[& \sin\theta\sin\phi + 2\sin\theta\sin\phi\,\frac{\delta r}{r} \\ & + 2\cos\theta\sin\phi\,\delta\theta + \sin\theta\cos\phi\,\delta\phi \end{aligned}$$

$$- \frac{1}{r} \cos\theta \sin\phi \frac{\partial(\delta r)}{\partial\theta} - \frac{\cos\phi}{r\sin\theta} \frac{\partial(\delta r)}{\partial\phi}$$

$$\left. + \sin\theta \sin\phi \frac{\partial(\delta\theta)}{\partial\theta} + \sin\theta\sin\phi \frac{\partial(\delta\phi)}{\partial\phi} \right]_{\mathbf{R}} dS\left(\mathbf{R}\right), \quad (7.14)$$

$$n_{\mathrm{p}}^{z}\, dS_{\mathrm{p}} = \left[\cos\theta + 2\cos\theta \frac{\delta r}{r} + \frac{1}{\sin\theta} \left(\cos^2\theta - \sin^2\theta\right) \delta\theta + \frac{1}{r} \sin\theta \frac{\partial(\delta r)}{\partial\theta} \right.$$

$$\left. + \cos\theta \frac{\partial(\delta\theta)}{\partial\theta} + \cos\theta \frac{\partial(\delta\phi)}{\partial\phi} \right]_{\mathbf{R}} dS\left(\mathbf{R}\right), \quad (7.15)$$

with $dS(\mathbf{R}) = R^2 \sin\theta d\theta d\phi$.

For the subsequent calculations, it is useful to pass on to the products $n_{\mathrm{p}}^{r} dS_{\mathrm{p}}$, $n_{\mathrm{p}}^{\theta} dS_{\mathrm{p}}$, $n_{\mathrm{p}}^{\phi} dS_{\mathrm{p}}$, where n_{p}^{r}, n_{p}^{θ}, n_{p}^{ϕ} are the components of the outward unit normal $\mathbf{n}_{\mathrm{p}}$ with respect to the local basis vectors $\partial/\partial r$, $\partial/\partial\theta$, $\partial/\partial\phi$. These components are related to the components n_{p}^{x}, n_{p}^{y}, n_{p}^{z} with respect to the basis vectors $\partial/\partial x$, $\partial/\partial y$, $\partial/\partial z$ by

$$\left.\begin{aligned} n_{\mathrm{p}}^{r} &= \frac{\partial R_{\mathrm{p}}}{\partial x_{\mathrm{p}}} n_{\mathrm{p}}^{x} + \frac{\partial R_{\mathrm{p}}}{\partial y_{\mathrm{p}}} n_{\mathrm{p}}^{y} + \frac{\partial R_{\mathrm{p}}}{\partial z_{\mathrm{p}}} n_{\mathrm{p}}^{z}, \\ n_{\mathrm{p}}^{\theta} &= \frac{\partial \theta_{\mathrm{p}}}{\partial x_{\mathrm{p}}} n_{\mathrm{p}}^{x} + \frac{\partial \theta_{\mathrm{p}}}{\partial y_{\mathrm{p}}} n_{\mathrm{p}}^{y} + \frac{\partial \theta_{\mathrm{p}}}{\partial z_{\mathrm{p}}} n_{\mathrm{p}}^{z}, \\ n_{\mathrm{p}}^{\phi} &= \frac{\partial \phi_{\mathrm{p}}}{\partial x_{\mathrm{p}}} n_{\mathrm{p}}^{x} + \frac{\partial \phi_{\mathrm{p}}}{\partial y_{\mathrm{p}}} n_{\mathrm{p}}^{y} + \frac{\partial \phi_{\mathrm{p}}}{\partial z_{\mathrm{p}}} n_{\mathrm{p}}^{z}. \end{aligned}\right\} \quad (7.16)$$

From the inverse of transformation formulae (7.7), one successively derives

$$\left.\begin{aligned} \frac{\partial R_{\mathrm{p}}}{\partial x_{\mathrm{p}}} &= (\sin\theta\cos\phi + \cos\theta\cos\phi\,\delta\theta - \sin\theta\sin\phi\,\delta\phi)_{\mathbf{R}}, \\ \frac{\partial R_{\mathrm{p}}}{\partial y_{\mathrm{p}}} &= (\sin\theta\sin\phi + \cos\theta\sin\phi\,\delta\theta + \sin\theta\cos\phi\,\delta\phi)_{\mathbf{R}}, \\ \frac{\partial R_{\mathrm{p}}}{\partial z_{\mathrm{p}}} &= (\cos\theta - \sin\theta\,\delta\theta)_{\mathbf{R}}\ ; \end{aligned}\right\} \quad (7.17)$$

$$\left.\begin{aligned}
\frac{\partial\theta_{\rm p}}{\partial x_{\rm p}} &= \left[\frac{1}{r^2}\left(r\cos\theta\cos\phi - \cos\theta\cos\phi\,\delta r\right.\right.\\
&\qquad\left.\left. -r\sin\theta\cos\phi\,\delta\theta - r\cos\theta\sin\phi\,\delta\phi\right)\right]_{\mathbf{R}},\\
\frac{\partial\theta_{\rm p}}{\partial y_{\rm p}} &= \left[\frac{1}{r^2}\left(r\cos\theta\sin\phi - \cos\theta\sin\phi\,\delta r\right.\right.\\
&\qquad\left.\left. -r\sin\theta\sin\phi\,\delta\theta + r\cos\theta\cos\phi\,\delta\phi\right)\right]_{\mathbf{R}},\\
\frac{\partial\theta_{\rm p}}{\partial z_{\rm p}} &= \left[-\frac{1}{r^2}\left(r\sin\theta - \sin\theta\,\delta r + r\cos\theta\,\delta\theta\right)\right]_{\mathbf{R}};
\end{aligned}\right\} \tag{7.18}$$

$$\left.\begin{aligned}
\frac{\partial\phi_{\rm p}}{\partial x_{\rm p}} &= \left[-\frac{1}{r^2\sin^2\theta}\left(r\sin\theta\sin\phi - \sin\theta\sin\phi\,\delta r\right.\right.\\
&\qquad\left.\left. -r\cos\theta\sin\phi\,\delta\theta + r\sin\theta\cos\phi\,\delta\phi\right)\right]_{\mathbf{R}},\\
\frac{\partial\phi_{\rm p}}{\partial y_{\rm p}} &= \left[\frac{1}{r^2\sin^2\theta}\left(r\sin\theta\cos\phi - \sin\theta\cos\phi\,\delta r\right.\right.\\
&\qquad\left.\left. -r\cos\theta\cos\phi\,\delta\theta - r\sin\theta\sin\phi\,\delta\phi\right)\right]_{\mathbf{R}},\\
\frac{\partial\phi_{\rm p}}{\partial z_{\rm p}} &= 0.
\end{aligned}\right\} \tag{7.19}$$

It results that

$$\left.\begin{aligned}
n_{\rm p}^r\,dS_{\rm p} &= \left[1 + 2\,\frac{\delta r}{r} + \frac{1}{\sin\theta}\,\frac{\partial}{\partial\theta}\left(\sin\theta\,\delta\theta\right) + \frac{\partial(\delta\phi)}{\partial\phi}\right]_{\mathbf{R}} dS\left(\mathbf{R}\right),\\
n_{\rm p}^\theta\,dS_{\rm p} &= \left[-\frac{1}{r^2}\,\frac{\partial(\delta r)}{\partial\theta}\right]_{\mathbf{R}} dS\left(\mathbf{R}\right),\\
n_{\rm p}^\phi\,dS_{\rm p} &= \left[-\frac{1}{r^2\sin^2\theta}\,\frac{\partial(\delta r)}{\partial\phi}\right]_{\mathbf{R}} dS\left(\mathbf{R}\right).
\end{aligned}\right\} \tag{7.20}$$

The two terms with the surface integrals in solution (1.67) of Poisson's equation can now be transformed as follows. In the first term, use of Taylor series (1.72) leads to

$$\begin{aligned}
&\frac{1}{4\pi}\int_{S_{\rm p}} \frac{1}{\left|\mathbf{r}'_{\rm p} - \mathbf{r}_{\rm p}\right|}\left(\frac{\partial\Phi}{\partial x^k}\right)_{\rm p}\left(\mathbf{r}'_{\rm p}\right) n_{\rm p}^k\left(\mathbf{r}'_{\rm p}\right) dS_{\rm p}\left(\mathbf{r}'_{\rm p}\right)\\
&\quad = \left(1 + \delta x^j\,\frac{\partial}{\partial x^j}\right)\left[\frac{1}{4\pi}\int_S \frac{1}{|\mathbf{r}' - \mathbf{r}|}\left(\frac{d\Phi}{dr}\right)(\mathbf{r}')\,dS\left(\mathbf{r}'\right)\right]
\end{aligned}$$

$$+\frac{1}{4\pi}\int_S \left(\delta x'^j \frac{\partial}{\partial x'^j}\left|\mathbf{r}'-\mathbf{r}\right|^{-1}\right)(\mathbf{r}')\left(\frac{d\Phi}{dr}\right)(\mathbf{r}')\,dS(\mathbf{r}')$$

$$+\frac{1}{4\pi}\int_S \frac{1}{|\mathbf{r}'-\mathbf{r}|}\left(\delta\frac{\partial\Phi}{\partial r}\right)(\mathbf{r}')\,dS(\mathbf{r}')$$

$$+\frac{1}{4\pi}\int_S \frac{1}{|\mathbf{r}'-\mathbf{r}|}\left(\frac{d\Phi}{dr}\right)(\mathbf{r}')$$

$$\left[2\frac{\delta r}{r}+\frac{1}{\sin\theta}\frac{\partial}{\partial\theta}(\sin\theta\,\delta\theta)+\frac{\partial(\delta\phi)}{\partial\phi}\right](\mathbf{r}')\,dS(\mathbf{r}')\,. \tag{7.21}$$

In the second term, the expansion of the function

$$\frac{\partial}{\partial x'^k_{\mathrm{p}}}\left|\mathbf{r}'_{\mathrm{p}}-\mathbf{r}_{\mathrm{p}}\right|^{-1}$$

into the Taylor series around the six Cartesian coordinates $x^1, x^2, x^3, x'^1, x'^2, x'^3$

$$\frac{\partial}{\partial x'^k_{\mathrm{p}}}\left|\mathbf{r}'_{\mathrm{p}}-\mathbf{r}_{\mathrm{p}}\right|^{-1}=\left(1+\delta x^j\frac{\partial}{\partial x^j}+\delta x'^j\frac{\partial}{\partial x'^j}\right)\frac{\partial}{\partial x'^k}\left|\mathbf{r}'-\mathbf{r}\right|^{-1} \tag{7.22}$$

and the substitution of this Taylor series lead to

$$-\frac{1}{4\pi}\int_{S_{\mathrm{p}}}\Phi_{\mathrm{p}}\left(\mathbf{r}'_{\mathrm{p}}\right)\left(\frac{\partial}{\partial x'^k_{\mathrm{p}}}\left|\mathbf{r}'_{\mathrm{p}}-\mathbf{r}_{\mathrm{p}}\right|^{-1}\right)\left(\mathbf{r}'_{\mathrm{p}}\right)n^k_{\mathrm{p}}\left(\mathbf{r}'_{\mathrm{p}}\right)dS_{\mathrm{p}}\left(\mathbf{r}'_{\mathrm{p}}\right)=\left(1+\delta x^j\frac{\partial}{\partial x^j}\right)$$

$$\left[-\frac{1}{4\pi}\int_S\Phi(\mathbf{r}')\left(\frac{\partial}{\partial x'^k}\left|\mathbf{r}'-\mathbf{r}\right|^{-1}\right)(\mathbf{r}')\,n^k(\mathbf{r}')\,dS(\mathbf{r}')\right]$$

$$-\frac{1}{4\pi}\int_S\Phi(\mathbf{r}')\left(\delta x'^j\frac{\partial}{\partial x'^j}\frac{\partial}{\partial r'}\left|\mathbf{r}'-\mathbf{r}\right|^{-1}\right)(\mathbf{r}')\,dS(\mathbf{r}')$$

$$-\frac{1}{4\pi}\int_S\delta\Phi(\mathbf{r}')\left(\frac{\partial}{\partial r'}\left|\mathbf{r}'-\mathbf{r}\right|^{-1}\right)(\mathbf{r}')\,dS(\mathbf{r}')$$

$$-\frac{1}{4\pi}\int_S\Phi(\mathbf{r}')\Bigg\{\left(\frac{\partial}{\partial r'}\left|\mathbf{r}'-\mathbf{r}\right|^{-1}\right)(\mathbf{r}')\left[2\frac{\delta r}{r}+\frac{1}{\sin\theta}\frac{\partial}{\partial\theta}(\sin\theta\,\delta\theta)+\frac{\partial(\delta\phi)}{\partial\phi}\right](\mathbf{r}')$$

$$+\left(\frac{\partial}{\partial\theta'}\left|\mathbf{r}'-\mathbf{r}\right|^{-1}\right)(\mathbf{r}')\left[-\frac{1}{r^2}\frac{\partial(\delta r)}{\partial\theta}\right](\mathbf{r}')$$

$$+\left(\frac{\partial}{\partial\phi'}\left|\mathbf{r}'-\mathbf{r}\right|^{-1}\right)(\mathbf{r}')\left[-\frac{1}{r^2\sin^2\theta}\frac{\partial(\delta r)}{\partial\phi}\right](\mathbf{r}')\Bigg\}dS(\mathbf{r}')\,. \tag{7.23}$$

In the sum of the two surface integrals, one has

$$\left(1 + \delta x^j \frac{\partial}{\partial x^j}\right) \left[\frac{1}{4\pi} \int_S \frac{1}{|\mathbf{r}' - \mathbf{r}|} \left(\frac{d\Phi}{dr}\right) (\mathbf{r}')\, dS\,(\mathbf{r}') \right.$$
$$\left. - \frac{1}{4\pi} \int_S \Phi(\mathbf{r}') \left(\frac{\partial}{\partial x'^k} |\mathbf{r}' - \mathbf{r}|^{-1}\right) (\mathbf{r}')\, n^k\,(\mathbf{r}')\, dS\,(\mathbf{r}') \right] = 0 \quad (7.24)$$

because of equality (2.29) and boundary condition (2.30). The other terms are developed by means of expansion (2.27). For a normal mode belonging to a spherical harmonic $Y_\ell^m(\theta, \phi)$, the use of orthogonality relation (Appendix D.4) and the property of the spherical harmonics given by Eq. (Appendix D.6) yields, for the first term,

$$\frac{1}{4\pi} \int_S \left(\delta x'^j \frac{\partial}{\partial x'^j} |\mathbf{r}' - \mathbf{r}|^{-1}\right) (\mathbf{r}') \left(\frac{d\Phi}{dr}\right) (\mathbf{r}')\, dS\,(\mathbf{r}')$$
$$= -\frac{\ell+1}{2\ell+1} \left(\frac{d\Phi}{dr}\right)_R \left[\xi_{\ell,m}(R) - \ell\, \frac{\eta_{\ell,m}(R)}{R}\right] \left(\frac{r}{R}\right)^\ell Y_\ell^m(\theta, \phi), \quad (7.25)$$

$$\frac{1}{4\pi} \int_S \frac{1}{|\mathbf{r}' - \mathbf{r}|} \left(\delta \frac{\partial \Phi}{\partial r}\right) (\mathbf{r}')\, dS\,(\mathbf{r}')$$
$$= \frac{1}{2\ell+1} \left[\left(\frac{d\Phi'_{\ell,m}}{dr}\right)_R - \frac{2}{R} \left(\frac{d\Phi}{dr}\right)_R \xi_{\ell,m}(R) + 4\pi G \rho(R)\, \xi_{\ell,m}(R)\right]$$
$$\frac{r^\ell}{R^{\ell-1}}\, Y_\ell^m(\theta, \phi), \quad (7.26)$$

$$\frac{1}{4\pi} \int_S \frac{1}{|\mathbf{r}' - \mathbf{r}|} \left(\frac{d\Phi}{dr}\right) (\mathbf{r}') \left[2\, \frac{\delta r}{r} + \frac{1}{\sin\theta} \frac{\partial}{\partial \theta} (\sin\theta\, \delta\theta) + \frac{\partial(\delta\phi)}{\partial \phi}\right] (\mathbf{r}')\, dS\,(\mathbf{r}')$$
$$= \frac{1}{2\ell+1} \left(\frac{d\Phi}{dr}\right)_R \left[2\, \xi_{\ell,m}(R) - \ell(\ell+1) \frac{\eta_{\ell,m}(R)}{R}\right] \left(\frac{r}{R}\right)^\ell Y_\ell^m(\theta, \phi), \quad (7.27)$$

and, for the second term,

$$-\frac{1}{4\pi} \int_S \Phi(\mathbf{r}') \left(\delta x'^j \frac{\partial}{\partial x'^j} \frac{\partial}{\partial r'} |\mathbf{r}' - \mathbf{r}|^{-1}\right) (\mathbf{r}')\, dS\,(\mathbf{r}')$$
$$= -\frac{\ell+1}{2\ell+1}\, \Phi(R) \left[(\ell+2)\, \xi_{\ell,m}(R) - \ell(\ell+1)\, \frac{\eta_{\ell,m}(R)}{R}\right]$$
$$\frac{r^\ell}{R^{\ell+1}}\, Y_\ell^m(\theta, \phi), \quad (7.28)$$

$$-\frac{1}{4\pi}\int_S \delta\Phi\left(\mathbf{r}'\right)\left(\frac{\partial}{\partial r'}\left|\mathbf{r}'-\mathbf{r}\right|^{-1}\right)\left(\mathbf{r}'\right)dS\left(\mathbf{r}'\right)$$
$$=\frac{\ell+1}{2\ell+1}\left[\Phi'_{\ell,m}(R)+\left(\frac{d\Phi}{dr}\right)_R \xi_{\ell,m}(R)\right]\frac{r^\ell}{R^\ell}\,Y_\ell^m(\theta,\phi), \tag{7.29}$$

$$-\frac{1}{4\pi}\int_S \Phi\left(\mathbf{r}'\right)\left\{\left(\frac{\partial}{\partial r'}\left|\mathbf{r}'-\mathbf{r}\right|^{-1}\right)\left(\mathbf{r}'\right)\right.$$
$$\left[2\,\frac{\delta r}{r}+\frac{1}{\sin\theta}\frac{\partial}{\partial\theta}\left(\sin\theta\,\delta\theta\right)+\frac{\partial(\delta\phi)}{\partial\phi}\right]\left(\mathbf{r}'\right)$$
$$+\left(\frac{\partial}{\partial\theta'}\left|\mathbf{r}'-\mathbf{r}\right|^{-1}\right)\left(\mathbf{r}'\right)\left[-\frac{1}{r^2}\frac{\partial(\delta r)}{\partial\theta}\right]\left(\mathbf{r}'\right)$$
$$\left.+\left(\frac{\partial}{\partial\phi'}\left|\mathbf{r}'-\mathbf{r}\right|^{-1}\right)\left(\mathbf{r}'\right)\left[-\frac{1}{r^2\sin^2\theta}\frac{\partial(\delta r)}{\partial\phi}\right]\left(\mathbf{r}'\right)\right\}dS\left(\mathbf{r}'\right)$$
$$=\frac{\ell+1}{2\ell+1}\,\Phi(R)\left[(\ell+2)\,\xi_{\ell,m}(R)-\ell(\ell+1)\,\frac{\eta_{\ell,m}(R)}{R}\right]\frac{r^\ell}{R^{\,\ell+1}}\,Y_\ell^m(\theta,\phi). \tag{7.30}$$

The sum of the two terms involving the surface integrals, which corresponds to a solution of Laplace's equation, then becomes

$$\frac{1}{4\pi}\int_{S_\mathrm{p}}\frac{1}{\left|\mathbf{r}'_\mathrm{p}-\mathbf{r}_\mathrm{p}\right|}\left(\frac{\partial\Phi}{\partial x^k}\right)_\mathrm{p}\left(\mathbf{r}'_\mathrm{p}\right)n^k_\mathrm{p}\left(\mathbf{r}'_\mathrm{p}\right)dS_\mathrm{p}\left(\mathbf{r}'_\mathrm{p}\right)$$
$$-\frac{1}{4\pi}\int_{S_\mathrm{p}}\Phi_\mathrm{p}\left(\mathbf{r}'_\mathrm{p}\right)\left(\frac{\partial}{\partial x'^k_\mathrm{p}}\left|\mathbf{r}'_\mathrm{p}-\mathbf{r}_\mathrm{p}\right|^{-1}\right)\left(\mathbf{r}'_\mathrm{p}\right)n^k_\mathrm{p}\left(\mathbf{r}'_\mathrm{p}\right)dS_\mathrm{p}\left(\mathbf{r}'_\mathrm{p}\right)$$
$$=\frac{1}{2\ell+1}\left[\left(\frac{d\Phi'_{\ell,m}}{dr}\right)_R+\frac{\ell+1}{R}\,\Phi'_{\ell,m}(R)+4\pi G\rho(R)\,\xi_{\ell,m}(R)\right]$$
$$\frac{r^\ell}{R^{\ell-1}}\,Y_\ell^m(\theta,\phi). \tag{7.31}$$

The conclusion is that the solution of Laplace's equation in the general integral solution of Poisson's differential equation is identically zero for the gravitational potential in a perturbed star because of boundary condition (5.97). Consequently, the general solution reduces to the solution given by Eq. (1.70).

The Eulerian perturbation of the gravitational potential is thus determined by Eq. (1.74) or (1.75). These solutions can be further developed by means of expansion (2.27) for $|\mathbf{r}-\mathbf{r}'|^{-1}$ in terms of spherical harmonics and the use of the orthogonality relation between spherical harmonics given by Eq. (Appendix D.4). One then again finds solution (7.2).

7.3 The Cowling Approximation

The Eulerian perturbation of the gravitational force is generally smaller than the buoyancy force of Archimedes and the pressure force in the perturbed equations of motion given by Eqs. (2.32). This is especially the case in the more superficial layers of stars, since star masses are often concentrated towards the centre. Therefore, the approximation of neglecting the perturbation of the gravitational force in the equations that govern linear, non-radial oscillations of stars has been adopted. Accordingly, it has been customary to leave out the Eulerian perturbation of the gravitational potential and its first derivative in the system of differential equations (6.15)–(6.17). Equations (6.15) and (6.16) then form a system of two first-order differential equations for the functions $u(r)$ and $y(r)$:

$$\frac{du}{dr} = \frac{g}{c^2}\,u + \left[\frac{\ell(\ell+1)}{\sigma^2} - \frac{r^2}{c^2}\right] y, \tag{7.32}$$

$$\frac{dy}{dr} = \left(\sigma^2 - N^2\right)\frac{u}{r^2} + \frac{N^2}{g}\,y. \tag{7.33}$$

The approximation is known as the approximation of Cowling, who adopted it in his fundamental paper on non-radial oscillations of polytropic equilibrium configurations (Cowling 1941). The approximation was used earlier by Emden (1907) (see also Kopal 1949, Ledoux & Walraven 1958).

By the introduction of the new functions

$$v(r) = f_1(r)\,u(r), \qquad w(r) = f_2(r)\,y(r), \tag{7.34}$$

with

$$f_1(r) = \exp\left[\int_0^r \left(-\frac{g}{c^2}\right) dr\right], \qquad f_2(r) = \exp\left[\int_0^r \left(-\frac{N^2}{g}\right) dr\right], \tag{7.35}$$

one brings Eqs. (7.32) and (7.33) into the form

$$\left.\begin{aligned} \frac{dv}{dr} &= aw, \\ \frac{dw}{dr} &= bv, \end{aligned}\right\} \tag{7.36}$$

where

$$a(r) = \left[\frac{\ell(\ell+1)}{\sigma^2} - \frac{r^2}{c^2}\right] h, \qquad b(r) = \left(\sigma^2 - N^2\right)\frac{1}{r^2 h}, \tag{7.37}$$

and $h(r) = f_1(r)/f_2(r)$ (Gabriel & Scuflaire 1979). Equations (7.36) are generalisations of equations given by Ledoux & Walraven (1958), in which the isentropic coefficient Γ_1 was considered to be constant.

By the neglect of Φ' and $d\Phi'/dr$ in Poisson's differential equation (5.92), it results that

$$\frac{d^2\Phi'}{dr^2} = 4\pi G\rho'. \tag{7.38}$$

Hence, the second derivative of the Eulerian perturbation of the gravitational potential is of the same order of magnitude as the Eulerian perturbation of the mass density, which is kept in the equations. Thus, the second derivative of the Eulerian perturbation of the gravitational potential cannot be neglected, in contrast with the perturbation of the gravitational potential itself and its first derivative.

An argument in support of the Cowling approximation is that $\Phi'(r)$ and $d\Phi'(r)/dr$ tend to zero for normal modes for which the Eulerian perturbation of the mass density displays a large number of nodes between the boundary points $r = 0$ and $r = R$ in solution (1.75) and for normal modes belonging to higher-degree spherical harmonics. In these cases, the Eulerian perturbation of the gravitational potential and its first derivative are expected to have only a slight effect on the eigenvalues. These expectations are seen to be usually met (Sauvenier-Goffin 1951, Robe 1968, Christensen-Dalsgaard 1991).

In the Cowling approximation, the system of governing differential equations is reduced from the fourth to the second order. This has the severe mathematical consequence that two particular solutions are kept out from the general solution. Therefore, it cannot be excluded that the Cowling approximation may have significant effects on particular normal modes (Christensen-Dalsgaard & Gough 2001).

Chapter 8
The Variational Principle of Hamilton

8.1 Introduction

The linearised equations of motion can also be derived as Euler–Lagrange equations of the variational principle of Hamilton

$$\delta \int_{t_1}^{t_2} L\,dt = 0, \tag{8.1}$$

where L is an appropriate Lagrangian of the star's displacement field[1]. The variations of the components of the Lagrangian displacement and their first derivatives are kept equal to zero at the times t_1 and t_2. An alternative variational principle is Rayleigh's principle, which is expressed in terms of the frequency. For this, the reader may be referred to Dahlen & Tromp (1998), especially to their Sect. 4.1.3.

The Lagrangian is determined as the difference between the kinetic energy of mass motion, $T_{\rm tot}$, and the potential energy of the star. In accordance with the procedure adopted in mechanics of continuous media, the potential energy of a star is defined as the sum of the internal energy, $U_{\rm tot}$, and the gravitational potential energy, Ω, of the star:

$$L = T_{\rm tot} - (U_{\rm tot} + \Omega) \tag{8.2}$$

(see, e.g., Mittag et al. 1968). The various energies of a star are given by

$$T_{\rm tot} = \int_V \frac{1}{2} v^2 \rho\, dV, \qquad U_{\rm tot} = \int_V U\rho\, dV, \qquad \Omega = \frac{1}{2}\int_V \Phi\rho\, dV, \tag{8.3}$$

where $\mathbf{v}$ is the macroscopic velocity of the matter, and U, the specific internal energy. The expression for the gravitational potential energy can be derived for a mass with spherical symmetry (Chandrasekhar 1939) and is adopted as an expression

[1] This chapter is partially a reproduction of Smeyers, P., Veugelen, P., Weigert, A.: Variational principles for the linear and adiabatic oscillations of a spherical star, and related definitions of canonical energy density and of canonical energy flux. Geophysical and Astrophysical Fluid Dynamics **16**, 55–71 (1980). With permission from the Taylor & Francis Group, http://www.informaworld.com.

P. Smeyers and T. Van Hoolst, *Linear Isentropic Oscillations of Stars: Theoretical Foundations*, Astrophysics and Space Science Library 371,
DOI 10.1007/978-3-642-13030-4_9,

applying also to the more general cases (Kellogg 1929). It corresponds to an extension, to continuous media, of the expression for the gravitational potential energy of a system of particles (Goldstein 1957).

The variations of the energies must be developed up to the second order in the components of the Lagrangian displacement (see Smeyers et al. 1980).

8.2 First- and Second-Order Energy Variations

The variation of an energy F of a star is given by

$$\delta F = \int_{V_{\rm p}} Q_{\rm p}\left(\mathbf{r}_{\rm p}\right) \rho_{\rm p}\left(\mathbf{r}_{\rm p}\right) dV_{\rm p}\left(\mathbf{r}_{\rm p}\right) - \int_{V} Q(\mathbf{r})\, \rho\,(\mathbf{r})\, dV(\mathbf{r}), \tag{8.4}$$

where Q is a scalar quantity. In accordance with the notations used in previous chapters, $V_{\rm p}$ is the volume occupied by the perturbed star, and V that occupied by the equilibrium star.

By virtue of the mass conservation of the moving elements expressed by equality (1.52), and of definition (1.23) of the Lagrangian perturbation of a scalar quantity Q at a point with position vector $\mathbf{r}$, definition (8.4) becomes

$$\delta F = \int_{V} \delta Q(\mathbf{r})\, \rho(\mathbf{r})\, dV(\mathbf{r}). \tag{8.5}$$

The part of the Lagrangian perturbation $\delta Q(\mathbf{r})$ of order i in the components of the Lagrangian displacement will be denoted by $\delta_i\, Q(\mathbf{r})$. Correspondingly, the notation

$$\delta_i\, F = \int_{V} \delta_i\, Q(\mathbf{r})\, \rho(\mathbf{r})\, dV(\mathbf{r}) \tag{8.6}$$

is introduced.

First, for the derivation of the variations $\delta_1 T_{\rm tot}$ and $\delta_2 T_{\rm tot}$ of the kinetic energy of a star, it is appropriate to start from the kinetic energy of mass motion, per unit mass,

$$T_{\rm kin} \equiv \frac{1}{2}\, v^2 = \frac{1}{2}\, g_{ij} \dot{q}^i \dot{q}^j . \tag{8.7}$$

Since there are supposed to be no mass motions in a quasi-static equilibrium star, one has, for the first variation of the kinetic energy,

$$\delta_1 \left(\frac{1}{2}\, v^2 \right) = 0, \tag{8.8}$$

and, for the second variation,

$$\delta_2 \left(\frac{1}{2}\, v^2 \right) = \frac{1}{2}\, g_{ij} \delta \dot{q}^i \delta \dot{q}^j \tag{8.9}$$

or, because of relation (1.21),

$$\delta_2 \left(\frac{1}{2} \, v^2 \right) = \frac{1}{2} \, g_{ij} \, \frac{\partial \left(\delta q^i \right)}{\partial t} \, \frac{\partial \left(\delta q^j \right)}{\partial t} . \tag{8.10}$$

It then follows

$$\delta_1 T_{\text{tot}} = 0 \tag{8.11}$$

and

$$\delta_2 T_{\text{tot}} = \int_V \frac{1}{2} \, \rho \, g_{ij} \, \frac{\partial \left(\delta q^i \right)}{\partial t} \, \frac{\partial \left(\delta q^j \right)}{\partial t} \, dV . \tag{8.12}$$

Secondly, the variations $\delta_1 U_{\text{tot}}$ and $\delta_2 U_{\text{tot}}$ of the internal energy are derived as follows. For isentropic perturbations, the specific internal energy only depends on the specific volume τ, so that its Taylor series takes the form

$$U \left[\tau_{\text{p}}(\mathbf{r}_{\text{p}}) \right] = U \left[\tau(\mathbf{r}) \right] + \left(\frac{\partial U}{\partial \tau} \right)_S \left[\tau_{\text{p}}(\mathbf{r}_{\text{p}}) - \tau(\mathbf{r}) \right] + \frac{1}{2} \left(\frac{\partial^2 U}{\partial \tau^2} \right)_S \left[\tau_{\text{p}}(\mathbf{r}_{\text{p}}) - \tau(\mathbf{r}) \right]^2 , \tag{8.13}$$

or, by use of relations (Appendix B.2) and (1.77),

$$\delta U = -P \, \delta \tau + \frac{1}{2} \, \frac{\Gamma_1 P}{\tau} \, (\delta \tau)^2 . \tag{8.14}$$

The Lagrangian perturbation of the specific volume τ can be expressed in terms of the components of the Lagrangian displacement by means of relation (1.61) and series expansion (1.57) of the determinant of the displacement gradients. One then obtains

$$\delta_1 U = -\frac{P}{\rho} \, \nabla_j \delta q^j , \tag{8.15}$$

$$\begin{aligned} \delta_2 U = \frac{1}{2 \rho} \Big[& -P \left(\nabla_i \, \delta q^i \right) \left(\nabla_j \, \delta q^j \right) + P \left(\nabla_j \, \delta q^i \right) \left(\nabla_i \, \delta q^j \right) \\ & + \Gamma_1 P \left(\nabla_i \, \delta q^i \right) \left(\nabla_j \, \delta q^j \right) \Big] . \end{aligned} \tag{8.16}$$

The first two terms in the right-hand member of the latter equality can be transformed as

$$\begin{aligned} P \left(\nabla_i \, \delta q^i \right) \left(\nabla_j \, \delta q^j \right) &= \nabla_j \left[P \left(\nabla_i \, \delta q^i \right) \delta q^j \right] \\ & -P \, \delta q^j \, \nabla_j \left(\nabla_i \, \delta q^i \right) - \left(\delta q^j \, \nabla_j P \right) \left(\nabla_i \, \delta q^i \right) , \end{aligned} \tag{8.17}$$

$$\begin{aligned} P \left(\nabla_j \, \delta q^i \right) \left(\nabla_i \, \delta q^j \right) &= \nabla_j \left[P \, \delta q^i \left(\nabla_i \, \delta q^j \right) \right] \\ & -P \, \delta q^i \, \nabla_i \left(\nabla_j \, \delta q^j \right) - \delta q^i \left(\nabla_j P \right) \left(\nabla_i \, \delta q^j \right) . \end{aligned} \tag{8.18}$$

Here the property is used that the order of taking a covariant derivative may be changed in an Euclidian space. Because of the two foregoing transformations, the expression for $\delta_2 U$ can be reduced to

$$\delta_2 U = \frac{1}{2\rho} \Big\{ -\nabla_j \left[P \left(\nabla_i \, \delta q^i \right) \delta q^j \right] + \nabla_j \left[P \, \delta q^i \left(\nabla_i \, \delta q^j \right) \right] + \left(\delta q^j \, \nabla_j P \right) \left(\nabla_i \, \delta q^i \right) - \delta q^i \left(\nabla_j P \right) \left(\nabla_i \, \delta q^j \right) + \Gamma_1 P \left(\nabla_i \, \delta q^i \right) \left(\nabla_j \, \delta q^j \right) \Big\}. \tag{8.19}$$

Equivalent expressions for $\delta_1 U$ and $\delta_2 U$ were derived by Friedman & Schutz (1978).

Substitution of the expressions for $\delta_1 U$ and $\delta_2 U$ into the definitions of the variations $\delta_1 U_{\text{tot}}$ and $\delta_2 U_{\text{tot}}$ of the internal energy yields the expressions for these quantities.

Thirdly, for the derivation of the variations $\delta_1 \Omega$ and $\delta_2 \Omega$ of the gravitational potential energy, an expression for the Lagrangian perturbation of the gravitational potential up to the second order in the components of the Lagrangian displacement is needed. To this end, solutions (1.69) and (1.70) of Poisson's differential equation, respectively, for an equilibrium star and for a perturbed star can be used. One takes into account the mass conservation of the moving elements, as expressed by equality (1.52), and uses the series expansion

$$\left| \mathbf{r}'_{\text{p}} - \mathbf{r}_{\text{p}} \right|^{-1} = \left[1 + \delta q^j \, \nabla_j + \delta q'^j \, \nabla_{\prime j} + \frac{1}{2} \left(\delta q^i \, \delta q^j \, \nabla_i \, \nabla_j + \delta q'^i \, \delta q'^j \, \nabla_{\prime i} \, \nabla_{\prime j} \right) + \delta q^i \, \delta q'^j \, \nabla_i \, \nabla_{\prime j} \right] \left| \mathbf{r}' - \mathbf{r} \right|^{-1}, \tag{8.20}$$

where ∇_j denotes a differentiation with respect to the generalised coordinate q^j of the point with position vector $\mathbf{r}$, and $\nabla_{\prime j}$, a differentiation with respect to the generalised coordinate q'^j of the point with position vector $\mathbf{r}'$. One then obtains

$$\delta_1 \Phi \left(\mathbf{r} \right) = -G \int_V \rho \left(\mathbf{r}' \right) \left(\delta q^j \, \nabla_j + \delta q'^j \, \nabla_{\prime j} \right) \left| \mathbf{r}' - \mathbf{r} \right|^{-1} dV(\mathbf{r}'), \tag{8.21}$$

$$\delta_2 \Phi \left(\mathbf{r} \right) = -G \int_V \rho \left(\mathbf{r}' \right) \left[\frac{1}{2} \left(\delta q^i \, \delta q^j \, \nabla_i \, \nabla_j + \delta q'^i \, \delta q'^j \, \nabla_{\prime i} \, \nabla_{\prime j} \right) + \delta q^i \, \delta q'^j \, \nabla_i \, \nabla_{\prime j} \right] \left| \mathbf{r}' - \mathbf{r} \right|^{-1} dV(\mathbf{r}'). \tag{8.22}$$

Similar expressions were derived by Kato & Unno (1967) and Van Hoolst (1992).

By virtue of solution (1.69) of Poisson's differential equation, equality (8.21) can be rewritten as

$$\delta_1 \Phi(\mathbf{r}) = \left(\delta q^j \, \nabla_j \Phi \right)(\mathbf{r}) - G \int_V \rho(\mathbf{r}') \, \delta q'^j \, \nabla_{\prime j} \left| \mathbf{r}' - \mathbf{r} \right|^{-1} dV(\mathbf{r}'). \tag{8.23}$$

Substitution into the definition of $\delta_1 \Omega$ yields

$$\delta_1 \Omega = \frac{1}{2} \int_V \rho \, (\delta q^j \, \nabla_j \Phi) \, dV + \frac{1}{2} \int_V \rho(\mathbf{r}') \, \delta q'^j \, \nabla_{\prime j} \left[-G \int_V \frac{\rho(\mathbf{r})}{|\mathbf{r}' - \mathbf{r}|} \, dV(\mathbf{r}) \right] dV(\mathbf{r}') \tag{8.24}$$

or, equivalently,

$$\delta_1 \Omega = \int_V \rho \, (\delta q^j \, \nabla_j \Phi) \, dV. \tag{8.25}$$

Furthermore, by virtue of solutions (1.69) and (1.74), equality (8.22) can be rewritten as

$$\delta_2 \Phi(\mathbf{r}) = \frac{1}{2} \left(\delta q^i \, \delta q^j \, \nabla_i \, \nabla_j \Phi \right)(\mathbf{r}) + \left(\delta q^i \, \nabla_i \Phi' \right)(\mathbf{r}) - \frac{1}{2} G \int_V \rho(\mathbf{r}') \, \delta q'^i \, \delta q'^j \, \nabla_{\prime i} \, \nabla_{\prime j} \, |\mathbf{r}' - \mathbf{r}|^{-1} \, dV(\mathbf{r}'). \tag{8.26}$$

Substitution into the definition of $\delta_2 \Omega$ yields

$$\delta_2 \Omega = \frac{1}{2} \left\{ \frac{1}{2} \int_V \rho(\mathbf{r}) \left(\delta q^i \, \delta q^j \, \nabla_i \, \nabla_j \Phi \right)(\mathbf{r}) \, dV(\mathbf{r}) + \int_V \rho(\mathbf{r}) \left(\delta q^i \, \nabla_i \Phi' \right)(\mathbf{r}) \, dV(\mathbf{r}) + \frac{1}{2} \int_V \rho(\mathbf{r}') \, \delta q'^i \, \delta q'^j \, \nabla_{\prime i} \, \nabla_{\prime j} \left[-G \int_V \frac{\rho(\mathbf{r})}{|\mathbf{r}' - \mathbf{r}|} \, dV(\mathbf{r}) \right] dV(\mathbf{r}') \right\}. \tag{8.27}$$

Use of solution (1.69) and addition of the first and the third term lead to

$$\delta_2 \Omega = \frac{1}{2} \left[\int_V \rho \left(\delta q^i \, \delta q^j \, \nabla_i \, \nabla_j \Phi \right) dV + \int_V \rho \left(\delta q^i \, \nabla_i \Phi' \right) dV \right]. \tag{8.28}$$

Equalities (8.25) and (8.28) express that the variations $\delta_1 \Omega$ and $\delta_2 \Omega$ of the gravitational potential energy of a star correspond to the opposite of the work of the gravitational force, respectively, at the first and at the second order, that is related to the Lagrangian displacements of the mass elements in the star (see, e.g., Cowling 1938).

By transformation of the first term in the right-hand member, equality (8.28) becomes

$$\delta_2 \Omega = \frac{1}{2} \int_V \rho \left[\delta q^i \, \nabla_i \left(\delta q^j \, \nabla_j \Phi \right) - \delta q^i \left(\nabla_j \Phi \right) \left(\nabla_i \, \delta q^j \right) + \delta q^i \, \nabla_i \Phi' \right] dV. \tag{8.29}$$

Relative to the first- and second-order variations of the gravitational potential energy, it is convenient to consider variations per unit mass as

$$\delta_1 \omega = \delta q^j \, \nabla_j \Phi, \tag{8.30}$$

$$\delta_2 \omega = \frac{1}{2} \left[\delta q^i \, \nabla_i \left(\delta q^j \, \nabla_j \Phi \right) - \delta q^i \left(\nabla_j \Phi \right) \left(\nabla_i \, \delta q^j \right) + \delta q^i \, \nabla_i \Phi' \right], \tag{8.31}$$

so that

$$\delta_k \Omega = \int_V \rho\, \delta_k \omega\, dV, \qquad k = 1, 2. \tag{8.32}$$

In the particular case of a purely radial displacement field, equality (8.19) can be reduced to

$$\delta_2 U = -\frac{1}{\rho\, r^2} \frac{\partial}{\partial r} \left[P\, r\, (\delta r)^2 \right] + \frac{1}{\rho} \frac{dP}{dr} \frac{(\delta r)^2}{r} + \frac{1}{2} \frac{\Gamma_1 P}{\rho} \left[\frac{1}{r^2} \frac{\partial}{\partial r} \left(r^2\, \delta r \right) \right]^2 . \tag{8.33}$$

Furthermore, in equality (8.31), $d^2\Phi/dr^2$ and $\partial\Phi'/\partial r$ can be eliminated by means of Eqs. (2.14) and (5.40). One then obtains

$$\delta_2 \omega = -\frac{G\, m(r)}{r^3} (\delta r)^2 . \tag{8.34}$$

Variations of the internal and gravitational potential energy for purely radial displacement fields were derived by Stothers & Frogel (1967) up to the third order in the radial displacement. In the more general case of non-radial displacement fields, expressions for the variations of the energies were derived by Däppen & Perdang (1985) to the third order in the components of the Lagrangian displacement, by Van Hoolst (1994), to the fourth order, and by Schenk et al. (2001), to the third order for rotating stars.

8.3 Equality of the Mean Kinetic and the Mean Potential Energy of Oscillation over a Period

For a normal isentropic mode of a static star, the mean values of the kinetic and the potential energy of oscillation, taken over a period, are equal to each other (Lee & Saio 1990). The kinetic and the potential energy of oscillation of a star correspond to the second-order variations of the kinetic and the potential energy derived up here. The equality of their mean values can be shown as follows.

Consider the normal mode

$$\delta q^k = F_n(t) \left(\delta q^k \right)_n (r, \theta, \phi) \tag{8.35}$$

with

$$F_n(t) = A_n \cos(\sigma_n t) + B_n \sin(\sigma_n t) , \tag{8.36}$$

A_n and B_n being arbitrary constants.

In terms of complex displacement fields, a star's kinetic energy of oscillation is defined, in accordance with Eq. (8.12), as

$$\delta_2 T_{\text{tot}} = \int_M \frac{1}{2}\, g_{ij}\, \frac{\overline{\partial(\delta q^i)}}{\partial t}\, \frac{\partial\left(\delta q^j\right)}{\partial t}\, dm, \tag{8.37}$$

so that

$$\delta_2 T_{\text{tot}} = \frac{1}{2}\left(\frac{dF_n}{dt}\right)^2 \int_M g_{ij}\, \overline{\left(\delta q^i\right)_n}\, \left(\delta q^j\right)_n\, dm. \tag{8.38}$$

On the other hand, a star's potential energy of oscillation is given by

$$\begin{aligned}
\delta_2 V_{\text{tot}} &= \delta_2 U_{\text{tot}} + \delta_2 \Omega \\
&= -\int_M \frac{1}{2}\frac{1}{\rho}\Big\{ \nabla_j \left[P\left(\nabla_i\, \delta q^i\right) \delta q^j\right] - \nabla_j \left[P\, \delta q^i \left(\nabla_i\, \delta q^j\right)\right] \\
&\qquad - \Gamma_1 P \left(\nabla_i\, \delta q^i\right)\left(\nabla_j\, \delta q^j\right) - \left(\delta q^j\, \nabla_j P\right)\left(\nabla_i\, \delta q^i\right) \\
&\qquad - \rho\, \delta q^i\, \nabla_i \left(\delta q^j\, \nabla_j \Phi\right) - \rho\, \delta q^i\, \nabla_i \Phi' \Big\}\, dm.
\end{aligned} \tag{8.39}$$

The first two terms of the right-hand member are equal to zero for stars in which the equilibrium pressure vanishes on the equilibrium surface. The third term of the integrand can be transformed as

$$-\Gamma_1 P \left(\nabla_i\, \delta q^i\right)\left(\nabla_j\, \delta q^j\right) = \nabla_j \left(\delta P\, \delta q^j\right) - \delta q^j\, \nabla_j \left(\delta P\right). \tag{8.40}$$

After integration over the star's mass, the first term of the right-hand member vanishes because of the boundary condition $\delta P = 0$ on the surface. Next, the fourth term of the integrand in the right-hand member of Eq. (8.39) can be written as

$$-\left(\delta q^j\, \nabla_j P\right)\left(\nabla_i\, \delta q^i\right) = \frac{\delta\rho}{\rho}\left(\delta q^j\, \nabla_j P\right). \tag{8.41}$$

The sum of the last two terms of the integrand in the right-hand member of Eq. (8.39) yields

$$-\rho\, \delta q^i\, \nabla_i \left(\delta q^j\, \nabla_j \Phi\right) - \rho\, \delta q^i\, \nabla_i \Phi' = -\rho\, \delta q^i\, \nabla_i \left(\delta\Phi\right). \tag{8.42}$$

In terms of complex displacement fields, a star's potential energy of oscillation can then be expressed as

$$\delta_2 V_{\text{tot}} = \frac{1}{2}\int_M \overline{\delta q^i}\left[\frac{1}{\rho}\, \nabla_i (\delta P) - \frac{\delta\rho}{\rho^2}\, \nabla_i P + \nabla_i \left(\delta\Phi\right)\right] dm. \tag{8.43}$$

With the use of Eq. (4.14) for $U_{ij}\, \delta q^j$ and Eq. (8.35) for the displacement field, the star's potential energy of oscillation becomes

$$\delta_2 V_{\text{tot}} = \frac{1}{2}\, F_n^2 \int_M \overline{\left(\delta q^i\right)_n}\, U_{ij}\, \left(\delta q^j\right)_n\, dm, \tag{8.44}$$

or, by means of wave equation (4.12),

$$\delta_2 V_{\text{tot}} = \frac{1}{2}\,\sigma_n^2\, F_n^2 \int_M g_{ij}\, \overline{\left(\delta q^i\right)_n}\, \left(\delta q^j\right)_n\, dm. \tag{8.45}$$

The mean values of the kinetic and the potential energy of oscillation, taken over a period $2\pi/\sigma_n$, are then

$$\left.\begin{aligned}
\frac{\sigma_n}{2\pi}\int_{-\pi/\sigma_n}^{\pi/\sigma_n} \delta_2 T_{\text{tot}}\, dt &= \frac{1}{4}\,\sigma_n^2\left(A_n^2 + B_n^2\right)\int_M g_{ij}\,\overline{\left(\delta q^i\right)_n}\,\left(\delta q^j\right)_n\, dm,\\
\frac{\sigma_n}{2\pi}\int_{-\pi/\sigma_n}^{\pi/\sigma_n} \delta_2 V_{\text{tot}}\, dt &= \frac{1}{4}\,\sigma_n^2\left(A_n^2 + B_n^2\right)\int_M g_{ij}\,\overline{\left(\delta q^i\right)_n}\,\left(\delta q^j\right)_n\, dm,
\end{aligned}\right\} \tag{8.46}$$

so that the equality follows

$$\frac{\sigma_n}{2\pi}\int_{-\pi/\sigma_n}^{\pi/\sigma_n} \delta_2 T_{\text{tot}}\, dt = \frac{\sigma_n}{2\pi}\int_{-\pi/\sigma_n}^{\pi/\sigma_n} \delta_2 V_{\text{tot}}\, dt. \tag{8.47}$$

8.4 First- and Second-Order Variational Principles

With the first- and second-order energies, Lagrangians of the corresponding orders can be constructed as

$$L_k = \delta_k T_{\text{tot}} - \left(\delta_k U_{\text{tot}} + \delta_k \Omega\right), \qquad k = 1, 2. \tag{8.48}$$

With them, Lagrangian densities $\mathscr{L}_k$ are associated:

$$L_k = \int_V \mathscr{L}_k\, dV. \tag{8.49}$$

8.4.1 First-Order Variational Principle

The Lagrangian of the first order is given by

$$L_1 = \int_V \left[P\left(\nabla_j\, \delta q^j\right) - \rho\left(\delta q^j\, \nabla_j \Phi\right)\right] dV. \tag{8.50}$$

By virtue of Hamilton's variational principle, the Lagrangian displacement field must satisfy the Euler–Lagrange-equations

$$\frac{\partial}{\partial t}\,\frac{\partial \mathscr{L}_1}{\partial\left[\partial\left(\delta q^i\right)/\partial t\right]} + \nabla_j\,\frac{\partial \mathscr{L}_1}{\partial\left(\nabla_j\, \delta q^i\right)} - \frac{\partial \mathscr{L}_1}{\partial\left(\delta q^i\right)} = 0, \qquad i = 1, 2, 3, \tag{8.51}$$

and the conditions on the surface of the equilibrium star

$$\frac{\partial \mathscr{L}_1}{\partial \left(\nabla_i \, \delta q^i\right)} n_j dS = 0, \qquad j = 1, 2, 3, \tag{8.52}$$

where n_j is the covariant component j of the outward unit normal **n** on the surface element dS [see Eqs. (Appendix I.7) and conditions (Appendix I.8)].

The Euler–Lagrange equations lead to the condition of hydrostatic equilibrium

$$\nabla_i P + \rho \nabla_i \Phi = 0, \qquad i = 1, 2, 3, \tag{8.53}$$

and the conditions on the surface of the equilibrium star, to the condition

$$P(R) = 0. \tag{8.54}$$

8.4.2 Second-Order Variational Principle

When the condition of hydrostatic equilibrium is fulfilled in the equilibrium star, the second-order Lagrangian is given by

$$\begin{aligned} L_2 = \int_V \frac{1}{2} \Big\{ & \rho \, g_{ij} \, \frac{\partial(\delta q^i)}{\partial t} \, \frac{\partial(\delta q^j)}{\partial t} + \nabla_j \left[P \left(\nabla_i \, \delta q^i \right) \delta q^j \right] \\ & - \nabla_j \left[P \, \delta q^i \left(\nabla_i \, \delta q^j \right) \right] - \Gamma_1 P \left(\nabla_i \, \delta q^i \right) \left(\nabla_j \, \delta q^j \right) \\ & - \left(\delta q^j \, \nabla_j P \right) \left(\nabla_i \, \delta q^i \right) - \rho \, \delta q^i \, \nabla_i \left(\delta q^j \, \nabla_j \Phi \right) - \rho \, \delta q^i \, \nabla_i \Phi' \Big\} \, dV. \end{aligned} \tag{8.55}$$

The second and the third integral in the right-hand member are equal to zero, when condition (8.54) is satisfied on the surface of the equilibrium star, and the displacement field is sufficiently regular there. The Lagrangian can therefore be reduced to

$$\begin{aligned} L_2^{(C)} = \int_V \frac{1}{2} \Big[& \rho \, g_{ij} \, \frac{\partial \left(\delta q^i\right)}{\partial t} \, \frac{\partial \left(\delta q^j\right)}{\partial t} - \Gamma_1 P \left(\nabla_i \, \delta q^i \right) \left(\nabla_j \, \delta q^j \right) \\ & - \left(\delta q^j \, \nabla_j P \right) \left(\nabla_i \, \delta q^i \right) - \rho \, \delta q^i \, \nabla_i \left(\delta q^j \, \nabla_j \Phi \right) - \rho \, \delta q^i \, \nabla_i \Phi' \Big] \, dV. \end{aligned} \tag{8.56}$$

The reduced Lagrangian $L_2^{(C)}$ corresponds to the functional derived by Chandrasekhar & Lebovitz (1964) on the basis of the linearised equations and considerations of symmetry in the components of the Lagrangian displacement. The procedure adopted by Chandrasekhar & Lebovitz in their construction of the variational principle is similar to that adopted by Detweiler & Ipser (1973) in a search for a variational principle relative to non-radial modes within the framework of general relativity. About their procedure, the latter authors noted:

> We use certain guides in our search for a variational principle. First of all, experience indicates that we should seek an expression for σ^2 in terms of volume ... integrals with integrands quadratic in the perturbation variables. In addition, ... the integrands should be symmetric in the perturbation variables if we are to expect to be able to demonstrate the stationary feature of the principle ... Surface integrals ... can arise from integrations by parts in volume integrals ... By the application of boundary conditions, we should be able to make the variations ... of any surface integrals vanish ... Another guide suggests that we can find the required quadratic expression by taking the inner product of the displacement 3-vector with the equations of motion ..., by integrating that inner product over the interior of a sphere ..., and by using integration by parts and other manipulations to force the expression into the desired symmetric form.

The Lagrangian density $\mathscr{L}_2^{(C)}$ defined as the integrand of the Lagrangian $L_2^{(C)}$ is related to the Lagrangian density $\mathscr{L}_2$ defined as the integrand of the Lagrangian L_2 as

$$\mathscr{L}_2^{(C)} = \mathscr{L}_2 - \frac{1}{2}\,\nabla_j \left[P\left(\nabla_i\,\delta q^i\right)\delta q^j\right] + \frac{1}{2}\,\nabla_j\left[P\,\delta q^i\left(\nabla_i\,\delta q^j\right)\right]. \tag{8.57}$$

It should be noted that the Lagrangian density $\mathscr{L}_2^{(C)}$ does *not* correspond to the difference between the local variation of the kinetic energy and the sum of the local variations of the internal and the gravitational potential energy.

The Lagrangian density $\mathscr{L}_2$, and thus also the Lagrangian density $\mathscr{L}_2^{(C)}$, contains not only the components of the Lagrangian displacement, their first derivatives with respect to time, and their covariant derivatives with respect to the generalised coordinates, but also the components of the gradient of the Eulerian perturbation of the gravitational potential. By virtue of solution (1.74) for this perturbation, the Lagrangian density $\mathscr{L}_2^{(C)}$ contains an integral extending over the whole displacement field. As noted also by Takata (2006a), this is a deviation from the standard formulation of Lagrangians in analytical mechanics, in which they are functionals of some functions and their first derivatives.

It is therefore convenient to split the Lagrangian density $\mathscr{L}_2^{(C)}$ into two parts by the introduction of the definitions

$$\mathscr{L}_2^{(C,1)} = \frac{1}{2}\Bigg[\rho\,g_{ij}\,\frac{\partial(\delta q^i)}{\partial t}\,\frac{\partial(\delta q^j)}{\partial t} - \Gamma_1\,P\left(\nabla_i\,\delta q^i\right)\left(\nabla_j\,\delta q^j\right) - \left(\delta q^j\,\nabla_j\,P\right)\left(\nabla_i\,\delta q^i\right) - \rho\,\delta q^i\,\nabla_i\left(\delta q^j\,\nabla_j\,\Phi\right)\Bigg], \tag{8.58}$$

and

$$\mathscr{L}_2^{(C,2)} = -\frac{1}{2}\,\rho\,\delta q^i\,\nabla_i \Phi'. \tag{8.59}$$

The term $\mathscr{L}_2^{(C,1)}$ involves the components of the Lagrangian displacement and their first derivatives, and the term $\mathscr{L}_2^{(C,2)}$, the Eulerian perturbation of the gravitational potential.

By means of solution (1.74) for the Eulerian perturbation of the gravitational potential, one derives the variation of the term $\mathscr{L}_2^{(C,2)}$ in terms of the variation of the Lagrangian displacement

$$-\delta \int_{t_1}^{t_2} dt \int_V \frac{1}{2}\,\rho\,\delta q^i\,\nabla_i \Phi'\,dV = \frac{G}{2}\,\delta \int_{t_1}^{t_2} dt$$
$$\int_V \int_V \rho\,(\mathbf{r})\,\rho\left(\mathbf{r}'\right)\delta q^i\,(\mathbf{r})\,\delta q^j\left(\mathbf{r}'\right)\nabla_i\,\nabla_{\prime j}\left|\mathbf{r}' - \mathbf{r}\right|^{-1} dV\left(\mathbf{r}'\right)dV\,(\mathbf{r}), \tag{8.60}$$

so that

$$-\delta \int_{t_1}^{t_2} dt \int_V \frac{1}{2}\,\rho\,\delta q^i\,\nabla_i \Phi'\,dV = -\int_{t_1}^{t_2} dt \int_V \rho\,\left(\nabla_i \Phi'\right)\delta\left(\delta q^i\right)dV \tag{8.61}$$

(Smeyers 1973).

The requirement that the Lagrangian displacement field satisfies Hamilton's variational principle then leads to the Euler–Lagrange equations

$$\frac{\partial}{\partial t}\,\frac{\partial \mathscr{L}_2^{(C,1)}}{\partial\left[\partial\left(\delta q^i\right)/\partial t\right]} + \nabla_j\,\frac{\partial \mathscr{L}_2^{(C,1)}}{\partial\left(\nabla_j\,\delta q^i\right)} - \frac{\partial \mathscr{L}_2^{(C,1)}}{\partial\left(\delta q^i\right)} + \rho\,\nabla_i \Phi' = 0, \quad i = 1,2,3, \tag{8.62}$$

and to the conditions on the surface of the equilibrium star

$$\frac{\partial \mathscr{L}_2^{(C,1)}}{\partial\left(\nabla_j\,\delta q^i\right)}\,n_j = 0, \quad i = 1,2,3. \tag{8.63}$$

The Euler–Lagrange equations take the form

$$\rho\,g_{ij}\,\frac{\partial^2\left(\delta q^j\right)}{\partial t^2} - \nabla_i\left(\Gamma_1 P\,\nabla_j\,\delta q^j\right) - \frac{1}{2}\,\nabla_i\left(\delta q^j\,\nabla_j P\right)$$
$$-\frac{1}{2}\,\nabla_j\left(\rho\,\delta q^j\,\nabla_i \Phi\right) + \frac{1}{2}\left(\nabla_i P\right)\left(\nabla_j\,\delta q^j\right) + \frac{1}{2}\,\rho\left(\nabla_j \Phi\right)\left(\nabla_i\,\delta q^j\right)$$
$$+\,\rho\,\delta q^j\,\nabla_i\,\nabla_j \Phi + \rho\,\nabla_i \Phi' = 0, \quad i = 1,2,3. \tag{8.64}$$

After transformation of the last but one term in the left-hand member, it follows

$$\begin{aligned}
&\rho\, g_{ij}\,\frac{\partial^2\left(\delta q^j\right)}{\partial t^2} - \nabla_i\left(\Gamma_1 P\,\nabla_j\,\delta q^j\right) - \frac{1}{2}\,\nabla_i\left(\delta q^j\,\nabla_j P\right)\\
&-\frac{1}{2}\,\nabla_j\left(\rho\,\delta q^j\,\nabla_i \Phi\right) + \frac{1}{2}\left(\nabla_i P\right)\left(\nabla_j\,\delta q^j\right) - \frac{1}{2}\,\rho\left(\nabla_j \Phi\right)\left(\nabla_i\,\delta q^j\right)\\
&+\rho\,\nabla_i\left(\delta q^j\,\nabla_j \Phi\right) + \rho\,\nabla_i \Phi' = 0, \qquad i = 1,2,3.
\end{aligned} \tag{8.65}$$

The fourth term in the left-hand member can be split as

$$-\frac{1}{2}\,\nabla_j\left(\rho\,\delta q^j\,\nabla_i \Phi\right) = -\frac{1}{2}\left(\rho\,\nabla_i \Phi\right)\left(\nabla_j\,\delta q^j\right) - \frac{1}{2}\,\delta q^j\,\nabla_j\left(\rho\,\nabla_i \Phi\right). \tag{8.66}$$

Use of the condition of hydrostatic equilibrium and change of the order of covariant differentiation in the second term of the right-hand member lead to

$$\begin{aligned}
-\frac{1}{2}\,\nabla_j\left(\rho\,\delta q^j\,\nabla_i \Phi\right) &= -\frac{1}{2}\left(\rho\,\nabla_i \Phi\right)\left(\nabla_j\,\delta q^j\right)\\
&\quad+\frac{1}{2}\,\nabla_i\left(\delta q^j\,\nabla_j P\right) - \frac{1}{2}\left(\nabla_j P\right)\left(\nabla_i\,\delta q^j\right).
\end{aligned} \tag{8.67}$$

Substitution into the Euler–Lagrange equations (8.65) yields

$$\begin{aligned}
&\rho\, g_{ij}\,\frac{\partial^2\left(\delta q^j\right)}{\partial t^2} - \nabla_i\left(\Gamma_1 P\,\nabla_j\,\delta q^j\right) - \left(\rho\,\nabla_i \Phi\right)\left(\nabla_j\,\delta q^j\right)\\
&+\rho\,\nabla_i\left(\delta q^j\,\nabla_j \Phi\right) + \rho\,\nabla_i \Phi\,' = 0, \qquad i = 1,2,3.
\end{aligned} \tag{8.68}$$

These Euler–Lagrange equations correspond to the perturbed equations of motion in Lagrangian form given by Eqs. (2.52).

Furthermore, boundary conditions (8.63) lead to

$$\frac{1}{2}\,\rho\left(\delta q^j\,\nabla_i \Phi\right) n_j + \left[\Gamma_1 P\left(\nabla_j\,\delta q^j\right) + \frac{1}{2}\left(\delta q^j \nabla_j P\right)\right] n_i = 0, \quad i = 1,2,3. \tag{8.69}$$

These conditions reduce to the usual boundary condition

$$\left[\Gamma_1 P\left(\nabla_j\,\delta q^j\right)\right]_{\mathrm{R}} = 0. \tag{8.70}$$

In the particular case of a purely radial displacement field, the Lagrangian density $\mathscr{L}_2^{(C)}$ can be simplified by the use of the expressions for $\delta_2 U$ and $\delta_2 \omega$ given by equalities (8.33) and (8.34), and the condition of hydrostatic equilibrium. It results that

$$\mathscr{L}_2^{(C)} = \frac{1}{2}\left\{\rho\left[\frac{\partial(\delta r)}{\partial t}\right]^2 - \Gamma_1 P\left[\frac{1}{r^2}\,\frac{\partial}{\partial r}\left(r^2\,\delta r\right)\right]^2 + 4\,\rho\,\frac{G\,m(r)}{r^3}\,(\delta r)^2\right\}. \tag{8.71}$$

The Lagrangian density $\mathscr{L}_2^{(C)}$ expressed in this form corresponds to a Lagrangian density given by Ledoux & Walraven (1958). No more than the Lagrangian density $\mathscr{L}_2^{(C)}$, the Lagrangian density of Ledoux & Walraven corresponds to the difference between the local variation of the kinetic energy and the sum of the local variations of the internal and the gravitational potential energy.

8.4.3 *Takata's Reformulation of the Second-Order Variational Principle*

In the second-order variational principle

$$\delta \int_{t_1}^{t_2} L_2^{(C)}\, dt = 0, \tag{8.72}$$

the variations of the components of the Lagrangian displacement and their derivatives are taken. This is particularly the case for the term

$$-\int_{t_1}^{t_2} dt \int_V \frac{1}{2}\, \rho\, \delta q^i\, \nabla_i \Phi'\, dV,$$

whose variation is determined by equality (8.61). Takata (2006a) observed that one obtains the same result by considering the term

$$\int_{t_1}^{t_2} dt \int_V \frac{1}{2} \left[-\left(\frac{g^{ij}\, (\nabla_i \Phi')\, (\nabla_j \Phi')}{4\pi G} + 2\, \rho\, \delta q^i\, \nabla_i \Phi' \right) \right] dV$$

and taking independently the variations of the Eulerian perturbation of the gravitational potential and the components of its gradient, besides the variations of the components of the Lagrangian displacement and their derivatives. When $\delta_{(\xi,\,\Phi')}$ denotes the operator for taking the variations of the Lagrangian displacement and the Eulerian perturbation of the gravitational potential independently, one has

$$\begin{aligned} \delta_{(\xi,\,\Phi')} & \int_{t_1}^{t_2} dt \int_V \frac{1}{2} \left[-\left(\frac{g^{ij}\, (\nabla_i \Phi')(\nabla_i \Phi')}{4\pi G} + 2\, \rho\, \delta q^i\, \nabla_i \Phi' \right) \right] dV \\ &= -\int_{t_1}^{t_2} dt \int_V \rho\, \delta q^i\, \nabla_i \Phi'\, dV - \int_{t_1}^{t_2} dt \int_V \left[\nabla_i \left(\frac{g^{ij}\, \nabla_j \Phi'}{4\pi G} + \rho\, \delta q^i \right) \right] \delta\Phi'\, dV. \end{aligned} \tag{8.73}$$

The first term of the right-hand member stems from the variation of the Lagrangian displacement and corresponds to the right-hand member of equality (8.61). The second term of the right-hand member stems from the variation of the Eulerian perturbation of the gravitational potential. It is identically zero, since

$$\nabla_i \left(\frac{g^{ij} \, \nabla_j \Phi'}{4\pi G} + \rho \, \delta q^i \right) = 0. \tag{8.74}$$

This equation follows from a combination of Poisson's perturbed differential equation (2.48) and the linearised continuity equation (2.33).

Takata then reformulated the second-order variational principle as

$$\begin{aligned}
&\delta_{(\xi,\,\Phi')} \int_{t_1}^{t_2} L_2^{(TAK)} \, dt \\
&\equiv \delta_{(\xi,\,\Phi')} \int_{t_1}^{t_2} dt \int_V \frac{1}{2} \left[\rho \, g_{ij} \, \frac{\partial \left(\delta q^i\right)}{\partial t} \, \frac{\partial \left(\delta q^j\right)}{\partial t} - \Gamma_1 \, P \left(\nabla_i \, \delta q^i\right) \left(\nabla_j \, \delta q^j\right) \right. \\
&\qquad - \left(\delta q^j \, \nabla_j \, P\right) \left(\nabla_i \, \delta q^i\right) - \rho \, \delta q^i \, \nabla_i \left(\delta q^j \, \nabla_j \Phi\right) \\
&\qquad \left. - \frac{g^{ij} \left(\nabla_i \Phi'\right) \left(\nabla_j \Phi'\right)}{4\pi G} - 2 \, \rho \, \delta q^i \, \nabla_i \Phi' \right] dV = 0.
\end{aligned} \tag{8.75}$$

The functional $L_2^{(TAK)}$ contains only the Lagrangian displacement and the Eulerian perturbation of the gravitational potential, and their derivatives, so that its form agrees with that of the Lagrangians considered in analytical mechanics.

8.4.4 The Lagrangian Density of Tolstoy

From the Lagrangian $L_2^{(C)}$, one passes on to another Lagrangian by transforming the expressions for $\delta_2 U$ and $\delta_2 \omega$.

First, the fourth term in the right-hand member of equality (8.19) can be transformed as

$$- \delta q^i \left(\nabla_j \, P\right) \left(\nabla_i \, \delta q^j\right) = -\nabla_i \left[\left(\delta q^j \, \nabla_j \, P\right) \delta q^i \right] + \delta q^i \, \nabla_i \left(\delta q^j \, \nabla_j \, P\right), \tag{8.76}$$

so that

$$\begin{aligned}
\delta_2 U = \frac{1}{2\,\rho} \Big\{ &-\nabla_j \left[P \left(\nabla_i \, \delta q^i\right) \delta q^j \right] + \nabla_j \left[P \, \delta q^i \left(\nabla_i \, \delta q^j\right) \right] \\
&- \nabla_i \left[\left(\delta q^j \, \nabla_j \, P\right) \delta q^i \right] + \Gamma_1 \, P \left(\nabla_i \, \delta q^i\right) \left(\nabla_j \, \delta q^j\right) \\
&+ \left(\delta q^j \, \nabla_j \, P\right) \left(\nabla_i \, \delta q^i\right) + \delta q^j \, \nabla_i \left(\delta q^i \, \nabla_j \, P\right) \Big\}.
\end{aligned} \tag{8.77}$$

Next, multiplication of equality (8.31) by ρ and rearrangement of the first two terms in the right-hand member yield

$$\rho\,\delta_2\omega = \frac{1}{2}\left[\delta q^j\,\nabla_i\left(\rho\,\delta q^i\,\nabla_j\Phi\right) - \left(\delta q^j\,\nabla_j\Phi\right)\nabla_i\left(\rho\,\delta q^i\right) + \rho\,\delta q^i\,\nabla_i\Phi'\right]. \quad (8.78)$$

When condition (8.53) of the hydrostatic equilibrium and boundary condition (8.54) on the surface of the equilibrium star are satisfied, the Lagrangian becomes

$$L_2^{(C)} = \int_V \frac{1}{2}\left\{\rho\,g_{ij}\,\frac{\partial\left(\delta q^i\right)}{\partial t}\,\frac{\partial\left(\delta q^j\right)}{\partial t} + \nabla_i\left[\left(\delta q^j\,\nabla_j P\right)\delta q^i\right]\right.$$
$$- \Gamma_1 P\left(\nabla_i\,\delta q^i\right)\left(\nabla_j\,\delta q^j\right) + 2\,\rho\left(\delta q^j\,\nabla_j\Phi\right)\left(\nabla_i\,\delta q^i\right)$$
$$\left. + \left(\delta q^j\,\nabla_j\Phi\right)\left(\delta q^i\,\nabla_i\rho\right) - \rho\,\delta q^i\,\nabla_i\Phi'\right\} dV. \quad (8.79)$$

When moreover ∇P or, equivalently, ρ vanishes on the surface of the equilibrium star, the second term in the right-hand member does not contribute to the Lagrangian, and the Lagrangian reduces to

$$L_2^{(T)} = \int_V \frac{1}{2}\left\{\rho\,g_{ij}\,\frac{\partial\left(\delta q^i\right)}{\partial t}\,\frac{\partial\left(\delta q^j\right)}{\partial t} - \Gamma_1 P\left(\nabla_i\,\delta q^i\right)\left(\nabla_j\,\delta q^j\right)\right.$$
$$\left. + 2\,\rho\left(\delta q^j\,\nabla_j\Phi\right)\left(\nabla_i\,\delta q^i\right) + \left(\delta q^j\,\nabla_j\Phi\right)\left(\delta q^i\,\nabla_i\rho\right) - \rho\,\delta q^i\,\nabla_i\Phi'\right\} dV. \quad (8.80)$$

The integrand of this Lagrangian, $\mathscr{L}_2^{(T)}$, corresponds to a Lagrangian density given by Tolstoy, apart from the last term, which does not appear in it (Tolstoy 1963, 1973). The reason for this difference is that the last term involves the gradient of the Eulerian perturbation of the gravitational potential, which is generally neglected in the context of atmospheric studies, as considered by Tolstoy.

The Lagrangian $L_2^{(T)}$ is equal to the difference between the variation of the kinetic energy and the sum of the variations of the internal and the gravitational potential energy in a perturbed star for which both the pressure and its gradient vanish on the surface of the equilibrium star. On the contrary, the Lagrangian density $\mathscr{L}_2^{(T)}$ does *not* correspond to the difference between the local variations of the energies. It is related to the Lagrangian density $\mathscr{L}_2^{(C)}$ as

$$\mathscr{L}_2^{(T)} = \mathscr{L}_2^{(C)} - \frac{1}{2}\,\nabla_i\left[\left(\delta q^j\,\nabla_j P\right)\delta q^i\right] \quad (8.81)$$

and can be decomposed as

$$\mathscr{L}_2^{(T)} = \mathscr{L}_2^{(T,1)} - \frac{1}{2}\,\rho\,\delta q^i\,\nabla_i\Phi'. \quad (8.82)$$

When the Lagrangian density $\mathscr{L}_2^{(C,1)}$ is replaced by the Lagrangian density $\mathscr{L}_2^{(T,1)}$, the Euler–Lagrange equations (8.62) still yield the perturbed equations of motion. The boundary conditions now lead to the condition

$$\left[\Gamma_1 P \left(\nabla_j \, \delta q^j\right) + \delta q^j \, \nabla_j P\right](R) = 0 \tag{8.83}$$

or, equivalently, to the condition

$$P'(R) = 0. \tag{8.84}$$

This is the condition to which the usual condition (8.70) reduces when ∇P is assumed to vanish on the surface of the equilibrium star.

8.5 Approximation Method of Rayleigh–Ritz

Successive approximations of eigenfrequencies and eigenfunctions of isentropic spheroidal normal modes in stars can be obtained by application of the approximation method of Rayleigh (1877) and Ritz (1909) to the Lagrangian $L_2^{(C)}$.

8.5.1 Convenient Form of the Lagrangian

For this purpose, it is desirable to express the Lagrangian $L_2^{(C)}$ in terms of products of spheroidal displacement fields that are associated with complex conjugate spherical harmonics. In order to derive the appropriate form of the Lagrangian, one can multiply the Euler–Lagrange equations (8.68) by $\overline{\delta q^i}$ and add them up. For normal spheroidal displacement fields depending on time by a factor $\exp(i\sigma t)$, it follows that

$$\begin{aligned} &\sigma^2 \rho \, g_{ij} \, \delta q^j \, \overline{\delta q^i} + \overline{\delta q^i} \, \nabla_i \left(\Gamma_1 P \, \nabla_j \, \delta q^j\right) - \left(\overline{\delta q^i} \, \nabla_i P\right) \left(\nabla_j \, \delta q^j\right) \\ &\quad - \rho \, \overline{\delta q^i} \, \nabla_i \left(\delta q^j \, \nabla_j \Phi\right) - \rho \, \overline{\delta q^i} \, \nabla_i \Phi' = 0. \end{aligned} \tag{8.85}$$

One obtains the convenient form of the Lagrangian after multiplication by a factor $1/2$, integration over the volume of the equilibrium star, and a partial integration:

$$\begin{aligned} L_2^{(C)} = \int_V \frac{1}{2} \Big[&\sigma^2 \rho \, g_{ij} \, \delta q^j \, \overline{\delta q^i} - \Gamma_1 P \left(\nabla_j \, \delta q^j\right) \left(\nabla_i \, \overline{\delta q^i}\right) \\ &- \left(\overline{\delta q^i} \, \nabla_i P\right) \left(\nabla_j \, \delta q^j\right) - \rho \, \overline{\delta q^i} \, \nabla_i \left(\delta q^j \, \nabla_j \Phi\right) - \rho \, \overline{\delta q^i} \, \nabla_i \Phi' \Big] \, dV. \end{aligned} \tag{8.86}$$

The Lagrangian is manifestly equal to zero when the eigenfrequency and the eigenfunctions of any normal mode are substituted into it.

The integrations over the angular variables θ and ϕ are carried out by use of the dependencies of the components of the spheroidal displacement fields on these variables that are given by definition (5.99).

For the first term, one has

$$\int_V \rho\, g_{ij}\, \delta q^j\, \overline{\delta q^i}\, dV = \int_V \rho(r) \Bigg\{ \xi^2(r)\, Y_\ell^m(\theta,\phi)\, \overline{Y_\ell^m}(\theta,\phi) + \frac{\eta^2(r)}{r^2} \left[\frac{\partial Y_\ell^m(\theta,\phi)}{\partial\theta}\, \frac{\partial \overline{Y_\ell^m}(\theta,\phi)}{\partial\theta} + \frac{1}{\sin^2\theta}\, \frac{\partial Y_\ell^m(\theta,\phi)}{\partial\phi}\, \frac{\partial \overline{Y_\ell^m}(\theta,\phi)}{\partial\phi} \right] \Bigg\}\, dV. \quad (8.87)$$

Use of properties (Appendix D.4) and (Appendix D.6) of the spherical harmonics yields

$$\int_V \rho\, g_{ij}\, \delta q^j\, \overline{\delta q^i}\, dV = N_{\ell m} \int_0^R \rho \left[\xi^2 + \frac{\ell(\ell+1)}{r^2}\, \eta^2 \right] r^2\, dr \equiv N_{\ell m}\, I_1. \quad (8.88)$$

By means of equality (5.93), one derives, for the second term,

$$-\int_V \Gamma_1 P \left(\nabla_j\, \delta q^j\right) \left(\nabla_i\, \overline{\delta q^i}\right) dV = -N_{\ell m} \int_0^R \Gamma_1 P \left[\frac{1}{r^2} \frac{d}{dr} \left(r^2\, \xi\right) - \frac{\ell(\ell+1)}{r^2}\, \eta \right]^2 r^2\, dr \equiv -N_{\ell m}\, I_2 \quad (8.89)$$

and, for the third term,

$$-\int_V \left(\overline{\delta q^i}\, \nabla_i P\right) \left(\nabla_j\, \delta q^j\right) dV = -N_{\ell m} \int_0^R \frac{dP}{dr}\, \xi \left[\frac{1}{r^2} \frac{d}{dr} \left(r^2\, \xi\right) - \frac{\ell(\ell+1)}{r^2}\, \eta \right] r^2\, dr \equiv -N_{\ell m}\, I_3. \quad (8.90)$$

The fourth term can be developed as

$$-\int_V \rho \left[\overline{\delta q^i}\, \nabla_i \left(\delta q^j\, \nabla_j \Phi\right) \right] dV = -N_{\ell m} \int_0^R \rho\, \xi \left[\frac{d}{dr} \left(\frac{d\Phi}{dr}\, \xi \right) + \frac{\ell(\ell+1)}{r^2}\, \frac{d\Phi}{dr}\, \eta \right] r^2\, dr \equiv -N_{\ell m}\, I_4. \quad (8.91)$$

For the last term, one has

$$-\int_V \rho\, \overline{\delta q^i}\, (\nabla_i \Phi')\, dV$$
$$= -N_{\ell m} \int_0^R \rho \left[\xi\, \frac{d\Phi'}{dr} + \frac{\ell(\ell+1)}{r^2}\, \eta\, \Phi' \right] r^2\, dr \equiv -N_{\ell m}\, I_5. \qquad (8.92)$$

The Lagrangian can then be expressed as

$$L_2^{(C)} = \left(\sigma^2 I_1 - I_2 - I_3 - I_4 - I_5\right) N_{\ell m} \qquad (8.93)$$

and is composed of integrals over the radial coordinate r extending from the boundary point $r = 0$ to the boundary point $r = R$. This form of the Lagrangian is convenient for the application of the method of Rayleigh–Ritz.

8.5.2 The Approximation Method

The approximation method of Rayleigh–Ritz makes it possible to derive consecutive approximations of eigenvalues and eigenfunctions by successive substitutions of more and more appropriate trial functions into a Lagrangian and the determination of the constants involved in these trial functions in such a way that the Lagrangian reaches an extremum (Courant & Hilbert 1968). The trial functions must satisfy the boundary conditions.

The method was applied to the determination of spheroidal non-radial modes in the equilibrium sphere of uniform mass density by Chandrasekhar (1964) and Chandrasekhar & Lebovitz (1964), in polytropic models by Robe & Brandt (1966), in an inhomogeneous gaseous mass consisting, under certain circumstances, of a convective core and a radiative envelope by Tassoul (1967), and in massive stars by Andrew (1968). It was also applied to the determination of higher-order spheroidal non-radial modes by Andrew (1967). The procedure is as follows. In Sect. 6.4.1, it is shown that the radial component of the Lagrangian displacement, $\xi(r)$, behaves as $\ell\, r^{\ell-1}$ as r $\to 0$, and the horizontal component, $\eta(r)$, as r^{ℓ}. Therefore, the trial functions

$$\xi(r) = C_1\, \ell\, r^{\ell-1}, \quad \eta(r) = C_1\, r^{\ell}, \qquad (8.94)$$

where C_1 is a general constant, can be substituted into the Lagrangian $L_2^{(C)}$. They define a divergence-free displacement field. The Eulerian perturbation of the gravitational potential, $\Phi'(r)$, and its first derivative, $d\Phi'/dr$, are determined by means of solutions (7.4) and (7.5) of Poisson's perturbed equation. The Lagrangian is then a homogeneous function of C_1^2. The requirement that the Lagrangian must be equal to zero leads to a first approximation of σ^2.

As second trial functions, one can use

$$\left.\begin{aligned} \xi(r) &= C_1\,\ell\,r^{\ell-1} + C_2\,r^{\ell+1}, \\ \eta(r) &= C_1\,r^{\ell} + C_3\,r^{\ell+2}, \end{aligned}\right\} \tag{8.95}$$

where the arbitrary constants C_2 and C_3 generally differ from each other. In order that the Lagrangian reaches an extremum, one imposes that

$$\frac{\partial L_2^{(C)}}{\partial C_1} = 0, \qquad \frac{\partial L_2^{(C)}}{\partial C_2} = 0, \qquad \frac{\partial L_2^{(C)}}{\partial C_3} = 0. \tag{8.96}$$

These conditions lead to a linear, homogeneous system of equations for the constants C_1, C_2, C_3. The necessary and sufficient condition for the existence of a non-trivial solution is that the determinant of the matrix of the coefficients be equal to zero. By means of this equation, a second approximation of σ^2 can be determined (Robe & Brandt 1966).

In illustration, we consider the particular case of the equilibrium sphere of uniform mass density. We restrict ourselves to the use of the first trial functions for $\xi(r)$ and $\eta(r)$.

For an equilibrium sphere with a uniform mass density ρ, the mass contained inside the sphere with radius r is given by

$$m(r) = \frac{4\pi r^3}{3}\,\rho, \tag{8.97}$$

so that substitution into equality (2.20) yields

$$g \equiv \frac{d\Phi}{dr} = \frac{4\pi G\rho}{3}\,r. \tag{8.98}$$

Since the displacement field defined by the trial functions is divergence-free, one has that

$$I_2 = 0, \qquad I_3 = 0. \tag{8.99}$$

From solution (7.4) of Poisson's perturbed differential equation, it results that

$$\Phi'(r) = -4\pi G\rho\, C_1\,\frac{\ell}{2\ell+1}\,r^{\ell}. \tag{8.100}$$

One then has

$$\left.\begin{aligned} I_1 &= C_1^2\,\rho\,\ell\,R^{2\ell+1}, \\ I_4 &= C_1^2\,\frac{4\pi G\rho^2}{3}\,\ell^2\,R^{2\ell+1}, \\ I_5 &= -C_1^2\,4\pi G\rho^2\,\frac{\ell^2}{2\ell+1}\,R^{2\ell+1}. \end{aligned}\right\} \tag{8.101}$$

The Lagrangian $L_2^{(C)}$ is equal to zero, and thus minimal, for each finite value of the constant C_1 that is different from zero, when

$$\sigma^2 = \frac{4\pi G\rho}{3} \frac{2\ell(\ell-1)}{2\ell+1}. \tag{8.102}$$

Taking into account that $\rho = \bar{\rho} = 3M/(4\pi R^3)$, one sees that the first approximations of the eigenvalues correspond to the squares of the eigenfrequencies of the Kelvin modes of the *incompressible* equilibrium sphere of uniform mass density, as given by Eq. (5.117) [see also Eq. (1)]. The cause of the agreement is that the trial functions used are here the exact eigenfunctions.

Since the term I_2 is the only term of the Lagrangian $L_2^{(C)}$ in which the compressibility of the medium is involved, and the term is equal to zero for divergence-free modes, the Kelvin modes must also be eigenmodes of the *compressible* equilibrium sphere of uniform mass density. This property was first noticed by Chandrasekhar (1964).

8.6 Weight Functions for Spheroidal Normal Modes

As noted above, the Lagrangian $L_2^{(C)}$ is equal to zero for any normal mode that satisfies the Euler–Lagrange equations. In particular for spheroidal normal modes, Eq. (8.93) can be transformed into an equation for the eigenvalue σ^2:

$$\sigma^2 = \frac{1}{I_1}\left(I_2 + I_3 + I_4 + I_5\right). \tag{8.103}$$

When the eigenfunctions are normalised such that $I_1 = 1$, it follows that

$$\begin{aligned}\sigma^2 = \int_0^R \Biggl\{ & c^2 \left[\frac{1}{r^2}\frac{d}{dr}\left(r^2\,\xi\right) - \frac{\ell(\ell+1)}{r^2}\,\eta\right]^2 + \left(4\pi G\rho - 4\,\frac{g}{r}\right)\xi^2 \\ & + 2\,\frac{\ell(\ell+1)}{r^2}\,g\,\xi\,\eta + \xi\,\frac{d\Phi'}{dr} + \frac{\ell(\ell+1)}{r^2}\,\eta\,\Phi' \Biggr\}\,\rho\, r^2\,dr\end{aligned} \tag{8.104}$$

The integrand can be regarded as a weight function whose variation gives an idea of the contribution of the various stellar layers to the eigenvalue of a spheroidal normal mode. The weight function allows one so to examine in which regions of a star various spheroidal normal modes are mainly formed.

Weight functions were used by Epstein (1950) and Pesnell (1987) for radial normal modes, and by Goossens & Smeyers (1974), Schwank (1976), and Van Hoolst & Smeyers (1991, see also Sect. 15.1), for spheroidal non-radial normal modes.

8.7 Energy Density and Energy Flux

From the Lagrangian L_k of order k, a Hamiltonian H_k can be derived by use of a Legendre transformation by which the roles of the velocity components $\partial\left(\delta q^i\right)/\partial t$ and their canonically conjugate components

$$p_i = \frac{\partial L_k}{\partial\left[\partial\left(\delta q^i\right)/\partial t\right]} \tag{8.105}$$

are interchanged (see, e.g., Goldstein 1950). The Hamiltonian of order k is then

$$H_k = -\left[L_k - \frac{\partial\left(\delta q^i\right)}{\partial t}\,\frac{\partial L_k}{\partial\left[\partial\left(\delta q^i\right)/\partial t\right]}\right]. \tag{8.106}$$

This functional corresponds to the star's energy of order k. Correspondingly, the energy density of the same order is defined as

$$\mathscr{H}_k = -\left[\mathscr{L}_k - \frac{\partial\left(\delta q^i\right)}{\partial t}\,\frac{\partial \mathscr{L}_k}{\partial\left[\partial\left(\delta q^i\right)/\partial t\right]}\right]. \tag{8.107}$$

One derives the first-order energy density by using the Lagrangian density $\mathscr{L}_1$, which is the integrand of the Lagrangian L_1 given by definition (8.50). By taking into account the condition of hydrostatic equilibrium, one has

$$\mathscr{H}_1 = -\nabla_j\left(P\,\delta q^j\right). \tag{8.108}$$

Similarly, one derives the second-order energy density by using the Lagrangian density $\mathscr{L}_2$, which is the integrand of the Lagrangian L_2 given by definition (8.55). The second-order energy density is then

$$\begin{aligned}\mathscr{H}_2 = \frac{1}{2}\Bigg\{&\rho\, g_{ij}\,\frac{\partial\left(\delta q^i\right)}{\partial t}\,\frac{\partial\left(\delta q^j\right)}{\partial t} - \nabla_j\left[P\left(\nabla_i\,\delta q^i\right)\delta q^j\right]\\ &+\nabla_j\left[P\,\delta q^i\left(\nabla_i\,\delta q^j\right)\right] + \Gamma_1 P\left(\nabla_i\,\delta q^i\right)\left(\nabla_j\,\delta q^j\right)\\ &+\left(\delta q^j\,\nabla_j P\right)\left(\nabla_i\,\delta q^i\right) + \rho\,\delta q^i\,\nabla_i\left(\delta q^j\,\nabla_j \Phi\right) + \rho\,\delta q^i\,\nabla_i \Phi'\Bigg\}.\end{aligned} \tag{8.109}$$

This energy density can be expressed in a different form. The fourth and the fifth term can be transformed by means of relation (1.30) between the Lagrangian and the Eulerian perturbation of a physical quantity and Eqs. (1.59) and (4.1). One then has

$$\Gamma_1 P\,\nabla_j \delta q^j + \delta q^j\,\nabla_j P = -P' \tag{8.110}$$

and

$$\nabla_i \, \delta q^i = -\frac{1}{\rho \, c^2} \left(P' + \frac{dP}{dr} \, \delta r \right), \tag{8.111}$$

so that

$$\left(\Gamma_1 \, P \, \nabla_j \delta q^j + \delta q^j \, \nabla_j \, P \right) \left(\nabla_i \, \delta q^i \right) = \frac{P'^2}{\rho \, c^2} + \frac{1}{\rho \, c^2} \frac{dP}{dr} \, P' \, \delta r. \tag{8.112}$$

The second term in the right-hand member of this equality can be transformed by means of Eq. (4.1) and definition (3.32) of N^2. It results

$$\left(\Gamma_1 \, P \, \nabla_j \delta q^j + \delta q^j \, \nabla_j \, P \right) \left(\nabla_i \, \delta q^i \right) = \frac{P'^2}{\rho \, c^2} - g \, \rho' \, \delta r + \rho \, N^2 \, (\delta r)^2 \, . \tag{8.113}$$

The second-order energy density is then given by

$$\begin{aligned} \mathscr{H}_2 = \frac{1}{2} \Bigg\{ & \rho \, g_{ij} \, \frac{\partial \left(\delta q^i \right)}{\partial t} \, \frac{\partial \left(\delta q^j \right)}{\partial t} - \nabla_j \left[P \left(\nabla_i \, \delta q^i \right) \delta q^j \right] \\ & + \nabla_j \left[P \, \delta q^i \left(\nabla_i \, \delta q^j \right) \right] + \frac{P'^2}{\rho \, c^2} - g \, \rho' \, \delta r + \rho \, N^2 \, (\delta r)^2 \\ & + \rho \, \delta q^i \, \nabla_i \left(\delta q^j \, \nabla_j \Phi \right) + \rho \, \delta q^i \, \nabla_i \Phi' \Bigg\} \, . \end{aligned} \tag{8.114}$$

Partial differentiation of the energy density with respect to time leads to an equation for the rate of change of this quantity and makes it possible to define the local energy flux.

The partial differentiation of equality (8.108) with respect to time yields

$$\frac{\partial \mathscr{H}_1}{\partial t} + \nabla_j \left[P \, \frac{\partial \left(\delta q^j \right)}{\partial t} \right] = 0. \tag{8.115}$$

The components of the first-order energy flux are then defined as

$$\mathscr{F}_1^j = P \, \frac{\partial \left(\delta q^j \right)}{\partial t}, \qquad j = 1, 2, 3. \tag{8.116}$$

Before differentiating partially the second-order energy density with respect to time, we make use of relation (8.57), so that this energy density is given by

$$\begin{aligned} \mathscr{H}_2 = & -\mathscr{L}_2^{(C)} - \frac{1}{2} \nabla_j \left[P \left(\nabla_i \, \delta q^i \right) \delta q^j \right] \\ & + \frac{1}{2} \nabla_j \left[P \, \delta q^i \left(\nabla_i \, \delta q^j \right) \right] + \frac{\partial \left(\delta q^i \right)}{\partial t} \, \frac{\partial \mathscr{L}_2^{(C)}}{\partial \left[\partial \left(\delta q^i \right) / \partial t \right]} . \end{aligned} \tag{8.117}$$

The partial differentiation with respect to time then yields

$$\frac{\partial \mathscr{H}_2}{\partial t} = -\frac{\partial \mathscr{L}_2^{(C)}}{\partial t} - \frac{1}{2} \nabla_j \frac{\partial}{\partial t} \left[P \left(\nabla_i \delta q^i \right) \delta q^j \right] + \frac{1}{2} \nabla_j \frac{\partial}{\partial t} \left[P \, \delta q^i \left(\nabla_i \delta q^j \right) \right]$$
$$+ \frac{\partial^2 \left(\delta q^i \right)}{\partial t^2} \frac{\partial \mathscr{L}_2^{(C)}}{\partial \left[\partial \left(\delta q^i \right) / \partial t \right]} + \frac{\partial \left(\delta q^i \right)}{\partial t} \frac{\partial}{\partial t} \frac{\partial \mathscr{L}_2^{(C)}}{\partial \left[\partial \left(\delta q^i \right) / \partial t \right]}. \tag{8.118}$$

The last term of the right-hand member can be eliminated by means of the Euler–Lagrange equations (8.62). It follows

$$\frac{\partial \mathscr{H}_2}{\partial t} = -\nabla_j \left\{ \frac{1}{2} \frac{\partial}{\partial t} \left[P \left(\nabla_i \delta q^i \right) \delta q^j \right] \right\} + \nabla_j \left\{ \frac{1}{2} \frac{\partial}{\partial t} \left[P \, \delta q^i \left(\nabla_i \delta q^j \right) \right] \right\}$$
$$+ \nabla_i \left[\frac{1}{2} \left(\delta q^j \, \nabla_j P \right) \frac{\partial \left(\delta q^i \right)}{\partial t} \right] + \nabla_i \left[\Gamma_1 P \left(\nabla_j \delta q^j \right) \frac{\partial \left(\delta q^i \right)}{\partial t} \right]$$
$$+ \frac{1}{2} \rho \frac{\partial \left(\delta q^i \right)}{\partial t} \nabla_i \left(\delta q^j \, \nabla_j \Phi \right) + \frac{1}{2} \rho \, \delta q^i \, \nabla_i \left(\frac{\partial \left(\delta q^j \right)}{\partial t} \nabla_j \Phi \right)$$
$$- \frac{\partial \left(\delta q^i \right)}{\partial t} \left[\nabla_j \left(-\frac{1}{2} \rho \, \delta q^j \, \nabla_i \Phi \right) + \frac{1}{2} \rho \left(\nabla_i \delta q^j \right) \left(\nabla_j \Phi \right) + \rho \, \delta q^j \, \nabla_i \nabla_j \Phi \right]$$
$$- \frac{1}{2} \rho \frac{\partial \left(\delta q^j \right)}{\partial t} \nabla_j \Phi' + \frac{1}{2} \rho \, \delta q^i \, \nabla_i \frac{\partial \Phi'}{\partial t}. \tag{8.119}$$

By transformation of the sixth term in the right-hand member and use of the condition of hydrostatic equilibrium, one obtains the equation

$$\frac{\partial \mathscr{H}_2}{\partial t} + \nabla_j \mathscr{F}_2^j = -\frac{1}{2} \left[\rho \frac{\partial \left(\delta q^i \right)}{\partial t} \nabla_i \Phi' - \rho \, \delta q^i \, \nabla_i \frac{\partial \Phi'}{\partial t} \right], \tag{8.120}$$

where the $\mathscr{F}_2^j$ are the components of the second-order energy flux defined as

$$\mathscr{F}_2^j = \frac{1}{2} \frac{\partial}{\partial t} \left[P \left(\nabla_i \delta q^i \right) \delta q^j \right] - \frac{1}{2} \frac{\partial}{\partial t} \left[P \, \delta q^i (\nabla_i \delta q^j) \right]$$
$$- \frac{1}{2} \left(\delta q^i \nabla_i P \right) \frac{\partial (\delta q^j)}{\partial t} + \frac{1}{2} \left(\frac{\partial (\delta q^i)}{\partial t} \nabla_i P \right) \delta q^j$$
$$- \Gamma_1 P \left(\nabla_i \delta q^i \right) \frac{\partial (\delta q^j)}{\partial t}, \qquad j = 1, 2, 3. \tag{8.121}$$

The sum of the terms in the right-hand member of Eq. (8.120) can be regarded as the amount of energy that is produced locally per unit time. This sum, as well as the sum of the third and the fourth term in the definition of the components of the second-order energy flux, is equal to zero for displacement fields depending harmonically on time by a factor $\cos(\sigma t)$ or $\sin(\sigma t)$ or depending exponentially on time by a factor $\exp(\pm \sigma t)$.

The components of the second-order energy flux vanish on the surface of the equilibrium star when boundary condition (8.70) is satisfied.

Over any spherical equipotential surface with radius r inside an equilibrium star, a mean second-order energy flux in the radial direction can be determined. For example, for displacement fields of the form

$$\left.\begin{aligned} \delta r &= \xi(r)\, P_\ell^m(\cos\theta)\, \cos(m\phi)\cos(\sigma t), \\ \delta\theta &= \frac{\eta(r)}{r^2}\, \frac{dP_\ell^m(\cos\theta)}{d\theta}\, \cos(m\phi)\, \cos(\sigma t), \\ \delta\phi &= -m\, \frac{\eta(r)}{r^2}\, \frac{P_\ell^m(\cos\theta)}{\sin^2\theta}\, \sin(m\phi)\, \cos(\sigma t), \end{aligned}\right\} \tag{8.122}$$

the mean radial second-order energy flux is given by

$$\begin{aligned} \overline{\mathscr{F}}_r(r) = -\frac{\sigma}{8\pi} &\left\{\int_{-1}^{1} \left[P_\ell^m(z)\right]^2 dz\right\} \left\{\int_0^{2\pi} \cos^2(m\phi)\, d\phi\right\} \\ &\left\{ P \left[\frac{2}{r}\,\xi^2 - \frac{\ell(\ell+1)}{r^2} \left(2\,\xi - \frac{\eta}{r}\right)\right] - \rho\, c^2\, \alpha\, \xi\right\} \sin(2\sigma t), \end{aligned} \tag{8.123}$$

where α is the divergence of the Lagrangian displacement determined by equality (5.93).

Some authors have adopted a wave energy density which corresponds to the energy density derived from Tolstoy's Lagrangian density $\mathscr{L}_2^{(T)}$. However, one should keep in mind that alternate forms of Lagrangian densities exist that are not defined as the difference between the local variation of the kinetic energy and the sum of the local variations of the internal and the gravitational potential energy but that nevertheless lead to the right Euler–Lagrange equations (see, e.g., Goldstein 1950, Tolstoy 1973). For such Lagrangian densities, the application of the Hamiltonian formalism does not lead to an energy density and an energy flux.

8.8 The Equations that Govern Linear, Isentropic Oscillations, as Canonical Equations

For the subsequent derivations, it is useful to express the Lagrangian $L_2^{(TAK)}$, defined by Eq. (8.75), in terms of spheroidal solutions that are associated with complex conjugate spherical harmonics. On account of Eq. (8.86) for the Lagrangian $L_2^{(C)}$ and the equality

$$2\,\rho\,\delta q^j\, \nabla_j \overline{\Phi'} = 2\,\rho\,\overline{\delta q^j}\, \nabla_j \Phi', \tag{8.124}$$

the Lagrangian $L_2^{(TAK)}$ can then be expressed as

$$L_2^{(TAK)} = \int_V \frac{1}{2}\Bigg[\sigma^2 \rho\, g_{ij}\, \delta q^j\, \overline{\delta q^i} - \Gamma_1 P\, (\nabla_j\, \delta q^j)\, \left(\nabla_i\, \overline{\delta q^i}\right)$$
$$- \left(\overline{\delta q^i}\, \nabla_i\, P\right) (\nabla_j\, \delta q^j) - \rho\, \overline{\delta q^i}\, \nabla_i\, (\delta q^j\, \nabla_j\, \Phi)$$
$$-g^{ij}\, \frac{(\nabla_i \Phi\,')\, (\nabla_j \overline{\Phi\,'})}{4\pi G} - 2\,\rho\, \delta q^j\, \nabla_j \overline{\Phi\,'}\,\Bigg] dV. \tag{8.125}$$

The first four integrals are already developed by means of Eqs. (8.88)–(8.91). The fifth and sixth integral can be developed as

$$-\int_V g^{ij}\, \frac{(\nabla_i \Phi\,')\, (\nabla_j \overline{\Phi\,'})}{4\pi G}\, dV = -\frac{N_{\ell m}}{4\pi G} \int_0^R \left[\left(\frac{d\Phi\,'}{dr}\right)^2 + \frac{\ell(\ell+1)}{r^2}\, \Phi\,'^2\right] r^2\, dr, \tag{8.126}$$

$$-\int_V 2\,\rho\, \delta q^j\, \nabla_j \overline{\Phi\,'}\, dV = -N_{\ell m} \int_0^R 2\,\rho \left[\xi\, \frac{d\Phi'}{dr} + \frac{\ell(\ell+1)}{r^2}\, \eta\, \Phi'\right] r^2\, dr. \tag{8.127}$$

Defining the Lagrangian density $\mathscr{L}_2^{(TAK)}$ as

$$L_2^{(TAK)} = \int_V \mathscr{L}_2^{(TAK)}\, dV, \tag{8.128}$$

one has

$$\mathscr{L}_2^{(TAK)} = \frac{1}{2}\Bigg\{\sigma^2 \rho \left[\xi^2 + \frac{\ell(\ell+1)}{r^2}\, \eta^2\right] - \Gamma_1 P \left[\frac{d\xi}{dr} + \frac{2}{r}\, \xi - \frac{\ell(\ell+1)}{r^2}\, \eta\right]^2$$
$$-\rho\, g\, \xi \left[\left(\frac{1}{g}\frac{dg}{dr} - \frac{2}{r}\right) \xi + 2\, \frac{\ell(\ell+1)}{r^2}\, \eta\right] - 2\,\rho \left[\xi\, \frac{d\Phi'}{dr} + \frac{\ell(\ell+1)}{r^2}\, \eta\, \Phi'\right]$$
$$-\frac{1}{4\pi G} \left[\left(\frac{d\Phi'}{dr}\right)^2 + \frac{\ell(\ell+1)}{r^2}\, \Phi\,'^2\right]\Bigg\}. \tag{8.129}$$

The Lagrangian density depends on the functions ξ, η, Φ' and the first derivatives $d\xi/dr$ and $d\Phi'/dr$, but not on the first derivative $d\eta/dr$. In order to use the analogy with mechanics, Takata replaced the function η by the solution of the variational problem. Since

$$\frac{\partial \mathscr{L}_2^{(TAK)}}{\partial \eta} = \rho c^2 \frac{\ell(\ell+1)}{r^2} \left\{ \left[\frac{\sigma^2}{c^2} - \frac{\ell(\ell+1)}{r^2} \right] \eta + \left[\frac{d\xi}{dr} + \left(\frac{2}{r} - \frac{g}{c^2} \right) \xi - \frac{\Phi'}{c^2} \right] \right\}, \tag{8.130}$$

the requirement that $\mathscr{L}_2^{(TAK)}$ is stationary with respect to arbitrary variations of η yields

$$\eta = - \left[\frac{\sigma^2}{c^2} - \frac{\ell(\ell+1)}{r^2} \right]^{-1} \left[\frac{d\xi}{dr} + \left(\frac{2}{r} - \frac{g}{c^2} \right) \xi - \frac{\Phi'}{c^2} \right]. \tag{8.131}$$

Substitution of the solution for η into Eq. (8.129) leads to a Lagrangian density $\mathscr{L}_2^{(TAK,1)}$ which only depends on the functions ξ and Φ', and their first derivatives $d\xi/dr$ and $d\Phi'/dr$.

By analogy with mechanics, Takata regarded the radius r as the independent variable, the functions ξ and Φ', as the coordinates of two particles, and the first derivatives $d\xi/dr$ and $d\Phi'/dr$, as the velocities of the particles. He determined the momentum variables, which are canonically conjugate to ξ and Φ', respectively, by

$$p_\xi = \frac{\partial \mathscr{L}_2^{(TAK,1)}}{\partial (d\xi/dr)}, \qquad p_{\Phi'} = \frac{\partial \mathscr{L}_2^{(TAK,1)}}{\partial \big(d\Phi'/dr\big)}. \tag{8.132}$$

The Hamiltonian density results from the Legendre transformation

$$\mathscr{H}^{(TAK)} = p_\xi \frac{d\xi}{dr} + p_{\Phi'} \frac{d\Phi'}{dr} - \mathscr{L}_2^{(TAK,1)} \tag{8.133}$$

[see also Eq. (8.107)]. The canonical equations are then given by

$$\left. \begin{aligned} \frac{d\xi}{dr} &= \frac{\partial \mathscr{H}^{(TAK)}}{\partial p_\xi}, \qquad & \frac{dp_\xi}{dr} &= -\frac{\partial \mathscr{H}^{(TAK)}}{\partial \xi}, \\ \frac{d\Phi'}{dr} &= \frac{\partial \mathscr{H}^{(TAK)}}{\partial p_{\Phi'}}, \qquad & \frac{dp_{\Phi'}}{dr} &= -\frac{\partial \mathscr{H}^{(TAK)}}{\partial \Phi'}, \end{aligned} \right\} \tag{8.134}$$

and form a fourth-order system of ordinary differential equations.

Chapter 9
Radial Propagation of Waves

9.1 Introduction

The spheroidal normal modes in stars are solutions of the eigenvalue problem defined by one of the fourth-order systems of linear, homogeneous differential equations that are presented in Chap. 6 and the associated regularity and boundary conditions. The eigenvalue problem is non-linear in the eigenvalue parameter σ^2. A general mathematical study of this eigenvalue problem has still to be made.

Understanding of the nature of various kinds of spheroidal normal modes can be acquired by a study of the possible wave propagation in the radial direction inside a star. This study starts from a *local* analysis of the solutions of the equations that govern linear, isentropic perturbations of a star in the Cowling approximation. Two local dispersion equations are derived: one valid in the whole star, and another one applying to the surface layers. On the basis of the first dispersion relation, the conditions for radial wave propagation in the stellar interior are analysed and illustrated by means of diagnostic diagrams. Next, the propagation diagram is introduced as a *global* representation of the radial wave propagation inside a star. Finally, the phase velocity and the group velocity of traveling waves in a star are considered.

9.2 Local Dispersion Equations

9.2.1 General Local Dispersion Equation

In the Cowling approximation, Eqs. (6.15) and (6.16) reduce to Eqs. (7.32) and (7.33). The coefficients of the equations are supposed to vary so slowly that they can be regarded as constant in the neigbourhood of the stellar layer considered. The supposition is not valid in the neighbourhood of the boundary points $r = 0$ and $r = R$, where some coefficients become indefinitely large in absolute value.

On the supposition made, solutions of Eqs. (7.32) and (7.33) can be sought of the form

$$u(r) = A \exp\left(-\mathrm{i}k_r r\right), \qquad y(r) = B \exp\left(-\mathrm{i}k_r r\right), \tag{9.1}$$

P. Smeyers and T. Van Hoolst, *Linear Isentropic Oscillations of Stars: Theoretical Foundations*, Astrophysics and Space Science Library 371,
DOI 10.1007/978-3-642-13030-4_10, © Springer-Verlag Berlin Heidelberg 2010

where A and B are yet undetermined constants, and k_r is the complex wavenumber represented as $k_r = k_r^{(r)} + \mathrm{i}k_r^{(i)}$. The requirement that a non-trivial solution exists for the constants A and B leads to the equation

$$\begin{aligned}\left(k_r + \frac{\mathrm{i}}{2}\frac{d\ln\rho}{dr}\right)^2 &= \frac{\sigma^2}{c^2} + \frac{\ell(\ell+1)}{r^2}\left(\frac{N^2}{\sigma^2} - 1\right) - \frac{1}{4}\left(\frac{d\ln\rho}{dr}\right)^2 \\ &\equiv F\left(\sigma^2, \ell, r\right)\end{aligned} \tag{9.2}$$

(Smeyers 1984).

Since the time has previously been separated from the equations by the introduction of a time dependence of the form $\exp(\mathrm{i}\sigma t)$ (see Sect. 4.3), the search for solutions of the form given by Eqs. (9.1) corresponds to a search for time-dependent solutions of the form

$$u(r,t) = A\exp\left[\mathrm{i}(\sigma t - k_r r)\right], \qquad y(r,t) = B\exp\left[\mathrm{i}(\sigma t - k_r r)\right]. \tag{9.3}$$

For a real value of the angular frequency σ, propagation of waves in the radial direction is then possible when the wavenumber k_r has a real part different from zero. The angular frequency is considered to be positive.

For the determination of the conditions under which the wavenumber k_r has a real part different from zero, it is convenient to concentrate on the right-hand member of Eq. (9.2). Two possibilities must be distinguished.

- Either $F\left(\sigma^2, \ell, r\right) > 0$. Then the left-hand member of the equation is positive, so that it can be regarded as the square of a real quantity. This implies that

$$k_r^{(i)} + \frac{1}{2}\frac{d\ln\rho}{dr} = 0 \tag{9.4}$$

  and $k_r^{(r)} \neq 0$.
- Or $F\left(\sigma^2, \ell, r\right) < 0$. Then the left-hand member of the equation is negative, so that it can be regarded as the square of a purely imaginary quantity. This implies that $k_r^{(r)} = 0$.

Hence, the wavenumber k_r has a real part when

$$F\left(\sigma^2, \ell, r\right) > 0. \tag{9.5}$$

When $k_r^{(r)} = 0$, and the exponential function $\exp\left(k_r^{(i)} r\right)$ is an exponentially decreasing function of r, the waves rapidly languish in the sense of the increasing values of r and are said to be evanescent.

In the circumstances in which radial propagation of waves is possible, Eq. (9.2) yields the local dispersion equation

$$\left(k_r^{(r)}\right)^2 = \frac{\sigma^2}{c^2} + \frac{\ell(\ell+1)}{r^2}\left(\frac{N^2}{\sigma^2} - 1\right) - \frac{1}{4}\left(\frac{d\ln\rho}{dr}\right)^2. \tag{9.6}$$

A similar dispersion equation was derived by Whitaker (1963) for a plane isothermal atmosphere in presence of a constant gravitational field and was adopted by Ulrich (1970) and Bahcall & Ulrich (1988) for spherically symmetric layers.

From the analogy of dispersion equation (9.6) with the dispersion equation valid for plane layers, it appears that the term $\sqrt{\ell(\ell+1)}/r$ plays the role of the local horizontal wavenumber of the perturbation in a spherically symmetric layer.

For the derivation of a dispersion equation for waves that propagate purely in the radial direction ($\ell = 0$), and that thus have a horizontal wavenumber equal to zero, the Eulerian perturbation of the gravitational potential does not need to be neglected. In this case, it is appropriate to return to Eqs. (6.21) and (6.22), and to eliminate $d\Phi'/dr$ by means of Eq. (6.25), so that

$$\frac{du}{dr} = \frac{g}{c^2}\,u - \frac{r^2}{c^2}\,y, \tag{9.7}$$

$$\frac{dy}{dr} = \left(\sigma^2 + 4\pi G\rho - N^2\right)\frac{u}{r^2} + \frac{N^2}{g}\,y. \tag{9.8}$$

Substitution of solutions (9.1) into these equations leads to

$$\left(k_r + \frac{\mathrm{i}}{2}\frac{d\ln\rho}{dr}\right)^2 = \frac{1}{c^2}\left(\sigma^2 + 4\pi G\rho\right) - \frac{1}{4}\left(\frac{d\ln\rho}{dr}\right)^2. \tag{9.9}$$

The additional term $4\pi G\rho/c^2$ appears in this equation in comparison with the terms found in Eq. (9.2) when ℓ is set equal to zero. In the regions of propagation of waves with a purely vertical wavenumber, the local dispersion equation takes the form

$$\left(k_r^{(r)}\right)^2 = \frac{1}{c^2}\left(\sigma^2 + 4\pi G\rho\right) - \frac{1}{4}\left(\frac{d\ln\rho}{dr}\right)^2. \tag{9.10}$$

9.2.2 *Local Dispersion Equation Applying to Surface Layers*

A local dispersion equation valid for the surface layers of the Sun and the stars was given by Deubner & Gough (1984). It is obtained in the supposition that the wavelength of the perturbation in the radial direction is much shorter than the radius of the Sun or the star considered, and that the local effects of the curvature may be neglected. Moreover, the gravity is considered to be constant.

The derivation can be started from Eqs. (6.5) and (6.7), in which $d\Phi'/dr$ and Φ' are neglected, and $\left(1/r^2\right)\,d(r^2\xi)/dr$ is replaced by $d\xi/dr$. Elimination of the function $\xi(r)$ by means of Eq. (6.2), from which the angular variables are separated, yields

$$-\frac{d}{dr}\frac{P'}{\rho} + \frac{\sigma^2}{g}\frac{P'}{\rho} = \frac{N^2-\sigma^2}{g}\,c^2\,\alpha\,, \tag{9.11}$$

$$-\frac{d}{dr}\frac{P'}{\rho} + \frac{\ell(\ell+1)}{\sigma^2 r^2}\, g\, \frac{P'}{\rho} = \frac{d}{dr}\left(c^2\,\alpha\right) - g\,\alpha. \tag{9.12}$$

Next, elimination of $(d/dr)\,(P'/\rho)$ leads to the equation

$$\frac{P'}{\rho} = -\frac{1}{X}\left[\frac{dv}{dr} + \left(\frac{\sigma^2}{g} + \frac{1}{\rho}\frac{d\rho}{dr}\right) v\right], \tag{9.13}$$

where

$$v = c^2\,\alpha, \qquad X = \frac{\sigma^2}{g} - \frac{\ell(\ell+1)}{\sigma^2 r^2}\, g. \tag{9.14}$$

This equation can be used for the elimination of P'/ρ from Eq. (9.11). Regarding the square of the horizontal wavenumber as constant in the differentiation of P'/ρ, one derives the second-order differential equation

$$\frac{d^2 v}{dr^2} + \frac{1}{\rho}\frac{d\rho}{dr}\frac{dv}{dr} + \left[\frac{\sigma^2}{c^2} + \frac{\ell(\ell+1)}{r^2}\left(\frac{N^2}{\sigma^2} - 1\right) + \frac{d}{dr}\left(\frac{1}{\rho}\frac{d\rho}{dr}\right)\right] v = 0. \tag{9.15}$$

When one passes on to the function $\Psi = \rho^{1/2}\, v$, the differential equation becomes

$$\frac{d^2\Psi}{dr^2} + \left[\frac{\sigma^2}{c^2} + \frac{\ell(\ell+1}{r^2}\left(\frac{N^2}{\sigma^2} - 1\right) - \frac{3}{4}\left(\frac{1}{\rho}\frac{d\rho}{dr}\right)^2 + \frac{1}{2}\frac{1}{\rho}\frac{d^2\rho}{dr^2}\right]\Psi = 0. \tag{9.16}$$

In the surface layers of the Sun and the stars where propagation of waves in the radial direction is possible, the differential equation leads to the dispersion equation

$$k_r^2 = \frac{\sigma^2}{c^2} + \frac{\ell(\ell+1}{r^2}\left(\frac{N^2}{\sigma^2} - 1\right) - \frac{3}{4}\left(\frac{1}{\rho}\frac{d\rho}{dr}\right)^2 + \frac{1}{2}\frac{1}{\rho}\frac{d^2\rho}{dr^2}, \tag{9.17}$$

which contains a term involving the second derivative of the mass density besides the term involving the square of the first derivative of that quantity.

One brings Eq. (9.16) into the form adopted by Deubner & Gough by introducing the density height scale

$$H_\rho \equiv \left(-\frac{1}{\rho}\frac{d\rho}{dr}\right)^{-1} \tag{9.18}$$

and the square of the Lamb frequency

$$S_\ell^2 \equiv \frac{\ell(\ell+1)\,c^2}{r^2}. \tag{9.19}$$

One then has

$$\frac{d^2\Psi}{dr^2} + \frac{1}{c^2}\left[\sigma^2 + S_\ell^2\left(\frac{N^2}{\sigma^2} - 1\right) - \frac{1}{4}\frac{c^2}{H_\rho^2}\left(1 - 2\frac{dH_\rho}{dr}\right)\right]\Psi = 0. \tag{9.20}$$

9.3 Local Radial Propagation of Waves

The possibilities of local radial propagation of waves in the interiors of stars can be examined by means of dispersion equation (9.2) on the analogy of analyses that are known for plane gaseous, isothermal layers (Hines 1960, Tolstoy 1963, 1973).

The main question is to determine under which conditions the function $F\left(\sigma^2, \ell, r\right)$ is positive in a stellar layer.

The function depends on three physical properties of the layer: the gravity, g, the isentropic sound velocity, c, which is related to the isentropic compressibility coefficient, $\kappa_S \equiv 1/\left(\Gamma_1 P\right)$, and the relative gradient of the mass density, $d \ln \rho / dr$.

Moreover, the function $F\left(\sigma^2, \ell, r\right)$ contains a term in which σ^2 appears in the numerator as well as a term in which it appears in the denominator. The first term tends to zero as $c^2 \rightarrow \infty$, the second term is equal to zero as $g = 0$. These properties are the basis for a division of the analysis into two parts: a first part relative to the radial propagation of waves in incompressible layers that are subject to gravity, and a second part relative to the radial propagation of waves in compressible layers that are not subject to gravity. In both cases, the layer is considered, first in absence, next in presence of a density stratification. Finally, the more general case of the radial propagation of waves in a compressible layer with a density stratification that is subject to gravity, is considered.

9.3.1 Radial Propagation of Waves in an Incompressible Layer Subject to Gravity

9.3.1.1 In Absence of Any Density Stratification

In the limiting case of an incompressible layer without density stratification that is subject to gravity, one has

$$c^2 \rightarrow \infty, \qquad \frac{1}{\rho}\frac{d\rho}{dr} = 0, \qquad N^2 = 0. \tag{9.21}$$

From definition (9.2) of the function $F\left(\sigma^2, \ell, r\right)$, it then follows

$$F\left(\sigma^2, \ell, r\right) = -\frac{\ell(\ell+1)}{r^2} < 0. \tag{9.22}$$

Hence, *no* radial propagation of waves is possible in an incompressible layer without density stratification that is subject to gravity.

9.3.1.2 In Presence of a Density Stratification

In the limiting case of an incompressible layer with density stratification that is subject to gravity, one has

$$c^2 \to \infty, \qquad N^2 = -g\,\frac{1}{\rho}\frac{d\rho}{dr}. \tag{9.23}$$

When $d\ln\rho/dr < 0$, it results that $N^2 > 0$.

From condition (9.5), it follows that radial propagation of waves is possible when

$$\sigma^2 < \frac{\dfrac{\ell(\ell+1)}{r^2}N^2}{\dfrac{\ell(\ell+1)}{r^2} + \dfrac{1}{4}\left(\dfrac{d\ln\rho}{dr}\right)^2}. \tag{9.24}$$

The radially traveling waves that satisfy this condition are called *internal gravity* waves. A deeper study of internal gravity waves in plane gaseous, isothermal layers was made, e.g., by Hines (1960).

According to a usual procedure in geophysics, it is convenient to use a diagnostic diagram for the visual representation of the conditions under which waves can propagate radially in a stellar layer. In such a diagram, the square of the angular frequency is plotted versus the square of the horizontal wavenumber. Diagnostic diagrams have been considered in the astrophysical literature mainly in connection with the propagation of waves in the surface layers of the Sun (Stein & Leibacher 1974).

Radial propagation of internal gravity waves is possible in the region of the diagnostic diagram that is delimited by the axis of the abscissae and the curve defined by the equation

$$\sigma^2 = \frac{\dfrac{\ell(\ell+1)}{r^2}N^2}{\dfrac{\ell(\ell+1)}{r^2} + \dfrac{1}{4}\left(\dfrac{d\ln\rho}{dr}\right)^2}. \tag{9.25}$$

The curve starts from the origin, for the value zero of the horizontal wavenumber, and tends asymptotically to the limiting value of N^2 for larger values of the horizontal wavenumber, which is determined as

$$\lim_{\ell(\ell+1)\to\infty} \frac{\dfrac{\ell(\ell+1)}{r^2}N^2}{\dfrac{\ell(\ell+1)}{r^2} + \dfrac{1}{4}\left(\dfrac{d\ln\rho}{dr}\right)^2} = N^2. \tag{9.26}$$

Hence, the value of N^2 is an upper limit for the squares of the angular frequencies of the internal gravity waves.

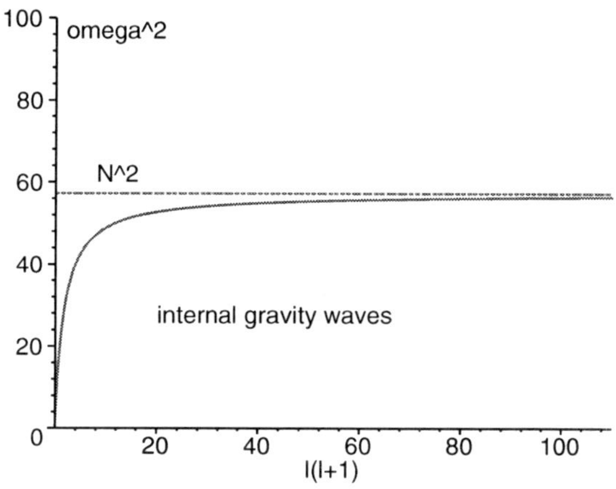

Fig. 9.1 The region of radial propagation of the internal gravity waves in the diagnostic diagram of the layer of the incompressible polytropic model with index $n = 3$ that is situated at the relative distance $r/R = 0.2833$ from the centre, as a function of $\ell(\ell+1)$

According to dispersion equation (9.6), the square of the angular frequency of an internal gravity wave with a radial wavenumber $k_r^{(r)}$ and a horizontal wavenumber $\sqrt{\ell(\ell+1)}/r$ is given by

$$\sigma_i^2 = \frac{\dfrac{\ell(\ell+1)}{r^2} N^2}{\left(k_r^{(r)}\right)^2 + \dfrac{\ell(\ell+1)}{r^2} + \dfrac{1}{4}\left(\dfrac{d\ln\rho}{dr}\right)^2}. \tag{9.27}$$

The angular frequency of an internal gravity wave is thus related to the local frequency of Brunt–Väisälä. For a given horizontal wavenumber $\sqrt{\ell(\ell+1)}/r$, it is smaller when the radial wavenumber $k_r^{(r)}$ is larger.

In Fig. 9.1 the diagnostic diagram is presented in which the region of possible radial propagation of the internal gravity waves is shown for the layer of the incompressible polytropic model with index $n = 3$ that is situated at the relative distance $r/R = 0.2833$ from the centre of the model. The layer is subject to gravity, has a density stratification, but is incompressible. The sphere inside it contains half of the total mass M of the model. The relevant physical quantities are

$$g = 6.230\,\frac{GM}{R^2}, \qquad \frac{1}{\rho}\frac{d\rho}{dr} = -9.192\,\frac{1}{R}, \qquad N^2 = 57.27\,\frac{GM}{R^3}.$$

In the figure, the square of the dimensionless angular velocity,

$$\omega^2 = \frac{GM}{R^3}\sigma^2, \tag{9.28}$$

is plotted versus the product $\ell(\ell+1)$, which is regarded as a continuous variable.

9.3.2 *Radial Propagation of Waves in a Compressible Layer not Subject to Gravity*

9.3.2.1 In Absence of Any Density Stratification

In an unstratified compressible layer that is not subject to gravity, one has

$$g = 0, \qquad \frac{1}{\rho}\frac{d\rho}{dr} = 0, \qquad N^2 = 0. \tag{9.29}$$

From condition (9.5), it follows that radial propagation of waves is possible when

$$\sigma^2 > \frac{\ell(\ell+1)}{r^2}c^2. \tag{9.30}$$

In the diagnostic diagram of a layer, radial propagation is possible in the region delimited by the axis of the ordinates and the straight line that is defined by the equation

$$\sigma^2 = \frac{\ell(\ell+1)}{r^2}c^2. \tag{9.31}$$

This straight line goes through the origin and has the slope c^2.

According to dispersion equation (9.6), the square of the angular frequency of a wave propagating with a radial wavenumber $k_r^{(r)}$ and a horizontal wavenumber $\sqrt{\ell(\ell+1)}/r$ is given by

$$\sigma_a^2 = c^2\left[\left(k_r^{(r)}\right)^2 + \frac{\ell(\ell+1)}{r^2}\right]. \tag{9.32}$$

This equation has the same form as Eq. (Appendix J.7) for the squares of the angular frequencies of acoustic waves in a uniform gaseous medium: the square of the angular frequency is equal to the product of the square of the isentropic sound velocity and the square of the total wavenumber. The waves whose angular frequencies satisfy relation (9.32) are *acoustic waves*. For a given horizontal wavenumber $\sqrt{\ell(\ell+1)}/r$, the angular frequency increases with the radial wavenumber $k_r^{(r)}$.

In the layer of the polytropic model with index $n = 3$ considered, one has, for $\Gamma_1 = 5/3$,

$$c^2 = 0.8472\,\frac{GM}{R}.$$

The layer is supposed to be compressible, but neither to be subject to gravity nor to have a density stratification. The region of possible radial propagation of acoustic waves in the diagnostic diagram is represented in Fig. 9.2.

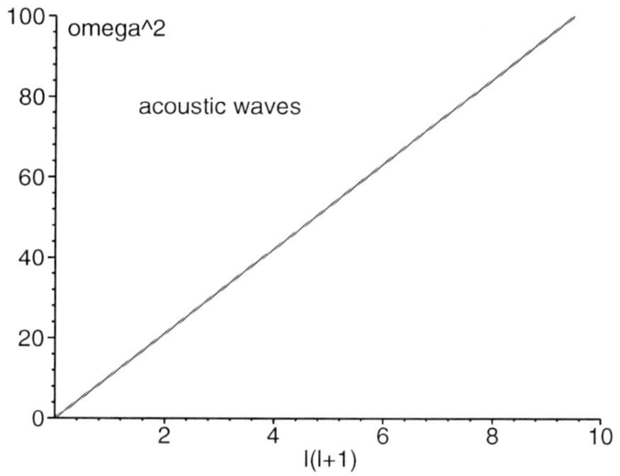

Fig. 9.2 Region of radial propagation of the acoustic waves in the diagnostic diagram of the layer of the compressible polytropic model with index $n = 3$ that is situated at the relative distance $r/R = 0.2833$ from the centre and is supposed neither to be subject to gravity nor to have a density stratification

9.3.2.2 In Presence of a Density Stratification

In a compressible layer that has a density stratification but is still not subject to gravity, one has

$$g = 0, \qquad N^2 = 0. \tag{9.33}$$

From condition (9.5), it follows that radial propagation of waves is possible when

$$\sigma^2 > c^2 \left[\frac{\ell(\ell+1)}{r^2} + \frac{1}{4} \left(\frac{d \ln \rho}{dr} \right)^2 \right]. \tag{9.34}$$

In the diagnostic diagram, the region of possible radial propagation is delimited by the axis of the ordinates and the straight line that is defined by the equation

$$\sigma^2 = c^2 \left[\frac{\ell(\ell+1)}{r^2} + \frac{1}{4} \left(\frac{d \ln \rho}{dr} \right)^2 \right]. \tag{9.35}$$

This straight line has the slope c^2 and goes through to the point on the axis of the ordinates whose ordinate is equal to the square of the cutoff angular frequency

$$\sigma_0 = (c/2) \, |d \ln \rho / dr|. \tag{9.36}$$

The cutoff frequency is the angular frequency associated with the horizontal wavenumber equal to zero ($\ell = 0$).

According to dispersion equation (9.6), the square of the angular frequency of a propagating wave with a radial wavenumber $k_r^{(r)}$ and a horizontal wavenumber $\sqrt{\ell(\ell+1)}/r$ is given by

$$\sigma_a^2 = c^2 \left[\left(k_r^{(r)} \right)^2 + \frac{\ell(\ell+1)}{r^2} + \frac{1}{4} \left(\frac{d \ln \rho}{dr} \right)^2 \right]. \tag{9.37}$$

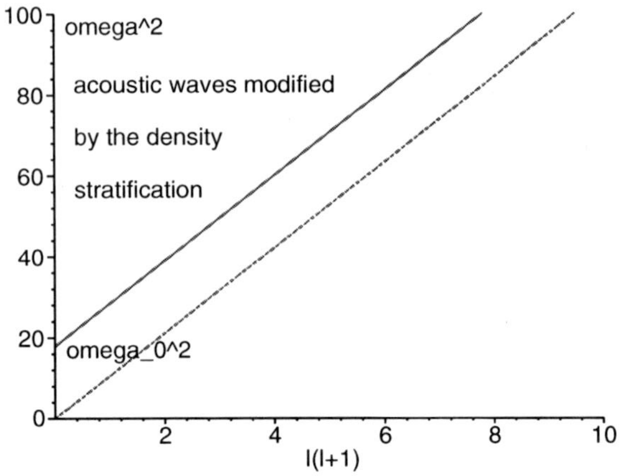

Fig. 9.3 Region of radial propagation of the acoustic waves modified by the density stratification in the diagnostic diagram of the layer of the compressible polytropic model with index $n = 3$ that is situated at the relative distance $r/R = 0.2833$ from the centre and is supposed not to be subject to gravity

The propagating waves are thus *acoustic waves modified by the density stratification*. The square of their angular frequency is increased by the amount σ_0^2 in comparison with the square of the angular frequency of an acoustic wave in absence of a density stratification.

For the layer considered in the compressible polytropic model with index $n = 3$, the square of the cutoff frequency is given by

$$\sigma_0^2 \equiv \omega_0^2 \frac{GM}{R^3} = 17.90 \frac{GM}{R^3}.$$

The diagnostic diagram of the layer is represented in Fig. 9.3.

In a star, the acoustic cutoff frequency varies considerably from the centre to the surface. As $r \to 0$, it tends to zero by virtue of power series (6.37) and the second power series (6.40). On the other hand, as $r \to R$, it becomes indefinitely large, when the behaviour of the mass density near $r = R$ can be represented by a power series of the form of power series (6.51). In physically realistic stellar models, the acoustic cutoff frequency may have a finite value at $r = R$.

The variation of the square of the acoustic cutoff frequency in the polytropic model with index $n = 3$ is represented in Fig. 9.4.

9.3.3 *Radial Propagation of Waves in a Compressible Layer with a Density Stratification that is Subject to Gravity*

On the basis of the preceding analysis, two types of radially propagating waves are expected to be possible in a compressible layer with a density stratification that is subject to gravity: radially propagating acoustic waves modified by the density stratification and the gravity, and radially propagating internal gravity waves modified by the compressibility of the medium.

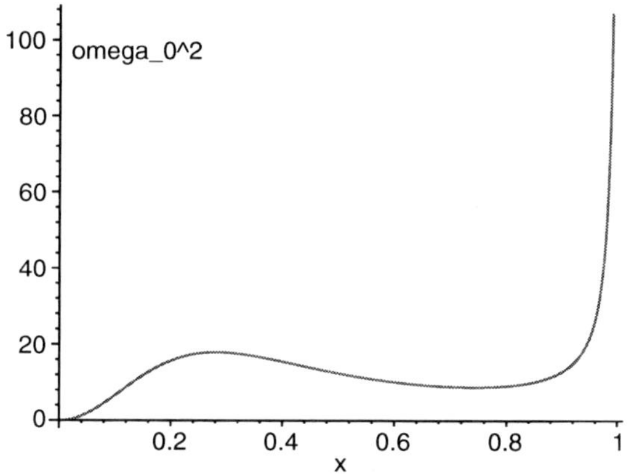

Fig. 9.4 Variation of the square of the dimensionless acoustic cutoff frequency, ω_0^2, in the polytropic model with index $n = 3$, as a function of the relative radial distance from the centre, $x = r/R$

The squares of the angular frequencies of the propagating waves are the roots of the quadratic equation stemming from dispersion equation (9.6)

$$\sigma^4 - \sigma^2 c^2 \left[\left(k_r^{(r)} \right)^2 + \frac{\ell(\ell+1)}{r^2} + \frac{1}{4} \left(\frac{d \ln \rho}{dr} \right)^2 \right] + \frac{\ell(\ell+1)\, c^2}{r^2} N^2 = 0. \quad (9.38)$$

The discriminant of this quadratic equation is

$$\Delta = \left\{ c^4 \left[\left(k_r^{(r)} \right)^2 + \frac{\ell(\ell+1)}{r^2} + \frac{1}{4} \left(\frac{d \ln \rho}{dr} \right)^2 \right]^2 - 4 \frac{\ell(\ell+1)\, c^2}{r^2} N^2 \right\}^{1/2} . \quad (9.39)$$

For a radial wavenumber $k_r^{(r)}$ and a horizontal wavenumber $\sqrt{\ell(\ell+1)}/r$, the square of the angular frequency of an acoustic wave is given by

$$\sigma_A^2 = \frac{1}{2} \left\{ c^2 \left[\left(k_r^{(r)} \right)^2 + \frac{\ell(\ell+1)}{r^2} + \frac{1}{4} \left(\frac{d \ln \rho}{dr} \right)^2 \right] + \Delta \right\} , \quad (9.40)$$

and the square of the angular velocity of an internal gravity wave, by

$$\sigma_I^2 = \frac{1}{2} \left\{ c^2 \left[\left(k_r^{(r)} \right)^2 + \frac{\ell(\ell+1)}{r^2} + \frac{1}{4} \left(\frac{d \ln \rho}{dr} \right)^2 \right] - \Delta \right\} . \quad (9.41)$$

The boundaries of the regions of possible radial propagation in a diagnostic diagram are determined by the condition that the radial wavenumber $k_r^{(r)}$ is equal to zero, respectively, in Eq. (9.40) and in Eq. (9.41). It follows, for acoustic waves,

$$\lim_{\ell(\ell+1) \to 0} \sigma_A^2 = \frac{c^2}{4} \left(\frac{d \ln \rho}{dr} \right)^2 \equiv \sigma_0^2 , \quad (9.42)$$

and, for internal gravity waves,

$$\lim_{\ell(\ell+1)\to 0} \sigma_I^2 = 0 \tag{9.43}$$

and

$$\lim_{\ell(\ell+1)\to\infty} \sigma_I^2 = \lim_{\ell(\ell+1)\to\infty} \frac{1}{2}\frac{\ell(\ell+1)\,c^2}{r^2}\left\{\left[1+\frac{(d\ln\rho/dr)^2/4}{\ell(\ell+1)/r^2}\right]\right.$$
$$\left. - \left\{\left[1+\frac{(d\ln\rho/dr)^2/4}{\ell(\ell+1)/r^2}\right]^2 - \frac{4N^2}{\ell(\ell+1)\,c^2/r^2}\right\}^{1/2}\right\} = N^2. \tag{9.44}$$

Hence, the angular frequency σ_0 determined by definition (9.36) is still a cutoff frequency for acoustic waves, and the Brunt–Väisälä frequency is still an upper limit for the angular frequencies of internal gravity waves. The cutoff frequency σ_0 corresponds to the acoustic cutoff frequency for acoustic waves that propagate vertically in a plane, compressible, isothermal atmosphere subject to gravity, as derived by Lamb (1908) (see Appendix J.2; see also Sutmann et al. 1998).

In any stellar layer, the angular cutoff frequency for acoustic waves is always larger than the Brunt–Väisälä frequency, i.e.,

$$\sigma_0^2 \equiv \frac{c^2}{4}\left(\frac{d\ln\rho}{dr}\right)^2 > N^2. \tag{9.45}$$

This property follows from the inequality

$$\left(\frac{c^2}{2}\frac{d\ln\rho}{dr} + \frac{g}{c^2}\right)^2 > 0. \tag{9.46}$$

Consequently, acoustic waves modified by the density stratification and the gravity, and internal gravity waves modified by the compressibility are locally two well separated types of radially propagating waves.

The diagnostic diagram of the layer of the polytropic model with index $n = 3$ that is situated at the relative distance $r/R = 0.2833$ from the centre is represented in Fig. 9.5, in the supposition that the layer is subject to gravity, is compressible, and displays a density stratification. Here the square of the Brunt–Väisälä frequency has the value $N^2 = 11.45\,GM/R^3$, which is approximately one fifth of the value given above in the supposition that the layer is incompressible.

In the particular case of the degree $\ell = 0$, waves can propagate only in the radial direction, and it is preferable to use dispersion equation (9.9), which is not affected by the Cowling approximation. Purely radial propagation of waves is then possible when

$$\sigma^2 > \frac{c^2}{4}\left(\frac{d\ln\rho}{dr}\right)^2 - 4\pi G\rho. \tag{9.47}$$

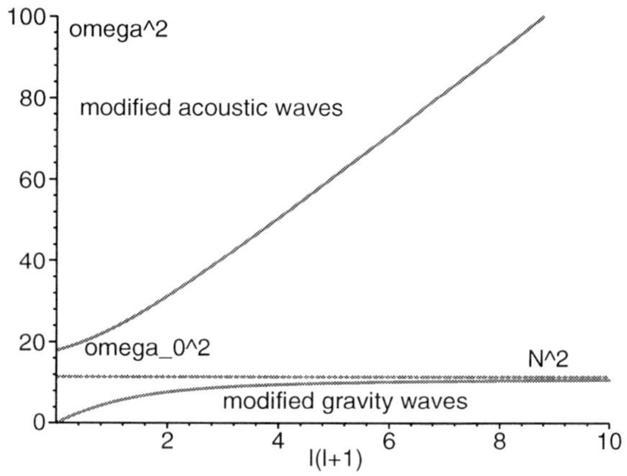

Fig. 9.5 Diagnostic diagram for the stratified layer of the compressible polytropic model with index $n = 3$ that is situated at the relative distance $r/R = 0.2833$ from the centre and is subject to gravity

These waves are manifestly acoustic waves, whose angular frequencies are given by

$$\sigma^2 = c^2 \left(k_r^{(r)}\right)^2 + \frac{c^2}{4}\left(\frac{d \ln \rho}{dr}\right)^2 - 4\pi G\rho. \tag{9.48}$$

The cutoff frequency is given by

$$\sigma_0^2 = \frac{c^2}{4}\left(\frac{d \ln \rho}{dr}\right)^2 - 4\pi G\rho. \tag{9.49}$$

9.4 Global Representation of the Radial Propagation of Waves

The radial propagation, throughout a star, of acoustic waves modified by the density gradient and the gravity and internal gravity waves modified by the compressibility is illustrated by means of a propagation diagram. Such a diagram is builded for the radial propagation of waves belonging to a given degree ℓ. In this, the square of the angular frequency is represented versus the radial distance from the star's centre.

A propagation diagram is generated as follows. From the diagnostic diagram of each individual stellar layer, both the square of the lowest possible frequency of the acoustic waves and the square of the highest possible frequency of the internal gravity waves, at the degree ℓ considered, are transferred to the propagation diagram at the radial distance of the layer. This is done layer by layer. Hereafter, the points that represent the squares of the lowest admissible frequencies of the acoustic waves are joined by a curve. Similarly, the points that represent the squares of the highest possible frequencies of the internal gravity waves are joined by another curve. The first curve delimitates the region of possible propagation of acoustic waves from below, and the second curve, the region of possible propagation of internal gravity waves from above. Correspondingly, the region of possible propagation of acoustic

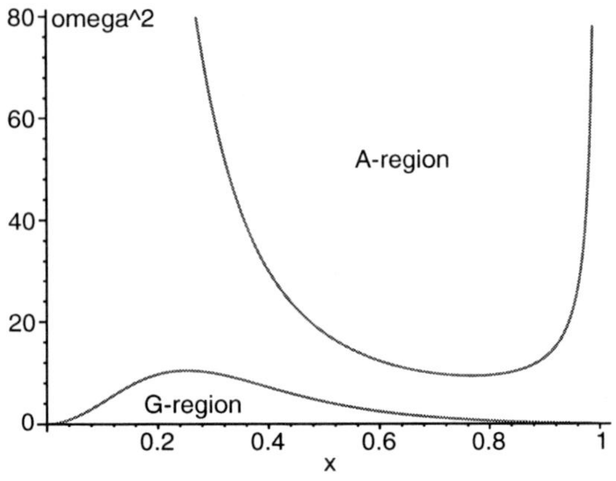

Fig. 9.6 Propagation diagram of the polytropic model with index $n = 3$ for the degree $\ell = 2$

waves is called the A-region or A-cavity, and that of possible propagation of internal gravity waves, the G-region or G-cavity.

The boundaries of the A- and G-region can also be determined in an analytical way. It suffices to set $k_r^{(r)} = 0$ in Eq. (9.38) and to solve the resulting quadratic equation for the square of the angular frequency, σ^2, as a function of the radial distance from the centre, r. As an example, the propagation diagram of the polytropic model with index $n = 3$, for the degree $\ell = 2$, is represented in Fig. 9.6. Here, the square of the dimensionless angular frequency, ω^2, is plotted versus the relative radial distance from the centre, $x = r/R$ (Smeyers 1984).

Propagation diagrams are more elaborate forms of the diagrams that were presented before by Scuflaire (1974) for polytropic models, by Unno (1975) for a solar model, and by Osaki (1975) for a main sequence star of $10 M_\odot$. In these diagrams, just the two curves

$$\sigma^2 = S_\ell^2 \equiv \ell(\ell+1)\, c^2/r^2, \qquad \sigma^2 = N^2$$

were displayed.

For larger values of σ^2, the inner boundary of the A-region in a propagation diagram is approximately determined by the equation

$$\sigma^2 = S_\ell^2, \tag{9.50}$$

and the outer boundary, by the equation

$$\sigma^2 = \sigma_0^2 \equiv \frac{c^2}{4}\left(\frac{d\ln\rho}{dr}\right)^2. \tag{9.51}$$

From Eq. (9.50), it follows that, for an increasing degree ℓ, the inner boundary of the A-region in the propagation diagram moves towards larger values of r. In illustration, the propagation diagram of the polytropic model with index $n = 3$, for the degree $\ell = 10$, is represented in Fig. 9.7.

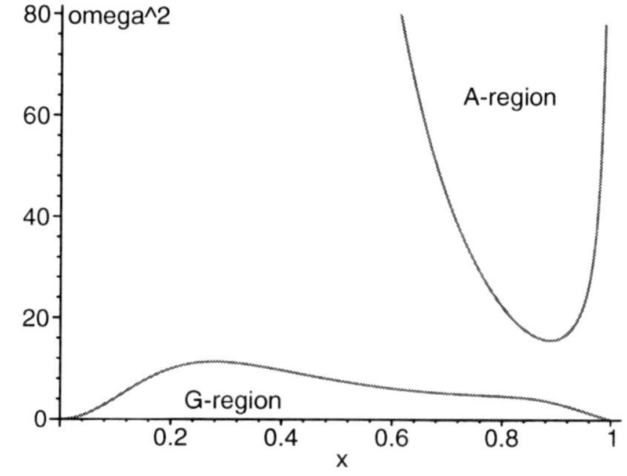

Fig. 9.7 Propagation diagram of the polytropic model with index $n = 3$ for the degree $\ell = 10$

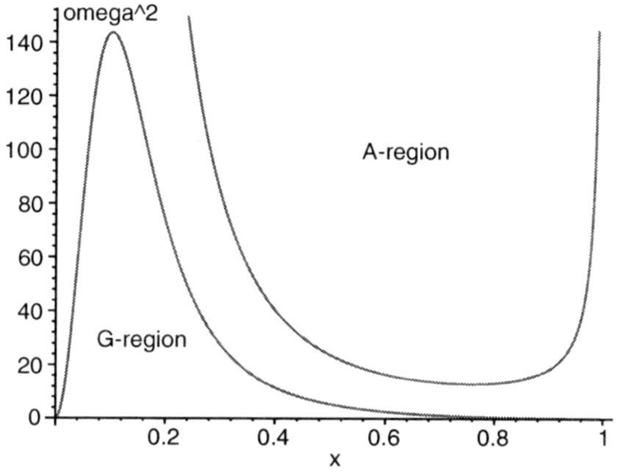

Fig. 9.8 Propagation diagram of the polytropic model with index $n = 4$ for the degree $\ell = 2$

In the propagation diagram of the polytropic model with index $n = 3$, for the degree $\ell = 2$ (Fig. 9.6), the top of the G-region has an ordinate that is only slightly larger than that of the lowest point of the A-region. However, for polytropic models with larger indexes, which display a more central mass condensation, the top of the G-region is situated at an ordinate largely higher than the ordinate of the lowest point of the A-region. By way of illustration, the propagation diagram of the polytropic model with index $n = 4$, for the degree $\ell = 2$, is represented in Fig. 9.8. This has for consequence that, in some interval of frequencies, radially propagating waves have a dual character: in the more central G-region of the polytropic model, they display the character of internal gravity waves, and in the more external A-region, the character of acoustic waves.

The propagation diagrams for waves that are associated with the degree $\ell = 0$ can be constructed on the basis of Eq. (9.48). In these diagrams, no G-region exists, and the A-region extends to the centre of the model. In Fig. 9.9, the propagation diagram of the polytropic model with index $n = 3$ is represented for the degree $\ell = 0$.

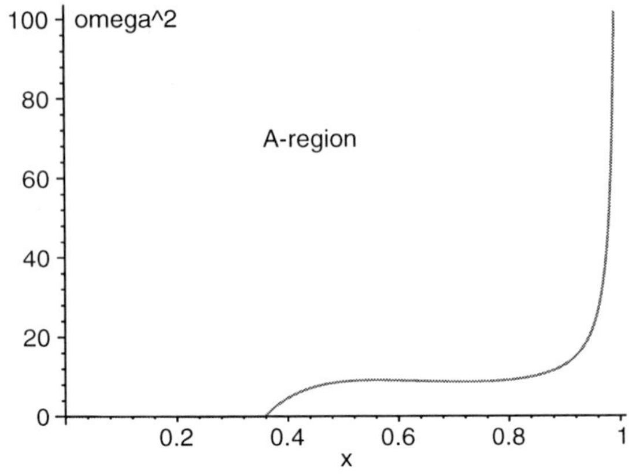

Fig. 9.9 Propagation diagram of the polytropic model with index $n = 3$ for the degree $\ell = 0$

The boundary of the region of propagation for acoustic waves associated with the degree $\ell = 0$ can also be determined by means of the second-order differential equation (6.27). By introducing the new independent variable

$$\tau(r) = \int_0^r \frac{dr'}{c(r')} \tag{9.52}$$

and applying the transformation $\psi(\tau) = f_1(r)\,\zeta(r)$, with $f_1(r) = \left(\rho c r^4\right)^{1/2}$, one derives the Schrödinger like equation

$$\frac{d^2\psi}{d\tau^2} + \left(\sigma^2 - \Omega_1^2\right)\psi = 0, \tag{9.53}$$

in which Ω_1^2 is the acoustic potential defined as

$$\Omega_1^2 \equiv -\frac{1}{\rho r}\frac{d}{dr}\left[(3\Gamma_1 - 4)\,P\right] + \frac{c^2}{f_1}\frac{d^2 f_1}{dr^2} + \frac{c}{f_1}\frac{df_1}{dr}\frac{dc}{dr} \tag{9.54}$$

(Christensen-Dalsgaard et al. 1983). The independent variable $\tau(r)$ corresponds to that introduced by Ledoux (1962) in his asymptotic representation of radial oscillation modes with higher frequencies (see Sect. 14.3) and is equal to the time needed by an acoustic wave for propagating from the star's centre to the radial distance r. It can therefore be considered as the acoustic height of the layer. By the use of this height as independent variable, the region near the star's surface is stretched, since $c \to 0$ as $r \to R$. This is illustrated by Fig. 9.10, in which the variation of the dimensionless acoustic height $\left(GM_\odot/R_\odot^3\right)^{1/2}\tau(r)$ is represented as a function of the relative radial distance from the centre, $x = r/R_\odot$, for the solar model of Christensen-Dalsgaard et al. (1993).

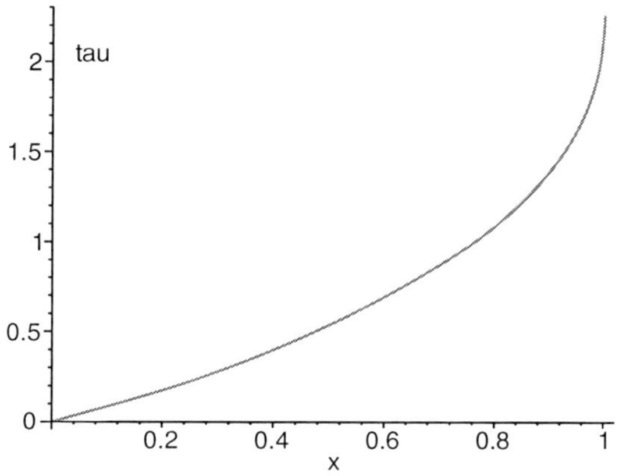

Fig. 9.10 Variation of the dimensionless acoustic height $\tau(r)$ as a function of the relative radial distance from the centre, $x = r/R_\odot$, for the solar model of Christensen-Dalsgaard et al. (1993)

Propagation of waves associated with the degree $\ell = 0$ is possible in stellar layers in which

$$\sigma^2 > \Omega_1^2. \tag{9.55}$$

This condition was used by Buchler et al. (1997) in a study of the propagation diagram for a model of a classical Cepheid. These authors observed that

> the sharpness of the hydrogen ionization front produces an immensely high barrier in the potential, effectively separating the star into two distinct regions. The barrier is ultimately caused by the rapid variation of the local sound speed, most by the adiabatic index.

Consequences of the existence of this high barrier in the potential are considered in Sect. 11.5.

A narrow potential well in surface layers was found earlier by Vorontsov & Zharkov (1989) for subphotospheric regions of a solar model. The potential was determined by means of a dispersion equation differing from the equation

$$\sigma^2 = \Omega_1^2 \tag{9.56}$$

by an additional term $4\pi G\rho$ in the right-hand member, which is due to the use of the Cowling approximation by the authors. The authors noted that

> the rapid variation of the potential ... corresponds to the outer superadiabatic layers of the convection zone. The peculiarity in the variation of the potential near $r \approx 0.98\ R_\odot$, where the stratification of the convection zone is close to the adiabatic stratification, corresponds to the region of second ionization of helium.

Chapter 10
Classification of the Spheroidal Normal Modes

10.1 Origin from Propagating Waves

Many spheroidal normal modes of a star originate from propagating waves that are reflected to-and-fro by the walls of the A- or G-cavity inside the star and become standing waves. This view about the origin of spheroidal normal modes has grown for the greater part from the study of the oscillation modes with periods around 5 min observed on the solar surface. In this connection, the following passage from Leibacher & Stein (1981) is particularly enlightening:

> Consider two identical, one-dimensional waves propagating in opposite directions. When the waves are in phase, that is, when the upward motions of both waves coincide, the resulting amplitude is twice that of one of the waves. One-quarter period later, each wave will have propagated one-quarter wavelength in opposite directions, and they will be out of phase and the sum will vanish everywhere. The period and wavelength of the resultant wave is the same as each of the two components, but it is stationary in space; the wave is said to be "standing". Where does one get two such nicely behaved waves, of the same period and amplitude but traveling in opposite directions? Reflecting a wave back upon itself is an obvious example, and it is the existence of reflections in nature that makes the consideration of interfering waves relevant ...
>
> Consider waves incident upon a rigid reflecting surface, with the reflected wave 180° out of phase with the incident wave. They always cancel at the surface and thus also every half integer wavelength away from the surface. An incident spectrum of wavelengths would give rise to a spectrum of standing waves with their zeros, or "nodes", at varying distances from the surface. (Of course, rigid surfaces are only one sort of boundary condition ...)
>
> Slightly more interesting things occur if one now slips a second reflecting surface into the mixture of oppositely propagating waves. For those wavelengths with a node where one placed the new surface, there is no change. Since they had zero amplitude where the surface is now located, they are unaffected. For other wavelengths, however, things get messy rapidly. The motion after a second reflection from the newly inserted wall is not in phase with the initial wave. In general, the distance between the two reflections is a nonintegral number of wavelengths, and the wave will interfere destructively with itself. Those waves with wavelengths that interfere constructively in this cavity are referred to as "its modes of oscillation", "eigenmodes", or just "modes". The requirement for the existence of such a mode is reflection at two boundaries; the propagation equations relate the motions at the two boundaries, and the eigenvalues of the resulting system prescribe the allowed motions within the cavity.
>
> In general, a cavity formed by two reflections will have many resonant modes, that is, many wavelengths that can satisfy the requirement that after two reflections they arrive

P. Smeyers and T. Van Hoolst, *Linear Isentropic Oscillations of Stars: Theoretical Foundations*, Astrophysics and Space Science Library 371,
DOI 10.1007/978-3-642-13030-4_11,

back at the starting point in phase. In addition to the fundamental mode, with a node at each reflecting surface

$$\text{wavelength} = 2 \times \text{separation},$$

there exists an infinity of shorter wavelength modes satisfying the following equation:

$$\text{wavelength} = \frac{2 \times \text{separation}}{\text{an integer}}.$$

These shorter wavelength modes are frequently called "higher modes" or "harmonics" ...

In a star, the reflections as well as the phase speed do depend on the wave period and wavelength. Hence, the size, and even the existence, of a cavity depends on the frequency of the wave and its wavenumber. This has the consequences that ... there is no simple integral relation between periods and wavelengths of harmonics, nor are the boundaries the same for all waves ...

Since each of the two running waves that produce the standing wave patterns has an equal but opposite energy flux, there is no net energy transfer in spite of the obvious mechanical energy density.

Many linear, isentropic, spheroidal normal modes of stars are standing waves resulting from interferences of running waves that travel in opposite directions (see also, e.g., Deubner & Gough 1984). These modes are almost imprisoned in their cavity, although they extend from the centre to the surface of the star, where they obey certain conditions. Spheroidal normal modes that originate from running waves are the radial modes and the non-radial p- and g^+-modes. On the other hand, f- and g^--modes have different origins.

10.2 The Radial Modes

Radial modes originate from interferences of acoustic waves that are associated with the degree $\ell = 0$ and travel to-and-fro in the acoustic cavity with a frequency fitting to the dimensions of the cavity. One gets an idea of the spectrum of the fitting frequencies by treating the eigenvalue problem of the radial oscillations as a Sturm–Liouville eigenvalue problem with singular end points, which is defined by the second-order differential equation (6.27), for the relative radial displacement $\zeta(r) = \xi(r)/r$,

$$\frac{d}{dr}\left(r^4 \Gamma_1 P \frac{d\zeta}{dr}\right) + \left\{\sigma^2 \rho r^4 + r^3 \frac{d}{dr}\left[(3\Gamma_1 - 4)\, P\right]\right\} \zeta = 0, \tag{10.1}$$

and the two regularity conditions

$$\zeta \text{ is finite at } r = 0, \tag{10.2}$$

$$\left(r\frac{d\zeta}{dr} + 3\zeta\right) \text{ is finite at } r = R. \tag{10.3}$$

These two regularity conditions stem from the conditions

$$\xi \equiv r\,\zeta = 0 \ \text{at} \ r = 0, \tag{10.4}$$

$$\delta P = -\frac{\Gamma_1 P}{r^2}\frac{d}{dr}\left(r^3\,\zeta\right) = 0 \ \text{at} \ r = R \tag{10.5}$$

(Ledoux & Walraven 1958).

A classic Sturm–Liouville eigenvalue problem is defined in an interval $a \leq x \leq b$ by a linear, homogeneous, second-order differential equation of the form

$$\frac{d}{dx}\left[k(x)\,\frac{dy}{dx}\right] + [\lambda\,g(x) - l(x)]y = 0, \tag{10.6}$$

where λ is the eigenvalue parameter, and two boundary conditions of the form

$$\left.\begin{aligned} \alpha_2 y(a) - \alpha_1 y'(a) &= 0, \\ \beta_2 y(b) + \beta_1 y'(b) &= 0. \end{aligned}\right\} \tag{10.7}$$

The coefficients $k(x)$, $g(x)$, $l(x)$ are supposed to be real, continuous functions of x in the interval considered and to be independent of the eigenvalue parameter. Without any limitation of the generality, it can be supposed that $k(x) > 0$ and $g(x) > 0$. Furthermore, α_1, α_2, β_1, β_2 are constant coefficients and are also independent of the eigenvalue parameter.

This eigenvalue problem admits of an infinite spectrum of increasing, discrete eigenvalues $\lambda_0, \lambda_1, \lambda_2, \ldots$ without upper limit. The associated eigenfunctions $y_0, y_1, y_2, \ldots$ have $0, 1, 2, \ldots$ nodes in the interval $a < x < b$. If moreover the conditions

$$l(x) \geq 0, \qquad \alpha_1\alpha_2 \geq 0, \qquad \beta_1\beta_2 \geq 0 \tag{10.8}$$

are satisfied, all eigenvalues are positive (Ince 1956).

The eigenvalue problem of the radial modes of a star differs from a classic Sturm–Liouville eigenvalue problem, since the end points $r = 0$ and $r = R$ are singular points of the differential equation, as noted by Ledoux & Walraven (1958). Therefore, besides the spectrum of the discrete eigenvalues, a continuous spectrum of eigenvalues may exist. In this context, Kemble (1937) observed in a footnote:

> The singular end-point problems are of course an old story to mathematicians, but the writer has not been able to discover any comprehensive treatment of a sufficiently elementary character to meet the needs of physicists. The powerful and elaborate work of H. Weyl, *Math. Ann.* **68**, 220 (1910); *Nachtrichten d. Kgl. Gesell. d. Wissenschaften zu Göttingen, Math.-phys. Klasse*, (1910, p. 1) covers most of the ground, but has for its primary purpose the study of the continuous spectrum and is not adapted to the requirements of beginners.

According to Ledoux & Walraven, the properties of a classic Sturm–Liouville eigenvalue problem remain valid, if the singular end points are regular singular points of the differential equation, and the regular parts of the solution satisfy the boundary conditions. By virtue of the analysis made in Sect. 6.3, this condition is satisfied for the end point $r = 0$. It is also satisfied for the end point $r = R$ when the mass density is analytic there and can be expanded in a power series with a positive exponent n_e.

Ledoux & Walraven made mention of an earlier investigation that had led to the existence of a continuous spectrum of eigenvalues for some models but presumed that the distributions of the mass density in these models had no physical meaning.

Since the discrete character of the spectrum of the eigenvalues has nearly generally been taken for granted. In a more recent mathematical study of this question by Beyer & Schmidt (1995), the conjecture proved to be true under certain conditions. For this study, the reader is referred to the subsequent Sect. 10.5.2.

From the similarity between the eigenvalue problem of the radial modes of a star and the Sturm–Liouville eigenvalue problem, one infers the existence of an infinite spectrum of increasing, discrete eigenvalues $\sigma_0^2, \sigma_1^2, \sigma_2^2, \ldots$, without upper limit. The associated eigenfunctions $\zeta_0, \zeta_1, \zeta_2, \ldots$ display respectively zero, one, two, ... nodes between $r = 0$ and $r = R$. These radial modes are successively called the fundamental radial mode, the first radial overtone, the second radial overtone,...

Conditions (10.8) for the eigenvalues σ^2 being positive are reduced to the single condition

$$-r^3 \frac{d}{dr}\left[(3\Gamma_1 - 4)\, P\right] \geq 0. \tag{10.9}$$

When Γ_1 is constant, the condition becomes simply

$$\Gamma_1 > \frac{4}{3}. \tag{10.10}$$

On the other hand, when Γ_1 is not constant, as this is the case in realistic stellar models, the condition is expressed in terms of an unspecified mean value $\overline{\Gamma_1}$ as

$$\overline{\Gamma_1} > \frac{4}{3}. \tag{10.11}$$

From the comparison between results obtained for different stellar models, it follows that the period of the fundamental radial mode generally increases when the ratio $\rho_c/\bar{\rho}$ decreases (Ledoux 1940).

The nodes in the eigenfunctions $\zeta(r)$ belonging to radial modes of successive orders are separated from each other in accordance with Sturm's fundamental theorem (see, e.g., Ince 1956).

A survey of early results of radial oscillations in stellar models is given in Ledoux & Walraven (1958). For the radial modes of the equilibrium sphere of uniform mass density, analytical solutions were obtained by Sterne (1937). These solutions are derived in Sect. 10.4.1. In Table 10.1, the dimensionless eigenvalues ω^2 of the first five radial modes of the polytropic model with index $n = 3$, and the relative radial distances from the centre, $x = r/R$, of the nodes in the eigenfunctions $\zeta(r)$ are given for $\Gamma_1 = 5/3$. For the fundamental mode, and the first and second overtone, the variation of the eigenfunction $\zeta(r)$ is represented in Fig. 10.1. The eigenfunctions are normalised to unity at $x = 1$.

Table 10.1 For the first five radial modes of the polytropic model with index $n = 3$, and $\Gamma_1 = 5/3$, the square of the dimensionless eigenvalue, ω^2, and the relative radial distances, $x = r/R$, of the nodes of the eigenfunction $\zeta(r)$

Radial mode	ω^2	Positions of the nodes of $\zeta(r)$			
Fundamental mode	9.255				
First overtone	16.98	0.6774			
Second overtone	28.48	0.5499	0.8310		
Third overtone	43.44	0.4737	0.7370	0.8946	
Fourth overtone	61.70	0.4212	0.6657	0.8237	0.9276

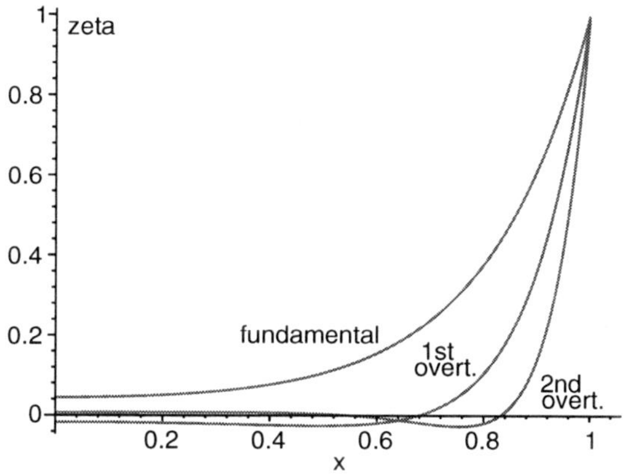

Fig. 10.1 The eigenfunctions $\zeta(r)$ for the fundamental radial mode, and the first and second overtone in the polytropic model with index $n = 3$, for $\Gamma_1 = 5/3$, as functions of the relative radial distance from the centre, $x = r/R$

10.3 Cowling's Classification of the Non-Radial Spheroidal Modes

10.3.1 The Non-Radial p- and g-Modes

The interferences of acoustic waves and internal gravity waves that are associated with a degree ℓ different from zero and travel to-and-fro in the A- or G-cavity of a star lead, for certain frequencies, to standing waves resulting in non-radial modes. Important insight into the types of spectra of eigenfrequencies is provided by an approach made by Cowling (1941). Cowling considered the eigenvalue problem of the non-radial oscillations of polytropic models for large as well for small values of $|\sigma^2|$. Neglecting the Eulerian perturbation of the gravitational potential, he derived an appropriate second-order differential equation for both cases and observed that the eigenvalue problem tends asymptotically to a Sturm–Liouville eigenvalue problem in each case. He also determined a few eigenfrequencies of non-radial modes belonging to the degree $\ell = 2$ for the polytropic model with index $n = 3$. Later Cowling's approach was extended to more general models by Ledoux & Walraven (1958).

To render Cowling's approach for general models, we start from Eqs. (7.32) and (7.33), in which the Eulerian perturbation of the gravitational potential and its first derivative are neglected. After elimination of the function $y(r)$, we derive two second-order differential equations for the function $u(r)$: an equation appropriate for large values of $|\sigma^2|$,

$$\frac{d}{dr}\left(\frac{\rho c^2}{r^2}\frac{du}{dr}\right) + \left[\frac{\sigma^2}{c^2} - \frac{\ell(\ell+1)}{r^2} + \frac{4g}{c^2 r} - \frac{4\pi G\rho}{c^2}\right]\frac{\rho c^2}{r^2}u$$
$$= \frac{S_\ell^2}{\sigma^2}\frac{1}{1 - S_\ell^2/\sigma^2}\frac{\rho c^2}{r^2}\left(\frac{c^2}{r^2}\frac{d}{dr}\frac{r^2}{c^2}\right)\left(\frac{du}{dr} - \frac{g}{c^2}u\right) - \frac{S_\ell^2}{\sigma^2}\frac{\rho N^2}{r^2}u, \quad (10.12)$$

and an equation appropriate for small values of $|\sigma^2|$,

$$\frac{d}{dr}\left(\rho\frac{du}{dr}\right) + \left[\frac{\ell(\ell+1)}{r^2}\left(\frac{N^2}{\sigma^2} - 1\right) - \frac{d}{dr}\frac{g}{c^2}\right]\rho u$$
$$= -\frac{\sigma^2}{S_\ell^2}\frac{1}{1 - \sigma^2/S_\ell^2}\rho\left(\frac{c^2}{r^2}\frac{d}{dr}\frac{r^2}{c^2}\right)\left(\frac{du}{dr} - \frac{g}{c^2}u\right) - \frac{\sigma^2}{c^2}\rho u. \quad (10.13)$$

In these equations, S_ℓ^2 is the square of the Lamb frequency, as defined by equality (9.19).

In Eq. (10.12), the coefficient $\rho c^2/r^2$ in the first term of the left-hand member becomes indefinitely large as $r \to 0$. Therefore, it is appropriate to pass on to an equation for the relative radial component of the displacement $\zeta(r) = u(r)/r^3$:

$$\frac{d}{dr}\left(\rho c^2 r^4 \frac{d\zeta}{dr}\right)$$
$$+\left[\frac{\sigma^2}{c^2} - \frac{\ell(\ell+1)}{r^2} + \frac{4g}{c^2 r} - \frac{4\pi G\rho}{c^2} + \frac{3}{\rho c^2 r}\frac{d(\rho c^2)}{dr}\right]\rho c^2 r^4 \zeta$$
$$= \frac{S_\ell^2}{\sigma^2}\frac{\rho c^2 r}{1 - S_\ell^2/\sigma^2}\left(\frac{c^2}{r^2}\frac{d}{dr}\frac{r^2}{c^2}\right)\left[\frac{d(r^3\zeta)}{dr} - \frac{g}{c^2}r^3\zeta\right]$$
$$-\frac{S_\ell^2}{\sigma^2}\rho^2 c^2 r^5 N^2 \zeta. \quad (10.14)$$

For sufficiently large values of $|\sigma^2|$, the right-hand member of the equation is small in comparison with the left-hand member, except in a limited interval in the neighbourhood of the boundary point $r = 0$ where the coefficient $1 - S_\ell^2/\sigma^2$ goes through zero and the ratio S_ℓ^2/σ^2 becomes indefinitely large as $r \to 0$. For increasing values of $|\sigma^2|$, the length of this interval decreases, and the equation tends to a Sturm–Liouville type equation. This analogy points towards the existence of an infinite spectrum of increasing, discrete eigenvalues σ^2 for each degree ℓ different from zero. If

$$T_1(r) \equiv \frac{\ell(\ell+1)}{r^2} - \frac{4g}{c^2 r} + \frac{4\pi G\rho}{c^2} - \frac{3}{\rho c^2 r}\frac{d(\rho c^2)}{dr} \geq 0 \quad (10.15)$$

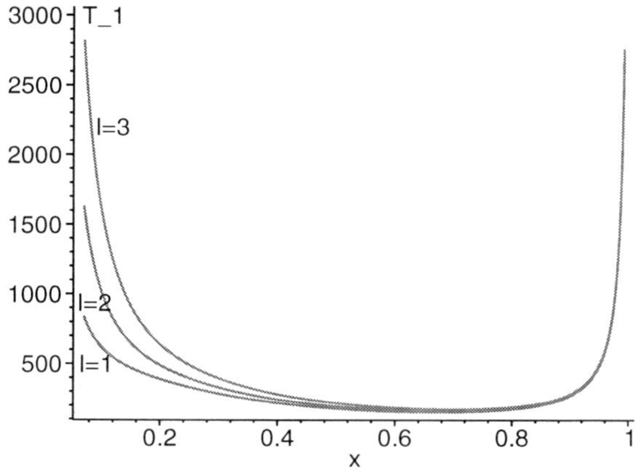

Fig. 10.2 The dimensionless function $T_1(r)$ for the polytropic model with index $n = 3$, and $\Gamma_1 = 5/3$, in the cases of the degrees $\ell = 1, 2, 3$

in the interval $0 < r < R$, the eigenvalues σ^2 are all positive. In Fig. 10.2, the dimensionless function $T_1(r)$ for the polytropic model with index $n = 3$, and $\Gamma_1 = 5/3$, is represented in the cases of the degrees $\ell = 1, 2, 3$. The function $T_1(r)$ appears to be positive in the interval $0 < r < R$ in these cases.

The modes involved arise from interferences of acoustic waves that are modified by the density stratification and the gravity, and propagate in the A-cavity of the star. Cowling called them p-modes (pressure modes) and classified them as the p_1-, the p_2-, the p_3-mode ... according to the number of nodes the eigenfunction $\zeta(r)$ displays in the interval $0 < r < R$. The eigenfunctions $\zeta(r)$ display one more node than the corresponding eigenfunctions in a classic Sturm–Liouville eigenvalue problem. The additional node must be attributed to the existence of the limited region near the boundary point $r = 0$ where the ratio S_ℓ^2/σ^2 becomes indefinitely large as $r \to 0$.

Next, we turn to Eq. (10.13). For sufficiently small values of $|\sigma^2|$, the right-hand member is small in comparison with the left-hand member, except in a limited interval in the neighbourhood of the boundary point $r = R$. There the ratio σ^2/S_ℓ^2 becomes indefinitely large in absolute value as $r \to R$. In case of positive values of σ^2, the coefficient $1 - \sigma^2/S_\ell^2$ also goes through zero. For decreasing values of σ^2, the length of the interval decreases, so that Eq. (10.13) too tends to a Sturm–Liouville type equation. This second analogy with the Sturm–Liouville eigenvalue problems points towards the existence of an infinite spectrum of increasing, discrete eigenvalues $1/\left|\sigma^2\right|$ for each degree ℓ different from zero. According to $N^2 > 0$ or $N^2 < 0$ at all points in the stellar model, the eigenvalues are $1/\sigma^2$ or $-1/\sigma^2$. In both cases, they are all positive if

$$T_2(r) \equiv \frac{\ell(\ell+1)}{r^2} + \frac{d}{dr}\frac{g}{c^2} > 0 \tag{10.16}$$

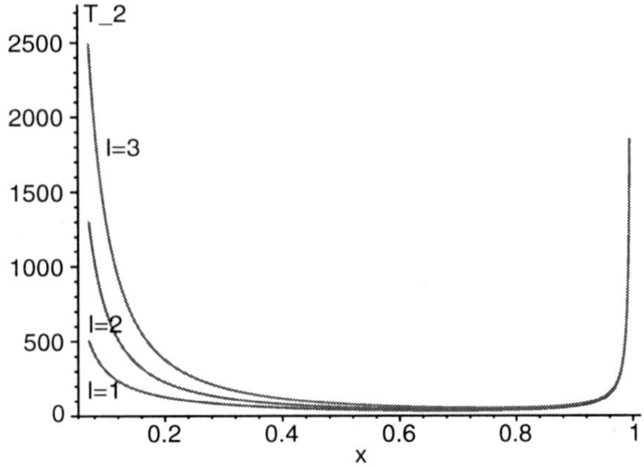

Fig. 10.3 The dimensionless function $T_2(r)$ for the polytropic model with index $n = 3$, and $\Gamma_1 = 5/3$, in the cases of the degrees $\ell = 1, 2, 3$

in the interval $0 < r < R$. The dimensionless function $T_2(r)$ is represented in Fig. 10.3 for the polytropic model with index $n = 3$, and $\Gamma_1 = 5/3$, in the cases of the degrees $\ell = 1, 2, 3$. It is seen to be positive in the interval $0 < r < R$ in these cases.

Hence, the eigenvalues σ^2 are positive as $N^2 > 0$ and negative as $N^2 < 0$. They decrease in absolute value and tend to the saturation point $\sigma^2 = 0$.

The modes associated with the spectra of the second type were called g-modes (gravity modes) by Cowling. Later Ledoux & Smeyers (1966) introduced the denominations g^+- and g^--modes, in order to distinguish between the g-modes associated with positive eigenvalues σ^2 and those associated with negative eigenvalues σ^2.

g^+-modes and g^--modes have different origins. g^+-modes result from propagations to-and-fro of gravity waves, modified by the compressibility of the medium, in the G-cavity of the star. On the other hand, g^--modes render the global reactions of the star to the local convective instabilities. The existence of a relation between g^--modes and starting convective motions in superadiabatic regions in stars has been stressed, probably for the first time in such an explicit way, by Ledoux (Ledoux 1949, Ledoux & Walraven (1958), Ledoux 1958).

The g-modes are classified either as modes $g_1^+, g_2^+, g_3^+, \ldots$, or as modes g_1^-, $g_2^-, g_3^-, \ldots$. The eigenfunction $\xi(r)$ of a mode g_k^+ displays k nodes in the interval $0 < r < R$, and the eigenfunction $\xi(r)$ of a mode g_k^-, $k - 1$ nodes. The eigenfunction $\xi(r)$ of a mode g_k^+ thus displays one more node than the corresponding eigenfunction of a Sturm–Liouville eigenvalue problem, while the eigenfunction $\xi(r)$ of a mode g_k^- has the same number of nodes.

In stars composed of a convectively stable region and a convectively unstable region, both an infinite spectrum of discrete positive eigenvalues σ^2_{g+} and an infinite spectrum of discrete negative eigenvalues σ^2_{g-} exist for each degree ℓ different from zero. Ledoux & Smeyers (1966) pointed out that the two types of spectra exist besides each other, by referring to the analogy between the eigenvalue problem of the non-radial oscillations of stars for small values of $|\sigma^2|$ and the Sturm–Liouville eigenvalue problems (see Ince 1956). They illustrated their conclusion by means of numerical results for an artificial model consisting of a core of uniform mass density and a polytropic envelope with index $n = 3$. For $\Gamma_1 = 5/3$, the core is convectively unstable, and the envelope, convectively stable (see also Smeyers 1966a).

For stellar models in which two convectively unstable regions are separated by a convectively stable region, Ledoux & Perdang (1980) showed that only one g^--spectrum associated with a degree ℓ different from zero exists, and not one spectrum for each convectively unstable region. Ledoux & Perdang showed this on the basis of an asymptotic approximation for g^--modes for which they adopted a tractable second-order differential equation and used the phase integral approach.

10.3.2 The Non-Radial f-Modes

In his study of the non-radial oscillations of polytropic models, Cowling distinguished an eigenvalue σ^2 situated between the eigenvalues $\sigma^2_{p_1}$ and $\sigma^2_{g_1^+}$ for the model with index $n = 3$ and the degree $\ell = 2$. The associated eigenfunctions $\xi(r)$ and $\rho'(r)$ display no node in the interval $0 < r < R$. Cowling denoted the mode as the fundamental non-radial mode or f-mode.

In the full treatment, in which the Eulerian perturbation of the gravitational potential is incorporated into the governing equations, an eigenvalue σ^2_f exists between the eigenvalues $\sigma^2_{p_1}$ and $\sigma^2_{g_1^+}$ for each degree $\ell > 1$. For $\ell = 1$, no eigenvalue σ^2_f is found between the eigenvalues $\sigma^2_{p_1}$ and $\sigma^2_{g_1^+}$, but $\sigma^2_f = 0$.

The f-modes can be regarded as transpositions of the Kelvin modes of the equilibrium sphere of uniform mass density to the compressible stellar models containing a density gradient. The Kelvin modes are derived above in Sect. 8.5.2 by application of the Rayleigh–Ritz method to Hamilton's variational principle, and their eigenvalues are given by Eq. (8.102).

As well as the Kelvin modes of degree $\ell = 1$ in the equilibrium sphere of uniform mass density, the f-modes of degree $\ell = 1$ in an arbitrary equilibrium stellar model with a density stratification represent linear combinations of translations of that equilibrium model. In contrast with the Kelvin modes belonging to the degrees $\ell > 1$, the f-modes belonging to the degrees $\ell > 1$ are neither divergence-free nor curl-free.

From a physical point of view, the Kelvin modes of higher degrees, and by extension the f-modes of higher degrees, are related to surface waves. With the use of the horizontal wave number at the surface, $k_h = \sqrt{\ell(\ell+1)}/R$, the eigenvalue equation of the Kelvin modes can be written as

$$\sigma^2 = \frac{GM}{R^2} \frac{k_h}{\sqrt{\ell(\ell+1)}} \frac{2\ell(\ell-1)}{2\ell+1}. \tag{10.17}$$

For larger degrees ℓ, this equation tends to

$$\sigma^2 = \frac{GM}{R^2} k_h, \tag{10.18}$$

what corresponds to the characteristic equation for surface-gravity waves in a homogeneous, incompressible half-space with a free surface in a constant gravity field (Tolstoy 1973).

10.4 Validity of Cowling's Classification

10.4.1 The Equilibrium Sphere of Uniform Mass Density

The spheroidal modes of the compressible equilibrium sphere of uniform mass density can be determined analytically. The study of these modes provides a guiding example for the classification of the spheroidal modes in more physical stellar models, despite the fact that the equilibrium sphere of uniform mass density is an unrealistic model because of its instability with respect to convection. The most important studies are those of Pekeris (1938) and Sauvenier-Goffin (1951) (see also Ledoux & Walraven 1958).

10.4.1.1 The Equilibrium Model

In the equilibrium sphere of uniform mass density, the mass contained inside the sphere with radius r is given by Eq. (8.97). The condition of hydrostatic equilibrium then takes the form

$$\frac{dP}{dr} = -\frac{4\pi G\rho^2}{3} r. \tag{10.19}$$

Integration and use of the boundary condition $P(R) = 0$ lead to the pressure distribution

$$P(r) = \frac{2\pi G\rho^2}{3} \left(R^2 - r^2\right). \tag{10.20}$$

Since $\rho = \bar{\rho} = 3M/\left(4\pi R^3\right)$, the pressure distribution can also be expressed as

$$P(r) = \frac{GM^2}{4\pi R^4} \frac{3}{2} \left(1 - \frac{r^2}{R^2}\right). \tag{10.21}$$

Furthermore, one has

$$\left.\begin{aligned} g(r) &= \frac{GM}{R^2}\frac{r}{R}, \qquad c^2(r) = \frac{GM}{R}\frac{\Gamma_1}{2}\left(1-\frac{r^2}{R^2}\right), \\ N^2(r) &= -\frac{GM}{R^3}\frac{2}{\Gamma_1}\frac{r^2/R^2}{1-r^2/R^2}, \end{aligned}\right\} \tag{10.22}$$

so that the equilibrium sphere is unstable against convection at all points.

10.4.1.2 The Oscillations of the Incompressible Equilibrium Sphere of Uniform Mass Density

According to a procedure of Rosseland (1932), the divergence-free Kelvin modes of the incompressible equilibrium sphere of uniform mass density can also be derived as solutions of the perturbed equation of motion given by Eq. (2.31) and the associated boundary conditions.

After separation of the time, Eq. (2.31) can be written in the vectorial form

$$\sigma^2\,\boldsymbol{\xi} = \nabla\chi \tag{10.23}$$

with $\chi = \Phi' + P'/\rho$. Since the equilibrium sphere is considered to be incompressible, the function $\chi(r,\theta,\phi)$ is harmonic, so that it can be expanded in terms of spherical harmonics as

$$\chi(r,\theta,\phi) = \sum_{\ell=0}^{+\infty}\sum_{m=-\ell}^{+\ell} A_{\ell,m}\left(\frac{r}{R}\right)^{\ell} Y_\ell^m(\theta,\phi), \tag{10.24}$$

where the $A_{\ell,m}$ are arbitrary constants. For $\sigma^2 \neq 0$, the solution for the radial component of the Lagrangian displacement then takes the form

$$\xi_r(r,\theta,\phi) = \frac{1}{\sigma^2}\sum_{\ell=0}^{+\infty}\sum_{m=-\ell}^{+\ell} A_{\ell,m}\frac{\ell}{R}\left(\frac{r}{R}\right)^{\ell-1} Y_\ell^m(\theta,\phi). \tag{10.25}$$

In accordance with the first Eq. (5.82), the solution can equivalently be expressed as

$$\xi_r(r,\theta,\phi) = \sum_{\ell=0}^{+\infty}\sum_{m=-\ell}^{+\ell} \xi_{\ell,m}(r)\, Y_\ell^m(\theta,\phi). \tag{10.26}$$

Since $\rho' = 0$, it follows from solution (7.2) of Poisson's differential equation that the Eulerian perturbation of the gravitational potential is exclusively determined by a (positive or negative) mass charge $\rho\,\xi(R)$ situated on the surface of the equilibrium sphere, per unit surface:

$$\Phi'(r,\theta,\phi) = -\sum_{\ell=0}^{+\infty}\sum_{m=-\ell}^{+\ell} \frac{4\pi G}{2\ell+1}\,\frac{\rho\,\xi_{\ell,m}(R)}{R^{\ell-1}}\, r^{\ell}\, Y_\ell^m(\theta,\phi). \tag{10.27}$$

Consequently, the Eulerian perturbation of the pressure can be expressed as

$$P'(r,\theta,\phi) = \rho \sum_{\ell=0}^{+\infty}\sum_{m=-\ell}^{+\ell} A_{\ell,m}\left(1+\frac{4\pi G\rho}{2\ell+1}\frac{\ell}{\sigma^2}\right)\left(\frac{r}{R}\right)^{\ell} Y_\ell^m(\theta,\phi), \tag{10.28}$$

and the Lagrangian perturbation of the pressure, as

$$\delta P(r,\theta,\phi) = \rho \sum_{\ell=0}^{+\infty}\sum_{m=-\ell}^{+\ell} A_{\ell,m}\left[\left(1+\frac{4\pi G\rho}{2\ell+1}\frac{\ell}{\sigma^2}\right)\frac{r}{R}-\frac{g}{\sigma^2}\frac{\ell}{R}\right]\left(\frac{r}{R}\right)^{\ell-1} Y_\ell^m(\theta,\phi). \tag{10.29}$$

The condition that the Lagrangian perturbation of the pressure vanishes on the surface of the equilibrium model leads to eigenvalue equation (8.102) for the divergence-free Kelvin modes.

No spheroidal modes other than the Kelvin modes are possible in an incompressible sphere of uniform mass density. Indeed, p-modes are excluded because of the absence of a compressibility in the medium, and g-modes, because of the absence of buoyancy $\left(N^2 = 0\right)$, due to the conjoint absence of a compressibility and a density stratification.

10.4.1.3 The Oscillations of the Compressible Equilibrium Sphere of Uniform Mass Density

In order to determine the oscillations of the compressible equilibrium sphere of uniform mass density, it is convenient to start from the fourth-order system of differential equations (6.8) and (6.10). The oscillations have also been determined by integration of the fourth-order system of differential equations (6.15)–(6.17) (Smeyers 1966b). Here we use the system of Eqs. (6.8) and (6.10).

For a constant value of Γ_1, the coefficients $K_1(r)$, $K_2(r)$, $K_3(r)$, $K_4(r)$ are given by

$$\left.\begin{aligned}
K_1(r) &= -\ell(\ell+1)\frac{GM}{R^3}\,\frac{2}{\Gamma_1}\,\frac{1}{R^2\left(1-r^2/R^2\right)},\\
K_2(r) &= \frac{2}{r}-\frac{4r}{R^2\left(1-r^2/R^2\right)},\\
K_3(r) &= -\frac{\ell(\ell+1)}{r^2}+2\left(\frac{4}{\Gamma_1}-3\right)\frac{1}{R^2\left(1-r^2/R^2\right)},\\
K_4(r) &= 0.
\end{aligned}\right\} \tag{10.30}$$

Equation (6.10) then reduces to a homogeneous second-order differential equation for the function $\alpha(r)$, as noted in Sect. 6.2.1. By passing on to the dimensionless radial coordinate $x = r/R$ and the dimensionless angular frequency $\omega = \left(GM/R^3\right)^{1/2} \sigma$, one transforms the equation into

$$\frac{d^2\alpha}{dx^2} + \frac{2-6x^2}{x(1-x^2)}\frac{d\alpha}{dx} + \left[\frac{C}{1-x^2} - \frac{\ell(\ell+1)}{x^2}\right]\alpha = 0, \tag{10.31}$$

where C is defined as

$$C = \frac{2}{\Gamma_1}\left(\omega^2 - \frac{\ell(\ell+1)}{\omega^2} + 4\right) - 6. \tag{10.32}$$

By the introduction of the dependent variable

$$v(r) = x(1-x^2)\,\alpha(r), \tag{10.33}$$

Eq. (10.31) becomes

$$\frac{d^2v}{dx^2} + \left[\frac{C+6}{1-x^2} - \frac{\ell(\ell+1)}{x^2}\right] v = 0. \tag{10.34}$$

Together with the regularity conditions that $\alpha(r)$ must be finite at the boundary points $r = 0$ and $r = R$, or, equivalently, that $v(r)$ must vanish at these points, the differential equation forms a Sturm–Liouville eigenvalue problem with singular end points, in which C is the eigenvalue parameter.

The transformation of Eq. (10.31) by means of

$$\alpha(r) = x^\ell\, w(z), \qquad x^2 = z \tag{10.35}$$

leads to the hypergeometric differential equation

$$z(1-z)\frac{d^2w}{dz^2} + \frac{1}{2}\left[2\ell+3-(2\ell+7)\,z\right]\frac{dw}{dz} + \frac{B}{4}\,w = 0 \tag{10.36}$$

with

$$B = C - 4\ell. \tag{10.37}$$

Comparison with the standard form of the hypergeometric differential equations

$$z(1-z)\frac{d^2w}{dz^2} + \left[c-(a+b+1)\,z\right]\frac{dw}{dz} - a\,b\,w = 0 \tag{10.38}$$

yields the identifications

$$c = \frac{1}{2}(2\ell+3), \qquad a+b = \ell+\frac{5}{2}, \qquad a\,b = -\frac{B}{4}. \tag{10.39}$$

The solution that remains finite at $z = 0$ is the hypergeometric series of Gauss

$$F(a,b;c;z) = \frac{\Gamma(c)}{\Gamma(a)\,\Gamma(b)} \sum_{j=0}^{\infty} \frac{\Gamma(a+j)\,\Gamma(b+j)}{\Gamma(c+j)} \frac{z^j}{j!} \equiv \sum_{j=0}^{\infty} C_{2j} z^j, \quad (10.40)$$

where $\Gamma(z)$ is the gamma function (Abramowitz & Stegun 1965). Because of the recurrence relations between gamma functions, the coefficients of successive terms in the series are related by

$$C_{2j+2} = C_{2j} \frac{2j(2\ell + 2j + 5) - B}{(2j+2)(2\ell + 2j + 3)}. \quad (10.41)$$

The series diverges at the boundary point $z = 1$, since $c - (a + b) = -1$. In order that it be finite at that point, the series must be broken of and reduced to a polynomial. Therefore, one must impose that

$$C_{2k+2} = 0, \qquad k = 0, 1, 2, \ldots, \quad (10.42)$$

or, more explicitly, that

$$2k(2k + 5 + 2\ell) - B = 0, \qquad k = 0, 1, 2, \ldots \quad (10.43)$$

This condition yields the eigenvalue equation for C, from which the quadratic eigenvalue equation for ω^2 follows

$$\omega^2 - \frac{\ell(\ell+1)}{\omega^2} = -4 + \Gamma_1[k(2k + 5 + 2\ell) + 3 + 2\ell] \equiv 2D_{\ell,k}, \quad (10.44)$$

with the roots

$$\omega^2_{\ell,k} = D_{\ell,k} \pm \left[D^2_{\ell,k} + \ell(\ell+1)\right]^{1/2}. \quad (10.45)$$

Consequently, for each degree ℓ different from zero, two infinite spectra of discrete eigenvalues σ^2 exist: a spectrum of indefinitely increasing positive eigenvalues and a spectrum of negative eigenvalues decreasing in absolute value towards the accumulation point $\sigma^2 = 0$. The first spectrum manifestly corresponds to that of the eigenvalues σ^2_p, the second spectrum, to that of the eigenvalues $\sigma^2_{g^-}$. The non-radial p- and g^--modes of a given degree ℓ different from zero are ordered from the parameter $k = 0$, while they are ordered from the radial order $n = 1$ in Cowling's classification. For $\ell = 0$, the spectrum of the eigenvalues σ^2_p is the only existing spectrum and corresponds to that of the radial modes.

The distributions of the dimensionless eigenvalues $\omega^2_{\ell,k}$ for the p- and g^--modes of the lowest degrees and lowest radial orders are presented respectively in Figs. 10.4 and 10.5, for $\Gamma_1 = 5/3$. From Fig. 10.4, it appears that the eigenvalue of the fundamental radial mode is in a direct line with the eigenvalues of the non-radial p_1-modes, the eigenvalue of the first radial overtone, in a direct line with the eigenvalues of the non-radial p_2-modes, ... Hence, the fundamental radial mode may

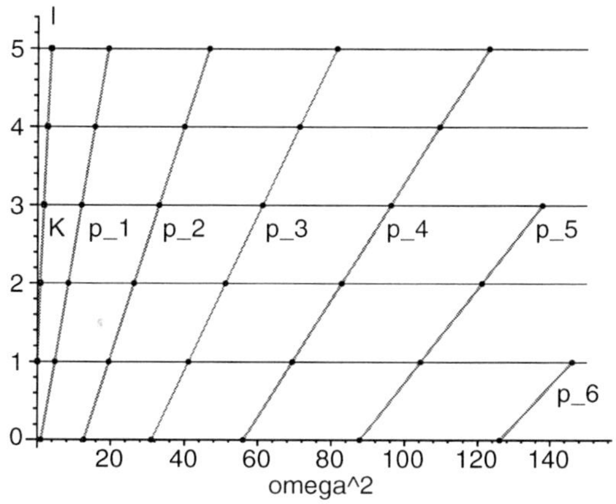

Fig. 10.4 Dimensionless positive eigenvalues ω^2 of radial modes, lowest-degrees Kelvin modes, and lowest-degrees p-modes in the compressible equilibrium sphere of uniform mass density, for $\Gamma_1 = 5/3$

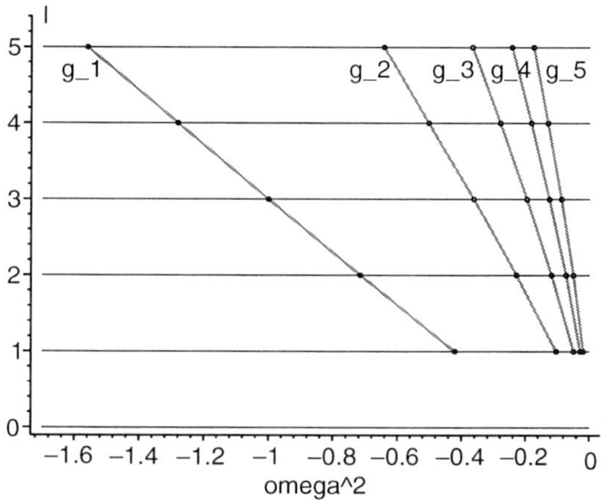

Fig. 10.5 Dimensionless negative eigenvalues ω^2 of lowest-degrees g^--modes in the compressible equilibrium sphere of uniform mass density, for $\Gamma_1 = 5/3$

be regarded as the p_1-mode belonging to $\ell = 0$, the first radial overtone, as the p_2-mode belonging to $\ell = 0, \ldots$

The eigenfunctions for the divergence of the Lagrangian displacement take the form

$$\alpha_{\ell,k}(r) = x^{\ell} \sum_{j=0}^{k} C_{2j} x^{2j}, \qquad k = 0, 1, 2, \ldots \tag{10.46}$$

From the relations existing between the coefficients C_{2j}, it results that the eigenfunctions $\alpha_{\ell,k}(r)$ of the p-mode and the g^--mode that belong to the same pair of values ℓ and k are identical. The eigenfunctions $\alpha_{\ell,k}(r)$ associated with $\ell = 2$ and $k = 0, 1, 2$ are represented as functions of x in Fig. 10.6. They are normalised such that $\alpha_{\ell,k}(r) = 1$ at $x = 1$. An eigenfunction $\alpha_{\ell,k}(r)$ displays k nodes in the interval $(0, 1)$, as expected for Sturm–Liouville eigenvalue problems.

In order to derive the eigenfunctions $\xi_{\ell,k}(r)$, one substitutes solution (10.46) for $\alpha_{\ell,k}(r)$ into Eq. (6.8) and solves the inhomogeneous differential equation by means

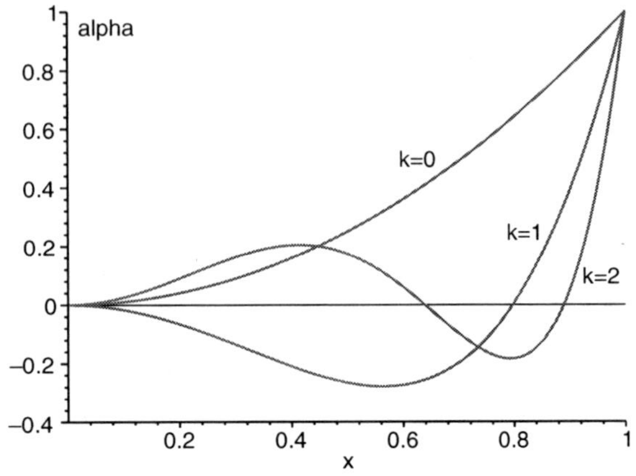

Fig. 10.6 The eigenfunctions $\alpha_{\ell,k}(r)$ in the compressible equilibrium sphere of uniform mass density for $\ell = 2$, $k = 0, 1, 2$, and $\Gamma_1 = 5/3$

of the method of the variation of the constants. The solution that remains finite at $r = 0$ can be written as

$$\frac{\xi_{\ell,k}(r)}{R} = \sum_{j=0}^{k} \left[(\ell + 2j + 2) + \frac{\ell(\ell+1)}{\omega_{\ell,k}^2} \right] \frac{C_{2j}}{(2j+2)(2\ell+2j+3)} x^{\ell+2j+1}$$
$$+ \left(1 - \delta_{\ell,0}\right) A_{\ell,k} x^{\ell-1}, \qquad (10.47)$$

where $\delta_{\ell,0}$ is a Kronecker delta, and $A_{\ell,k}$, an arbitrary constant.

For the transformation of this solution, it is useful to express recurrence relation (10.41) as

$$C_{2j+2} = C_{2j} \left[1 - \frac{B + 4\ell + 6}{(2j+2)(2\ell+2j+3)} \right], \qquad (10.48)$$

or, also as

$$\frac{C_{2j}}{(2j+2)(2\ell+2j+3)} = -\frac{1}{B + 4\ell + 6} \left(C_{2j+2} - C_{2j}\right) \qquad (10.49)$$

(Sauvenier-Goffin 1951). Solution (10.47) can then be rewritten as

$$\frac{\xi_{\ell,k}(r)}{R} = -\frac{1}{B + 4\ell + 6}$$
$$\left\{ \frac{d}{dx} \left[\left(1 - x^2\right) \alpha_{\ell,k} \right] + \frac{\ell(\ell+1)}{\omega_{\ell,k}^2} \frac{1 - x^2}{x} \alpha_{\ell,k} \right\} + E_{\ell,k} x^{\ell-1}, \qquad (10.50)$$

where $E_{\ell,k}$ is a constant defined as

$$E_{\ell,k} = \frac{C_0}{B + 4\ell + 6}\left[\ell + \frac{\ell(\ell+1)}{\omega^2}\right] + \left(1 - \delta_{\ell,0}\right) A_{\ell,k}. \tag{10.51}$$

The constant is fixed by means of boundary condition (5.97). For this, the eigensolution for the Eulerian perturbation of the gravitational potential is needed.

Solution (7.2) takes here the form

$$\Phi'_{\ell,k}(r) = \frac{4\pi G\rho R^2}{2\ell+1}\left[x^{-(\ell+1)}\int_0^x \alpha_{\ell,k}\left(r'\right) x'^{(\ell+2)}\, dx' \right.$$
$$\left. + x^{\ell}\int_x^1 \alpha_{\ell,k}\left(r'\right) x'^{-(\ell-1)}\, dx' - \frac{\xi_{\ell,k}(R)}{R}x^{\ell}\right]. \tag{10.52}$$

By substitution of solution (10.46) for $\alpha_{\ell,k}(r)$ and integration, it results that

$$\Phi'_{\ell,k}(r) = \frac{4\pi G\rho R^2}{2\ell+1}\left[\sum_{j=0}^{k}\left(\frac{1}{2\ell+2j+3} - \frac{1}{2j+2}\right) C_{2j}x^{\ell+2j+2}\right.$$
$$\left. + \sum_{j=0}^{k}\frac{C_{2j}}{2j+2}x^{\ell} - \frac{\xi_{\ell,k}(R)}{R}x^{\ell}\right]. \tag{10.53}$$

The first term inside the square brackets in the right-hand member can be transformed as

$$\sum_{j=0}^{k}\left(\frac{1}{2\ell+2j+3} - \frac{1}{2j+2}\right) C_{2j}x^{\ell+2j+2}$$
$$= -(2\ell+1)\sum_{j=0}^{k}\frac{C_{2j}}{(2\ell+2j+3)(2j+2)}x^{\ell+2j+2} \tag{10.54}$$

or, by means of equality (10.49), as

$$\sum_{j=0}^{k}\left(\frac{1}{2\ell+2j+3} - \frac{1}{2j+2}\right) C_{2j}x^{\ell+2j+2}$$
$$= \frac{2\ell+1}{B+4\ell+6}\left[\left(\alpha_{\ell,k} - C_0 x^{\ell}\right) - x^2\alpha_{\ell,k}\right]. \tag{10.55}$$

Next, the second term is transformed. From equality (10.49), it also follows that

$$\sum_{j=0}^{k}\frac{C_{2j}}{2j+2} = -\frac{1}{B+4\ell+6}\sum_{j=0}^{k}(2\ell+2j+3)\left(C_{2j+2} - C_{2j}\right), \tag{10.56}$$

so that

$$\sum_{j=0}^{k} \frac{C_{2j}}{2j+2} = -\frac{1}{B+4\ell+6}\left[\frac{d}{dx}\sum_{j=0}^{k}\left(C_{2j+2}-C_{2j}\right)x^{2\ell+2j+3}\right]_{x=1}, \tag{10.57}$$

or

$$\sum_{j=0}^{k} \frac{C_{2j}}{2j+2} = -\frac{1}{B+4\ell+6}\left\{\frac{d}{dx}\left[\left(1-x^2\right)x^{\ell+1}\,\alpha_{\ell,k} - C_0\,x^{2\ell+1}\right]\right\}_{x=1}. \tag{10.58}$$

It results that

$$\sum_{j=0}^{k} \frac{C_{2j}}{2j+2} = \frac{1}{B+4\ell+6}\left[2\sum_{j=0}^{k} C_{2j} + (2\ell+1)\,C_0\right]. \tag{10.59}$$

Solution (10.53) for $\Phi'_{\ell,k}(r)$ then becomes

$$\Phi'_{\ell,k}(r) = \frac{4\pi G\rho R^2}{B+4\ell+6}\left(1-x^2\right)\alpha_{\ell,k} - \frac{4\pi G\rho R^2}{2\ell+1}E_{\ell,k}x^{\ell}. \tag{10.60}$$

From boundary condition (5.97), it follows that $E_{\ell,k} = 0$ for all values of ℓ, so that

$$\left(1-\delta_{\ell,0}\right)A_{\ell,k} = -\frac{C_0}{B+4\ell+6}\left[\ell + \frac{\ell(\ell+1)}{\omega_{\ell,k}^2}\right]. \tag{10.61}$$

Consequently, solution (10.50) for $\xi_{\ell,k}(r)$ and solution (10.60) for $\Phi'_{\ell,k}(r)$ take the form

$$\frac{\xi_{\ell,k}(r)}{R} = -\frac{1}{B+4\ell+6}\left\{\frac{d}{dx}\left[\left(1-x^2\right)\alpha_{\ell,k}\right] + \frac{\ell(\ell+1)}{\omega_{\ell,k}^2}\frac{1-x^2}{x}\alpha_{\ell,k}\right\}, \tag{10.62}$$

$$\Phi'_{\ell,k}(r) = \frac{4\pi G\rho R^2}{B+4\ell+6}\left(1-x^2\right)\alpha_{\ell,k}. \tag{10.63}$$

These eigenfunctions were first derived by Sauvenier-Goffin (1951).

The eigenfunctions $\xi_{\ell,k}(r)/R$ for the second-degree modes p_1, p_2, p_3 and g_1^-, g_2^-, g_3^- are represented respectively in Figs. 10.7 and 10.8, for $\Gamma_1 = 5/3$, as functions of x. For the p-modes, the eigenfunction $\xi_{\ell,k}(r)$ displays one more node in the interval $0 < r < R$ than the eigenfunction $\alpha_{\ell,k}(r)$, while, for the g^--modes, both eigenfunctions have the same number of nodes.

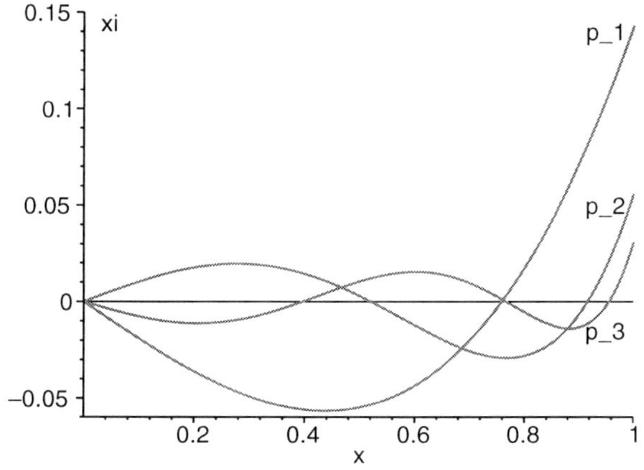

Fig. 10.7 The eigenfunctions $\xi_{\ell,k}(r)/R$ for the second-degree modes p_1, p_2, p_3 of the compressible equilibrium sphere of uniform mass density, for $\Gamma_1 = 5/3$

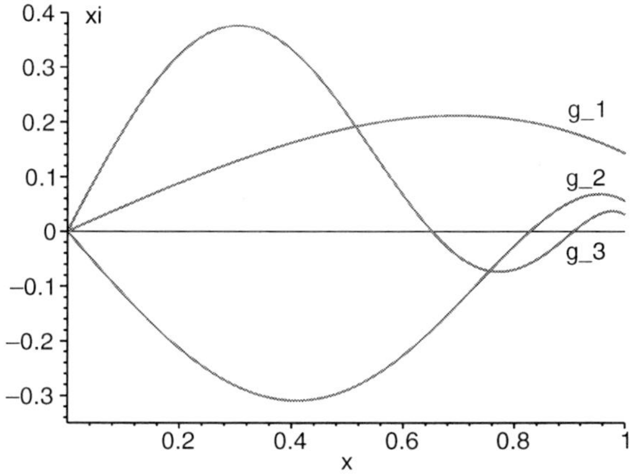

Fig. 10.8 The eigenfunctions $\xi_{\ell,k}(r)/R$ for the second-degree modes g_1^-, g_2^-, g_3^- of the compressible equilibrium sphere of uniform mass density, for $\Gamma_1 = 5/3$

In the particular case of the purely radial oscillations ($\ell = 0$), eigenvalue equation (10.44) reduces to a first-degree equation. The eigenfunctions are still given by solutions (10.46), (10.62), and (10.63). In illustration, the eigenfunctions $\xi_{0,k}(r)/R$ for the first four radial modes are presented as

$$\left.\begin{aligned} \frac{\xi_{0,0}(r)}{R} &= \frac{1}{3}x, \\ \frac{\xi_{0,1}(r)}{R} &= \frac{1}{3}\left(x - \frac{7}{5}x^3\right), \\ \frac{\xi_{0,2}(r)}{R} &= \frac{1}{3}\left(x - \frac{18}{5}x^3 + \frac{99}{35}x^5\right), \\ \frac{\xi_{0,3}(r)}{R} &= \frac{1}{3}\left(x - \frac{33}{5}x^3 + \frac{429}{35}x^5 - \frac{143}{21}x^7\right), \end{aligned}\right\} \tag{10.64}$$

in which $C_0 = 1$. The eigenfunctions correspond to those derived by Sterne (1937).

To the radial, and the non-radial p- and g^--modes, the Kelvin modes must be added. As divergence-free modes, the Kelvin modes are associated with solutions of Eq. (10.31) that are identically zero. Since such solutions were presumed to be trivial, no attention was paid to them for a long time.

The eigenfunctions for the Kelvin modes can readily be derived from Eqs. (2.52), (2.53), and (3.24), written in a Lagrangian form, and the related boundary conditions. We here consider the time to be separated from the equations.

From Eqs. (2.53) and (3.24), it follows that

$$\delta\rho = 0, \qquad \delta P = 0, \tag{10.65}$$

so that Eqs. (2.52) can be reduced to the vectorial form

$$\sigma^2\, \boldsymbol{\xi} = \nabla(\delta\Phi). \tag{10.66}$$

By taking the divergence of both members, one has

$$\nabla^2(\delta\Phi) = 0. \tag{10.67}$$

Hence, the Lagrangian perturbation of the gravitational potential is solution of Laplace's equation. The solution for the radial part $\delta\Phi(r)$ that remains finite at $r = 0$ has the form

$$\delta\Phi(r) = Ar^{\ell}, \tag{10.68}$$

where A is an arbitrary constant. The radial component of Eq. (10.66) can then be written as

$$\sigma^2\, \xi(r) = A\ell r^{\ell-1}. \tag{10.69}$$

Use of the relation between the Eulerian and the Lagrangian perturbation of a scalar quantity and multiplication by σ^2 yield

$$\sigma^2\, \Phi'(r) = A\left(\sigma^2 - \frac{4\pi G\rho}{3}\ell\right) r^{\ell}. \tag{10.70}$$

By imposing boundary condition (5.97), one obtains eigenvalue equation (8.102).

As is seen in Fig. 10.4, the eigenvalue ω^2 of the fundamental radial mode is close to that of the second-degree Kelvin mode for the value $\Gamma_1 = 5/3$ considered. From eigenvalue equations (8.102) and (10.45), it follows that the eigenvalues of the fundamental radial mode and the second-degree Kelvin mode are equal for $\Gamma_1 = 8/5$. Then, the eigenvalue problem of the linear, isentropic oscillations of the spherically symmetric equilibrium sphere of uniform mass density admits of a double root for σ^2, in accordance with the possibility mentioned in Sect. 4.6, and is accidently degenerate. This degeneracy was found by Chandrasekhar & Lebovitz (1962a) for polytropic models (see also Chandrasekhar & Lebovitz 1962b, Chandrasekhar & Lebovitz 1963).

10.4.2 Polytropic and Physical Models

Cowling illustrated his classification of non-radial modes by means of numerical results for the eigenfrequencies of a few second-degree low-order modes in the polytropic model with index $n = 3$. He did this in the approximation afterwards named after him, in which the Eulerian perturbation of the gravitational potential is neglected. Using the same approximation, Mullan & Ulrich (1988) and Mullan (1989) determined a large number of high-precision eigenfrequencies of p- and g-modes of several polytropic models.

Cowling's classification remains valid even when the Eulerian perturbation of the gravitational potential is incorporated into the governing equations, at least for polytropic models with a not too high central mass condensation. In this case, accurate eigenfrequencies of polytropic models were determined by Christensen-Dalsgaard & Mullan (1994).

In Table 10.2, the dimensionless eigenvalues ω^2 of the lowest-order second-degree non-radial modes in the polytropic models with index $n = 0, 1, 1.5, 2, 3$ are presented for $\Gamma_1 = 5/3$. The eigenvalues ω^2 of the g-modes are negative for the

Table 10.2 Dimensionless eigenvalues ω^2 of the lowest-order second-degree non-radial modes in the polytropic models with index $n = 0, 1, 1.5, 2, 3$

	$n = 0$	$n = 1$	$n = 1.5$	$n = 2$	$n = 3$
$\rho_c/\bar{\rho}$	1.00	3.290	5.991	10.50	54.18
p_{10}	412.7	311.9	285.0	264.3	233.6
p_9	341.0	259.4	237.7	221.0	196.7
p_8	276.0	211.5	194.5	181.5	163.0
p_7	217.7	168.3	155.5	145.7	132.3
p_6	166.0	129.9	120.6	113.7	104.9
p_5	121.0	96.11	89.92	85.51	80.57
p_4	82.74	67.08	63.47	61.13	59.43
p_3	51.12	42.84	41.30	40.63	41.47
p_2	26.23	23.49	23.51	24.07	26.72
p_1	8.382	9.308	10.29	11.55	15.26
f	0.8000	1.497	2.120	3.113	8.175
g_1	−0.7158	−0.3029	0	0.5633	4.915
g_2	−0.2288	−0.1383	0	0.2967	2.828
g_3	−0.1174	−0.08021	0	0.1839	1.822
g_4	−0.07252	−0.05272	0	0.1254	1.271
g_5	−0.04957	−0.03742	0	0.09113	0.9364
g_6	−0.03614	−0.02798	0	0.06928	0.7189
g_7	−0.02756	−0.02174	0	0.05449	0.5695
g_8	−0.02174	−0.01739	0	0.04401	0.4624
g_9	−0.01759	−0.01424	0	0.03630	0.3830
g_{10}	−0.01454	−0.01187	0	0.03046	0.3225

polytropic models with index $n = 0$ and $n = 1$, in which $N^2 < 0$, equal to zero for the polytropic model with index $n = 1.5$, in which $N^2 = 0$, and positive for the polytropic models with index $n = 2$ and $n = 3$, in which $N^2 > 0$.

With regard to the polytropic model with index $n = 3$, the dimensionless eigenvalues ω^2 of the lowest-order non-radial modes and the relative radial distances $x = r/R$ of the nodes of the associated eigenfunctions $\xi(r)$ are presented in Tables 10.3–10.5, for the degrees $\ell = 1, 2, 3$ and $\Gamma_1 = 5/3$. For the lowest-order modes belonging to $\ell = 2$, the eigenvalues ω^2 and the positions of the nodes of the eigenfunctions $\xi(r)$ are represented in the propagation diagram given by Fig. 10.9.

Table 10.3 Lowest-order first-degree non-radial modes in the polytropic model with index $n = 3$, for $\Gamma_1 = 5/3$, dimensionless eigenvalues ω^2, relative distances $x = r/R$ of the nodes of the eigenfunctions $\xi(r)$

Non-radial mode	ω^2	Positions of the nodes of $\xi(r)$				
g_5^+	0.3699	0.08868	0.1598	0.2519	0.3789	0.7019
g_4^+	0.5178	0.1055	0.2009	0.3248	0.6492	
g_3^+	0.7757	0.1362	0.2598	0.5844		
g_2^+	1.286	0.1771	0.5051			
g_1^+	2.516	0.4082				
f	0					
p_1	11.40	0.2550				
p_2	21.55	0.2444	0.7629			
p_3	34.92	0.2241	0.6545	0.8658		
p_4	51.50	0.2053	0.5785	0.7837	0.9122	
p_5	71.29	0.1889	0.5211	0.7177	0.8495	0.9378

Table 10.4 Lowest-order second-degree non-radial modes in the polytropic model with index $n = 3$, for $\Gamma_1 = 5/3$, dimensionless eigenvalues ω^2, relative distances $x = r/R$ of the nodes of the eigenfunctions $\xi(r)$

Non-radial mode	ω^2	Positions of the nodes of $\xi(r)$				
g_5^+	0.9364	0.1061	0.1781	0.2674	0.3995	0.7237
g_4^+	1.271	0.1263	0.2185	0.3496	0.6736	
g_3^+	1.822	0.1574	0.2901	0.6094		
g_2^+	2.828	0.2135	0.5237			
g_1^+	4.915	0.3948				
f	8.175					
p_1	15.26	0.6041				
p_2	26.72	0.4735	0.8165			
p_3	41.47	0.3985	0.7190	0.8888		
p_4	59.43	0.3481	0.6458	0.8155	0.9246	
p_5	80.57	0.3111	0.5881	0.7542	0.8680	0.9452

Table 10.5 Lowest-order third-degree non-radial modes in the polytropic model with index $n = 3$, for $\Gamma_1 = 5/3$, dimensionless eigenvalues ω^2, relative distances $x = r/R$ of the nodes of the eigenfunctions $\xi(r)$

Non-radial mode	ω^2	Positions of the nodes of $\xi(r)$				
g_5^+	1.587	0.1207	0.1925	0.2813	0.4137	0.7482
g_4^+	2.093	0.1425	0.2343	0.3655	0.7045	
g_3^+	2.887	0.1757	0.3086	0.6486		
g_2^+	4.236	0.2356	0.5737			
g_1^+	6.767	0.4547				
f	9.414					
p_1	18.44	0.6956				
p_2	31.26	0.5750	0.8465			
p_3	47.32	0.4962	0.7585	0.9035		
p_4	66.57	0.4391	0.6896	0.8370	0.9331	
p_5	89.00	0.3954	0.6337	0.7798	0.8812	0.9506

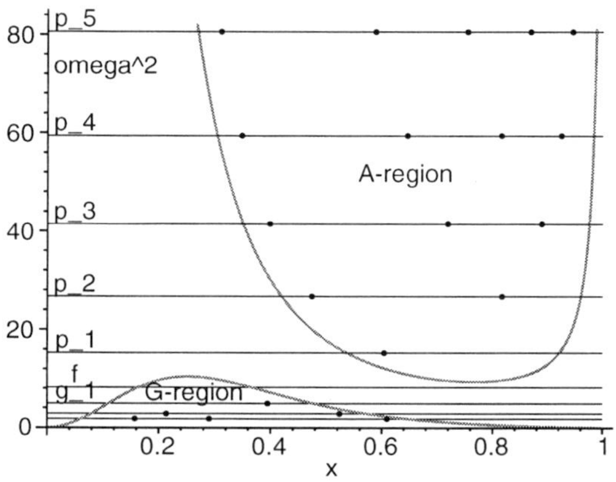

Fig. 10.9 Propagation diagram of the polytropic model with index $n = 3$, for $\Gamma_1 = 5/3$, in which the dimensionless eigenvalues ω^2 and the positions of the nodes of the eigenfunctions $\xi(r)$ are represented for the lowest-order second-degree non-radial modes

The ordinates of the horizontal lines correspond to the eigenvalues of the modes. The horizontal lines pass across the A-cavity or the G-cavity. Moreover, the nodes of the eigenfunctions $\xi(r)$ lie in the cavity with which the mode is associated. This illustrates that the p-modes are formed in the A-cavity, and the g^+-modes, in the G-cavity. The oscillatory behaviour of the eigenfunctions $\xi(r)$ in their cavity is a mathematical reflection of the physical reality that the mode originates from waves propagating to-and-fro in that cavity.

Cowling's classification is also valid for physical models of stars with a not too high central mass condensation. This was shown for models of massive stars consisting of a convective core in adiabatic equilibrium and a radiative envelope by Smeyers (1963), Van der Borght & Wan Fook Sun (1965), Smeyers (1967) and has afterwards been confirmed for many other physical models.

In the spheroidal p-, g-, and f-modes, the Lagrangian displacement field consists of a curl-free longitudinal part and a divergence-free poloidal part, in accordance with the resolution given by Eq. (5.63), in which the toroidal part is here equal to zero. A quantity that measures the relative importance of the two parts is the ratio of the radial amplitude of the divergence to that of the horizontal component of the curl:

$$\begin{aligned}\frac{\alpha(r)}{(\nabla \times \boldsymbol{\xi})_h\,(r)} &= \left[\frac{1}{r^2}\frac{d\left(r^2\,\xi\right)}{dr} - \frac{\ell(\ell+1)}{r^2}\eta\right] \Big/ \left(\xi - \frac{d\eta}{dr}\right)\\ &= \sigma^2 \frac{g(r)}{c^2(r)\,N^2(r)} \equiv \sigma^2\,W(r).\end{aligned} \tag{10.71}$$

Wolff (1979) derived the expression in the Cowling approximation, but this expression remains valid when the Eulerian perturbation of the gravitational potential is included.

The equality shows that the divergence of the Lagrangian displacement field is dominant where $\left|\sigma^2\right|$ is sufficiently larger than $\left|c^2(r)\,N^2(r)/g(r)\right|$, and that the horizontal component of the curl is dominant where $\left|\sigma^2\right|$ is sufficiently smaller than $\left|c^2(r)\,N^2(r)/g(r)\right|$. The p-modes with their larger eigenvalues are therefore largely divergence-dominated modes, and the g-modes with their smaller eigenvalues, largely curl-dominated modes.

For a given eigenvalue σ^2, the ratio of the radial amplitude of the divergence to that of the horizontal component of the curl varies from layer to layer as the function $W(r) = g(r)/\left[c^2(r)\,N^2(r)\right]$. By way of illustration, the variation of the function $W(r)$ in the polytropic model with index $n = 3$ is represented in Fig. 10.10. In this model, the ratio is large in a small region near the centre, decreases rapidly, and then increases slightly towards the surface. Hence, for any mode, the ratio of the amplitude of the divergence to that of the horizontal component of the curl is much larger near the centre than it is farther away from that point.

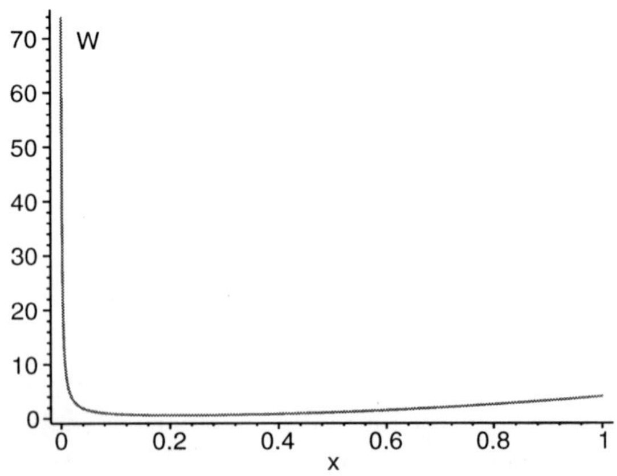

Fig. 10.10 Variation of the ratio of the divergence to the horizontal component of the curl of the Lagrangian displacement field in the polytropic model with index $n = 3$, for $\Gamma_1 = 5/3$, as a function of the relative radial distance from the centre, $x = r/R$

In conclusion, because of the possible existence of p- and g^+-modes, side by side, a vibrating star is comparable to a double oscillator that depends on the degree ℓ: the oscillator with the eigenfrequencies σ_p, which are associated with the A-cavity, and the oscillator with the eigenfrequencies σ_{g^+}, which are associated with the G-cavity (Aizenman & Smeyers 1977).

In stellar models with a lower central mass condensation, the cavities have no, or almost no, common interval of frequencies. See, for example, the polytropic model with index $n = 3$, whose propagation diagram for $\ell = 2$ is represented in Fig. 10.9. In these models, the oscillators can be regarded as almost uncoupled. However, in stellar models with a higher central mass condensation, the A- and G-cavity have a common domain of frequencies. See, for example, the polytropic model with index $n = 4$, whose propagation diagram for $\ell = 2$ is represented in Fig. 9.8. In such models, interactions between low-order g^+-modes, on one side, and the f-mode and low-order p-modes of the same degree, on the other side, appear to occur. The interactions of modes of the same degree and their implications for the Cowling classification of non-radial modes in stars are discussed in the next chapter.

Even in stellar models with a larger central mass condensation, the higher-order p- and g^+-modes of a same degree remain uncoupled. Therefore, asymptotic representations of higher-order p- and g^+-modes can be developed separately, as is done in Chap. 14 and following chapters.

10.5 Beyer's Study on the Nature of the Oscillation Spectra

Beyer has investigated the nature of the spectra of the radial and non-radial isentropic stellar oscillations within the framework of functional analysis (Beyer 1994, Beyer 1995a, Beyer 1995b). The results of his investigations were synthesised in a research note by Beyer & Schmidt (1995). The subsequent summary is based on this research note.

10.5.1 System of Equations

Beyer & Schmidt (1995) started from a vectorial wave equation, which, after separation of the time by a factor $\exp(i\sigma t)$, can be written in a form equivalent to Eq. (5.4)

$$\sigma^2 \boldsymbol{\xi} = \nabla \left(g\, \xi_r - c^2\, \alpha \right) + \frac{N^2}{g}\, c^2\, \alpha\, \mathbf{1}_r + \nabla \Phi'. \tag{10.72}$$

For spheroidal normal modes associated with a spherical harmonic $Y_\ell^m(\theta, \phi)$, the equation has the radial and transverse components

$$\left.\begin{aligned}\sigma^2 \xi &= \frac{d}{dr}\left\{g\,\xi - c^2\left[\frac{1}{r^2}\frac{d}{dr}\left(r^2\,\xi\right) - \frac{\ell(\ell+1)}{r^2}\eta\right]\right\}\\ &\quad + \frac{N^2}{g}c^2\left[\frac{1}{r^2}\frac{d}{dr}\left(r^2\,\xi\right) - \frac{\ell(\ell+1)}{r^2}\eta\right] + \frac{d\Phi'}{dr},\\ \sigma^2\eta &= g\,\xi - c^2\left[\frac{1}{r^2}\frac{d}{dr}\left(r^2\,\xi\right) - \frac{\ell(\ell+1)}{r^2}\eta\right] + \Phi'.\end{aligned}\right\} \tag{10.73}$$

By passing on to the relative radial component of the Lagrangian displacement $\zeta(r) = \xi(r)/r$ and to the function $\chi = c_\ell\,\eta/r^2$, in which $c_\ell = \sqrt{\ell(\ell+1)}$, one obtains the system of equations

$$\left.\begin{aligned}\sigma^2\zeta &= -\frac{1}{\rho\,r^4}\frac{d}{dr}\left(r^4\,\Gamma_1\,P\frac{d\zeta}{dr}\right)\\ &\quad - \left\{\frac{1}{\rho\,r}\frac{d}{dr}\left[(3\Gamma_1 - 4)\,P\right]\right\}\zeta + 4\pi G\rho\,\zeta\\ &\quad + c_\ell\left[\frac{d}{dr}\left(\frac{c^2}{r}\chi\right) - \frac{c^2}{r}\left(\frac{N^2}{g} - \frac{1}{r}\right)\chi\right] + \frac{1}{r}\frac{d\Phi'}{dr},\\ \sigma^2\chi &= -c_\ell\left[\frac{\Gamma_1 P}{\rho}\frac{1}{\rho\,r^5}\frac{d}{dr}\left(\rho\,r^4\,\zeta\right)\right.\\ &\quad \left. + \frac{\Gamma_1 P}{\rho}\frac{1}{r}\left(\frac{N^2}{g} - \frac{1}{r}\right)\zeta\right] + \frac{c_\ell^2}{r^2}\frac{\Gamma_1 P}{\rho}\chi + \frac{c_\ell}{r^2}\Phi'.\end{aligned}\right\} \tag{10.74}$$

In these equations, Φ' and $d\Phi'/dr$ can be expressed in terms of $\zeta(r)$ and $\chi(r)$ by means of solutions (7.4) and (7.5) of Poisson's equation. The system then corresponds to that of Eqs. (8) and (9) of Beyer & Schmidt (1995), apart from the fact that the time is not yet separated from the authors' equations.

10.5.2 The Radial Modes

For $\ell = 0$, $c_\ell = 0$, and $\chi = 0$, the first equation (10.74) corresponds to Eq. (6.28). As noted in Sect. 10.2, this second-order differential equation is of the type of the differential equation considered in the Sturm–Liouville eigenvalue problems but has singular end points at $r = 0$ and $r = R$. In this connection, Beyer & Schmidt (1995) first recalled the theory of the singular Sturm–Liouville eigenvalue problems developed by Weyl (1910a, 1910b, 1950):

> The new feature that arises for singular problems is that the eigenfunctions do not necessarily form a complete set and thus there may be a continuous part of the spectrum. This is similar to quantum mechanics where for example the bound states of the hydrogen atom do not form a basis of the Hilbert space. Additional scattering states are needed for a complete description. These are connected with the "continuous part of the spectrum". Singular coefficients in the differential equation can act like an infinite system. *To exclude a*

> *continuous part of the spectrum for the case of stellar oscillations is very important because all linear stability arguments depend crucially on the completeness of the eigenfunctions.*
>
> Weyl's theory states that quite generally, at a boundary point, there are two possibilities concerning boundary conditions, depending on the behaviour of the solutions of the differential equation near the singular boundary point: either one has to prescribe a boundary condition – the limit circle case – or the problem is already fixed by the condition that the solution is square integrable – the limit point case ... In both cases, selfadjoint extensions can be constructed and a generalisation of the regular Sturm–Liouville theory is available. In both cases, however, the spectrum may have a continuous part.

For the distinction between the limit circle case and the limit point case, the reader may be referred to Richtmyer (1978).

Beyer (1994) brought Eq. (6.28) into the form

$$\sigma^2 f = -\frac{d^2 f}{dx^2} + V(x) f, \quad 0 < x < x_0, \tag{10.75}$$

which is similar to the form of Eq. (9.53). He considered polytropic equilibrium models with finite radius, i.e., $0 < n < 5$, for which Γ_1 is constant. For such models, he determined the asymptotic behaviour of the potential $V(x)$ near the end points $x = 0$ and $x = x_0$ in terms of the Lane-Emden function of the background model. This asymptotic behaviour determines whether one is dealing with the limit point case or with the limit circle case.

Beyer & Schmidt (1995) drew the conclusion:

> *at the center one has the limit point case for all* n, $0 < n < 5$. *At the surface ... one has the limit point case for* $0 < n < 1$ *and the limit point case for* $1 < n < 5$. In the limit point case the square integrable solution satisfies the boundary condition
>
> $$\delta P(t, r) \to 0 \quad \text{for} \quad r \to R.$$
>
> In the limit circle case the boundary condition selects a unique self-adjoint extension. In both cases a theorem by Friedrichs (1948) which uses the asymptotic form of V shows that *the spectrum is "purely discrete"*. This means the spectrum has no accumulation points and consists only of eigenvalues of finite multiplicity and hence there is a basis of eigenvectors of the *Hilbert space*. Hence finally *one obtains a complete confirmation of what was conjectured by Ledoux & Walraven* (1958) and taken for granted in most of the astrophysical literature.
>
> By the same technique one can also ensure a purely discrete spectrum in the case of (possibly different) polytropic background models only near the center and the boundary and if Γ_1 is only constant near the center and the boundary of the star but otherwise quite arbitrary.

10.5.3 The Non-Radial Modes

For non-radial modes, a fourth-order system of equations must be considered, but no extension of Weyl's theory to such systems seems to be available. Beyer (1995a, 1995b) analysed the system of Eqs. (10.74) and, again in the wording of Beyer & Schmidt (1995), showed that the system is symmetric in the Hilbert space of pairs, ζ, χ, of square integrable functions with respect to the measure ρr^4. Therefore the Hilbert space is fixed ... *The Cowling approximation already determines the domain of the self-adjoint extensions* for the non-radial pulsation problem. To construct all these extensions Beyer generalised Weyl's theory, which holds only for one second-order differential equation, to systems of the form of the Cowling approximation. Again one has the limit-point, limit-circle alternative for this class of operators.

In the particular case of a polytropic equation of state for the background and Γ_1 =const. one has the same situation as for the radial case ...

Having determined the correct self-adjoint extensions, we have to investigate the nature of the spectrum. Unfortunately the spectrum can change if we pass from the Cowling approximation to the full problem ... Therefore the full system has to be analysed to find the spectrum.

No general theorems seem to be available to determine the nature of the spectrum. So the only possibility is to determine it directly ...

Beyer & Schmit (1995) presented the conclusion:

The spectrum is countable and ... there exists a countable collection of eigenvectors which span the Hilbert space, and the corresponding eigenspaces are orthogonal. The essential difference to the case of a "purely discrete" spectrum is that the eigenvalues of a "pure point spectrum" may have accumulation points and this actually happens in our case. We conjecture in agreement with numerical calculation that, for non-radial stellar oscillations, $\sigma^2 = 0$ is the only accumulation point, but we cannot prove it.

Chapter 11
Classification of the Spheroidal Normal Modes (continued)

11.1 Additional Nodes for Models with a Larger Central Mass Condensation

In a study of second-degree non-radial modes of polytropic models with index $n = 3.25, 3.50, 3.75$, in the Cowling approximation, Owen (1957) observed that, with an increasing central mass condensation, one low-radial order mode after the other ceases to exist: this is the case first for the f-mode, then for the neighbouring p_1- and g_1^+-mode, and so on. He inferred that, in stellar models with a larger central mass condensation, Cowling's classification of the non-radial oscillations does not remain valid for the modes of the lowest radial orders.

Robe (1968) redid Owen's study by integrating the governing equations for various modes of polytropic models with increasing index, with inclusion of the Eulerian perturbation of the gravitational potential. He came to the conclusion that no modes cease to exist, but that, for polytropic models with index $n \geq 3.37$, additional nodes appear in the eigenfunction $\xi(r)$ for the radial component of the Lagrangian displacement (see also Christensen-Dalsgaard & Gough 2001).

The dimensionless eigenvalues ω^2 of the lowest-order second-degree modes in the polytropic models with index $n = 3,\ 3.25,\ 3.50,\ 4$ are presented in Table 11.1, for $\Gamma_1 = 5/3$. In Table 11.2, the numbers of nodes of the eigenfunctions $\xi(r)$ are given for the modes contained in Tables 10.2 and 11.1. Moreover, the variations of the dimensionless eigenvalues ω^2 of the lowest-order second-degree modes in the polytropic models with index $n = 0,\ 1,\ 1.5,\ 2,\ 3,\ 3.25,\ 3.50,\ 4$ are represented in Fig. 11.1.

In Table 11.2, one observes an increase of the number of nodes in the eigenfunctions $\xi(r)$ of lower-order modes in polytropic models with a larger central mass condensation ($n > 3.25$). This increase has its origin in the fact that, in the propagation diagram, the top of the G-cavity is situated at a higher frequency than the lower limit of the A-cavity. Eigenfrequencies of g^+-modes may therefore acquire values that are comparable to, and even larger than, the eigenfrequencies of the f-mode and the lowest-order p-modes. Consequently, an interval of eigenfrequencies exists in which the modes show a dual character: a character of g^+-mode in the G-cavity and a character of p-mode in the A-cavity. The eigenfunctions $\xi(r)$ of

P. Smeyers and T. Van Hoolst, *Linear Isentropic Oscillations of Stars: Theoretical Foundations*, Astrophysics and Space Science Library 371,
DOI 10.1007/978-3-642-13030-4_12,

Table 11.1 Dimensionless eigenvalues ω^2 of the lowest-order second-degree non-radial modes in the polytropic models with index $n = 3,\ 3.25,\ 3.5,\ 4$, for $\Gamma_1 = 5/3$

	n = 3	n = 3.25	n = 3.5	n = 4
$\rho_c/\bar{\rho}$	54.18	88.15	152.9	622.4
p_{10}	233.6	227.8	222.7	215.9
p_9	196.7	192.3	188.5	184.0
p_8	163.0	159.7	157.1	154.8
p_7	132.3	130.2	128.5	128.4
p_6	104.9	103.6	102.9	105.3
p_5	80.57	80.14	80.14	87.47
p_4	59.43	59.66	60.29	76.69
p_3	41.47	42.22	43.41	62.88
p_2	26.72	27.88	29.76	50.82
p_1	15.26	16.97	20.93	42.14
f	8.175	11.24	16.17	34.34
g_1^+	4.915	7.970	12.10	27.60
g_2^+	2.828	4.857	8.546	23.01
g_3^+	1.822	3.171	5.741	18.00
g_4^+	1.271	2.224	4.063	15.36
g_5^+	0.9364	1.644	3.017	12.77
g_6^+	0.7189	1.265	2.327	10.09
g_7^+	0.5695	1.003	1.848	8.092
g_8^+	0.4624	0.8155	1.504	6.617
g_9^+	0.3830	0.6759	1.247	5.505
g_{10}^+	0.3225	0.5695	0.8983	4.649

these modes then display not only more nodes, but also nodes in both cavities. The dual character of a number of second-degree lower-order modes is illustrated by Fig. 11.2 for the polytropic model with index $n = 4$: in addition to the dimensionless eigenvalues ω^2 of the modes, the positions of the nodes of the eigenfunctions $\xi(r)$ are shown.

The dual character of some non-radial modes was first observed by Dziembowski (1971) for a model of a 7 $M_\odot$ Cepheid. In models for stars with a larger central mass condensation and a rarefied atmosphere, as classical Cepheids and RR Lyrae stars, dual modes are found with only a few nodes in the A-region and many more nodes in the G-region. The separation in frequency between two modes whose numbers of nodes differ by one node in the G-region is then much smaller than the separation in frequency between two modes whose numbers of nodes differ by one node in the A-region (Dziembowski 1977, Van Hoolst et al. 1998).

Scuflaire (1974) observed that the additional nodes appear in pairs: initially a double node appears and then the two nodes move away from each other for polytropic models with increasing indexes. For example, the eigenfunction $\xi(r)$ of

Table 11.2 The numbers of nodes in the eigenfunction $\xi(r)$ for the lowest second-degree modes in polytropic models with increasing index, for $\Gamma_1 = 5/3$

	n = 0	n = 1	n = 2	n = 3	n = 3.25	n = 3.5	n = 4
p_{10}	10	10	10	10	10	10	10
p_9	9	9	9	9	9	9	9
p_8	8	8	8	8	8	8	8
p_7	7	7	7	7	7	7	7
p_6	6	6	6	6	6	6	6
p_5	5	5	5	5	5	5	5
p_4	4	4	4	4	4	4	4
p_3	3	3	3	3	3	3	5
p_2	2	2	2	2	2	2	4
p_1	1	1	1	1	1	1	3
f	0	0	0	0	0	2	4
g_1	0	0	1	1	1	1	3
g_2	1	1	2	2	2	2	4
g_3	2	2	3	3	3	3	5
g_4	3	3	4	4	4	4	4
g_5	4	4	5	5	5	5	5
g_6	5	5	6	6	6	6	6
g_7	6	6	7	7	7	7	7
g_8	7	7	8	8	8	8	8
g_9	8	8	9	9	9	9	9
g_{10}	9	9	10	10	10	10	10

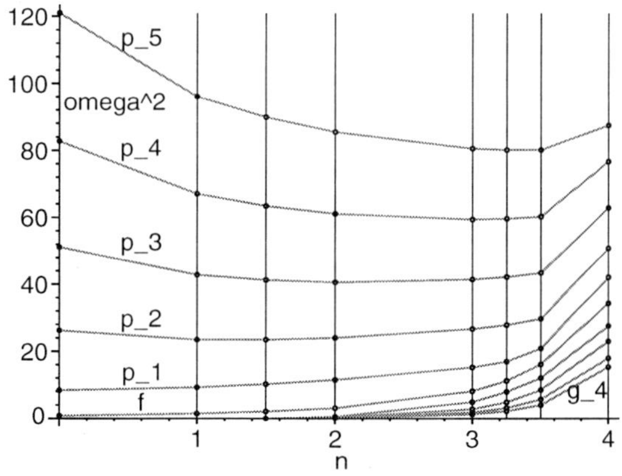

Fig. 11.1 The variations of the dimensionless eigenvalues ω^2 of the lowest-order second-degree modes in the polytropic models with index $n = 0,\ 1,\ 1.5,\ 2,\ 3,\ 3.25,\ 3.50,\ 4$, for $\Gamma_1 = 5/3$

the second-degree f-mode acquires an additional double node in the polytropic model with index $n = 3.4213$ ($\rho_c/\bar{\rho} = 127.52$). The additional nodes in the eigenfunctions $\xi(r)$ of the neighbouring p- and g^+-modes appear in polytropic models with larger indexes.

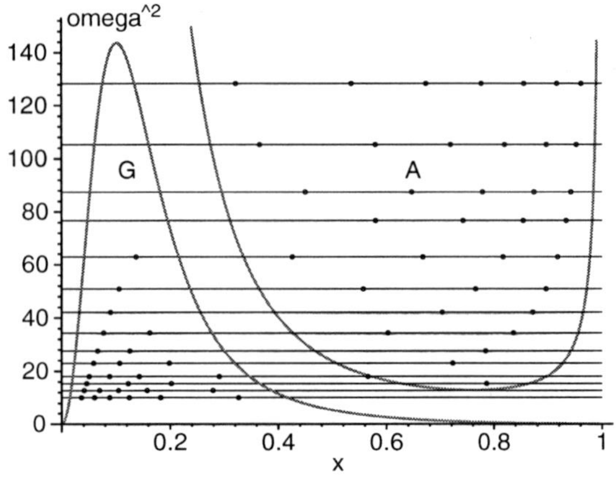

Fig. 11.2 Propagation diagram of the polytropic model with index $n = 4$, for $\Gamma_1 = 5/3$, in which the dimensionless eigenvalues ω^2 and the positions of the nodes of the eigenfunctions $\xi(r)$ are represented for the second-degree lowest-order non-radial modes

Both Scuflaire (1974) and Osaki (1975) drew attention to the fact that, even for a mode of dual character, the radial order can be determined on the basis of the number of the nodes of the eigenfunction $\xi(r)$, if these nodes are counted in an appropriate way. For modes with a singular character, the rule holds that the radial order of a p-/g^+-mode is equal to the number of nodes of the eigenfunction $\xi(r)$ in the A-/G-cavity; for modes with a dual character, this rule is modified: the radial order of a p-/g^+-mode is equal to the number of nodes of the eigenfunction $\xi(r)$ in the A-/G-cavity minus the number of nodes in the other cavity.

Scuflaire (1974) and Osaki (1975) also proposed a practical way of counting the nodes on the analogy of a procedure used in geophysics (Eckart 1960). The rule has been elaborated in the Cowling approximation. The radial order of a mode is determined by the behaviour of the solution in a phase diagram, in which the radial component of the Lagrangian displacement, $\xi(r)$, is displayed in abscissa, and the Eulerian perturbation of the pressure, $P'(r)$, in ordinate. The nodes of the eigenfunction $\xi(r)$ are considered as g^+-nodes or as p-nodes according to the phase point (ξ, P') cuts the axis $\xi = 0$ clockwise or anticlockwise for increasing radial distances r. The modes are classified in terms of the difference k between the number of p-nodes, N_p, and the number of g^+-nodes, N_{g^+}: $k = N_p - N_{g^+}$. For p-modes, k is positive, for g-modes, negative, and for f-modes, equal to zero. A mathematical basis for the scheme proposed by Scuflaire (1974) and Osaki (1975) was developed by Gabriel & Scuflaire (1979).

In the full problem, in which the Eulerian perturbation of the gravitational potential is taken into account, the classification scheme of Scuflaire (1974) and Osaki (1975) fails for first-degree modes in δ Scuti stars (Lee 1985), in evolving 1 $M_\odot$ models (Guenther 1991), and in polytropic models with larger central mass condensations (Christensen-Dalsgaard & Mullan 1994). The reason is that the Eulerian perturbation of the gravitational potential is not small in the cores of stars with a larger central mass condensation.

Takata (2006b), on his part, presented a scheme for the classification of the first-degree modes in which the Eulerian perturbation of the gravitational potential is included. The scheme is based on a second-order system of two ordinary differential equations derived by a suitable choose of the dependent variables and a use of the first integral specific to first-degree modes. It seems to solve the question of the classification of the first-degree modes (Dogan et al. 2008).

11.2 Mode Bumping in Models with a Larger Central Mass Condensation

In sequences of stellar models with an increasing central mass condensation, modes with a dual character are seen to bump against other modes. A typical example is the mode bumping in a sequence of 14 models of an evolving 10 $M_{\odot}$ star described by Osaki (1975). For each model, the evolution time and the ratio $\rho_c/\bar{\rho}$ are given in Table 11.3. In Fig. 11.3, the variations of the dimensionless eigenvalues ω^2 of the second-degree modes g_2^+, g_1^+, f, p_1, p_2 are represented. The g_1^+-mode bumps against the f-mode, the f-mode, against the p_1-mode, and the p_1-mode, against the p_2-mode. At a later stage in the star's evolution, the g_2^+-mode bumps against the g_1^+-mode, the g_1^+-mode, against the f-mode, and the f-mode, against the p_1-mode. A striking phenomenon is that the eigenfrequency of a mode against which another mode has bumped, rapidly increases, while the bumping mode almost adopts the eigenfrequency of the mode against which it bumped.

Shibahashi & Osaki (1976) and Aizenman et al. (1977) independently pointed out the close analogy between the phenomenon of mode bumping and that of the avoided crossing of modes of a double oscillator. The latter authors considered the phenomenon of mode bumping in a sequence of models of an evolving 16 $M_{\odot}$ star and referred to the example of avoided crossings of modes for a string vibrating in a sphere, which has been treated in Morse & Feshbach (1953).

Table 11.3 Properties of the successive models for an evolving 10 $M_{\odot}$ star, as considered by Osaki (1975)

Model number	Evolution time (10^6 year)	$\rho_c/\bar{\rho}$	Model number	Evolution time (10^6 year)	$\rho_c/\bar{\rho}$
1	0	38.9	8	10.6	124
2	3.03	46.6	9	11.6	155
3	4.62	55.6	10	12.5	194
4	5.65	62.5	11	13.3	244
5	6.83	71.3	12	13.9	310
6	8.16	84.2	13	14.5	395
7	9.42	101	14	14.7	471

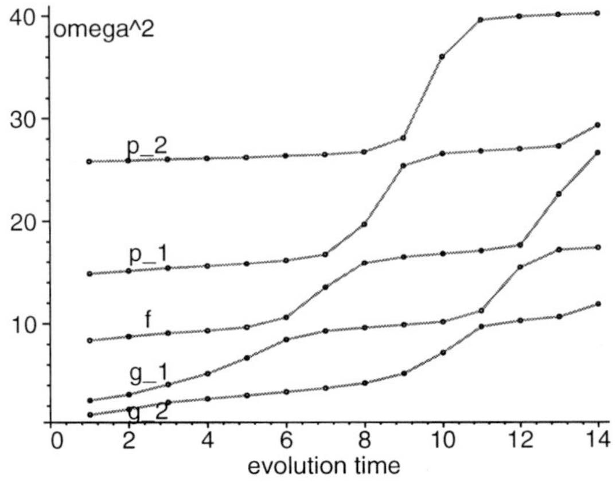

Fig. 11.3 The dimensionless eigenvalues ω^2 of the second-degree modes g_2^+, g_1^+, f, p_1, p_2 in a sequence of models for an evolving $10\,M_\odot$ star, as considered by Osaki (1975)

11.3 Theory of the Avoided Crossings of Modes

A theoretical study of the phenomenon of the avoided crossings of spheroidal modes was made by Gabriel (1980) for the adiabatic (isentropic) case and by Christensen-Dalsgaard (1981) for the non-adiabatic case. Both studies are based on an earlier study of von Neumann & Wigner (1929) on the behaviour of eigenvalues in processes examined in quantum mechanics.

Consider a sequence of stellar models with the same total mass M, whose structure depends on a parameter ε. This parameter may be, for example, the polytropic index for a sequence of polytropic models, or the evolution time for a sequence of models of an evolving star. Since the models are spherically symmetric, a relation between the radial distance r from the centre and the mass contained inside the sphere with that radius exists in each model. For a model associated with a given parameter ε, the radial distance, the pressure, the mass density, ...can then be expressed in the form

$$r(m,\varepsilon), \quad P[r(m,\varepsilon)], \quad \rho[r(m,\varepsilon)], \quad \ldots$$

so that wave equation (4.23) takes the form

$$\sigma^2(\varepsilon)\, g_{ij}[r(m,\varepsilon),\theta,\phi]\,\delta q^j[r(m,\varepsilon),\theta,\phi]$$
$$-U_{ij}[r(m,\varepsilon),\theta,\phi]\,\delta q^j[r(m,\varepsilon),\theta,\phi] = 0, \quad i = 1,2,3. \qquad (11.1)$$

Suppose that the two eigenfrequencies $\sigma_1(\varepsilon)$ and $\sigma_2(\varepsilon)$ are close to each other for the model with a given value of ε, in the sense that

$$\left|\sigma_2^2(\varepsilon) - \sigma_1^2(\varepsilon)\right| << \left|\sigma_2^2(\varepsilon) - \sigma_\mu^2(\varepsilon)\right|, \quad \mu = 3,4,5,\ldots$$

Be the components of the associated displacement fields

$$\delta q_1^j[r(m,\varepsilon),\theta,\phi],\quad \delta q_2^j[r(m,\varepsilon),\theta,\phi],\quad j=1,2,3.$$

For the stellar model associated with $\varepsilon = 0$, it is convenient to set

$$\sigma_1^2(0)=\sigma_1^2,\quad \sigma_2^2(0)=\sigma_2^2, \tag{11.2}$$

$$\delta q_1^j[r(m,0),\theta,\phi]=\delta q_1^j(0),\quad \delta q_2^j[r(m,0),\theta,\phi]=\delta q_2^j(0),\quad j=1,2,3. \tag{11.3}$$

Consider that $\sigma_1^2 < \sigma_2^2$.

Wave equation (11.1) is defined in the volume occupied by the stellar model associated with the value of ε. Stellar models associated with different values of ε occupy different volumes. These differences are taken into account by a transformation of wave equation (11.1) into a wave equation which is defined in the volume occupied by the stellar model associated with $\varepsilon = 0$. The parameter ε is here assumed to be small. The transformation is performed in two steps.

First, the operators U_{ij} in the wave equation applying to the model that is associated with the given value of ε are transformed from operators valid at the point with spherical coordinates $r(m,\varepsilon)$, θ, ϕ into operators valid at the point with spherical coordinates r_0, θ_0, ϕ_0. The transformation is done by means of the coordinate transformation

$$r(m,\varepsilon)=r_0+\varepsilon r_1(r_0),\quad \theta=\theta_0,\quad \phi=\phi_0, \tag{11.4}$$

where $r_0 = r(m,0)$, and series expansions of the form

$$\left.\begin{aligned} P[r(m,\varepsilon)] &= P_0(r_0)+\varepsilon P_1(r_0),\\ \rho[r(m,\varepsilon)] &= \rho_0(r_0)+\varepsilon\rho_1(r_0). \end{aligned}\right\} \tag{11.5}$$

It results that

$$U_{ij}[r(m,\varepsilon),\theta,\phi]=U_{ij}(r_0,\theta_0,\phi_0)+\varepsilon V_{ij}(r_0,\theta_0,\phi_0)\equiv U_{ij}(0)+\varepsilon V_{ij}(0). \tag{11.6}$$

For the components of the metric tensor, the Taylor series are used

$$g_{ij}[r(m,\varepsilon),\theta,\phi]=g_{ij}(0)+\varepsilon\left(\frac{\partial g_{ij}}{\partial r}\right)_0 r_1(r_0). \tag{11.7}$$

Secondly, the Lagrangian displacement of a mass element at the point with spherical coordinates $r(m,\varepsilon)$, θ, ϕ is subject to a parallel transfer along the radius, from this point to the point with spherical coordinates r_0, θ_0, ϕ_0. The components of the transferred Lagrangian displacement are

$$\widetilde{\delta q}^j(0;\varepsilon) = \delta q^j[r(m,\varepsilon),\theta,\phi] \\ +\varepsilon \Gamma^j_{1k}[r(m,\varepsilon),\theta,\phi]\, r_1(r_0)\, \delta q^k[r(m,\varepsilon),\theta,\phi], \quad j = 1,2,3. \tag{11.8}$$

The lower index 1 of the Christoffel symbols of the second kind stands for the radial coordinate. In the approximation linear in the small parameter ε, one has equivalently

$$\delta q^j[r(m,\varepsilon),\theta,\phi] = \widetilde{\delta q}^j(0;\varepsilon) - \varepsilon \Gamma^j_{1k}(0)\, r_1(r_0)\, \widetilde{\delta q}^k(0;\varepsilon), \quad j = 1,2,3. \tag{11.9}$$

Substitution into wave equation (11.1) applying to a mode λ yields the transformed wave equation at the point with spherical coordinates r_0, θ_0, ϕ_0 in the volume V_0 of the stellar model that is associated with $\varepsilon = 0$:

$$\left[\sigma^2_\lambda(\varepsilon)\, g_{ij}(0) - U_{ij}(0) - \varepsilon V_{ij}(0)\right] \widetilde{\delta q}^j_\lambda(0;\varepsilon) \\ -\varepsilon\, \sigma^2_\lambda\, r_1(r_0) \left[g_{ij}(0)\, \Gamma^j_{1k}(0)\, \widetilde{\delta q}^k_\lambda(0;\varepsilon) - \left(\frac{\partial g_{ij}}{\partial r}\right)_0 \widetilde{\delta q}^j_\lambda(0;\varepsilon)\right] \\ +\varepsilon\, U_{ij}(0) \left[\Gamma^j_{1k}(0)\, r_1(r_0)\, \widetilde{\delta q}^k_\lambda(0;\varepsilon)\right] = 0, \quad i = 1,2,3. \tag{11.10}$$

In the term of order ε, $\sigma^2_\lambda(\varepsilon)$ is approximated by $\sigma^2_\lambda \equiv \sigma^2_\lambda(0)$.

Consider the two modes $\lambda = 1$ and $\lambda = 2$ that are coupled during their encounter. In the stellar model associated with $\varepsilon = 0$, the Lagrangian displacements have respectively the components $\delta q^j_1(0)$ and $\delta q^j_2(0)$, with $j = 1,2,3$. The two displacement fields are orthogonal to each other and are supposed to be normalised such that

$$\left.\begin{aligned} \int_{V_0} \rho(0)\, g_{ij}(0)\, \overline{\delta q^i_1}(0)\, \delta q^j_1(0)\, dV(0) = 1, \\ \int_{V_0} \rho(0)\, g_{ij}(0)\, \overline{\delta q^i_2}(0)\, \delta q^j_2(0)\, dV(0) = 1. \end{aligned}\right\} \tag{11.11}$$

For the transferred displacement fields $\widetilde{\delta q}^j_1(0;\varepsilon)$ and $\widetilde{\delta q}^j_2(0;\varepsilon)$, combined solutions of the following form are adopted:

$$\widetilde{\delta q}^j_\lambda(0;\varepsilon) = C_{\lambda,1}(\varepsilon)\, \delta q^j_1(0) + C_{\lambda,2}(\varepsilon)\, \delta q^j_2(0), \quad j = 1,2,3, \tag{11.12}$$

where $C_{\lambda,1}(\varepsilon)$ and $C_{\lambda,2}(\varepsilon)$ are yet undetermined constants depending on the parameter ε. No interactions with other modes are considered.

Substitution of the linear combinations into Eq. (11.10), multiplication of both members by

$$\rho(0)\left(\overline{\delta q_1^i}(0) - \varepsilon \Gamma^i_{1k}(0)\, r_1\,(r_0)\, \overline{\delta q_1^k}(0)\right),$$

and integration over the volume V_0 of the stellar model associated with $\varepsilon = 0$ yield

$$\left[\sigma_\lambda^2(\varepsilon) - \sigma_1^2 - \varepsilon\left(M_{11} - \sigma_\lambda^2\, N_{11}\right)\right] C_{\lambda,1}(\varepsilon) - \varepsilon\left(M_{12} - \sigma_\lambda^2\, N_{12}\right) C_{\lambda,2}(\varepsilon) = 0, \tag{11.13}$$

where

$$M_{1s} = \int_{V_0} \rho(0)\left\{\overline{\delta q_1^i}(0)\, V_{ij}(0)\, \delta q_s^j(0) - \Gamma^i_{1k}(0)\, r_1\,(r_0)\, \overline{\delta q_1^k}(0)\, U_{ij}(0)\, \delta q_s^j(0)\right.$$
$$\left. - \overline{\delta q_1^i}(0)\, U_{ij}(0)\left[\Gamma^j_{1k}(0)\, r_1\,(r_0)\, \delta q_s^k(0)\right]\right\} dV(0), \tag{11.14}$$

$$N_{1s} = -\int_{V_0} \rho(0)\, r_1\,(r_0)\left[g_{ij}(0)\, \Gamma^j_{1k}(0)\, \overline{\delta q_1^i}(0)\, \delta q_s^k(0) + g_{ij}(0)\, \Gamma^i_{1k}(0)\, \overline{\delta q_1^k}(0)\, \delta q_s^j(0)\right.$$
$$\left. - \left(\frac{\partial g_{ij}}{\partial r}\right)_0 \overline{\delta q_1^i}(0)\, \delta q_s^j(0)\right] dV(0), \quad s = 1, 2. \tag{11.15}$$

The coefficients N_{1s} can also be expressed as

$$N_{1s} = \int_{V_0} \rho(0)\, r_1\,(r_0)\, \overline{\delta q_1^i}(0)\, \delta q_{(s)}^j(0)$$
$$\left[\left(\frac{\partial g_{ij}}{\partial r}\right)_0 - g_{in}(0)\, \Gamma^n_{1j}(0) - g_{nj}(0)\, \Gamma^n_{1i}(0)\right] dV(0). \tag{11.16}$$

They are identically zero, since the sum of the terms inside the square brackets corresponds to the covariant derivative of the component g_{ij} of the metric tensor with respect to the radial coordinate, which is identically zero (see, e.g., Lawden 1962). This property of the components of the metric tensor is also known as Ricci's theorem (see, e.g., Adler et al. 1965).

Next, Eq. (11.10) is considered for the complex conjugate of the mode λ:

$$\left[\sigma_\lambda^2(\varepsilon)\, g_{ij}(0) - U_{ij}(0) - \varepsilon V_{ij}(0)\right] \overline{\widetilde{\delta q}_\lambda^j}(0;\varepsilon)$$
$$-\varepsilon\, \sigma_\lambda^2\, r_1\,(r_0)\left[g_{ij}(0)\, \Gamma^j_{1k}(0)\, \overline{\widetilde{\delta q}_\lambda^k}(0;\varepsilon) - \left(\frac{\partial g_{ij}}{\partial r}\right)_0 \overline{\widetilde{\delta q}_\lambda^j}(0;\varepsilon)\right]$$
$$+\varepsilon\, U_{ij}(0)\left[\Gamma^j_{1k}(0)\, r_1\,(r_0)\, \overline{\widetilde{\delta q}_\lambda^k}(0;\varepsilon)\right] = 0, \quad i = 1, 2, 3. \tag{11.17}$$

Substitution of the linear combination given by Eq. (11.12), multiplication by

$$\rho(0)\left(\delta q_2^i(0) - \varepsilon \Gamma^i_{1k}(0)\, r_1(r_0)\, \delta q_2^k(0)\right),$$

and integration over the volume V_0 of the stellar model associated with $\varepsilon = 0$ yield

$$-\varepsilon\left(M_{21} - \sigma_\lambda^2 N_{21}\right) C_{\lambda,1}(\varepsilon) + \left[\sigma_\lambda^2(\varepsilon) - \sigma_2^2 - \varepsilon\left(M_{22} - \sigma_\lambda^2 N_{22}\right)\right] C_{\lambda,2}(\varepsilon) = 0, \tag{11.18}$$

where

$$M_{2s} = \int_{V_0} \rho(0)\Big\{\delta q_2^i(0)\, V_{ij}(0)\, \overline{\delta q_s^j}(0) - \Gamma^i_{1k}(0)\, r_1(r_0)\, \delta q_2^k(0)\, U_{ij}(0)\, \overline{\delta q_s^j}(0)$$
$$-\delta q_2^i(0)\, U_{ij}(0)\left[\Gamma^j_{1k}(0)\, r_1(r_0)\, \overline{\delta q_s^k(0)}\right]\Big\}\, dV(0), \tag{11.19}$$

$$N_{2s} = -\int_{V_0} \rho(0)\, r_1(r_0)\Bigg[g_{ij}(0)\, \Gamma^j_{1k}(0)\, \delta q_2^i(0)\, \overline{\delta q_s^k}(0) + g_{ij}(0)\, \Gamma^i_{1k}(0)\, \delta q_2^k(0)\, \overline{\delta q_s^j}(0)$$
$$-\left(\frac{\partial g_{ij}}{\partial r}\right)_0 \delta q_2^i(0)\, \overline{\delta q_s^j}(0)\Bigg]\, dV(0), \quad s = 1, 2. \tag{11.20}$$

The coefficients N_{2s} too are identically zero.

The two equations (11.13) and (11.18) form a homogeneous system of equations for the two constants $C_{\lambda,1}(\varepsilon)$ and $C_{\lambda,2}(\varepsilon)$. When one sets

$$\sigma_1^2 = \sigma_0^2 - \frac{1}{2}\,\Delta\sigma^2(0), \quad \sigma_2^2 = \sigma_0^2 + \frac{1}{2}\,\Delta\sigma^2(0) \tag{11.21}$$

with

$$\sigma_0^2 = \frac{1}{2}\left(\sigma_1^2 + \sigma_2^2\right), \quad \Delta\sigma^2(0) = \sigma_2^2 - \sigma_1^2, \tag{11.22}$$

the system becomes

$$\left.\begin{aligned}
&\left[\sigma_\lambda^2(\varepsilon) - \sigma_0^2 + \frac{1}{2}\,\Delta\sigma^2(0) - \varepsilon M_{11}\right] C_{\lambda,1}(\varepsilon)\\
&\quad -\varepsilon M_{12}\, C_{\lambda,2}(\varepsilon) = 0,\\
&-\varepsilon M_{21}\, C_{\lambda,1}(\varepsilon)\\
&\quad + \left[\sigma_\lambda^2(\varepsilon) - \sigma_0^2 - \frac{1}{2}\,\Delta\sigma^2(0) - \varepsilon M_{22}\right] C_{\lambda,2}(\varepsilon) = 0.
\end{aligned}\right\} \tag{11.23}$$

This system is comparable to that consisting of Eqs. (7) of von Neumann & Wigner (1929).

The requirement that the system should admit of a non-trivial solution for the two constants $C_{\lambda,1}(\varepsilon)$ and $C_{\lambda,2}(\varepsilon)$ leads to the quadratic equation for $\sigma_\lambda^2(\varepsilon)$

$$\sigma_\lambda^4(\varepsilon) - \left[2\,\sigma_0^2 + \varepsilon\,(M_{11}+M_{22})\right]\sigma_\lambda^2(\varepsilon) + \sigma_0^4 - \frac{1}{4}\left(\Delta\sigma^2(0)\right)^2 + \sigma_0^2\,\varepsilon\,(M_{11}+M_{22})$$
$$+\,\frac{1}{2}\,\Delta\sigma^2(0)\,\varepsilon\,(M_{11}-M_{22}) + \varepsilon^2\,(M_{11}\,M_{22} - M_{12}\,M_{21}) = 0, \tag{11.24}$$

whose discriminant is

$$D(\varepsilon) = \left[\Delta\sigma^2(0) - \varepsilon\,(M_{11}-M_{22})\right]^2 + 4\,\varepsilon^2\,M_{12}\,M_{21}. \tag{11.25}$$

The property holds that

$$M_{21} = M_{12}. \tag{11.26}$$

It can be verified as follows. The symmetry of the tensorial operator U, expressed by equality (4.32), applies to the stellar model associated with the value ε of the parameter, so that

$$\int_V \rho[r(m,\varepsilon)]\,\delta q_2^i[r(m,\varepsilon),\theta,\phi]\,U_{ij}[r(m,\varepsilon),\theta,\phi]\,\overline{\delta q_1^j}[r(m,\varepsilon),\theta,\phi]\,dV$$
$$= \int_V \rho[r(m,\varepsilon)]\,\overline{\delta q_1^i}[r(m,\varepsilon),\theta,\phi]\,U_{ij}[r(m,\varepsilon),\theta,\phi]\,\delta q_2^j[r(m,\varepsilon),\theta,\phi]\,dV. \tag{11.27}$$

By the use of the mass conservation and Eqs. (11.6), (11.9), and (11.12), the left-hand member can be transformed into

$$\int_V \rho[r(m,\varepsilon)]\,\delta q_2^i[r(m,\varepsilon),\theta,\phi]\,U_{ij}[r(m,\varepsilon),\theta,\phi]\,\overline{\delta q_1^j}[r(m,\varepsilon),\theta,\phi]\,dV$$
$$= \int_{V_0} \rho(0)\left\{\left[C_{2,1}(\varepsilon)\,\delta q_1^i(0) + C_{2,2}(\varepsilon)\,\delta q_2^i(0)\right]\right.$$
$$U_{ij}(0)\left[C_{1,1}(\varepsilon)\,\overline{\delta q_1^j}(0) + C_{1,2}(\varepsilon)\,\overline{\delta q_2^j}(0)\right]$$
$$+\,\varepsilon\left[C_{2,1}(\varepsilon)\,\delta q_1^i(0) + C_{2,2}(\varepsilon)\,\delta q_2^i(0)\right]$$
$$V_{ij}(0)\left[C_{1,1}(\varepsilon)\,\overline{\delta q_1^j}(0) + C_{1,2}(\varepsilon)\,\overline{\delta q_2^j}(0)\right]$$
$$-\,\varepsilon\,\Gamma_{1k}^i(0)\,r_1\,(r_0)\left[C_{2,1}(\varepsilon)\,\delta q_1^k(0) + C_{2,2}(\varepsilon)\,\delta q_2^k(0)\right]$$

$$U_{ij}(0)\left[C_{1,1}(\varepsilon)\,\overline{\delta q_1^j}(0)+C_{1,2}(\varepsilon)\,\overline{\delta q_2^j}(0)\right]$$

$$-\,\varepsilon\left[C_{2,1}(\varepsilon)\,\delta q_1^i(0)+C_{2,2}(\varepsilon)\,\delta q_2^i(0)\right]$$

$$U_{ij}(0)\Big\{\Gamma^j_{1k}(0)\,r_1\,(r_0)\left[C_{1,1}(\varepsilon)\,\overline{\delta q_1^k}(0)+C_{1,2}(\varepsilon)\,\overline{\delta q_2^k}(0)\right]\Big\}\Bigg\}\,dV(0). \tag{11.28}$$

The right-hand member can similarly be transformed into

$$\int_V \rho[r(m,\varepsilon)]\,\overline{\delta q_1^i}[r(m,\varepsilon),\theta,\phi]\,U_{ij}[r(m,\varepsilon),\theta,\phi]\,\delta q_2^j[r(m,\varepsilon),\theta,\phi]\,dV$$

$$=\int_{V_0}\rho(0)\Bigg\{\left[C_{1,1}(\varepsilon)\,\overline{\delta q_1^i}(0)+C_{1,2}(\varepsilon)\,\overline{\delta q_2^i}(0)\right]$$

$$U_{ij}(0)\left[C_{2,1}(\varepsilon)\,\delta q_1^j(0)+C_{2,2}(\varepsilon)\,\delta q_2^j(0)\right]$$

$$+\,\varepsilon\left[C_{1,1}(\varepsilon)\,\overline{\delta q_1^i}(0)+C_{1,2}(\varepsilon)\,\overline{\delta q_2^i}(0)\right]$$

$$V_{ij}(0)\left[C_{2,1}(\varepsilon)\,\delta q_1^j(0)+C_{2,2}(\varepsilon)\,\delta q_2^j(0)\right]$$

$$-\,\varepsilon\Gamma^i_{1k}(0)\,r_1\,(r_0)\left[C_{1,1}(\varepsilon)\,\overline{\delta q_1^k}(0)+C_{1,2}(\varepsilon)\,\overline{\delta q_2^k}(0)\right]$$

$$U_{ij}(0)\left[C_{2,1}(\varepsilon)\,\delta q_1^j(0)+C_{2,2}(\varepsilon)\,\delta q_2^j(0)\right]$$

$$-\,\varepsilon\left[C_{1,1}(\varepsilon)\,\overline{\delta q_1^i}(0)+C_{1,2}(\varepsilon)\,\overline{\delta q_2^i}(0)\right]U_{ij}(0)\Big\{\Gamma^j_{1k}(0)\,r_1\,(r_0)$$

$$\left[C_{2,1}(\varepsilon)\,\delta q_1^k(0)+C_{2,2}(\varepsilon)\,\delta q_2^k(0)\right]\Big\}\Bigg\}\,dV(0). \tag{11.29}$$

The left-hand member and the right-hand member are equal for any value of ε. This implies in particular that, in these members, the sums of the coefficients of the product $\varepsilon\,C_{1,1}(\varepsilon)\,C_{2,2}(\varepsilon)$ are equal. Equality (11.26) then follows.

Consequently, the discriminant $D(\varepsilon)$ of the quadratic equation (11.24) consists of a sum of two squares and is positive for all values of ε. The roots of the equation are

$$\left.\begin{aligned}\sigma_1^2(\varepsilon)&=\frac{1}{2}\left[2\,\sigma_0^2+\varepsilon\,(M_{11}+M_{22})-D^{1/2}(\varepsilon)\right],\\ \sigma_2^2(\varepsilon)&=\frac{1}{2}\left[2\,\sigma_0^2+\varepsilon\,(M_{11}+M_{22})+D^{1/2}(\varepsilon)\right].\end{aligned}\right\} \tag{11.30}$$

In order to determine the nature of the curves described by the two roots $\sigma_1^2(\varepsilon)$ and $\sigma_2^2(\varepsilon)$ in the plane (ε, σ^2), it is convenient to rewrite the quadratic equation as

$$\sigma_\lambda^4(\varepsilon) - (M_{11} + M_{22})\,\sigma_\lambda^2(\varepsilon)\,\varepsilon + (M_{11}\,M_{22} - M_{12}\,M_{21})\,\varepsilon^2 - 2\,\sigma_0^2\,\sigma_\lambda^2(\varepsilon)$$
$$+\left[\sigma_0^2\,(M_{11} + M_{22}) + \frac{1}{2}\,\Delta\sigma^2(0)\,(M_{11} - M_{22})\right]\varepsilon + \sigma_0^4 - \frac{1}{4}\left(\Delta\sigma^2(0)\right)^2 = 0. \tag{11.31}$$

This equation is of the second degree in σ_λ^2 and ε, and represents a hyperbola, since the discriminant of the second-degree terms is negative:

$$\delta \equiv -\frac{1}{4}\left[(M_{11} - M_{22})^2 + 4\,M_{12}^2\right] < 0. \tag{11.32}$$

Hence, each root describes a branch of a hyperbola. Equation (11.31) is similar to Eq. (12) of Gabriel (1980).

The difference between the two roots associated with a certain value of ε is given by

$$\Delta\sigma^2(\varepsilon) \equiv \sigma_2^2(\varepsilon) - \sigma_1^2(\varepsilon) = D^{1/2}(\varepsilon). \tag{11.33}$$

The value of ε for which the difference is minimal, ε_{min}, is solution of the equation

$$\frac{d}{d\varepsilon}\,\Delta\sigma^2(\varepsilon) = 0, \tag{11.34}$$

so that

$$\varepsilon_{\text{min}} = \Delta\sigma^2(0)\,\frac{M_{11} - M_{22}}{(M_{11} - M_{22})^2 + 4\,M_{12}\,M_{21}}. \tag{11.35}$$

From Eqs. (11.23), it follows that, for a certain value of ε, the ratios of the coefficients in the combined solutions (11.12) are given by

$$\frac{C_{1,1}(\varepsilon)}{C_{1,2}(\varepsilon)} = \frac{\varepsilon M_{12}}{\sigma_1^2(\varepsilon) - \sigma_0^2 + \Delta\sigma^2(0)/2 - \varepsilon M_{11}}, \tag{11.36}$$

$$\frac{C_{2,1}(\varepsilon)}{C_{2,2}(\varepsilon)} = \frac{\varepsilon M_{21}}{\sigma_2^2(\varepsilon) - \sigma_0^2 - \Delta\sigma^2(0)/2 - \varepsilon M_{22}}, \tag{11.37}$$

or, after elimination of $\sigma_1^2(\varepsilon)$ and $\sigma_2^2(\varepsilon)$ by means of Eqs. (11.30), equivalently by

$$\frac{C_{1,1}(\varepsilon)}{C_{1,2}(\varepsilon)} = \frac{2\,\varepsilon M_{12}}{-\varepsilon\,(M_{11} - M_{22}) + \Delta\sigma^2(0) - D^{1/2}(\varepsilon)}, \tag{11.38}$$

$$\frac{C_{2,1}(\varepsilon)}{C_{2,2}(\varepsilon)} = \frac{-2\,\varepsilon M_{21}}{-\varepsilon\,(M_{11} - M_{22}) + \Delta\sigma^2(0) - D^{1/2}(\varepsilon)}. \tag{11.39}$$

For larger values of $|\varepsilon|$, one has

$$D^{1/2}(\varepsilon) \simeq \pm 2\,\varepsilon M_{12}\left(1+y^2\right)^{1/2} \tag{11.40}$$

with $y = (M_{11} - M_{22})/(2\,M_{12})$. For $D^{1/2}(\varepsilon)$ being real, the plus sign must be used as $\varepsilon > 0$, and the minus sign, as $\varepsilon < 0$. Neglecting $\Delta\sigma^2(0)$ in the denominators of the right-hand members of Eqs. (11.38) and (11.39), one then verifies that

$$\frac{C_{1,1}(\varepsilon)}{C_{1,2}(\varepsilon)} \simeq \left\{-y - \left[\pm\left(1+y^2\right)^{1/2}\right]\right\}^{-1}, \tag{11.41}$$

$$\frac{C_{2,1}(\varepsilon)}{C_{2,2}(\varepsilon)} \simeq -\left\{-y - \left[\pm\left(1+y^2\right)^{1/2}\right]\right\}^{-1}. \tag{11.42}$$

When one associates the value $\varepsilon = 0$ with the stellar model for which the difference $M_{11} - M_{22}$ is smallest and considers y to be small, one sees that, for $\varepsilon < 0$, $C_{1,1}(\varepsilon)/C_{1,2}(\varepsilon)$ is close to 1 and $C_{2,1}(\varepsilon)/C_{2,2}(\varepsilon)$, close to -1, and that, for $\varepsilon > 0$, $C_{11}(\varepsilon)/C_{12}(\varepsilon)$ is close to -1 and $C_{21}(\varepsilon)/C_{22}(\varepsilon)$, close to 1. Consequently, the coupling between two modes whose eigenfrequencies have a close encounter leads to an exchange of properties of the Lagrangian displacement fields, although the eigenfrequencies avoid to cross each other.

Aizenman et al. (1977) examined the avoided crossings of non-radial modes in a sequence of models for an evolving 16 $M_\odot$ star. The authors artificially decoupled the modes of the different types by removing the term $\ell(\ell+1)/\left(C\omega^2\right)$, on the one side, and the term $C\omega^2$, on the other side, from the system of differential equations (6.31)–(6.34) used in their numerical integrations. By the removal of the term $\ell(\ell+1)/\left(C\omega^2\right)$, the g^+-modes are filtered out, and by the removal of the term $C\omega^2$, the f-mode and the p-modes. The authors called the uncoupled modes respectively π-, γ-, and ϕ-modes. In following the π-, γ-, and ϕ-modes in the sequence of models, they observed, for example, that the eigenfrequency of the γ_1-mode successively crosses the eigenfrequency of the ϕ-mode, the eigenfrequency of the π_1-mode, the eigenfrequency of the π_2-mode, the eigenfrequency of the π_3-mode, … Hence, the eigenfrequencies of the g^+-modes really cross the eigenfrequencies of the f-mode and the p-modes when these modes are decoupled from each other.

The phenomenon of the avoided crossings of non-radial stellar oscillation modes was studied further by Roth & Weigert (1979).

11.4 Implications of Avoided Crossings of Non-Radial Modes for Mode Identifications

In the identifications of non-radial spheroidal modes in stars with a larger central mass condensation, the transfers of properties in avoided crossings of modes must be taken into consideration.

An illustrative example is the identification of second-degree modes in polytropic models with a larger central mass condensation. In Sect. 11.1, it is mentioned that the eigenfunction $\xi(r)$ of the second-degree f-mode acquires an additional double node in the polytropic model with index $n = 3.4213$ and displays two nodes in the polytropic models with larger indexes. Actually the g_1^+-mode has there an avoided crossing with the f-mode. Because of the transfer of properties, the mode in the polytropic model with index $n = 3.5$ that is associated with the dimensionless eigenvalue $\omega^2 = 16.17$ must be identified as the g_1^+-mode, and the mode that is associated with the dimensionless eigenvalue $\omega^2 = 12.10$, as the f-mode (see Table 11.1). From Table 11.2, it appears that, after the avoided crossing, the eigenfunction $\xi(r)$ of the g_1^+-mode has two nodes, and the eigenfunction $\xi(r)$ of the f-mode, one node. Consequently, for both modes, the eigenfunction $\xi(r)$ has acquired one additional node by the avoided crossing.

Adopting as a rule that the eigenfunction $\xi(r)$ of a mode acquires one additional node by an avoided crossing with another mode, one comes to the modified identifications of the second-degree, lower-order modes of the polytropic model with index $n = 4$ that are presented in Table 11.4. In subscript, the number of avoided crossings ("ac") to which the mode has been subject, is mentioned after the radial order of the mode according to the Cowling classification. Notice that the sum of the radial order and the number of avoided crossings corresponds to the number of nodes of the eigenfunction $\xi(r)$. The comparison of the mode identifications given in Table 11.4 with those given in Table 11.1 is instructive. The identifications given in Table 11.4 correspond more with the physical characters of the modes than do the formal identifications given in Table 11.1, in which the identification of a mode does not always agree with the main properties of the mode.

From the propagation diagram for the polytropic model with index $n = 4$, represented in Fig. 11.2, it appears that, for the g_1^+-mode, which has had an avoided crossing successively with the f-, the p_1-, the p_2-, and the p_3-mode, the eigenfunction

Table 11.4 Identification of the lower-order second-degree modes of the polytropic model with index $n = 4$ on the basis of the rule that the eigenfunction $\xi(r)$ acquires an additional node by each avoided crossing of the mode with another mode

Mode	ω^2	Number of nodes of $\xi(r)$	Mode	ω^2	Number of nodes of $\xi(r)$
p_{10}	215.9	10	$p_{1,\,2\mathrm{ac}}$	27.60	3
p_9	184.0	9	$g^+_{3,\,1\mathrm{ac}}$	23.01	4
p_8	154.8	8	$g^+_{4,\,1\mathrm{ac}}$	18.00	5
p_7	128.4	7	$f_{0,\,4\mathrm{ac}}$	15.36	4
p_6	105.3	6	g_5^+	12.77	5
p_5	87.47	5	g_6^+	10.09	6
p_4	76.69	4	g_7^+	8.092	7
$g^+_{1,\,4\mathrm{ac}}$	62.88	5	g_8^+	6.617	8
$p_{3,\,1\mathrm{ac}}$	50.82	4	g_9^+	5.505	9
$p_{2,\,1\mathrm{ac}}$	42.14	3	g_{10}^+	4.649	10
$g^+_{2,\,2\mathrm{ac}}$	34.34	4			

$\xi(r)$ acquired four additional nodes in the A-cavity; for the p_2- and p_3-mode, which have had an avoided crossing with the g_1^+-mode, the eigenfunction $\xi(r)$ acquired one additional node in the G-cavity; for the g_2^+-mode, which has had an avoided crossing with the f- and the p_1-mode, the eigenfunction $\xi(r)$ acquired two additional nodes in the A-cavity; for the p_1-mode, which has had an avoided crossing with the g_1^+- and the g_2^+-mode, the eigenfunction $\xi(r)$ acquired two additional nodes in the G-cavity; for the g_3^+- and the g_4^+-mode, which have had an avoided crossing with the f-mode, the eigenfunction $\xi(r)$ acquired one additional node: in the case of the g_3^+-mode, the node is situated in the A-cavity, in the case of the g_4^+-mode, it is situated on the boundary of this cavity; finally, for the f-mode, which has had an avoided crossing with the g_1^+-, the g_2^+-, the g_3^+-, and the g_4^+-mode, the eigenfunction $\xi(r)$ acquired three additional nodes in the G-cavity, and one additional node in the A-cavity.

The variations of the dimensionless eigenvalues ω^2 that are obtained for the lowest-order second-degree modes in the polytropic models with index $n = 0, 1, 1.5, 2, 3, 3.25, 3.50, 4$, for $\Gamma_1 = 5/3$, are represented in Fig. 11.4, in the supposition that the implications of the avoided crossings for the identifications of the modes are taken into account.

A similar analysis can be made for the mode identification in the sequence of models of an evolving 10 $M_\odot$ star considered by Osaki (1975). Table 11.5 contains the numbers of nodes of the eigenfunctions $\xi(r)$ of the lowest-order second-degree modes for stellar model 13. The mode associated with the dimensionless eigenvalue $\omega^2 = 40.07$ is the mode p_3, for which the eigenfunction $\xi(r)$ acquired an additional node by an avoided crossing with the mode g_1^+, and the mode associated with the dimensionless eigenvalue $\omega^2 = 27.24$ is the mode p_2, for which the eigenfunction $\xi(r)$ also acquired an additional node by an avoided crossing with the mode g_1^+. The eigenfunction $\xi(r)$ of the mode associated with the dimensionless eigenvalue $\omega^2 = 22.59$ is surprisingly mentioned to display only two nodes, although the mode seems to be the mode g_2^+, which has undergone an avoided crossing with the mode f and the mode p_1. The mode associated with the dimensionless eigenvalue

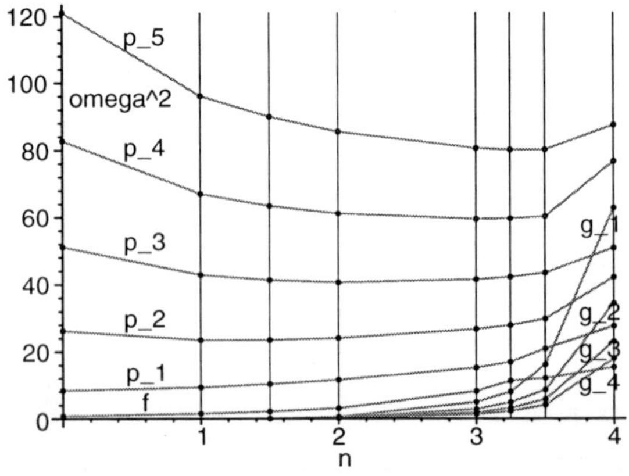

Fig. 11.4 The variations of the dimensionless eigenvalues ω^2 that are obtained for the lowest-order second-degree modes in the polytropic models with index $n = 0, 1, 1.5, 2, 3, 3.25, 3.50, 4$, for $\Gamma_1 = 5/3$, when the implications of the avoided crossings for the identifications of the modes are taken into account

Table 11.5 The numbers of nodes of the eigenfunctions $\xi(r)$ for the lowest-order second-degree modes in model 13 in the sequence of models considered by Osaki (1975)

Mode	ω^2	Number of nodes of $\xi(r)$
$p_{3,\,1\mathrm{ac}}$	40.07	4
$p_{2,\,1\mathrm{ac}}$	27.24	3
$g^+_{2,\,2\mathrm{ac}}$	22.59	2
$p_{1,\,2\mathrm{ac}}$	17.18	3
$f_{0,\,2\mathrm{ac}}$	10.62	2

$\omega^2 = 17.18$ is the mode p_1, for which the eigenfunction $\xi(r)$ acquired two additional nodes by an avoided crossing with the mode g_1^+ and an avoided crossing with the mode g_2^+, and the mode associated with the dimensionless eigenvalue $\omega^2 = 10.62$ is the mode f, for which the eigenfunction $\xi(r)$ also acquired two additional nodes by an avoided crossing with the mode g_1^+ and an avoided crossing with the mode g_2^+.

Avoided crossings of oscillation modes are a frequently occurring phenomenon in stars. For example, in a study of the properties of oscillations of low degrees $\ell = 0, \ldots, 6$ and low frequencies from 100 to 2000 μHz in six solar models, Provost et al. (2000) found oscillations of a dual character. They investigated the variation of the normalised kinetic energy density $\rho\, r^2\, \boldsymbol{\xi}\, \overline{\boldsymbol{\xi}}$ as a function of the solar radius for several modes in their models. Above the maximum of the Brunt–Väsälä frequency at about 450 μHz, the modes are pure p-modes, and below about 200 μHz, pure g^+-modes. Between them, low-radial order g^+-modes, a f-mode, and a p_1-mode are found, of which some have a dual character. For example, for $\ell = 1$ in the first model, the mode g_3^+ with frequency 151 μHz displays three peaks in the amplitude of the energy density that are concentrated in the solar core ($r \leq 0.2\, R_\odot$), and the mode g_2^+ with frequency 189 μHz, two peaks; the mode g_1^+ with frequency 260 μHz displays one peak in the solar core and a slight peak in the solar envelope ($r \geq 0.75\, R_\odot$); the mode p_1 with frequency 284 μHz displays a peak in the solar core and a larger peak in the solar envelope; the mode p_2 with frequency 448 μHz displays two peaks in the amplitude of the energy density that are concentrated in the solar envelope, and the mode p_3 with frequency 597 μHz, three peaks (see Fig. 1 of Provost et al.). The little additional peak in the solar envelope in the case of the mode denoted as g_1^+ and the additional peak in the solar core in the case of the mode denoted as p_1 seem to show that the two modes have been subject to an avoided crossing with exchange of their properties. Therefore they may more appropriately be identified as the mode $p_{1,\,1\mathrm{ac}}$ with frequency 260 μHz, and the mode $g^+_{1,\,1\mathrm{ac}}$ with frequency 284 μHz.

11.5 Strange Radial Modes

The denomination strange radial modes was first attributed to some strongly non-adiabatic, radial modes in models of R Coronae Borealis stars by Cox et al. (1980).

In a sequence of models of these stars as a function of the increasing effective temperature, the eigenfrequencies of some modes vary more strongly than those of the most other modes, and the modes themselves can hardly be related to modes determined in the isentropic approximation.

Strange radial modes were found later between the non-adiabatic modes of stellar models of very different types (Saio et al. 1998). They display remarkably small amplitudes in the interior of the stellar envelope.

For some time, the origin of the strange radial modes was ascribed to the strongly non-adiabatic character of the modes. However, in a surprising way, strange radial modes were also found in models of classic Cepheids, although the modes are only slightly non-adiabatic in these models (Buchler et al. 1997). They were even found in the isentropic approximation (Kiriakidis et al. 1993, Gautschy 1993).

The enigma of the strange radial modes was elucidated by Buchler et al. (1997) with the help of a simple model for isentropic oscillations. The modes in this model are essentially acoustic surface modes such as those existing in models of Cepheids, in which the zone of the partial ionisation of hydrogen acts as a potential barrier and divides the star between an internal region and a surface region. The importance of an ionisation zone had already been stressed by Zalewski (1992), who had ascribed the existence of strange radial modes to the presence of a strong peak of entropy in such a zone.

Buchler et al. (1997) developed an illustrative model by using a Schrödinger-like equation. In order to reduce the potential barrier to a thin strip, they set the acoustic potential equal to zero, except at a given point with the coordinate $\tau = \tau_0$, where they set it equal to a Dirac delta function. The equation then takes the form

$$\frac{d^2\psi}{d\tau^2} + \left[\sigma^2 - V_0\,\sigma\,\delta\,(\tau - \tau_0)\right]\psi = 0, \tag{11.43}$$

and is considered in the interval $0 \leq \tau \leq 1$. The coefficient V_0 determines the height of the potential barrier. The authors adopted a boundary condition for a closed pipe on the internal side and a boundary condition for an open pipe on the external side:

$$\psi(0) = 0, \quad \psi'(1) = 0. \tag{11.44}$$

The eigenvalue problem can be solved analytically. The solutions for $\tau \leq \tau_0$ are

$$\psi(\tau) = B_1 \sin(\sigma\tau), \tag{11.45}$$

and those for $\tau \geq \tau_0$,

$$\psi(\tau) = B_2 \cos[\sigma(1 - \tau)], \tag{11.46}$$

where B_1 and B_2 are arbitrary constants.

At the point $\tau = \tau_0$, Buchler et al. imposed the conditions

$$\left.\begin{aligned} \psi\left(\tau_0^+\right) &= \psi\left(\tau_0^-\right), \\ \psi'\left(\tau_0^+\right) &= \psi'\left(\tau_0^-\right) + V_0\,\sigma\,\psi\left(\tau_0\right), \end{aligned}\right\} \qquad (11.47)$$

or, more explicitly,

$$\left.\begin{aligned} B_2\,\cos\left[\sigma(1-\tau_0)\right] &= B_1\,\sin\left(\sigma\tau_0\right), \\ B_2\,\sin\left[\sigma(1-\tau_0)\right] &= B_1\,\cos\left(\sigma\tau_0\right) + B_2\,V_0\,\cos\left[\sigma(1-\tau_0)\right]. \end{aligned}\right\} \qquad (11.48)$$

These conditions lead to the equation for the eigenfrequencies

$$\cos\sigma + V_0\,\left[\sin\left(\sigma\tau_0\right)\right]\cos\left[\sigma(1-\tau_0)\right] = 0. \qquad (11.49)$$

When the solutions are normalised by the condition that $\psi(1) = 1$, it results that $B_2 = 1$. From the first condition (11.48), it follows that

$$B_1 = \frac{\cos\left[\sigma(1-\tau_0)\right]}{\sin\left(\sigma\tau_0\right)}. \qquad (11.50)$$

Relative to the resolution of Eq. (11.49), two limiting cases can be distinguished. As $\tau_0 \to 0$, the internal region is reduced to zero, and the spectrum of the eigenfrequencies is determined by

$$\cos\sigma = 0, \qquad (11.51)$$

so that the eigenfrequencies are given by

$$\sigma_k = (2k+1)\frac{\pi}{2}, \quad k = 0, 1, 2, \ldots, \qquad (11.52)$$

and correspond to those of an open pipe. The associated modes are *external* modes. On the other side, as $\tau_0 \to 1$, the external region is reduced to zero. For larger values of V_0, i.e. for higher potential barriers, the spectrum of the eigenfrequencies is approximately determined by

$$\sin\sigma \simeq 0, \qquad (11.53)$$

so that the eigenfrequencies are nearly given by

$$\sigma \simeq (n+1)\pi, \quad n = 0, 1, 2, \ldots, \qquad (11.54)$$

and correspond almost to those of a closed pipe. The associated modes are *internal* modes.

Buchler et al. considered the case $V_0 = 10$. For this case, the variations of the eigenfrequencies σ for a number of lowest-order internal and external modes are represented as functions of $1 - \tau_0$ in Fig. 11.5. The quantity $1 - \tau_0$ is a measure for the depth of the position of the potential barrier from the surface. For larger depths, only external modes are found, and for smaller depths, only internal modes, as shown above. For the intermediate depths, both types of modes appear, and the eigenfrequencies of modes of different types display avoided crossings. The eigenfrequencies of the internal modes increase as the depth from the surface at which the potential barrier is located increases, while those of the external modes increase as that depth decreases.

Buchler et al. entered into more details for the case $\tau_0 = 0.96$, since this case is close to the situation encountered in Cepheids. On analogy, the case $\tau_0 = 0.942$ is examined here. In Fig. 11.5, a vertical line is drawn at the depth from the surface $1 - \tau_0 = 0.058$. For this location of the potential barrier, the first seven eigenfrequencies are those of the internal modes $n = 0, \ldots, 6$, and the eighth eigenfrequency is the external mode $k = 0$, which has been subject to an avoided crossing which each of the first seven internal modes. The ninth and the tenth mode are again internal modes. The external mode $k = 0$ appears here as a strange mode between the internal modes $n = 6$ and $n = 7$.

Still relative to the model for $V_0 = 10$ and $\tau_0 = 0.942$, the eigenfrequencies σ_n of the nine internal modes $n = 0, \ldots, 8$ are given in Table 11.6. The eigenfunctions $\psi(\tau)$ of the seven internal modes $n = 0, \ldots, 6$ are represented in Figs. 11.6

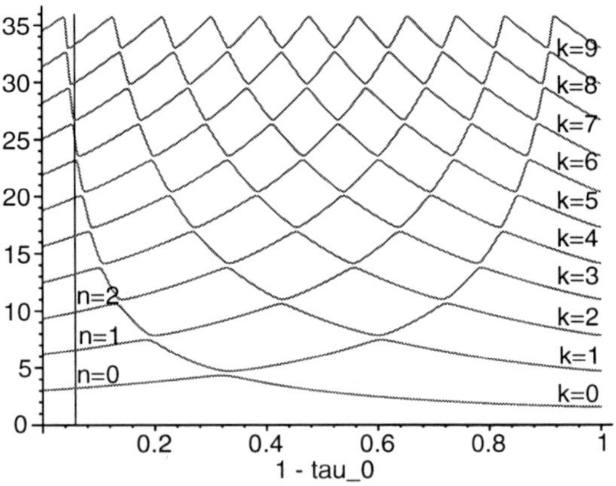

Fig. 11.5 The eigenfrequencies σ of the lowest-order internal and external modes as functions of the depth $1 - \tau_0$ of the potential barrier from the surface, for the model of Buchler et al. (1997) in which $V_0 = 10$

Table 11.6 Eigenfrequencies σ_n of the first nine internal modes for the model of Buchler et al. (1997) for $V_0 = 10$ and $\tau_0 = 0.942$

n	Eigenfrequency σ_n	n	Eigenfrequency σ_n
0	3.23	5	18.8
1	6.56	6	22.0
2	9.89	7	26.7
3	13.2	8	29.9
4	15.7		

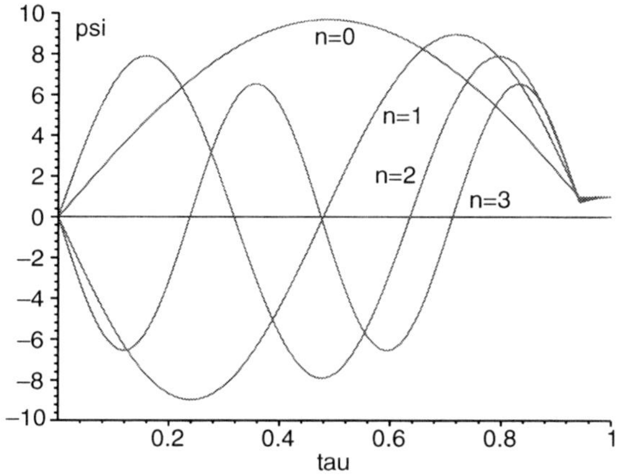

Fig. 11.6 The eigenfunctions $\psi(\tau)$ of the internal modes $n = 0, \ldots, 3$ in the model of Buchler et al. (1997) for $V_0 = 10$ and $\tau_0 = 0.948$

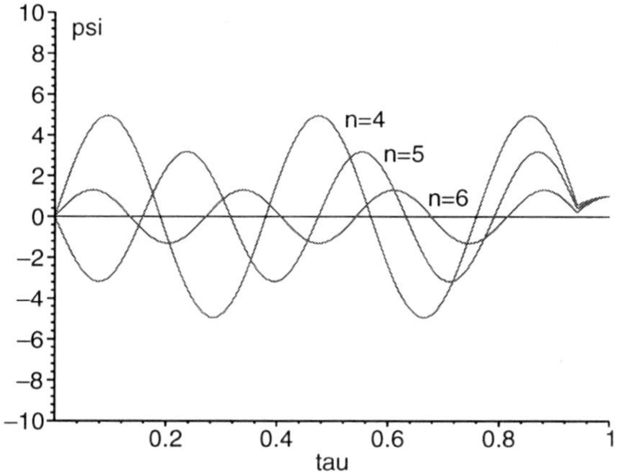

Fig. 11.7 The eigenfunctions $\psi(\tau)$ of the internal modes $n = 4, 5, 6$ in the model of Buchler et al. (1997) for $V_0 = 10$ and $\tau_0 = 0.948$

and 11.7. They display respectively $0, \ldots, 6$ nodes in the internal region. The external mode $k = 0$ has the eigenfrequency $\sigma_{k=0} = 25.3$, and its eigenfunction $\psi(\tau)$ is represented in Fig. 11.8. Noteworthy is that the amplitude of the eigenfunction is larger in the external region than in the internal region and that the eigenfunction displays seven nodes resulting from the avoided crossings with the seven internal modes $n = 0, \ldots, 6$. The eigenfunctions $\psi(\tau)$ of the two internal modes $n = 7$ and $n = 8$ are represented in Fig. 11.9. Both eigenfunctions have acquired one additional node by the avoided crossing with the external mode $k = 0$, so that they display respectively eight and nine nodes. As a result, the external mode $k = 0$ is intercalated between the internal modes $n = 6$ and $n = 7$ without any interference with the regularity of the increase of the number of nodes of the internal modes.

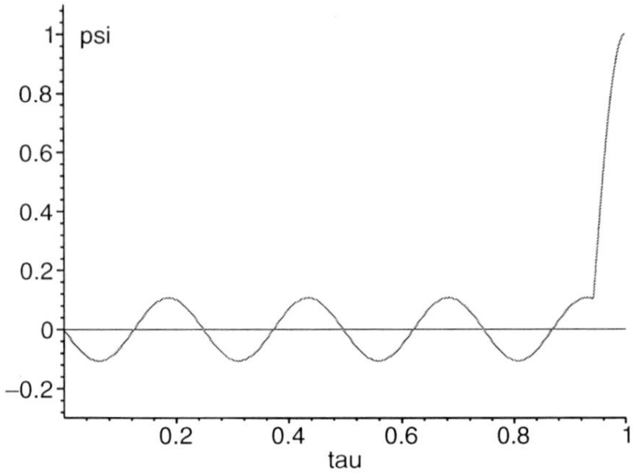

Fig. 11.8 The eigenfunction $\psi(\tau)$ of the external mode $k = 0$ in the model of Buchler et al. (1997) for $V_0 = 10$ and $\tau_0 = 0.948$

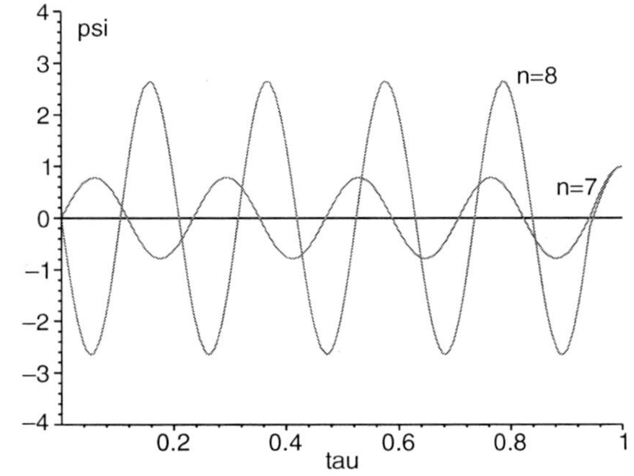

Fig. 11.9 The eigenfunctions $\psi(\tau)$ of the internal modes $n = 7$ and $n = 8$ in the model of Buchler et al. (1997) for $V_0 = 10$ and $\tau_0 = 0.948$

11.6 The First-Degree f-Modes

As mentioned in Sect. 10.3.2, a f-mode with an eigenvalue σ^2 between the eigenvalues $\sigma^2_{g_1^+}$ and $\sigma^2_{p_1}$ exists for each degree $\ell > 1$, but the f-mode belonging to $\ell = 1$ has the eigenvalue $\sigma^2 = 0$. Aizenman et al. (1977) examined the transition from the eigenvalue of the f-mode for $\ell = 2$ to the value zero for $\ell = 1$ by treating the degree ℓ as a non-integer and gradually decreasing its value in the system of equations that governs the linear, isentropic oscillations of a quasi-static star. They pointed at the fact that, after the separation of the angular variables by means of spherical harmonics, the system of equations contains the degree ℓ as a parameter and is well defined for any arbitrary ℓ. From their analysis for one of their evolutionary models, the authors concluded that the eigenvalue gradually shifts towards the value $\sigma^2 = 0$ while the "f-mode" undergoes avoided crossings with the "g^+-modes".

Another question concerning the first-degree f-modes is related to the property that, in the Cowling approximation, the eigenvalue σ_f^2 of the degree $\ell = 1$ is situated between the eigenvalues $\sigma^2_{g_1^+}$ and $\sigma^2_{p_1}$ as the eigenvalues σ_f^2 of the degrees $\ell > 1$. As noted by Christensen-Dalsgaard & Gough (2001), "this ... raises the question of the relation between the $\ell = 1$ oscillation spectra in the Cowling approximation and in the full case".

Christensen-Dalsgaard & Gough investigated the question by introducing a gravitational dilution factor λ in their equations. When the factor is introduced into Eqs. (6.15)–(6.17), these equations become

$$\frac{du}{dr} = \frac{g}{c^2}\,u + \left[\frac{\ell(\ell+1)}{\sigma^2} - \frac{r^2}{c^2}\right] y + \lambda\,\frac{\ell(\ell+1)}{\sigma^2}\,\Phi\,', \tag{11.55}$$

$$\frac{dy}{dr} = \left(\sigma^2 - N^2\right)\frac{u}{r^2} + \frac{N^2}{g}\,y - \lambda\,\frac{d\Phi\,'}{dr}, \tag{11.56}$$

$$\frac{1}{r^2}\frac{d}{dr}\left(r^2\,\frac{d\Phi\,'}{dr}\right) - \frac{\ell(\ell+1)}{r^2}\,\Phi\,' = 4\pi G\rho\left(\frac{N^2}{g}\,\frac{u}{r^2} + \frac{1}{c^2}\,y\right). \tag{11.57}$$

They are supplemented by the usual boundary conditions. The authors assumed that $\rho = 0$ at $r = R$.

When $\lambda = 1$, the equations correspond to those that govern the full problem. When $\lambda = 0$, the Eulerian perturbation of the gravitational potential and its first derivative are eliminated from Eqs. (11.55) and (11.56), so that these equations reduce to Eqs. (7.32) and (7.33) valid in the Cowling approximation. By varying λ from 0 to 1, one makes a gradual transition from the Cowling approximation to the full treatment.

Christensen-Dalsgaard & Gough concluded that the effect of increasing λ is to decrease the eigenfrequencies belonging to the degree $\ell = 1$. They also noted:

> However, the most striking aspect of the results is the segment of the solution that converges towards $\sigma^2 = 0$ as $\lambda \to 1$. Where it joins with the other 'normal' segments the frequencies undergo *avoided crossings*, approaching each other quite closely as λ varies without actually crossing.

Chapter 12
Completeness of the Linear, Isentropic Normal Modes

12.1 Status Questionis

The question whether the inner-product-space of the linear, isentropic normal modes of a quasi-static star, as considered in Sect. 4.4, forms a complete space has been worded by Dyson & Schutz (1979) in the following terms:

> In order to have confidence in the results of a normal mode analysis one needs to know whether the modes are complete: can any perturbation be expressed as a superposition of normal modes?

In case of completeness of the space of the linear, isentropic normal modes, the time-dependent evolution of any perturbation in a quasi-static star can be represented as

$$\boldsymbol{\xi}\,(\mathbf{r},t) = \sum_{\kappa=0}^{\infty} c_\kappa \,\exp(\mathrm{i}\sigma_\kappa t)\,\boldsymbol{\xi}_\kappa\,(\mathbf{r}). \tag{12.1}$$

The possibility of approximating any linear, isentropic perturbation by an infinite series of normal modes is of main importance for the analysis of the dynamic stability of a quasi-star. An initial perturbation remains bounded at all times, if no eigenfrequency σ_κ of a linear, isentropic normal mode κ contains a *negative imaginary part* (see, e.g., Chandrasekhar 1961). Since σ^2 is the parameter in the eigenvalue problem, the investigation of the dynamic stability of a star reduces to the question whether or not all eigenvalues σ^2 are positive.

When the space of the linear, isentropic normal modes is complete, any linear, isentropic perturbation $\boldsymbol{\xi}\,(\mathbf{r})$ at a certain time, which is square integrable and has a sufficiently regular behaviour, can arbitrarily closely be approximated in L^2-sense by an infinite superposition of linear, isentropic normal modes $\boldsymbol{\xi}_\kappa\,(\mathbf{r})$ at that time as

$$\boldsymbol{\xi}\,(\mathbf{r}) = \sum_{\kappa'=0}^{\infty} c'_\kappa\,\boldsymbol{\xi}_{\kappa'}\,(\mathbf{r}). \tag{12.2}$$

Such an approximation of any linear, isentropic perturbation by an infinite series of normal modes is important with regard to the use of perturbation methods for the

P. Smeyers and T. Van Hoolst, *Linear Isentropic Oscillations of Stars: Theoretical Foundations*, Astrophysics and Space Science Library 371,
DOI 10.1007/978-3-642-13030-4_13,

evaluation of smaller effects on normal modes, as there are the non-adiabatic energy exchanges, and the effects of axial rotations and magnetic fields.

The completeness of the linear, isentropic normal modes of a quasi-static star is usually (tacitly) assumed (see, e.g., Cox 1980, Unno et al. 1989, Reisenegger 1994). For the completeness, the toroidal modes have to be taken into consideration in addition to the spheroidal modes.

The question of the completeness of the linear, isentropic normal modes of a quasi-static star was approached by Eisenfeld (1969) for stellar models with a vanishing surface density. Eisenfeld considered the integro-differential operator involved in the system of governing equations for the functions $r^2\,\xi(r)$ and $\eta(r)$ that follows from Eqs. (5.88)–(5.89). He attempted to prove that the eigenvectors of the operator form a complete orthonormal basis of a Hilbert space. For this purpose, he showed that the operator is symmetric. He also gave a proof of the property that the operator has a non-empty resolvent, but it is questionable whether the boundary conditions on the surface are satisfied. The main point, however, is that Eisenfeld's treatment has been vitiated by an invalid conclusion.

The completeness is closely connected to the question whether the tensorial operator U, which is involved in vectorial wave equation (4.24), is not merely symmetric but admits a self-adjoint extension, i.e. an extension for which the domain is identical to the domain of the adjoint. Kaniel & Kovetz (1967) and Dyson & Schutz (1979) adopted the point of view that the quadratic associated with the tensorial operator U is bounded from below. This means that the function $c\,(\boldsymbol{\xi})$ of the Lagrangian displacement field given by

$$c\,(\boldsymbol{\xi}) = \frac{\displaystyle\int_V \rho\,\delta q^i\,\left(U_{ij}\,\delta q^j\right) dV}{\displaystyle\int_V \rho\,g_{ij}\,\delta q^i\,\delta q^j\,dV} \tag{12.3}$$

is bounded from below as a function on the domain of U. The operator U can then be extended, on the basis of the theorem of Stone–Friedrichs (Riesz & Sz.-Nagy 1955), to a self-adjoint operator $\widetilde{\mathsf{U}}$ whose spectrum has the same lower bound as the function $c\,(\boldsymbol{\xi})$.

For the determination of the lower bound of the function $c\,(\boldsymbol{\xi})$, Dyson & Schutz observed that, due to the symmetry of the operator U, it suffices to consider real displacement fields $\boldsymbol{\xi}$.

In the numerator in the right-hand member of the equality (12.3), Kaniel & Kovetz and Dyson & Schutz decomposed the covariant vector components $\rho\,U_{ij}\,\delta q^j$, with $i = 1,2,3$, into two parts by means of Eq. (4.33). They proved that the first term of $c\,(\boldsymbol{\xi})$, which is determined by the part $R_{ij}\,\delta q^i$, is bounded from below. Relative to the second term of $c\,(\boldsymbol{\xi})$, which is determined by the part $\rho\,\nabla_i \Phi'$, Kaniel & Kovetz simply observed that this term is clearly bounded from below. On their side, Dyson & Schutz referred to a derivation of Hunter (1977), in which the star is considered to be enclosed in a cube, and Fourier series are introduced. In this way, Hunter was able to show that the second term of $c(\boldsymbol{\xi})$ is also bounded from below. However, when the Eulerian perturbation of the gravitational potential, Φ', is expressed in terms of the solution of Poisson's perturbed differential equation that is valid in a spherically symmetric star, one obtains an additional term for $c(\boldsymbol{\xi})$ for which the existence of a lower bound remains to be established. A conclusive proof of the existence of a lower bound for the tensorial operator U has thus still to be given.

If the operator U is nevertheless bounded from below, use can be made of the self-adjointness of the extended operator $\widetilde{\mathsf{U}}$. On this ground, Kaniel & Kovetz and Dyson & Schutz gave a spectral theorem, which enabled them to formulate an expansion theorem for a general time-dependent, linear, isentropic displacement field $\xi(\mathbf{r}, t)$ in a quasi-static star, in terms of normal modes.

In this chapter, Eisenfeld's approach is presented. Next, the derivation of the lower bound of the tensorial operator U is reconsidered. Both Hunter's derivation and the derivation based on the use of the solution of Poisson's perturbed differential equation for the Eulerian perturbation of the gravitational potential are given. Finally, the spectral and expansion theorems of Kaniel & Kovetz and Dyson & Schutz are described.

12.2 Approach of Eisenfeld

Eisenfeld's approach concerns the spheroidal normal modes of a spherically symmetric star that belong to a spherical harmonic of degree $\ell \geq 1$.

In the equilibrium star, the mass density is supposed to be positive everywhere, except at the boundary point $r = R$, where it vanishes. It is also supposed to have at least two derivatives in the interval $[0, R]$ and to be analytic at the boundary points $r = 0$ and $r = R$. Moreover, the local criterion for stability against convection, $N^2(r) \geq 0$, is supposed to be satisfied at all points. It is also required that $N^2(r)$ does not vanish identically in any subinterval.

12.2.1 Eisenfeld's Operator T

Eisenfeld started from the system of Eqs. (5.88)–(5.89), which result from the separation of the angular variables in the components of the perturbed equation of motion by means of a spherical harmonic $Y_\ell^m(\theta, \phi)$:

$$\sigma^2 \xi = \frac{d\Phi'}{dr} - \frac{\rho'}{\rho^2}\frac{dP}{dr} + \frac{1}{\rho}\frac{dP'}{dr}, \tag{12.4}$$

$$\sigma^2 \eta = \Phi' + \frac{P'}{\rho}. \tag{12.5}$$

These equations correspond to Eisenfeld's Eqs. (2.6) and (2.7), in which $\psi = r^2 \xi$ and $d\chi/dr = \ell(\ell+1)\,\eta$. The Eulerian perturbations $\rho'(r)$, $P'(r)$, and $\Phi'(r)$ can be expressed in terms of the two functions $\xi(r)$ and $\eta(r)$ by means of Eqs. (5.90) and (5.91), and integral solution (7.4) for the Eulerian perturbation of the gravitational potential. The solutions must satisfy boundary conditions: $\xi(r)$ remains finite as $r \to 0$, and $P'(r) = 0$ at $r = R$. The two equations and the boundary conditions define an eigenvalue problem for σ^2.

The system of Eqs. (12.4)–(12.5) can also be written as

$$\sigma^2 \left(r^2 \xi\right) = r^2 \left(\frac{d\Phi'}{dr} - \frac{\rho'}{\rho^2}\frac{dP}{dr} + \frac{1}{\rho}\frac{dP'}{dr}\right), \tag{12.6}$$

$$\sigma^2 \eta = \Phi' + \frac{P'}{\rho}. \tag{12.7}$$

This form of the equations leads to the consideration of vectors of the type

$$u = \begin{pmatrix} r^2 \xi \\ \eta \end{pmatrix}. \tag{12.8}$$

Therefore, Eisenfeld worked in the Hilbert space $\mathscr{H}$ of the vector valued functions

$$g = \begin{pmatrix} g_1 \\ g_2 \end{pmatrix} \tag{12.9}$$

such that $\rho^{1/2}\, r\, g_1$ and $\rho^{1/2}\, g_2$ are vectors in the function space $L_2(0, R)$. A function space $L_2(0, 1)$ is defined as the set of all complex-valued functions $f(t)$ on $0 \le t \le 1$ that are Lebesgue-measurable and satisfy the condition that the square of their absolute value is integrable in the sense of Lebesgue:

$$\int_0^1 |f|^2\, dt < \infty.$$

One may think of the continuous, or even piecewise continuous, functions on $0 \le t \le 1$, which are included in the class of functions considered (Coddington & Levinson 1955).

The inner product for $f, g \in \mathscr{H}$ is defined as

$$(f, g) = \int_0^R \left[\frac{f_1\, \overline{g_1}}{r^2} + \ell(\ell + 1)\, f_2\, \overline{g_2}\right] \rho\, dr. \tag{12.10}$$

This definition agrees with the definition of the inner product of two spheroidal modes given by Eq. (5.139).

In addition, Eisenfeld introduced the operator T on $\mathscr{H}$ by

$$Tu = v, \tag{12.11}$$

where the vector v is supposed to be in $\mathscr{H}$ and to have the components

$$\left.\begin{aligned} v_1 &= r^2\left(\frac{d\Phi'}{dr} - \frac{\rho'}{\rho^2}\frac{dP}{dr} + \frac{1}{\rho}\frac{dP'}{dr}\right),\\ v_2 &= \Phi' + \frac{P'}{\rho}. \end{aligned}\right\} \tag{12.12}$$

The system of Eqs. (12.6)–(12.7) can then be written in the form

$$\sigma^2 u = Tu. \tag{12.13}$$

About the domain of the operator T, $D(T)$, Eisenfeld specified only that it consists of all vectors u in $\mathscr{H}$ for which Tu is defined in the given sense.

12.2.2 Symmetry of the Operator T

The operator T is symmetric, if the property

$$\left(Tu^{(\alpha)}, u^{(\beta)}\right) = \left(u^{(\alpha)}, Tu^{(\beta)}\right) \tag{12.14}$$

is valid for arbitrary vectors $u^{(\alpha)}$ and $u^{(\beta)}$ in $D(T)$. Eisenfeld noted that Chandrasekhar (1964) already showed the validity of the property for any pair of eigenvectors $u^{(\alpha)}$ and $u^{(\beta)}$.

For the proof, Eisenfeld considered the decomposition

$$Tu = T_0\, u + Gu, \tag{12.15}$$

where T_0 is a differential operator, and G an integro-differential operator, respectively defined by

$$T_0\, u = \begin{pmatrix} r^2\left(\dfrac{1}{\rho}\dfrac{dP'}{dr} + \dfrac{\rho'}{\rho}\dfrac{d\Phi}{dr}\right) \\ \dfrac{P'}{\rho} \end{pmatrix}, \quad Gu = \begin{pmatrix} r^2\dfrac{d\Phi'}{dr} \\ \Phi' \end{pmatrix}. \tag{12.16}$$

According to definition (12.10), one has that

$$\left(T_0\, u^{(\alpha)}, u^{(\beta)}\right) = \int_0^R \left[\left(\frac{1}{\rho} \frac{dP'^{(\alpha)}}{dr} + \frac{\rho'^{(\alpha)}}{\rho} \frac{d\Phi}{dr} \right) \xi^{(\beta)} + \frac{\ell(\ell+1)}{r^2} \frac{P'^{(\alpha)}}{\rho} \eta^{(\beta)} \right] \rho\, r^2\, dr. \tag{12.17}$$

By partial integration of the first term in the right-hand member and use of the conditions that $r^2\, P'^{(\alpha)}\, \xi^{(\beta)} = 0$ at $r = 0$ and $r = R$, one obtains

$$\left(T_0\, u^{(\alpha)}, u^{(\beta)}\right) = \int_0^R \left[-P'^{(\alpha)} \frac{d\left(r^2\, \xi^{(\beta)}\right)}{dr} + r^2 \frac{d\Phi}{dr}\, \rho'^{(\alpha)}\, \xi^{(\beta)} + \ell(\ell+1)\, P'^{(\alpha)}\, \eta^{(\beta)} \right] dr. \tag{12.18}$$

The elimination of η^{β}, $P'^{(\alpha)}$, and $\rho'^{(\alpha)}$ by means of Eqs. (5.90), (5.91), and (5.93) leads to the equality

$$\left(T_0\, u^{(\alpha)}, u^{(\beta)}\right) = \int_0^R \left[c^2\, \alpha^{(\alpha)}\, \alpha^{(\beta)} - g\, \xi^{(\alpha)}\, \alpha^{(\beta)} - g\, \xi^{(\beta)}\, \alpha^{(\alpha)} - g\, \frac{1}{\rho} \frac{d\rho}{dr}\, \xi^{(\alpha)}\, \xi^{(\beta)} \right] \rho\, r^2\, dr, \tag{12.19}$$

from which the symmetry of the differential operator T_0 for any pair of vectors $u^{(\alpha)}$ and $u^{(\beta)}$ in $D(T)$ is apparent.

Next, one has that

$$\left(G u^{(\alpha)}, u^{(\beta)}\right) = \int_0^R \left[\frac{d\Phi'^{(\alpha)}}{dr}\, \xi^{(\beta)} + \frac{\ell(\ell+1)}{r^2}\, \Phi'^{(\alpha)}\, \eta^{(\beta)} \right] \rho\, r^2\, dr. \tag{12.20}$$

Partial integration of the first term in the right-hand member, use of the conditions that $\rho\, r^2\, \Phi'^{(\alpha)}\, \xi^{(\beta)} = 0$ at $r = 0$ and $r = R$, and use of Eqs. (5.90) and (5.93) yield

$$\left(G u^{(\alpha)}, u^{(\beta)}\right) = \int_0^R \Phi'^{(\alpha)}\, \rho'^{(\beta)}\, r^2 dr. \tag{12.21}$$

By elimination of $\rho'^{(\beta)}$ by means of Eq. (5.92) and a partial integration, one obtains

$$\left(G u^{(\alpha)}, u^{(\beta)}\right) = \frac{1}{4\pi G}\left[R^2\, \Phi'^{(\alpha)}(R)\left(\frac{d\Phi'^{(\beta)}}{dr}\right)_R - \int_0^R \frac{d\Phi'^{(\alpha)}}{dr}\,\frac{d\Phi'^{(\beta)}}{dr}\, r^2\, dr \right.$$
$$\left. -\ell(\ell+1)\int_0^R \Phi'^{(\alpha)}\, \Phi'^{(\beta)}\, dr\right]. \tag{12.22}$$

Because of boundary condition (5.97), this equality becomes

$$\left(G u^{(\alpha)}, u^{(\beta)}\right) = \frac{1}{4\pi G}\left[-(\ell+1)\, R\, \Phi'^{(\alpha)}(R)\, \Phi'^{(\beta)}(R) - \int_0^R \frac{d\Phi'^{(\alpha)}}{dr}\,\frac{d\Phi'^{(\beta)}}{dr}\, r^2\, dr \right.$$
$$\left. -\ell(\ell+1)\int_0^R \Phi'^{(\alpha)}\, \Phi'^{(\beta)}\, dr\right], \tag{12.23}$$

so that the integro-differential operator G too is symmetric in the vectors $u^{(\alpha)}$ and $u^{(\beta)}$.

12.2.3 Existence of a Non-Empty Resolvent Set for the Operator T ?

As recalled by Eisenfeld, the resolvent set $\rho(L)$ of a linear operator L is the set of the complex numbers λ such that the operator $L - \lambda I$, where I is the identic operator, has a bounded inverse operator $(L - \lambda I)^{-1}$ on the entire space (see also Dunford & Schwartz 1957, Richtmyer 1978). For a complex number λ in the resolvent set $\rho(L)$, the operator

$$R_\lambda \equiv (L - \lambda)^{-1}$$

is called the resolvent operator of L.

An operator $L : D(L) \to \mathcal{H}$ is said to have a bounded inverse operator when a bounded operator $K : \mathcal{H} \to \mathcal{H}$ exists such that, for all u in $\mathcal{H}$, $K\,u$ belongs to $D(L)$ and $L(K\,u) = u$, and that, for all u in $D(L)$, $K(L\,u) = u$. Note that a linear operator L that has a bounded inverse operator, does not have zero as eigenvalue. Indeed, suppose that $L\,u = 0$, then $0 = K(L\,u) = u$, so that also $u = 0$. Conversely, an operator that does not have zero as eigenvalue, cannot necessarily be inverted into a bounded operator. When a linear operator L does have a bounded inverse operator K, this operator is uniquely determined and is denoted as L^{-1} (Quaegebeur 2002).

The operator T, which is considered here, has the eigenvalue zero for $\ell = 1$ with the associated eigenfunctions $\xi(r) = C$ and $\eta(r) = C\,r$, where C is a constant [see Sect. 5.8.3]. The modes correspond to uniform translations of the star. Eisenfeld eliminated these solutions by supposing that the star's mass center has been fixed [see his comment below his Eq. (3.6)].

The determination of the resolvent set of the operator T thus requires the knowledge of the complex numbers σ^2 for which the inhomogeneous equation

$$R_{\sigma^2}\, f \equiv (T - \sigma^2)^{-1}\, f = u$$

admits of a solution for u in $D(T)$ and for all f in the Hilbert space $\mathscr{H}$. This is equivalent to the determination of the solutions of the inhomogeneous equation

$$(T - \sigma^2)\, u = f$$

or, more explicitly,

$$\left.\begin{aligned} &r^2 \left(\frac{1}{\rho}\frac{dP'}{dr} + \frac{\rho'}{\rho}\frac{d\Phi}{dr} + \frac{d\Phi'}{dr} \right) - \sigma^2 \left(r^2\, \xi \right) = r^2\, f_1, \\ &\left(\frac{P'}{\rho} + \Phi' \right) - \sigma^2\, \eta = f_2. \end{aligned}\right\} \tag{12.24}$$

This system of equations corresponds to Eisenfeld's system of Eqs. (5.1)–(5.2), apart from the fact that Eisenfeld's function $f_1(r)$ has been replaced by $r^2\, f_1(r)$, and Eisenfeld's function $f_2(r)/[\ell(\ell+1)]$, by $f_2(r)$. The functions $f_1(r)$ and $f_2(r)$ are assumed to be continuous in [0,R] and the function $f_2(r)$ is assumed to have a continuous first derivative.

Eisenfeld determined the solutions for values of σ^2 different from zero and from the eigenvalues. To this end, he passed on to an associated inhomogeneous system of equations, which is derived as follows.

From the second equation (12.24), P' is eliminated by means of Eqs. (5.90) and (5.91), so that the equation becomes

$$\frac{d\xi}{dr} - \left(\frac{g}{c^2} - \frac{2}{r} \right) \xi - \left[\frac{\ell(\ell+1)}{r^2} - \frac{\sigma^2}{c^2} \right] \eta - \frac{1}{c^2}\, \Phi' = -\frac{1}{c^2}\, f_2. \tag{12.25}$$

Next, the first equation (12.24) is rewritten as

$$\frac{d}{dr}\left(\frac{P'}{\rho} + \Phi' \right) + \frac{P'}{\rho^2}\frac{d\rho}{dr} + \frac{\rho'}{\rho}\frac{d\Phi}{dr} - \sigma^2\, \xi = f_1, \tag{12.26}$$

or, by means of the second equation (12.24), as

$$\sigma^2 \frac{d\eta}{dr} + \frac{P'}{\rho^2}\frac{d\rho}{dr} + \frac{\rho'}{\rho}\frac{d\Phi}{dr} - \sigma^2\, \xi = f_1 - \frac{df_2}{dr}. \tag{12.27}$$

Elimination of P', ρ', and $d\xi/dr$ leads to

$$\frac{d\eta}{dr} - \left(1 - \frac{N^2}{\sigma^2}\right)\xi - \frac{N^2}{g}\eta + \frac{1}{\sigma^2}\frac{N^2}{g}\Phi' = \frac{1}{\sigma^2}\left(f_1 + \frac{N^2}{g}f_2 - \frac{df_2}{dr}\right). \tag{12.28}$$

To this equation and Eq. (12.25), Eisenfeld added Poisson's perturbed differential equation (5.92). After elimination of ρ' and $d\xi/dr$, the equation can be decomposed into two first-order differential equations as

$$\frac{d\Phi'}{dr} - \Psi = 0, \tag{12.29}$$

$$\frac{d\Psi}{dr} - 4\pi G\rho\frac{N^2}{g}\xi - \sigma^2\frac{4\pi G\rho}{c^2}\eta - \left[\frac{\ell(\ell+1)}{r^2} - \frac{4\pi G\rho}{c^2}\right]\Phi' + \frac{2}{r}\Psi = \frac{4\pi G\rho}{c^2}f_2. \tag{12.30}$$

Equations (12.25) and (12.28)–(12.30) form a fourth-order inhomogeneous system of ordinary differential equations. When one uses the four dependent variables $w_1(r)$, $w_2(r)$, $w_3(r)$, $w_4(r)$ defined by Eqs. (6.69), the inhomogeneous system takes the form

$$\frac{dw}{dr} - A(r)\,w = h(r), \tag{12.31}$$

where the matrix $A(r)$ of the coefficients corresponds to that of the homogeneous system of ordinary differential equations (6.70), and the vector $h(r)$ of the inhomogeneous terms is given by

$$h(r) = \begin{pmatrix} -\dfrac{r}{c^2}f_2 \\[2ex] \dfrac{1}{\sigma^2}\left(f_1 + \dfrac{N^2}{g}f_2 - \dfrac{df_2}{dr}\right) \\[2ex] 0 \\[2ex] \dfrac{4\pi G\rho}{c^2}rf_2 \end{pmatrix}. \tag{12.32}$$

Note that the four-component vector $w(r)$ adopted by Eisenfeld differs from the one used here. The relations between the components are:

$$(w_1)_E = \rho w_1, \quad (w_2)_E = \rho w_2, \quad (w_3)_E = -w_3, \quad (w_4)_E = -w_4. \tag{12.33}$$

By his Eqs. (4.18), Eisenfeld also introduced the functions $\hat{w}_1, \hat{w}_2, \hat{w}_3, \hat{w}_4$, for which the relations hold

$$\hat{w}_1 = w_1, \quad \hat{w}_2 = w_2, \quad \hat{w}_3 = -w_3, \quad \hat{w}_4 = -w_4. \tag{12.34}$$

The four-component vector $h(r)$ given by Eq. (12.32) agrees with the four-component vector $h(r)$ given by Eisenfeld's Eqs. (5.4)–(5.6) apart from the component $h_4(r)$ and apart from the fact that Eisenfeld's components $h_1(r)$ and $h_2(r)$ contain an additional factor ρ, which must be related to the factor ρ incorporated into the definitions of the functions $(w_1)_E$ and $(w_2)_E$.

A general solution of the linear, inhomogeneous system of differential equations (12.31) consists of the sum of a general solution of the homogeneous system and a particular solution of the inhomogeneous system.

For the construction of a general solution of the homogeneous system, we suppose, as in Chap. 6, that the mass density is analytic at the singular boundary points $r = 0$ and $r = R$, and that the layers near the boundary point $r = R$ are characterised by an effective polytropic index $n_e = 3$. At distances sufficiently large from the boundary points, a fundamental solution matrix of the homogeneous system is given by the matrix whose determinant $\Delta\left(\sigma^2\right)$ determines eigenvalue equation (6.98), since σ^2 is here supposed to be different from the eigenvalues. This matrix, denoted as the matrix $\tau(r)$, consists of the solution vectors $w^{(c,1)}(r)$ and $w^{(c,2)}(r)$, constructed from the singular boundary point $r = 0$, and the solution vectors $w^{(s,1)}(r)$ and $w^{(s,2')}(r)$, constructed from the singular boundary point $r = R$.

A particular solution of the inhomogeneous system can be constructed by means of the method of the variation of the constants (see, e.g., Coddington & Levinson 1955).

A general solution of the inhomogeneous system is then given by

$$w(r) = \tau(r) \int_{R/2}^{r} \tau^{-1}\left(r'\right) h\left(r'\right) dr' + A_1\, w^{(c,1)}(r) + A_2\, w^{(c,2)}(r)$$
$$+ B_1\, w^{(s,1)}(r) + B_2\, w^{(s,2')}(r), \tag{12.35}$$

where A_1, A_2, B_1, B_2 are arbitrary constants. With the definition of the vector components

$$k_i(r) = \left[\tau^{-1}(r)\, h(r)\right]_i, \quad i = 1, \ldots, 4, \tag{12.36}$$

the general solution can be rewritten as

$$w(r) = \left(\int_{R/2}^{r} k_1\left(r'\right) dr' + A_1\right) w^{(c,1)}(r) + \left(\int_{R/2}^{r} k_2\left(r'\right) dr' + A_2\right) w^{(c,2)}(r)$$
$$+ \left(\int_{R/2}^{r} k_3\left(r'\right) dr' + B_1\right) w^{(s,1)}(r) + \left(\int_{R/2}^{r} k_4\left(r'\right) dr' + B_2\right) w^{(s,2')}(r). \tag{12.37}$$

In order that the general solution *may* satisfy the boundary conditions at $r = 0$ and $r = R$, Eisenfeld imposed that

$$\left.\begin{aligned} \int_{R/2}^{R} k_1\left(r'\right) dr' + A_1 = 0, \quad \int_{R/2}^{R} k_2\left(r'\right) dr' + A_2 = 0\,, \\ \int_{R/2}^{0} k_3\left(r'\right) dr' + B_1 = 0, \quad \int_{R/2}^{0} k_4\left(r'\right) dr' + B_2 = 0. \end{aligned}\right\} \tag{12.38}$$

The solution then takes the form

$$\begin{aligned} w(r) = & -w^{(c,1)}(r) \int_{r}^{R} k_1\left(r'\right) dr' - w^{(c,2)}(r) \int_{r}^{R} k_2\left(r'\right) dr' \\ & + w^{(s,1)}(r) \int_{0}^{r} k_3\left(r'\right) dr' + w^{(s,2')}(r) \int_{0}^{r} k_4\left(r'\right) dr'. \end{aligned} \tag{12.39}$$

Next, Eisenfeld verified whether the solution actually satisfies the boundary conditions. In our subsequent analysis, we restrict ourselves to the degree $\ell = 2$.

12.2.3.1 For $r \to 0$

In the fundamental solution matrix $\tau(r)$, the solution vectors $w^{(s,1)}(r)$ and $w^{(s,2')}(r)$ have been constructed from the singular boundary point $r = R$ but their behaviours near the singular boundary point $r = 0$ are unknown. Near $r = 0$, these solution vectors must be regarded as two independent linear combinations of the four solution vectors $w^{(c,1)}(r)$, $w^{(c,2)}(r)$, $w^{(c,3)}(r)$, $w^{(c,4)}(r)$ constructed from that point. Therefore, the fundamental solution matrix $\tau(r)$ is represented as the product of the fundamental solution matrix $\tau_c(r)$ of the homogeneous system of Eqs. (6.70), which is constructed from $r = 0$, with the matrix

$$Q_c = \begin{pmatrix} 1 & 0 & D_1 & E_1 \\ 0 & 1 & D_2 & E_2 \\ 0 & 0 & D_3 & E_3 \\ 0 & 0 & D_4 & E_4 \end{pmatrix}. \tag{12.40}$$

In this matrix, D_1, D_2, D_3, D_4, E_1, E_2, E_3, E_4 are constants, which satisfy the condition that $D_3\, E_4 - E_3\, D_4 \neq 0$. One then has

$$\tau(r) = \tau_c(r)\, Q_c. \tag{12.41}$$

With the leading terms in the components of the solution vectors $w^{(c,1)}(r)$, $w^{(c,2)}(r)$, $w^{(c,3)}(r)$, $w^{(c,4)}(r)$, given by Eqs. (6.75)–(6.78), the leading terms of the elements in the matrix $\tau_c(r)$ have the form

$$\tau_c(r) = \begin{pmatrix} 2\,r^2 & a_{12}\,r^4 & -3\,r^{-3} & a_{14}\,r^{-1} \\ r^2 & a_{22}\,r^4 & r^{-3} & a_{24}\,r^{-1} \\ a_{31}\,r^4 & r^2 & a_{33}\,r^{-1} & r^{-3} \\ a_{41}\,r^4 & 2\,r^2 & a_{34}\,r^{-1} & -3\,r^{-3} \end{pmatrix}. \tag{12.42}$$

Consequently, the leading terms of the elements in the matrix $\tau(r)$ have the form

$$\tau(r) = \begin{pmatrix} 2\,r^2 & a_{12}\,r^4 & -3\,D_3\,r^{-3} & -3\,E_3\,r^{-3} \\ r^2 & a_{22}\,r^4 & D_3\,r^{-3} & E_3\,r^{-3} \\ a_{31}\,r^4 & r^2 & D_4\,r^{-3} & E_4\,r^{-3} \\ a_{41}\,r^4 & 2\,r^2 & -3\,D_4\,r^{-3} & -3\,E_4\,r^{-3} \end{pmatrix}. \tag{12.43}$$

In the inverse matrix $\tau^{-1}(r)$, the leading terms of the elements are of the orders in r

$$\tau^{-1}(r) = \begin{pmatrix} O\left(r^{-2}\right) & O\left(r^{-2}\right) & O(1) & O(1) \\ O(1) & O(1) & O\left(r^{-2}\right) & O\left(r^{-2}\right) \\ O\left(r^3\right) & O\left(r^3\right) & O\left(r^3\right) & O\left(r^3\right) \\ O\left(r^3\right) & O\left(r^3\right) & O\left(r^3\right) & O\left(r^3\right) \end{pmatrix}. \tag{12.44}$$

The behaviour of the elements of the inverse matrix $\tau^{-1}(r)$ corresponds to that given by Eisenfeld in his Eq. (5.11), except for the elements $\left(\tau^{-1}\right)_{1,3}$, $\left(\tau^{-1}\right)_{1,4}$, $\left(\tau^{-1}\right)_{2,1}$, $\left(\tau^{-1}\right)_{2,2}$.

In the supposition that $f_1(r) = f_{1,c} + O(r)$ and $f_2(r) = f_{2,c} + O(r)$, it follows that the components of the vector $h(r)$ behave as

$$h(r) = \begin{pmatrix} -\dfrac{f_{2,c}}{c_c^2}\, r \\[2ex] \dfrac{f_{1,c}}{\sigma^2} \\[2ex] 0 \\[2ex] \dfrac{4\pi G \rho_c}{c_c^2}\, f_{2,c}\, r \end{pmatrix}. \tag{12.45}$$

These behaviours agree with those of the components of Eisenfeld's vector $h(r)$, except for the fourth component [see Eisenfeld's first Eq. (5.13)].

The components of the four-vector $k(r)$ are then of the orders in r

$$k(r) = \begin{pmatrix} O\left(r^{-2}\right) \\[2ex] O\left(r^{-1}\right) \\[2ex] O\left(r^{3}\right) \\[2ex] O\left(r^{3}\right) \end{pmatrix}. \tag{12.46}$$

These orders in r agree with those given by Eisenfeld, except for the second component [see Eisenfeld's first Eq. (5.15)].

When the integrals $\int_r^R k_1\left(r'\right) dr'$ and $\int_r^R k_2\left(r'\right) dr'$ are evaluated as

$$\int_r^R k_1\left(r'\right) dr' = O\left(r^{-1}\right), \quad \int_r^R k_2\left(r'\right) dr' = O(\ln r), \tag{12.47}$$

the solution vector $w(r)$, determined by Eq. (12.39), behaves asymptotically as

$$w(r) = O\left(r^{-1}\right) \begin{pmatrix} O\left(r^{2}\right) \\[2ex] O\left(r^{2}\right) \\[2ex] O\left(r^{4}\right) \\[2ex] O\left(r^{4}\right) \end{pmatrix} + O(\ln r) \begin{pmatrix} O\left(r^{4}\right) \\[2ex] O\left(r^{4}\right) \\[2ex] O\left(r^{2}\right) \\[2ex] O\left(r^{2}\right) \end{pmatrix}$$

$$+O\left(r^{4}\right)\begin{pmatrix} O\left(r^{-3}\right) \\ O\left(r^{-3}\right) \\ O\left(r^{-3}\right) \\ O\left(r^{-3}\right) \end{pmatrix} + O\left(r^{4}\right)\begin{pmatrix} O\left(r^{-3}\right) \\ O\left(r^{-3}\right) \\ O\left(r^{-3}\right) \\ O\left(r^{-3}\right) \end{pmatrix}. \tag{12.48}$$

Hence, the functions $\xi(r)$, $\eta(r)$, $\Phi'(r)$, $d\Phi'(r)/dr$ remain bounded as $r \rightarrow 0$. A more detailed analysis is needed to see whether these functions satisfy the boundary conditions at $r = 0$.

12.2.3.2 For $r \rightarrow R$

A fundamental solution matrix of the homogeneous system of Eqs. (6.70) valid from the singular boundary point $r = R$ is given by Eq. (6.92) and consists of the four solution vectors $w^{(s,1)}(r)$, $w^{(s,2)}(r)$, $w^{(s,3)}(r)$, $w^{(s,4)}(r)$, where the solution vector $w^{(s,4)}(r)$ is defined by Eq. (6.86):

$$\tau_s(r) = \begin{pmatrix} w_1^{(s,1)}(r) & w_1^{(s,2)}(r) & w_1^{(s,3)}(r) & w_1^{(s,4)}(r) \\ w_2^{(s,1)}(r) & w_2^{(s,2)}(r) & w_2^{(s,3)}(r) & w_2^{(s,4)}(r) \\ w_3^{(s,1)}(r) & w_3^{(s,2)}(r) & w_3^{(s,3)}(r) & w_3^{(s,4)}(r) \\ w_4^{(s,1)}(r) & w_4^{(s,2)}(r) & w_4^{(s,3)}(r) & w_4^{(s,4)}(r) \end{pmatrix}. \tag{12.49}$$

With the use of the functions $\hat{w}_1(r)$, $\hat{w}_2(r)$, $\hat{w}_3(r)$, $\hat{w}_4(r)$, the characteristic roots of Eisenfeld's homogeneous system of equations are the threefold root 0 and the singular root -3, as in Sect. 6.4.2 [see below Eisenfeld's Eq. (4.19)].

Because of boundary condition (5.97) relative to the Eulerian perturbation of the gravitational potential at $r = R$, the solution vector $w^{(s,2)}(r)$ has been replaced by the solution vector $w^{(s,2')}(r)$, which is a linear combination of the solution vectors $w^{(s,2)}(r)$ and $w^{(s,3)}(r)$, as is expressed by Eq. (6.97). The fundamental solution matrix $\tau_s(r)$, valid from the singular boundary point $r = R$, is hereafter regarded as consisting of the solution vectors $w^{(s,3)}(r)$, $w^{(s,4)}(r)$, $w^{(s,1)}(r)$, $w^{(s,2')}(r)$. Equation (6.92) is therefore rewritten as

$$\tau_s(r) = \begin{pmatrix} w_1^{(s,3)}(r) & w_1^{(s,4)}(r) & w_1^{(s,1)}(r) & w_1^{(s,2')}(r) \\ w_2^{(s,3)}(r) & w_2^{(s,4)}(r) & w_2^{(s,1)}(r) & w_2^{(s,2')}(r) \\ w_3^{(s,3)}(r) & w_3^{(s,4)}(r) & w_3^{(s,1)}(r) & w_3^{(s,2')}(r) \\ w_4^{(s,3)}(r) & w_4^{(s,4)}(r) & w_4^{(s,1)}(r) & w_4^{(s,2')}(r) \end{pmatrix}. \tag{12.50}$$

The solution vectors $w^{(c,1)}(r)$ and $w^{(c,2)}(r)$, constructed from the singular boundary point $r = 0$, are regarded as two independent linear combinations of the solution vectors $w^{(s,3)}(r)$, $w^{(s,4)}(r)$, $w^{(s,1)}(r)$, $w^{(s,2')}(r)$. The fundamental solution matrix $\tau(r)$ can then be represented as

$$\tau(r) = \tau_s(r)\, Q_s, \tag{12.51}$$

where the matrix Q_s is defined as

$$Q_s = \begin{pmatrix} D_1' & E_1' & 0 & 0 \\ D_2' & E_2' & 0 & 0 \\ D_3' & E_3' & 1 & 0 \\ D_4' & E_4' & 0 & 1 \end{pmatrix}. \tag{12.52}$$

D_1', D_2', D_3', D_4', E_1', E_2', E_3', E_4' are constants, for which $D_1'\, E_2' - E_1'\, D_2' \neq 0$. Note that Eisenfeld considered the two solution vectors $w^{(c,1)}(r)$ and $w^{(c,2)}(r)$ as linear combinations of the solution vectors $w^{(s,3)}(r)$ and $w^{(s,4)}(r)$ [see below his Eq. (5.13)].

With the leading terms in the components of the solution vectors $w^{(s,3)}(r)$, $w^{(s,4)}(r)$, $w^{(s,1)}(r)$, $w^{(s,2')}(r)$, given by Eqs. (6.82)–(6.84) and Eq. (6.93), the leading terms in the elements of the matrix $\tau_s(r)$ are of the form

$$\tau_s(r) = \begin{pmatrix} a_{11}\, z & a_{12}\, z^{-3} & a_{13} & a_{14} \\ a_{21}\, z & a_{22}\, z^{-3} & 1 & a_{24}\, z \\ a_{31}\, z & a_{32}\, z^{-2} & a_{33}\, z^4 & 1 \\ 1 & a_{42}\, z^{-3} & a_{43}\, z^3 & -3 \end{pmatrix}. \tag{12.53}$$

The leading terms in the elements of the matrix $\tau(r)$ are of the form

$$\tau(r) = \begin{pmatrix} \dfrac{D_2' a_{12}}{z^3} & \dfrac{E_2' a_{12}}{z^3} & a_{13} & a_{14} \\ \dfrac{D_2' a_{22}}{z^3} & \dfrac{E_2' a_{22}}{z^3} & 1 & a_{24}\, z \\ \dfrac{D_2' a_{32}}{z^2} & \dfrac{E_2' a_{32}}{z^2} & a_{33}\, z^4 & 1 \\ \dfrac{D_2' a_{42}}{z^3} & \dfrac{E_2' a_{42}}{z^3} & a_{43}\, z^3 & -3 \end{pmatrix}. \tag{12.54}$$

Because of the overwhelming influence of the solution vector $w^{(\mathrm{s},4)}(r)$, the first two solution vectors in the fundamental solution matrix $\tau(r)$ tend to become proportional to each other for $z \to 0$.

In the inverse matrix $\tau^{-1}(r)$, all elements are of the order $O(1)$:

$$\tau^{-1}(r) = \begin{pmatrix} O(1) & O(1) & O(1) & O(1) \\ O(1) & O(1) & O(1) & O(1) \\ O(1) & O(1) & O(1) & O(1) \\ O(1) & O(1) & O(1) & O(1) \end{pmatrix}. \tag{12.55}$$

In the supposition that $f_1(r) = f_1(R) + O(z)$ and $f_2(r) = f_2(R) + O(z)$, it follows that the components of the vector $h(r)$ behave as

$$h(r) = \begin{pmatrix} -\dfrac{R}{c_\mathrm{s}^2}\, f_2(R)\, z^{-1} \\ \dfrac{1}{\sigma^2}\, \dfrac{N_\mathrm{s}^2}{g_\mathrm{s}}\, f_2(R)\, z^{-1} \\ 0 \\ \dfrac{4\pi G\rho_\mathrm{s}}{c_\mathrm{s}^2}\, R\, f_2(R)\, z^2 \end{pmatrix}. \tag{12.56}$$

The components of the vector $k(r)$ then behave as

$$k(r) = \begin{pmatrix} O\left(z^{-1}\right) \\ O\left(z^{-1}\right) \\ O\left(z^{-1}\right) \\ O\left(z^{-1}\right) \end{pmatrix}. \tag{12.57}$$

In conclusion, the functions $\xi(r)$, $\eta(r)$, $\Phi'(r)$, $d\Phi'(r)/dr$ can hardly remain bounded as $r \to R$, in contrast with Eisenfeld's deduction.

12.2.4 Eisenfeld's Conclusion

From the foregoing study, Eisenfeld arrived at the conclusion that the resolvent set of the symmetric operator T is non-empty and inferred that the operator is (strictly) self-adjoint:

> ...if $\rho(L)$ is non-empty then L is closed and has no proper extension ...In particular, a symmetric operator with a non-empty resolvent set is closed, and in fact, self-adjoint.

However, Eisenfeld's proposition that a closed operator has no extension is invalid. The following proposition holds here: if T is a closed symmetric operator, four and only four possibilities for the spectrum $\sigma(T)$ of T are possible.

1. $\sigma(T) = \mathbf{C}$, i.e., $\sigma(T)$ contains all complex numbers
2. $\sigma(T) = \{\lambda \in \mathbf{C} \mid \text{Im}\,\lambda \geq 0\}$
3. $\sigma(T) = \{\lambda \in \mathbf{C} \mid \text{Im}\,\lambda \leq 0\}$
4. $\sigma(T) \subseteq \mathbf{R}$

In case 1, the operator T can still be extended, possibly but not necessarily, to a self-adjoint operator. In cases 2 and 3, the operator T is maximally symmetric and *not* self-adjoint; it cannot be extended without loosing its property of symmetry. Case 4 occurs when, and only when, the operator T is self-adjoint (Quaegebeur 2002).

12.3 Lower Bound of the Tensorial Operator U

12.3.1 Expression for the Lower Bound

For the derivation of a lower bound for the tensorial operator U, Kaniel & Kovetz and Dyson & Schutz substituted the decomposition of the covariant vector components $\rho\, U_{ij}\, \delta q^j$, with $i = 1, 2, 3$, given by Eq. (4.33), into definition (12.3) of the function $c\,(\boldsymbol{\xi})$.

When, in the term $R_{ij}\,\delta q^j$, the Eulerian perturbations ρ' and P' are expressed in terms of the components of the Lagrangian displacement, and the term is multiplied by δq^i, one obtains

$$\delta q^i\left(R_{ij}\,\delta q^j\right) = -\nabla_i\left[\left(\rho\,c^2\,\nabla_j\,\delta q^j + \delta q^j\,\nabla_j\,P\right)\delta q^i\right] + \rho\,c^2\left(\nabla_i\,\delta q^i\right)\left(\nabla_j\,\delta q^j\right)$$
$$+\,2\left(\delta q^i\,\nabla_i\,P\right)\left(\nabla_j\,\delta q^j\right) + \frac{1}{\rho}\left(\delta q^i\,\nabla_i\,P\right)\left(\delta q^j\,\nabla_j\rho\right). \qquad (12.58)$$

By transformation of the right-hand member and use of the condition of hydrostatic equilibrium, it follows

$$\delta q^i\left(R_{ij}\,\delta q^j\right) = -\nabla_i\left[\left(\rho\,c^2\,\nabla_j\,\delta q^j + \delta q^j\,\nabla_j\,P\right)\delta q^i\right]$$
$$+\left[\left(\rho\,c^2\right)^{1/2}\left(\nabla_j\,\delta q^j\right) + \frac{\delta q^i\,\nabla_i\,P}{\left(\rho\,c^2\right)^{1/2}}\right]^2$$
$$-\rho\left(\delta q^i\,\nabla_i\Phi\right)\left(\frac{1}{c^2}\,\delta q^j\,\nabla_j\Phi + \frac{1}{\rho}\,\delta q^j\,\nabla_j\rho\right). \qquad (12.59)$$

After integration over the equilibrium star's volume, the first term of the right-hand member can be transformed into a surface integral, which is equal to zero since both the pressure and the gradient of pressure are considered to vanish on the star's surface. It results that

$$\int_V \delta q^i\left(R_{ij}\,\delta q^j\right)dV \geq -\int_V \rho\left(\delta q^i\,\nabla_i\Phi\right)\left(\frac{1}{c^2}\,\delta q^j\,\nabla_j\Phi + \frac{1}{\rho}\,\delta q^j\,\nabla_j\rho\right)dV$$
$$= \int_V \rho\,N^2\,|\delta r|^2\,dV. \qquad (12.60)$$

In a physically acceptable stellar model, the function $N^2(r)$ does not reach an infinite negative value, so that it has a finite minimum value. Consequently, one has

$$\int_V \delta q^i\left(R_{ij}\,\delta q^j\right)dV \geq N^2_{\min}\int_V \rho\,|\delta r|^2\,dV. \qquad (12.61)$$

Relative to the lower bound of the second term in the function $c\,(\xi)$ defined by Eq. (12.3), Dyson & Schutz referred to the derivation of Hunter (1977), from which it follows that

$$\frac{\int_V \delta q^i \left(W_{ij}\, \delta q^j\right) dV}{\int_V \rho\, g_{ij}\, \delta q^i\, \delta q^j\, dV} \geq -4\pi G \rho_{\max}. \tag{12.62}$$

Here $\rho_{\max}$ is the maximum mass density in the stellar model.

12.3.2 Hunter's Derivation

Hunter regarded the star as enclosed in a cube with edge $2L$ $\left(-L \leq x^j \leq L, j = 1, 2, 3\right)$. By partial integration and use of the equation for the mass conservation, one has

$$W(\xi) \equiv \int_V \rho\, \overline{\delta q^i} \left(\nabla_i \Phi'\right) dV = \int_V \Phi'\, \overline{\rho'}\, dV, \tag{12.63}$$

where V is the volume of the cube. The Eulerian perturbation of the gravitational potential is determined by Poisson's integral solution (1.74) or, equivalently, by

$$\Phi'(\mathbf{r}) = G \int_V \rho\left(\mathbf{r}'\right) \delta x^j\left(\mathbf{r}'\right) \left(\frac{\partial}{\partial x^j} \frac{1}{|\mathbf{r}' - \mathbf{r}|}\right) dV\left(\mathbf{r}'\right). \tag{12.64}$$

Hence, the Eulerian perturbation of the gravitational potential can be expressed as the opposite of the divergence of a vector potential $\mathbf{F}$:

$$\Phi'(\mathbf{r}) = -\frac{\partial F^j}{\partial x^j} \equiv -\nabla \cdot \mathbf{F} \tag{12.65}$$

with

$$\mathbf{F}(\mathbf{r}) = -G \int_V \frac{\rho\left(\mathbf{r}'\right) \boldsymbol{\xi}\left(\mathbf{r}'\right)}{|\mathbf{r}' - \mathbf{r}|}\, dV\left(\mathbf{r}'\right). \tag{12.66}$$

Moreover, Hunter assumed that the vector potential can be represented as a Fourier series of the form

$$\mathbf{F}(\mathbf{r}) = \sum_{\mathbf{k}} \mathbf{A}(\mathbf{k}) \exp\left(\mathrm{i}\pi\, \frac{\mathbf{k} \cdot \mathbf{r}}{L}\right) \tag{12.67}$$

in the cube with edge $2L$. By means of equality (12.65), the Fourier series for the Eulerian perturbation of the gravitational potential can then be derived as

$$\Phi'(\mathbf{r}) = -\frac{\mathrm{i}\pi}{L} \sum_{\mathbf{k}} \left[\mathbf{A}(\mathbf{k}) \cdot \mathbf{k}\right] \exp\left(\mathrm{i}\pi\, \frac{\mathbf{k} \cdot \mathbf{r}}{L}\right). \tag{12.68}$$

In addition, Hunter derived a Fourier series for the displacement field $\rho\,\boldsymbol{\xi}$, by considering the vector potential $\mathbf{F}(\mathbf{r})$ as a solution of the Poisson's equation

$$\rho\,\boldsymbol{\xi} = \frac{1}{4\pi G}\,\nabla^2\mathbf{F} \equiv \frac{1}{4\pi G}\left(\mathbf{1}_x\,\nabla^2 F_x + \mathbf{1}_y\,\nabla^2 F_y + \mathbf{1}_z\,\nabla^2 F_z\right) \tag{12.69}$$

and differentiating, term by term, the Fourier series of the right-hand member of this equation. It follows that

$$\rho\,\boldsymbol{\xi} = -\frac{\pi}{4GL^2}\sum_{\mathbf{k}}\mathbf{A}(\mathbf{k})\,|\mathbf{k}|^2\,\exp\left(\mathrm{i}\pi\,\frac{\mathbf{k}\cdot\mathbf{r}}{L}\right). \tag{12.70}$$

A Fourier series for ρ' is derived by use of continuity equation (1.60) and differentiation, term by term, of the Fourier series for $\rho\,\boldsymbol{\xi}$. Here Hunter noted that the differentiation can be performed, since $\rho\,\boldsymbol{\xi}$ vanishes at the end points $x^j = \pm L$. One then has

$$\rho' = \frac{\mathrm{i}\pi^2}{4GL^3}\sum_{\mathbf{k}}\left[\mathbf{A}(\mathbf{k})\cdot\mathbf{k}\right]|\mathbf{k}|^2\,\exp\left(\mathrm{i}\pi\,\frac{\mathbf{k}\cdot\mathbf{r}}{L}\right), \tag{12.71}$$

so that the complex conjugate is given by

$$\overline{\rho'} = -\frac{\mathrm{i}\pi^2}{4GL^3}\sum_{\mathbf{k}}\overline{\mathbf{A}(\mathbf{k})\cdot\mathbf{k}}\,|\mathbf{k}|^2\,\exp\left(-\mathrm{i}\pi\,\frac{\mathbf{k}\cdot\mathbf{r}}{L}\right). \tag{12.72}$$

Substitution of Fourier series (12.68) and (12.72) into the right-hand member of equality (12.63) yields

$$\begin{aligned}
&\int_V \Phi'\,\overline{\rho'}\,dV \\
&= -\frac{\pi^3}{4GL^3}\sum_{\mathbf{k}}\sum_{\mathbf{k}'}\left[\mathbf{A}(\mathbf{k})\cdot\mathbf{k}\right]\left[\mathbf{A}(\mathbf{k}')\cdot\mathbf{k}'\right]|\mathbf{k}'|^2\int_V \exp\left[\frac{\mathrm{i}\pi}{L}\left(\mathbf{k}-\mathbf{k}'\right)\cdot\mathbf{r}\right]dV \\
&= -\frac{2\pi^3}{GL}\sum_{\mathbf{k}}\left|\mathbf{A}(\mathbf{k})\cdot\mathbf{k}\right|^2\,|\mathbf{k}|^2.
\end{aligned} \tag{12.73}$$

On the other hand, by means of Fourier series (12.70), one derives that

$$\int_V \rho^2\,\boldsymbol{\xi}\cdot\overline{\boldsymbol{\xi}}\,dV = \frac{\pi^2}{2G^2L}\sum_{\mathbf{k}}\left|\mathbf{A}(\mathbf{k})\right|^2|\mathbf{k}|^4. \tag{12.74}$$

Since

$$-\left|\mathbf{A}(\mathbf{k})\cdot\mathbf{k}\right|^2 \geq -\left|\mathbf{A}(\mathbf{k})\right|^2|\mathbf{k}|^2, \tag{12.75}$$

the inequality holds

$$\int_V \Phi' \overline{\rho'} \, dV \geq -4\pi G \int_V \rho^2 \, \boldsymbol{\xi} \cdot \overline{\boldsymbol{\xi}} \, dV \geq -4\pi G \rho_{\max} \int_V \rho \, \boldsymbol{\xi} \cdot \overline{\boldsymbol{\xi}} \, dV. \tag{12.76}$$

Hence, it results that

$$\frac{\displaystyle\int_V \Phi' \overline{\rho'} \, dV}{\displaystyle\int_V \rho \, \boldsymbol{\xi} \cdot \overline{\boldsymbol{\xi}} \, dV} \geq -4\pi G \rho_{\max}. \tag{12.77}$$

12.3.3 Alternative Derivation

In this section, an inequality is derived for $W(\boldsymbol{\xi})$ by the use of expressions known for spheroidal modes of a spherically symmetric star. For any spheroidal mode belonging to a spherical harmonic $Y_\ell^m(\theta, \phi)$, the function $W(\boldsymbol{\xi})$ can be expressed as

$$\begin{aligned} W(\boldsymbol{\xi}) &\equiv \int_V \rho \, \overline{\delta q^i} \left(\nabla_i \Phi'\right) dV \\ &= N_{\ell m} \int_0^R \rho(r) \left[\xi(r) \, \frac{d\Phi'}{dr} + \ell(\ell+1) \, \frac{\eta(r)}{r^2} \, \Phi'(r) \right] r^2 \, dr, \end{aligned} \tag{12.78}$$

where $N_{\ell m}$ is defined by equality (Appendix D.5). Substitution of solution (7.4) of Poisson's perturbed differential equation and its first derivative, given by Eq. (7.5), leads to

$$\begin{aligned} W(\boldsymbol{\xi}) = {} & -4\pi G N_{\ell m} \int_0^R \rho^2(r) \, \xi^2(r) \, r^2 \, dr + 4\pi G N_{\ell m} \, \frac{\ell(\ell+1)}{2\ell+1} \\ & \int_0^R \rho(r) \Bigg\{ \left[r^{-(\ell+2)} \, \xi(r) - \ell \, r^{-(\ell+3)} \, \eta(r) \right] \\ & \qquad \left[\int_0^r \rho\left(r'\right) r'^{\ell} \left[r' \, \xi\left(r'\right) + (\ell+1) \, \eta\left(r'\right) \right] dr' \right] \\ & \qquad + \left[r^{\ell-1} \, \xi(r) + (\ell+1) \, r^{\ell-2} \, \eta(r) \right] \\ & \qquad \left[\int_r^R \rho\left(r'\right) r'^{-(\ell+1)} \left[r' \, \xi\left(r'\right) - \ell \, \eta\left(r'\right) \right] dr' \right] \Bigg\} \, r^2 \, dr. \end{aligned} \tag{12.79}$$

In the right-hand member, three types of terms can be distinguished: the terms involving only the function $\xi(r)$, the terms involving only the function $\eta(r)$, and the terms involving both the functions $\xi(r)$ and $\eta(r)$.

The sum of the terms involving only the function $\xi(r)$ is given by

$$T_1(\boldsymbol{\xi}) \equiv -4\pi G N_{\ell m} \int_0^R \rho^2(r)\,\xi^2(r)\,r^2\,dr + 4\pi G N_{\ell m} \frac{\ell(\ell+1)}{2\ell+1} \int_0^R \rho(r)\,\xi(r)\,dr$$
$$\left[r^{-\ell} \int_0^r \rho(r')\,r'^{\ell+1}\,\xi(r')\,dr' + r^{\ell+1} \int_r^R \rho(r')\,r'^{-\ell}\,\xi(r')\,dr' \right]. \tag{12.80}$$

Transformation of the second term in the right-hand member yields

$$T_1(\boldsymbol{\xi}) = -4\pi G N_{\ell m} \int_0^R \rho^2(r)\,\xi^2(r)\,r^2\,dr + 2\pi G N_{\ell m} \frac{\ell(\ell+1)}{2\ell+1}$$
$$\left\{ \int_0^R r^{-(2\ell+1)} \left\{ \frac{d}{dr} \left[\int_0^r \rho(r')\,r'^{\ell+1}\,\xi(r')\,dr' \right]^2 \right\} dr \right.$$
$$\left. - \int_0^R r^{2\ell+1} \left\{ \frac{d}{dr} \left[\int_r^R \rho(r')\,r'^{-\ell}\,\xi(r')\,dr' \right]^2 \right\} dr \right\}. \tag{12.81}$$

By performing partial integrations and taking into account that, for $\ell > 0$, $\xi(r) \propto r^{\ell-1}$ as $r \to 0$, one obtains

$$T_1(\boldsymbol{\xi}) = -4\pi G N_{\ell m} \int_0^R \rho^2(r)\,\xi^2(r)\,r^2\,dr + 2\pi G N_{\ell m} \frac{\ell(\ell+1)}{2\ell+1}$$
$$\left\{ (2\ell+1) \int_0^R r^{-2(\ell+1)} \left[\int_0^r \rho(r')\,r'^{\ell+1}\,\xi(r')\,dr' \right]^2 dr \right.$$
$$+ (2\ell+1) \int_0^R r^{2\ell} \left[\int_r^R \rho(r')\,r'^{-\ell}\,\xi(r')\,dr' \right]^2 dr$$
$$\left. + R^{-(2\ell+1)} \left[\int_0^R \rho(r')\,r'^{\ell+1}\,\xi(r')\,dr' \right]^2 \right\}, \tag{12.82}$$

so that

$$T_1(\boldsymbol{\xi}) \geq -4\pi G N_{\ell m} \int_0^R \rho^2(r)\,\xi^2(r)\,r^2\,dr. \tag{12.83}$$

Secondly, the sum of the terms involving only the function $\eta(r)$ is given by

$$T_2(\xi) \equiv -4\pi G N_{\ell m} \frac{[\ell(\ell+1)]^2}{2\ell+1} \int_0^R \rho(r)\,\eta(r) \left[r^{-(\ell+1)} \int_0^r \rho(r')\, r'^{\ell}\, \eta(r')\, dr' + r^{\ell} \int_r^R \rho(r')\, r'^{-(\ell+1)}\, \eta(r')\, dr' \right] dr. \tag{12.84}$$

This sum can be transformed into

$$T_2(\xi) = -4\pi G N_{\ell m} \frac{[\ell(\ell+1)]^2}{2\ell+1} \int_0^R \rho(r)\,\eta(r) \left[\frac{r^{-(\ell+1)}}{\ell+1} \int_0^r \rho(r') \frac{dr'^{(\ell+1)}}{dr'} \eta(r')\, dr' - \frac{r^{\ell}}{\ell} \int_r^R \rho(r') \frac{dr'^{-\ell}}{dr'} \eta(r')\, dr' \right] dr. \tag{12.85}$$

Partial integration leads to

$$\begin{aligned} T_2(\xi) = & -4\pi G N_{\ell m}\, \ell(\ell+1) \int_0^R \rho^2(r)\, \eta^2(r)\, dr \\ & -4\pi G N_{\ell m} \frac{\ell(\ell+1)}{2\ell+1} \left\{ -\ell \int_0^R \rho(r)\,\eta(r)\, r^{-(\ell+1)} \left\{ \int_0^r r'^{\ell+1} \left[\frac{d}{dr'} \left(\rho(r')\,\eta(r')\right) \right] dr' \right\} dr \right. \\ & \left. + (\ell+1) \int_0^R \rho(r)\,\eta(r)\, r^{\ell} \left\{ \int_r^R r'^{-\ell} \left[\frac{d}{dr'} \left(\rho(r')\,\eta(r')\right) \right] dr' \right\} dr \right\} \end{aligned} \tag{12.86}$$

or, equivalently, to

$$\begin{aligned} T_2(\xi) = & -4\pi G N_{\ell m}\, \ell(\ell+1) \int_0^R \rho^2(r)\, \eta^2(r)\, dr - 4\pi G N_{\ell m} \frac{\ell(\ell+1)}{2\ell+1} \\ & \left\{ \int_0^R \rho(r)\,\eta(r) \frac{dr^{-\ell}}{dr} \left\{ \int_0^r r'^{\ell+1} \left[\frac{d}{dr'} \left(\rho(r')\,\eta(r')\right) \right] dr' \right\} dr \right. \\ & \left. + \int_0^R \rho(r)\,\eta(r) \frac{dr^{\ell+1}}{dr} \left\{ \int_r^R r'^{-\ell} \left[\frac{d}{dr'} \left(\rho(r')\,\eta(r')\right) \right] dr' \right\} dr \right\}. \end{aligned} \tag{12.87}$$

By partial integration of the second term in the right-hand member and use of the property that $\eta(r) \propto r^{\ell}$ as $r \to 0$, one obtains

$$T_2(\boldsymbol{\xi}) = -4\pi G N_{\ell m}\, \ell(\ell+1) \int_0^R \rho^2(r)\, \eta^2(r)\, dr + 4\pi G N_{\ell m} \frac{\ell(\ell+1)}{2\ell+1}$$
$$\left\{ \int_0^R r^{-\ell} \left[\frac{d}{dr}\left(\rho(r)\,\eta(r)\right)\right] \left\{ \int_0^r r'^{\ell+1} \left[\frac{d}{dr'}\left(\rho\left(r'\right)\eta\left(r'\right)\right)\right] dr' \right\} dr \right.$$
$$\left. + \int_0^R r^{\ell+1} \left[\frac{d}{dr}(\rho(r)\,\eta(r))\right] \left\{ \int_r^R r'^{-\ell} \left[\frac{d}{dr'}\left(\rho\left(r'\right)\eta\left(r'\right)\right)\right] dr' \right\} dr \right\} \tag{12.88}$$

or, equivalently,

$$T_2(\boldsymbol{\xi}) = -4\pi G N_{\ell m}\, \ell(\ell+1) \int_0^R \rho^2(r)\, \eta^2(r)\, dr + 2\pi G N_{\ell m} \frac{\ell(\ell+1)}{2\ell+1}$$
$$\left\{ \int_0^R r^{-(2\ell+1)} \left\{ \frac{d}{dr} \left[\int_0^r r'^{\ell+1} \frac{d}{dr'} \left[\rho\left(r'\right)\eta\left(r'\right)\right] dr' \right]^2 \right\} dr \right.$$
$$\left. - \int_0^R r^{2\ell+1} \left\{ \frac{d}{dr} \left[\int_r^R r'^{-\ell} \frac{d}{dr'} \left[\rho\left(r'\right)\eta\left(r'\right)\right] dr' \right]^2 \right\} dr \right\}. \tag{12.89}$$

A third partial integration yields

$$T_2(\boldsymbol{\xi}) = -4\pi G N_{\ell m}\, \ell(\ell+1) \int_0^R \rho^2(r)\, \eta^2(r)\, dr + 2\pi G N_{\ell m} \frac{\ell(\ell+1)}{2\ell+1}$$
$$\left\{ (2\,\ell+1) \int_0^R r^{-2(\ell+1)} \left[\int_0^r r'^{\ell+1} \frac{d}{dr'} \left[\rho\left(r'\right)\eta\left(r'\right)\right] dr' \right]^2 dr \right.$$
$$+ (2\ell+1) \int_0^R r^{2\ell} \left[\int_r^R r'^{-\ell} \frac{d}{dr'} \left[\rho\left(r'\right)\eta\left(r'\right)\right] dr' \right]^2 dr$$
$$\left. + \frac{1}{R^{2\ell+1}} \left[\int_0^R r^{\ell+1} \frac{d}{dr} [\rho(r)\,\eta(r)]\, dr \right]^2 \right\}, \tag{12.90}$$

so that

$$T_2(\boldsymbol{\xi}) \geq -4\pi G N_{\ell m}\, \ell(\ell+1) \int_0^R \rho^2(r)\, \eta^2(r)\, dr. \tag{12.91}$$

By adding up inequalities (12.83) and (12.91), member by member, one obtains

$$T_1(\boldsymbol{\xi}) + T_2(\boldsymbol{\xi}) \geq -4\pi G N_{\ell m} \int_0^R \rho^2(r) \left[\xi^2(r) + \frac{\ell(\ell+1)}{r^2}\, \eta^2(r) \right] r^2\, dr, \quad (12.92)$$

so that

$$T_1(\boldsymbol{\xi}) + T_2(\boldsymbol{\xi}) \geq -4\pi G N_{\ell m}\, \rho_{\max} \int_0^R \rho(r) \left[\xi^2(r) + \frac{\ell(\ell+1)}{r^2}\, \eta^2(r) \right] r^2\, dr \quad (12.93)$$

or, equivalently,

$$\frac{T_1(\boldsymbol{\xi}) + T_2(\boldsymbol{\xi})}{\displaystyle\int_V \rho\, \boldsymbol{\xi} \cdot \overline{\boldsymbol{\xi}}\, dV} \geq -4\pi G \rho_{\max}. \quad (12.94)$$

Hunger's lower boundary for the function $W(\boldsymbol{\xi})$ of the Lagrangian displacement is here thus obtained as a lower boundary for the normalised sum of the functions $T_1(\boldsymbol{\xi})$ and $T_2(\boldsymbol{\xi})$.

However, in addition, the sum of the remaining terms in equality (12.79) for the function $W(\boldsymbol{\xi})$ must be considered:

$$\begin{aligned} T_3(\boldsymbol{\xi}) \equiv\ & 4\pi G N_{\ell m} \frac{\ell(\ell+1)}{2\ell+1} \\ & \left\{ \int_0^R \rho(r)\, \xi(r) \left[(\ell+1)\, r^{-\ell} \int_0^r \rho(r')\, r'^{\ell}\, \eta(r')\, dr' \right.\right. \\ & \left. -\ell\, r^{\ell+1} \int_r^R \rho(r')\, r'^{-(\ell+1)}\, \eta(r')\, dr' \right] dr \\ & - \int_0^R \rho(r)\, \eta(r) \left[\ell\, r^{-(\ell+1)} \int_0^r \rho(r')\, r'^{\ell+1}\, \xi(r')\, dr' \right. \\ & \left.\left. -(\ell+1)\, r^{\ell} \int_r^R \rho(r')\, r'^{-\ell}\, \xi(r')\, dr' \right] dr \right\}. \end{aligned} \quad (12.95)$$

The existence of a lower bound of the normalised function $T_3(\boldsymbol{\xi})$ remains to be established.

For conclusion, inequality (12.77) becomes

$$\frac{\displaystyle\int_V \Phi'\, \overline{\rho'}\, dV}{\displaystyle\int_V \rho\, \boldsymbol{\xi} \cdot \overline{\boldsymbol{\xi}}\, dV} \geq -4\pi G \rho_{\max} + \frac{T_3(\boldsymbol{\xi})}{\displaystyle\int_V \rho\, \boldsymbol{\xi} \cdot \overline{\boldsymbol{\xi}}\, dV}. \quad (12.96)$$

12.4 Spectral and Expansion Theorems

Kaniel & Kovetz and Dyson & Schutz used the property that, because of its self-adjointness, the extended operator $\widetilde{\mathsf{U}}$ admits a spectral decomposition

$$\widetilde{\mathsf{U}} = \int_{-M}^{\infty} \mu \, dE_{\mu}, \tag{12.97}$$

where $-M$ is a lower boundary for the spectrum of the operator $\widetilde{\mathsf{U}}$, and $\{E_{\mu}\}_{\mu \in \mathbf{R}}$ is a resolution of the identity, i.e. an increasing family of orthogonal projections such that

$$\int_{-M}^{\infty} dE_{\mu} = \mathsf{I}. \tag{12.98}$$

An eigenvalue μ stands here for σ^2. Notice that, in case the spectrum of $\widetilde{\mathsf{U}}$ is discrete, E_{μ} is the projection onto the closed subspace spanned by the eigenvectors that are associated with the eigenvalues smaller than or equal to μ.

From Eq. (12.97), it follows that, for any integer k,

$$\widetilde{\mathsf{U}}^k = \int_{-M}^{\infty} \mu^k \, dE_{\mu}. \tag{12.99}$$

Therefore, the spectral decomposition is particularly relevant for the expansion of time-dependent, linear, isentropic displacement fields in a quasi-static star in terms of linear, isentropic normal modes. Consider a general linear, isentropic displacement field $\boldsymbol{\xi}(\mathbf{r}, t)$ that is solution of the initial-value problem defined by the wave equation

$$\frac{\partial^2 \boldsymbol{\xi}(\mathbf{r}, t)}{\partial t^2} + \mathsf{U}(\mathbf{r})\, \boldsymbol{\xi}(\mathbf{r}, t) = 0 \tag{12.100}$$

and the initial conditions

$$\boldsymbol{\xi}(\mathbf{r}, 0) = \boldsymbol{\xi}_0, \quad \left(\frac{\partial \boldsymbol{\xi}(\mathbf{r}, t)}{\partial t}\right)_0 = \left(\frac{\partial \boldsymbol{\xi}}{\partial t}\right)_0. \tag{12.101}$$

If the displacement field is analytical in time at the initial instant $t = 0$, the Taylor series holds

$$\boldsymbol{\xi}(\mathbf{r}, t) = \boldsymbol{\xi}_0 + \frac{t}{1!}\left(\frac{\partial \boldsymbol{\xi}(\mathbf{r}, t)}{\partial t}\right)_0 + \frac{t^2}{2!}\left(\frac{\partial^2 \boldsymbol{\xi}(\mathbf{r}, t)}{\partial t^2}\right)_0 + \ldots \tag{12.102}$$

The coefficients can be determined in terms of ξ_0 and $(\partial\xi/\partial t)_0$ by means of the equations

$$\left(\frac{\partial^2\xi(\mathbf{r},t)}{\partial t^2}\right)_0 = -\widetilde{\mathsf{U}}\,\xi_0, \quad \left(\frac{\partial^3\xi(\mathbf{r},t)}{\partial t^3}\right)_0 = -\widetilde{\mathsf{U}}\left(\frac{\partial\xi}{\partial t}\right)_0, \quad \ldots, \tag{12.103}$$

so that the Taylor series becomes

$$\xi(\mathbf{r},t) = \left[\mathsf{I} - \frac{t^2}{2!}\widetilde{\mathsf{U}} + \frac{t^4}{4!}\widetilde{\mathsf{U}}^2 - \ldots\right]\xi_0 + \left[t\,\mathsf{I} - \frac{t^3}{3!}\widetilde{\mathsf{U}} + \frac{t^5}{5!}\widetilde{\mathsf{U}}^2 - \ldots\right]\left(\frac{\partial\xi}{\partial t}\right)_0. \tag{12.104}$$

In virtue of the spectral theorem and the appropriate convergence theorems, it can be rewritten as

$$\xi(\mathbf{r},t) = \left[\int_{-M}^{\infty}\left(1 - \frac{t^2}{2!}\mu + \frac{t^4}{4!}\mu^2 - \ldots\right)dE_\mu\right]\xi_0 + \left[\int_{-M}^{\infty}\left(t - \frac{t^3}{3!}\mu + \frac{t^5}{5!}\mu^2 - \ldots\right)dE_\mu\right]\left(\frac{\partial\xi}{\partial t}\right)_0. \tag{12.105}$$

Modes with a frequency equal to zero represent displacement fields that are linear functions of the time.

The Taylor series can be expressed in the shorter form

$$\xi(\mathbf{r},t) = \frac{1}{2}\left\{\int_{-M}^{\infty}\left[\exp\left(\mathrm{i}\,\mu^{1/2}t\right) + \exp\left(-\mathrm{i}\,\mu^{1/2}t\right)\right]dE_\mu\right\}\xi_0 + \frac{1}{2\mathrm{i}}\left\{\int_{-M}^{\infty}\mu^{-1/2}\left[\exp\left(\mathrm{i}\,\mu^{1/2}t\right) - \exp\left(-\mathrm{i}\,\mu^{1/2}t\right)\right]dE_\mu\right\}\left(\frac{\partial\xi}{\partial t}\right)_0 \tag{12.106}$$

or, after a rearrangement of the terms,

$$\xi(\mathbf{r},t) = \frac{1}{2}\left\{\int_{-M}^{\infty}\exp\left(\mathrm{i}\,\mu^{1/2}t\right)dE_\mu\left[\xi_0 + \frac{1}{\mathrm{i}\,\mu^{1/2}}\left(\frac{\partial\xi}{\partial t}\right)_0\right]\right\} + \frac{1}{2}\left\{\int_{-M}^{\infty}\exp\left(-\mathrm{i}\,\mu^{1/2}t\right)dE_\mu\left[\xi_0 - \frac{1}{\mathrm{i}\,\mu^{1/2}}\left(\frac{\partial\xi}{\partial t}\right)_0\right]\right\}. \tag{12.107}$$

As noted by Dyson & Schutz, the integrals over μ with respect to dE_μ correspond to linear superpositions of normal modes. Consequently, the Taylor series represents an expansion of a time-dependent displacement field $\boldsymbol{\xi}(\mathbf{r},t)$ in terms of normal modes, in which each normal mode is determined by the particular choice of the initial conditions $\boldsymbol{\xi}_0$ and $(\partial\boldsymbol{\xi}/\partial t)_0$. In this connection, Dyson & Schutz emphasised the distinction between eigensolutions and normal modes: normal modes are eigensolutions to which initial conditions are added.

With each eigenvalue μ, the two eigenfrequencies $\mu^{1/2}$ and $-\mu^{1/2}$ are associated. The solutions depend on time by $\exp\left(\mathrm{i}\,\mu^{1/2}t\right)$ and $\exp\left(-\mathrm{i}\,\mu^{1/2}t\right)$ and evolve from the initial conditions $\boldsymbol{\xi}_0$ and $(\partial\boldsymbol{\xi}/\partial t)_0$. For a certain eigenvalue μ, the time-dependency $\exp\left(-\mathrm{i}\,\mu^{1/2}t\right)$ can be removed by a choice of the initial conditions such that $(\partial\boldsymbol{\xi}/\partial t)_0 = \mathrm{i}\,\mu^{1/2}\,\boldsymbol{\xi}_0$.

Because of the existence of a lower bound for the operator U, Taylor series (12.107) imposes an upper bound on the rate of the time-dependent evolution of an arbitrary perturbation.

Chapter 13
$N^2(r)$ Nowhere Negative as Condition for Non-Radial Modes with Real Eigenfrequencies

13.1 Introduction

This chapter is devoted to the relation between the condition that $N^2(r)$ is greater than or equal to zero at all points in a star and the dynamic stability of the star with respect to non-radial oscillations. The existence of the relation between the convective stability in a star and the dynamic stability of the star was suggested by Ledoux (1949):

> Il semble ... que sous son aspect le plus général, la question de la stabilité dynamique se rattache à celle de la stabilité vis-à-vis des courants de convection à laquelle est associée le critère de K. Schwarzschild ...
>
> Si cette condition n'est pas satisfaite dans une région de l'étoile, des courants de convection doivent y apparaître. Dans la méthode habituelle d'établir ce critère, on ne considère qu'un seul petit élément de matière déplacé de sa position d'équilibre et on néglige complètement les effets de cette perturbation sur le reste de l'étoile. Si on cherche à tenir compte de ces effets et du mouvement des autres éléments de la couche instable, le problème devient très complexe.[1]

For equilibrium stars whose mass density vanishes on their surface, Lebovitz (1965a) showed that the eigenfrequencies σ of the non-radial, spheroidal normal modes are real when $N^2(r) \geq 0$ everywhere in the star. As noted by him, this implies that the condition is a *sufficient* condition for the dynamic stability of the star with respect to linear, non-radial, isentropic normal modes, if the linear, isentropic normal modes are complete.

Lebovitz (1966) showed moreover that, if $N^2 < 0$ in an interval between $r = 0$ and $r = R$, a negative eigenvalue σ^2 exists. In other words, when the

[1] It seems ... that in its most general form the question of the dynamic stability is related to the stability with respect to convective currents, with which the criterion of K. Schwarzschild is associated ...

If this condition is not satisfied in a region of the star, convective currents must appear there. In the usual way to establish this criterion, one considers only a single small element of matter that is displaced from its equilibrium position and one completely neglects the effects of this perturbation on the rest of the star. When one tries to take account of these effects and the motion of the other elements in the unstable layer, the problem becomes very complex.

P. Smeyers and T. Van Hoolst, *Linear Isentropic Oscillations of Stars: Theoretical Foundations*, Astrophysics and Space Science Library 371, DOI 10.1007/978-3-642-13030-4_14, © Springer-Verlag Berlin Heidelberg 2010 eigenfrequencies of the non-radial, spheroidal normal modes of a star are all real, the condition $N^2 \geq 0$ is satisfied everywhere in the star. The condition $N^2(r) \geq 0$ at all points in a star is thus also a *necessary* condition for the eigenfrequencies of the non-radial, spheroidal normal modes of a star being real.

13.2 $N^2(r)$ Nowhere Negative as Sufficient Condition

Lebovitz (1965a) derived his proof from an equation for the eigenvalues σ^2 of the spheroidal normal modes established by Chandrasekhar (1964), in which the eigenvalue is expressed in terms of integrals of the eigenfunctions over the star's mass. We first derive Chandrasekhar's equation and next present Lebovitz's proof.

13.2.1 Chandrasekhar's Equation for the Eigenvalue of a Spheroidal Normal Mode

In Eq. (8.103), the sum of the integrals I_3 and I_4, and the integral I_5 are transformed as follows.

From definition (8.91) of the integral I_4, the function $\eta(r)$ is eliminated by means of Eq. (5.93). When furthermore the condition of hydrostatic equilibrium is used, it results that

$$I_4 = \int_0^R \left[\frac{dP}{dr}\, \xi\, \alpha + \frac{1}{r^2}\frac{d}{dr}\left(r^2 \rho\, g\, \xi^2\right) - g\,\frac{d\rho}{dr}\,\xi^2 \right] r^2\, dr. \tag{13.1}$$

By use of Eq. (5.93) in definition (8.90) of the integral I_3, and integration of a term in the sum $I_3 + I_4$, one derives

$$I_3 + I_4 = \int_0^R \left(2\,\frac{dP}{dr}\,\xi\,\alpha - g\,\frac{d\rho}{dr}\,\xi^2 \right) r^2\, dr + GM\,\rho(R)\,\xi^2(R). \tag{13.2}$$

Next, definition (8.92) of the integral I_5 is transformed by use of solutions (7.2) and (7.3) of Poisson's differential equation for the Eulerian perturbation of the gravitational potential and its first derivative. One then has

$$I_5 = -\frac{4\pi G}{2\ell + 1}$$

$$\int_0^R \rho(r) \Big\{ \Big[-(\ell+1)\, r^{-(\ell+2)}\, \xi(r) + \ell(\ell+1)\, r^{-(\ell+3)}\, \eta(r) \Big]$$

$$\int_0^r \rho'\left(r'\right) r'^{(\ell+2)}\, dr'$$

$$+\left[\ell\, r^{\ell-1}\,\xi(r)+\ell(\ell+1)\,r^{\ell-2}\,\eta(r)\right]\int_r^R \rho'\left(r'\right) r'^{-(\ell-1)}\,dr'$$
$$+\frac{\rho(R)\,\xi(R)}{R^{\ell-1}}\left[\ell\, r^{\ell-1}\,\xi(r)+\ell(\ell+1)\,r^{\ell-2}\,\eta(r)\right]\Bigg\}\, r^2\,dr. \tag{13.3}$$

Elimination of the function $\eta(r)$ from the first two terms of the integrand by means of Eqs. (5.93) and (5.90) yields

$$I_5=-\frac{4\pi G}{2\ell+1}$$
$$\int_0^R\Bigg\{\left[\rho'(r)\,r^{-(\ell-1)}+\frac{d}{dr}\left(\rho(r)\,r^{-(\ell-1)}\,\xi(r)\right)\right]\int_0^r \rho'\left(r'\right) r'^{(\ell+2)}\,dr'$$
$$+\left[\rho'(r)\,r^{\ell+2}+\frac{d}{dr}\left(\rho(r)\,r^{\ell+2}\,\xi(r)\right)\right]\int_r^R \rho'\left(r'\right) r'^{-(\ell-1)}\,dr'$$
$$+\frac{\rho(R)\,\xi(R)}{R^{\ell-1}}\,\rho(r)\left[\ell\, r^{\ell+1}\,\xi(r)+\ell(\ell+1)\,r^{\ell}\,\eta(r)\right]\Bigg\}\,dr. \tag{13.4}$$

Partial integration in the first and the second term inside the braces leads to

$$I_5=-\frac{4\pi G}{2\ell+1}\Bigg\{\int_0^R \rho'(r)\,r^{-(\ell-1)}\left[\int_0^r \rho'\left(r'\right) r'^{(\ell+2)}\,dr'\right]dr$$
$$+\int_0^R \rho'(r)\,r^{\ell+2}\left[\int_r^R \rho'\left(r'\right) r'^{-(\ell-1)}\,dr'\right]dr$$
$$+\frac{\rho(R)\,\xi(R)}{R^{\ell-1}}\int_0^R \rho'\left(r'\right) r'^{(\ell+2)}\,dr'$$
$$+\frac{\rho(R)\,\xi(R)}{R^{\ell-1}}\int_0^R \rho(r)\left[\ell\, r^{\ell+1}\,\xi(r)+\ell(\ell+1)\,r^{\ell}\,\eta(r)\right]dr\Bigg\}. \tag{13.5}$$

In the first term inside the braces, the integration domain of the double integral extends from 0 to r for r', and from 0 to R for r. It can also be regarded as the integration domain that extends from r' to R for r, and from 0 to R for r'. It follows that

$$\int_0^R\int_0^r \rho'(r)\,r^{-(\ell-1)}\,\rho'\left(r'\right) r'^{(\ell+2)}\,dr'\,dr$$
$$=\int_0^R\int_{r'}^R \rho'(r)\,r^{-(\ell-1)}\,\rho'\left(r'\right) r'^{(\ell+2)}\,dr\,dr'. \tag{13.6}$$

Next, by an interchange of the coordinates r and r', the first two terms inside the braces in the right-hand member of equality (13.5) are seen to be equal. One then has

$$\int_0^R \rho'(r)\, r^{-(\ell-1)} \left[\int_0^r \rho'\left(r'\right) r'^{(\ell+2)}\, dr'\right] dr$$
$$+\int_0^R \rho'(r)\, r^{\ell+2} \left[\int_r^R \rho'\left(r'\right) r'^{-(\ell-1)}\, dr'\right] dr$$
$$= -\int_0^R r^{2\ell+1} \left\{\frac{d}{dr}\left[\int_r^R \rho'\left(r'\right) r'^{-(\ell-1)}\, dr'\right]^2\right\} dr \tag{13.7}$$

and, after partial integration,

$$\int_0^R \rho'(r)\, r^{-(\ell-1)} \left[\int_0^r \rho'\left(r'\right) r'^{(\ell+2)}\, dr'\right] dr$$
$$+\int_0^R \rho'(r)\, r^{\ell+2} \left[\int_r^R \rho'\left(r'\right) r'^{-(\ell-1)}\, dr'\right] dr$$
$$= (2\ell+1)\int_0^R r^{2\ell} \left[\int_r^R \rho'\left(r'\right) r'^{-(\ell-1)}\, dr'\right]^2 dr. \tag{13.8}$$

By substitution into equality (8.103) and use of the condition that $\rho(R) = 0$, one obtains Chandrasekhar's equation for the eigenvalue of a spheroidal normal mode in terms of integrals of the eigenfunctions over the star's mass:

$$\sigma^2 \int_0^R \left[\xi^2 + \frac{\ell(\ell+1)}{r^2}\,\eta^2\right] \rho\, r^2\, dr$$
$$= \int_0^R \left[\rho\, c^2\, \alpha^2 - \rho\, g \left(2\,\xi\,\alpha + \frac{1}{\rho}\frac{d\rho}{dr}\,\xi^2\right)\right] r^2\, dr$$
$$-4\pi G \int_0^R r^{2\ell} \left[\int_r^R \rho'\left(r'\right) r'^{-(\ell-1)}\, dr'\right]^2 dr. \tag{13.9}$$

13.2.2 The Sufficient Condition

Lebovitz' procedure started from Eq. (13.9). By eliminating the function $\alpha(r)$ by means of Eq. (2.53), and the function $\xi(r)$ by means of the relation between the Lagrangian and the Eulerian perturbation of the mass density, and using definition (3.32) of $N^2(r)$, one has

$$\rho\, c^2\, \alpha^2 - 2\,\rho\, g\, \xi\, \alpha - g\,\frac{d\rho}{dr}\,\xi^2 = -\frac{c^2}{d\rho/dr}\,\frac{N^2}{g}\,(\delta\rho)^2 - \frac{g}{d\rho/dr}\,\rho'^2. \tag{13.10}$$

Equation (13.9) then becomes

$$\sigma^2 \int_0^R \left[\xi^2 + \frac{\ell(\ell+1)}{r^2} \eta^2 \right] \rho r^2 dr = \int_0^R \frac{r^2 c^2}{-d\rho/dr} \frac{N^2}{g} (\delta\rho)^2(r)\, dr$$
$$+ \int_0^R \left\{ \frac{r^2 g\, \rho'^2(r)}{-d\rho/dr} - 4\pi G r^{2\ell} \left[\int_r^R \rho'\left(r'\right) r'^{-(\ell-1)}\, dr' \right]^2 \right\} dr. \quad (13.11)$$

Since $d\rho/dr < 0$, the first term of the right-hand member is non-negative when $N^2(r) \geq 0$. The second term of the right-hand member is non-negative, if

$$\int_0^R \left\{ \frac{r^2 g\, \rho'^2(r)}{-d\rho/dr} - 4\pi G r^{2\ell} \left[\int_r^R \rho'\left(r'\right) r'^{-(\ell-1)}\, dr' \right]^2 \right\} dr \geq 0 \quad (13.12)$$

for any admissible function $\rho'(r)$. This condition can also be expressed as

$$\int_0^R 4\pi G r^{2\ell} \left[\int_r^R \rho'\left(r'\right) r'^{-(\ell-1)}\, dr' \right]^2 (\Lambda - 1)\, dr \geq 0 \quad (13.13)$$

with

$$\Lambda = \frac{\int_0^R r^2 g\, \rho'^2(r)/(-d\rho/dr)\, dr}{4\pi G \int_0^R r^{2\ell} \left[\int_r^R \rho'\left(r'\right) r'^{-(\ell-1)}\, dr' \right]^2 dr}. \quad (13.14)$$

The condition is then equivalent to the condition that

$$\Lambda \geq 1 \quad (13.15)$$

for any admissible function $\rho'(r)$.

By the introduction of the function

$$f(r) = \int_r^R \rho'\left(r'\right) r'^{-(\ell-1)}\, dr', \quad (13.16)$$

Λ can be written as

$$\Lambda = \frac{\int_0^R \frac{g r^{2\ell}}{-d\rho/dr} \left(\frac{df}{dr} \right)^2 dr}{4\pi G \int_0^R r^{2\ell} f^2(r)\, dr}. \quad (13.17)$$

It follows that

$$\int_0^R \left[\frac{g r^{2\ell}}{-d\rho/dr} \left(\frac{df}{dr} \right)^2 - \Lambda\, 4\pi G\, r^{2\ell}\, f^2(r) \right] dr = 0. \tag{13.18}$$

A necessary condition for the existence of a minimum of Λ is given by the Euler–Lagrange equation

$$\frac{d}{dr} \left(\frac{g r^{2\ell}}{-d\rho/dr} \frac{df}{dr} \right) + \Lambda\, 4\pi G\, r^{2\ell}\, f = 0. \tag{13.19}$$

By means of the function

$$h(r) = \frac{df}{dr} \left(\frac{d\rho}{dr} \right)^{-1} \tag{13.20}$$

with

$$f(r) = -\int_r^R \frac{d\rho}{dr'}\, h\left(r'\right) dr', \tag{13.21}$$

the Euler–Lagrange equation can be transformed into

$$\frac{d}{dr} \left(g r^{2\ell}\, h \right) + \Lambda\, 4\pi G r^{2\ell} \int_r^R \frac{d\rho}{dr'}\, h\left(r'\right) dr' = 0. \tag{13.22}$$

The function $h(r)$ can be more explicitly written as

$$h(r) = -\frac{\dfrac{\rho'}{\rho} r^{-(\ell-1)}}{\dfrac{1}{\rho} \dfrac{d\rho}{dr}} \tag{13.23}$$

and remains finite at $r = 0$ and $r = R$.

Multiplication of Eq. (13.22) by $r^{-2\ell}$ and differentiation with respect to r lead to the second-order differential equation for $h(r)$

$$\frac{d}{dr} \left[r^{-2\ell} \frac{d}{dr} \left(g r^{2\ell}\, h \right) \right] - \Lambda\, 4\pi G\, \frac{d\rho}{dr}\, h = 0, \tag{13.24}$$

which can be transformed into the self-adjoint form

$$\frac{d}{dr} \left(g^2 r^{2\ell} \frac{dh}{dr} \right) + g r^{2\ell} \left[\frac{d^2 g}{dr^2} + 2\,\ell \frac{1}{r} \left(\frac{dg}{dr} - \frac{g}{r} \right) - 4\pi G \Lambda\, \frac{d\rho}{dr} \right] h = 0. \tag{13.25}$$

By eliminating d^2g/dr^2 and passing on from the gravity, $g(r)$, to the mean mass density inside the sphere with radius r, $\bar{\rho}(r)$, by means of the relation

$$g(r) = \frac{4\pi G r}{3}\,\bar{\rho}(r),$$

one obtains Lebovitz' equation

$$\frac{d}{dr}\left(r^{2\ell+2}\,\bar{\rho}^2(r)\,\frac{dh}{dr}\right) + r^{2\ell+1}\,\bar{\rho}(r)\left[2(\ell-1)\,\frac{d\bar{\rho}(r)}{dr} + 3(1-\Lambda)\,\frac{d\rho}{dr}\right] h = 0. \qquad (13.26)$$

Multiplication of all terms by $h(r)$ and integration from 0 to R lead to the equality

$$\left[r^{2\ell+2}\,\bar{\rho}^2(r)\,h\,\frac{dh}{dr}\right]_0^R - \int_0^R r^{2\ell+2}\,\bar{\rho}^2(r)\left(\frac{dh}{dr}\right)^2 dr + 2(\ell-1)\int_0^R r^{2\ell+1}\,\bar{\rho}(r)\,\frac{d\bar{\rho}(r)}{dr}\,h^2\,dr + 3(1-\Lambda)\int_0^R r^{2\ell+1}\,\bar{\rho}(r)\,\frac{d\rho}{dr}\,h^2\,dr = 0. \qquad (13.27)$$

In the first term, the contribution stemming from the lower boundary is equal to zero, since it follows from definition (13.23) that $h(r)$ remains finite and $dh/dr \propto r$ as $r \to 0$. For the determination of the contribution stemming from the upper boundary, it is appropriate to derive $(dh/dr)_R$ from Eq. (13.22). Since

$$\left(\frac{dh}{dr}\right)_R + \frac{2(\ell-1)}{R}\,h(R) = 0, \qquad (13.28)$$

the contribution is

$$-2(\ell-1)\,R^{2\ell+1}\,\bar{\rho}^2(R)\,h^2(R).$$

Hence, equality (13.27) becomes

$$-2(\ell-1)\,R^{2\ell+1}\,\bar{\rho}^2(R)\,h^2(R) - \int_0^R r^{2\ell+2}\,\bar{\rho}^2(r)\left(\frac{dh}{dr}\right)^2 dr + 2(\ell-1)\int_0^R r^{2\ell+1}\,\bar{\rho}(r)\,\frac{d\bar{\rho}(r)}{dr}\,h^2\,dr + 3(1-\Lambda)\int_0^R r^{2\ell+1}\,\bar{\rho}(r)\,\frac{d\rho}{dr}\,h^2\,dr = 0. \qquad (13.29)$$

Suppose from here on that $\ell > 0$. Apart from the exceptional case in which $\ell = 1$, and $dh/dr = 0$ at all points in the star, the first three terms in the left-hand member are all negative, so that the fourth term must be positive, and

$$\Lambda > 1. \tag{13.30}$$

In the exceptional case,

$$\Lambda = 1. \tag{13.31}$$

Moreover, it follows from definition (13.21) that

$$\frac{df}{dr} = \text{constant} \cdot \frac{d\rho}{dr} \tag{13.32}$$

and, from definition (13.16), that

$$\frac{df}{dr} = -\rho', \tag{13.33}$$

so that the Eulerian perturbation of the mass density is related to the first derivative of the equilibrium mass density as

$$\rho' = -\text{constant} \cdot \frac{d\rho}{dr}. \tag{13.34}$$

Hence, the exceptional case corresponds to that of a uniform translation of the star, for which the radial component ξ of the Lagrangian displacement is independent of the radial coordinate r, as shown in Sect. 4.7. In other words, the exceptional case occurs for the f-modes associated with $\ell = 1$.

One then reaches the conclusion that, for all admissible solutions $\rho'(r)$, condition (13.15) is satisfied, so that the eigenvalues σ^2 of the non-radial, spheroidal normal modes are positive or equal to zero in stars in which $N^2(r) \geq 0$ at all points.

13.3 $N^2(r)$ Nowhere Negative as Necessary Condition

Lebovitz (1966) applied an energy principle of Laval et al. (1965) to the symmetric tensorial operator U, defined in Sect. 4.2: if for a displacement field $\boldsymbol{\xi}$ that is not identically zero, the inequality holds

$$-\frac{(\boldsymbol{\xi}, \mathsf{U}\,\boldsymbol{\xi})}{(\boldsymbol{\xi}, \boldsymbol{\xi})} \equiv \nu^2 > 0, \tag{13.35}$$

initial conditions $\boldsymbol{\xi}\,(\boldsymbol{r}, 0)$ and $(\partial\boldsymbol{\xi}/\partial t)\,(\boldsymbol{r}, 0)$ exist for which the solution of the vectorial wave equation (4.15) satisfies the inequality

$$\left(\frac{\partial\boldsymbol{\xi}}{\partial t}, \frac{\partial\boldsymbol{\xi}}{\partial t}\right) \geq C \exp(\nu t), \qquad \text{with} \quad C > 0. \tag{13.36}$$

Lebovitz proved that a displacement field exists for which condition (13.35) is satisfied when $N^2 < 0$ in some layers of a star. After multiplication of definition (4.14) for $U_{ij}\,\delta q^j$ by $-\rho\,\overline{\delta q^i}$, integration over the star's volume, and partial integration of the first and the third term, one has

$$-\left(\overline{\xi}, \mathsf{U}\,\xi\right) = -\int_S \rho\,\delta\Phi\,\overline{\xi}_r\,dS - \int_V \left[\frac{c^2}{\rho}\,\overline{\delta\rho}\,\delta\rho + g\,\overline{\xi}_r\,\delta\rho + \overline{\rho'}\,\delta\Phi\right] dV. \quad (13.37)$$

When, in the volume integral, use is made of the relation between the Lagrangian perturbation and the Eulerian perturbation of the gravitational potential, and $\overline{\xi}_r$ is eliminated by means of the relation between the Lagrangian perturbation and the Eulerian perturbation of the mass density, it results that

$$\begin{aligned} -\left(\overline{\xi}, \mathsf{U}\,\xi\right) = {} & -\int_S \rho\,\delta\Phi\,\overline{\xi}_r\,dS - \int_V \left\{\frac{c^2}{\rho}\,\overline{\delta\rho}\,\delta\rho\right. \\ & \left. + \frac{g}{d\rho/dr}\left[\delta\rho\left(\overline{\delta\rho} - \overline{\rho'}\right) + \overline{\rho'}\left(\delta\rho - \rho'\right)\right] + \overline{\rho'}\,\Phi'\right\} dV. \end{aligned} \quad (13.38)$$

By the use of definition (3.32) of $N^2(r)$, one obtains

$$\begin{aligned} -\left(\overline{\xi}, \mathsf{U}\xi\right) = {} & -\int_S \rho\,\delta\Phi\,\overline{\xi}_r\,dS \\ & + \int_V \left\{\frac{c^2}{d\rho/dr}\,\frac{N^2}{g}\,\overline{\delta\rho}\,\delta\rho + \left[\frac{g}{d\rho/dr}\,\overline{\rho'}\,\rho' - \overline{\rho'}\,\Phi'\right]\right\} dV. \end{aligned} \quad (13.39)$$

Let $N^2(r)$ be negative in a subinterval (α, β) of the interval $(0, R)$. Lebovitz considered a perturbation for which $\rho'(r) = 0$ in the interval $(0, R)$, so that, for a perturbation associated with a spherical harmonic $Y_\ell^m(\theta, \phi)$, it follows that

$$\delta\rho\,(\mathbf{r}) = \frac{d\rho}{dr}\,\xi(r)\,Y_\ell^m(\theta, \phi). \quad (13.40)$$

The function $\xi(r)$ is supposed to be sufficiently regular, not to become identically zero in the subinterval (α, β), but to be identically zero outside it. In the supposition that also $\xi(R) = 0$, equality (13.39) reduces to

$$-\left(\overline{\xi}, \mathsf{U}\,\xi\right) = N_{\ell m} \int_\alpha^\beta c^2\,\frac{d\rho}{dr}\,\frac{N^2}{g}\,\xi^2\,r^2\,dr. \quad (13.41)$$

Since, in the right-hand member, the integrand is non-singular, and both $N^2(r)$ and $d\rho/dr$ are negative, it thus follows that

$$-\left(\overline{\xi}, \mathsf{U}\,\xi\right) > 0. \quad (13.42)$$

Scuflaire (1971), on his side, also showed that the condition $N^2(r) \geq 0$ at all points in a star is a necessary condition for stability. His approach is based on the use of a principle of minimal energy given by Ledoux (1958) and requires only a small number of assumptions. In particular, the fact that the equilibrium configuration is spherically symmetric is not used.

Chapter 14
Asymptotic Representation of Low-Degree, Higher-Order p-Modes

14.1 State of the Art

Asymptotic representations of linear, isentropic, spheroidal normal modes of stars lead to rather simple equations for the eigenfrequencies and to semi-analytic solutions for the eigenfunctions. Therefore, they largely contribute to the gain of insight in the influence exerted by various stellar properties on distinct types of modes, and provide tools for mode identifications, which are complementary to numerical integrations of the governing equations.

The asymptotic theory of the linear, isentropic, spheroidal normal modes has had its beginning with Ledoux' asymptotic representation of *radial* modes with a high angular frequency σ, or, equivalently, radial modes of a higher radial order (Ledoux 1962). Ledoux started from the linear, homogeneous second-order differential equation (10.1) for the relative radial displacement and considered the square of the angular frequency as a large parameter. From each of the singular boundary points $r = 0$ and $r = R$, he constructed an asymptotic solution in terms op Bessel functions by applying a method of Langer (1935). By imposing the asymptotic solutions and their first derivatives to be continuous at a point in their common domain of validity, he derived a first-order approximation of the eigenfrequencies. For the polytropic model with index $n = 3$, the relative errors of the asymptotic eigenfrequencies of the modes with radial orders $n = 8, 12, 40$ appeared to be respectively $3.5\%, 2\%, 0.5\%$.

In order to improve the approximation, Tassoul & Tassoul (1968) developed the asymptotic representation of radial modes with a high angular frequency up to the second order. To this end, they applied a method of Olver (1956), which involves the use of series of Bessel functions (see also Olver 1974).

For *low-degree non-radial* normal modes associated with either a high or a low angular frequency, attempts were made to develop similarly asymptotic representations from a linear, homogeneous second-order differential equation. However, only for the compressible equilibrium sphere of uniform mass density, an exact second-order differential equation is available, i.e. Eq. (6.10) for the function $\alpha(r)$, in which the coefficient $K_4(r) = 0$ in this case. The equation was used by Iweins &

P. Smeyers and T. Van Hoolst, *Linear Isentropic Oscillations of Stars: Theoretical Foundations*, Astrophysics and Space Science Library 371, DOI 10.1007/978-3-642-13030-4_15,

Smeyers (1968) for the construction of a first-order asymptotic representation of low-degree p-modes associated with a large positive eigenvalue σ^2 and for low-degree g^--modes associated with a small negative eigenvalue σ^2.

For models different from the compressible equilibrium sphere of uniform mass density, second-order differential equations have been adopted that were derived from the system of Eqs. (6.15)–(6.17) in the Cowling approximation. So Eq. (10.12) for the function $u(r)$ has been used for the construction of an asymptotic representation of low-degree p-modes associated with a large positive eigenvalue σ^2. The right-hand member of the differential equation was considered to be negligibly small and was even set equal to zero. However, the differential equation is inadequate in a small region near the boundary point $r = 0$, since its right-hand member becomes there indefinitely large: the factor $(1 - S_\ell^2/\sigma^2)$, which appears in the denominator of one of the terms, goes through zero as $r \to 0$. In this connection, the term mobile singularity is often used, since the position of the point where the singularity occurs, i.e. here the point where $S_\ell^2 = \sigma^2$, depends on the value of σ^2.

Likewise Eq. (10.13) for the function $u(r)$ has been used for the construction of an asymptotic representation of low-degree g-modes associated with an eigenvalue σ^2 that is small in absolute value. Here too the right-hand member of the differential equation was set equal to zero. The differential equation is now inadequate in a small region near the boundary point $r = R$, since its right-hand member is there no more negligible. For small positive values of σ^2, a mobile singularity also appears at the point where $S_\ell^2 = \sigma^2$.

Iweins (1964) and Vandakurov (1967) independently got around the difficulties resulting from the existence of excluded regions by passing on to an adequate second-order differential equation for the function $y(r)$ in these regions. Tassoul (1980) adopted the idea of Iweins and Vandakurov in a careful development of asymptotic representations of low-degree non-radial modes in the Cowling approximation. By applying asymptotic methods of Olver (1974) to the second-order differential equations used and imposing the functions $u(r)$ and $y(r)$ to be continuous at a point of their common domain of validity, she constructed first- and second-order asymptotic representations of p-modes associated with a large eigenvalue σ^2 and of g^+- and g^--modes associated with an eigenvalue σ^2 small in absolute value. Tassoul's eigenvalue equation for higher-order p-modes has been of main importance in a large number of helioseismological analyses (see, e.g., Bahcall & Ulrich 1988). Tassoul's study was done over again in a more transparant way by Smeyers & Tassoul (1987).

A major progress in the development of the asymptotic theory of low-degree p-modes with higher frequencies was made by Tassoul (1990). In this investigation, Tassoul applied asymptotic methods of Olver to the fourth-order system composed of the two differential equations (6.8) and (6.10), which were derived by Pekeris (1938) for the functions $\alpha(r)$ and $\xi(r)$ but were almost fallen into oblivion. These differential equations are valid in the whole interval $[0, R]$, so that difficulties related to mobile singularities and excluded subintervals do no longer occur. Another important point is that the asymptotic representations were constructed from the full fourth-order system of differential equations, without any neglect of the Eulerian perturbation of the gravitational potential, in contrast with what had been done before.

A next significant progress in the asymptotic theory of the spheroidal normal modes of stars has been made by the introduction of asymptotic methods that are adequate for singular perturbation problems, particularly two-variable expansion procedures and boundary-layer theory (Kevorkian & Cole 1981, 1996). The two methods are commonly described for single second-order differential equations, but their use has been extended to the system of the two differential equations (6.8) and (6.10). The application of these methods has the advantage that the asymptotic solutions are constructed at each level without any foregoing introduction of a series of Bessel functions, as it is the case in the use of Olver's method for higher-order asymptotic approximations. These methods have also the advantage of being more closely related to the physical reality that p- and g^+-modes originate from waves propagating to-and-fro in a resonant cavity of the star, as described in Chap. 10.

The asymptotic methods mentioned have been applied in a series of papers: on low-degree higher-order g^+-modes by Smeyers et al. (1995), on low-degree higher-order p-modes to the second approximation by Smeyers et al. (1996), on low-degree higher-order g^+-modes in stars composed of a convective core and a radiative envelope by Willems et al. (1997), on p-modes of intermediate degree by Smeyers (2003), and on low-degree higher-order g^+-modes in stars containing a convective core by Smeyers & Moya (2007).

Hereafter the second-order asymptotic representation of low-degree p-modes of higher radial orders developed by Smeyers (2006) is described. This asymptotic representation is an improved version of the asymptotic representation by Smeyers et al. (1996), which was partly obscured by the fact that the authors adopted boundary-layer coordinates identical to the fast independent variables used in the two-variable expansions at larger distances from the boundary points.

14.2 Appropriate Equations

After introduction of the parameter

$$\varepsilon = \frac{1}{|\sigma|}, \tag{14.1}$$

the fourth-order system of differential equations (6.10) and (6.8) takes the form

$$\frac{d^2\alpha}{dr^2} + K_2(r)\,\frac{d\alpha}{dr} + \left[\frac{1}{\varepsilon^2}\,\frac{1}{c^2} + K_3(r) + \varepsilon^2\,K_1(r)\right]\alpha = -K_4(r)\,\frac{d\xi}{dr}, \tag{14.2}$$

$$\frac{d^2\xi}{dr^2} + \frac{4}{r}\,\frac{d\xi}{dr} - \frac{\ell(\ell+1)-2}{r^2}\,\xi = \frac{d\alpha}{dr} + \left[\frac{2}{r} - \varepsilon^2\,\frac{c^2}{g}\,K_1(r)\right]\alpha. \tag{14.3}$$

The solutions must satisfy the usual boundary and regularity conditions: the divergence $\alpha(r)$ and the radial component $\xi(r)$ of the Lagrangian displacement must remain finite at $r = 0$ and $r = R$, and the Eulerian perturbation of the gravitational potential, $\Phi'(r)$, must satisfy condition (5.97) at $r = R$.

The equations and boundary conditions are made dimensionless. Therefore, the time t, the radial coordinate r, the pressure $P(r)$, the mass density $\rho(r)$, the gravity $g(r)$, the isentropic sound velocity $c(r)$, the gravitational potential $\Phi(r)$, and both the radial component $\xi(r)$ and the transverse component $\eta(r)$ of the Lagrangian displacement are expressed respectively in the unit $\left[R^3/(GM)\right]^{1/2}$, R, $GM^2/\left(4\pi R^4\right)$, $M/\left(4\pi R^3\right)$, GM/R^2, $(GM/R)^{1/2}$, GM/R, R. With these units, ε is a dimensionless parameter which corresponds to the ratio of the oscillation period to 2π times the star's dynamic time scale and is supposed to be small.

The interval $[0, R]$ is divided into three subintervals: a subinterval distant from the singular boundary points $r = 0$ and $r = R$, a subinterval near the singular boundary point $r = 0$, and a subinterval near the singular boundary point $r = R$. The first subinterval may be thought of as the resonant acoustic cavity of the star. It is supposed that, at the high angular frequencies considered, the lower and the upper boundary of the resonant acoustic cavity are situated so close respectively to the boundary point $r = 0$ and the boundary point $r = R$ that the subintervals outside the cavity are thin boundary layers.

In the three subintervals, a similar procedure is followed. From differential equation (14.2), a *homogeneous* second-order differential equation is derived for the lowest-order asymptotic approximation of the divergence $\alpha(r)$ of the Lagrangian displacement, while from differential equation (14.3), an *inhomogeneous* second-order differential equation is derived for the lowest-order asymptotic approximation of the radial component $\xi(r)$ of the Lagrangian displacement. Next, inhomogeneous second-order differential equations are derived for the second-order asymptotic approximations of $\alpha(r)$ and $\xi(r)$.

14.3 Two-Variable Expansions at Larger Distances from the Boundary Points

The procedure followed in the subinterval situated at larger distances from the boundary points is mainly that of Smeyers et al. (1996)[1]. Differential equation (14.2) is compared to a second-order differential equation that governs the motion of a linear oscillator with small damping. Following an example given by Kevorkian & Cole (1981, 1996), Smeyers et al. used a two-variable expansion procedure.

[1] This section is a partial reproduction of Smeyers, P., Vansimpsen, T., De Boeck, I., Van Hoolst, T.: Asymptotic representation of high-frequency, low-degree p-modes in stars and in the Sun. Astronomy & Astrophysics **307**, 105–120 (1996). With permission from Astronomy & Astrophysics, http://www.aanda.org.

A fast independent variable $\tau(r)$ is defined on the ground of the assumption that the asymptotic solutions for $\alpha(r)$ oscillate so rapidly that their second derivatives $d^2\alpha/dr^2$ are of the same order of magnitude as the term $\alpha/\left(\varepsilon^2\, c^2\right)$, which contains the large parameter. By setting

$$\left(\frac{d\tau}{dr}\right)^2 = \frac{1}{\varepsilon^2\, c^2(r)} \tag{14.4}$$

and imposing that $\tau(0) = 0$, one obtains the positive variable

$$\tau(r) = \frac{1}{\varepsilon}\int_0^r \frac{dr'}{c\,(r')}. \tag{14.5}$$

The integral corresponds to the time needed by an acoustic wave to propagate from the star's centre to the radial distance r and was first introduced by Ledoux (1962) in his asymptotic representation of radial modes with a high angular frequency.

In addition, the radial coordinate r is used as a slow independent variable.

The operators in differential equations (14.2) and (14.3) are transformed according to the chain rule as

$$\left.\begin{aligned} \frac{d}{dr} &= \frac{d\tau}{dr}\frac{\partial}{\partial\tau} + \frac{\partial}{\partial r}, \\ \frac{d^2}{dr^2} &= \left(\frac{d\tau}{dr}\right)^2 \frac{\partial^2}{\partial\tau^2} + 2\,\frac{d\tau}{dr}\,\frac{\partial^2}{\partial\tau\,\partial r} + \frac{d^2\tau}{dr^2}\,\frac{\partial}{\partial\tau} + \frac{\partial^2}{\partial r^2}, \end{aligned}\right\} \tag{14.6}$$

and, for the functions $\alpha(r)$ and $\xi(r)$, asymptotic expansions in terms of the two independent variables are introduced as

$$\left.\begin{aligned} \alpha^{(o)}(r;\varepsilon) &= \alpha_0^{(o)}(\tau,r) + \varepsilon\,\alpha_1^{(o)}(\tau,r) + \varepsilon^2\,\alpha_2^{(o)}(\tau,r) \\ &\quad + \varepsilon^3\,\alpha_3^{(o)}(\tau,r) + O\left(\varepsilon^4\right), \\ \xi^{(o)}(r;\varepsilon) &= \nu(\varepsilon)\Big[\,\xi_0^{(o)}(\tau,r) + \varepsilon\,\xi_1^{(o)}(\tau,r) + \varepsilon^2\,\xi_2^{(o)}(\tau,r) \\ &\quad + \varepsilon^3\,\xi_3^{(o)}(\tau,r) + O\left(\varepsilon^4\right)\Big]. \end{aligned}\right\} \tag{14.7}$$

The function $\nu(\varepsilon)$ makes it possible, if necessary, to take account of the fact that the oscillatory parts in $\alpha^{(o)}(r;\varepsilon)$ and $\xi^{(o)}(r;\varepsilon)$ are of different orders in ε.

Equation (14.2) then takes the form

$$\frac{1}{\varepsilon^2}\frac{1}{c^2}\left(\frac{\partial^2\alpha_0^{(o)}}{\partial\tau^2} + \alpha_0^{(o)}\right) + O\left(\varepsilon^{-1}\right) = \nu(\varepsilon)\left[-\frac{1}{\varepsilon}\,\frac{K_4}{c}\,\frac{\partial\xi_0^{(o)}}{\partial\tau} + O\left(\varepsilon^0\right)\right]. \tag{14.8}$$

The lowest-order differential equation is homogeneous when $\nu(\varepsilon)$ is at least of the order ε^0, and is given by

$$\frac{\partial^2 \alpha_0^{(o)}}{\partial \tau^2} + \alpha_0^{(o)} = 0. \tag{14.9}$$

Its general solution

$$\alpha_0^{(o)}(\tau, r) = A_0^{(o)}(r) \cos \tau + B_0^{(o)}(r) \sin \tau \tag{14.10}$$

involves the two yet undetermined functions $A_0^{(o)}(r)$ and $B_0^{(o)}(r)$ of the slow variable.

Smeyers et al. extended the procedure of Kevorkian & Cole, which is described for a single second-order differential equation, by applying it also to the second-order differential equation (14.3). This equation then takes the form

$$\nu(\varepsilon) \left[\frac{1}{\varepsilon^2} \frac{1}{c^2} \frac{\partial^2 \xi_0^{(o)}}{\partial \tau^2} + O\left(\varepsilon^{-1}\right) \right] = \frac{1}{\varepsilon} \frac{1}{c} \frac{\partial \alpha_0^{(o)}}{\partial \tau} + O\left(\varepsilon^0\right). \tag{14.11}$$

Here the lowest-order differential equation is inhomogeneous when $\nu(\varepsilon) = \varepsilon$, so that

$$\frac{\partial^2 \xi_0^{(o)}}{\partial \tau^2} = c \, \frac{\partial \alpha_0^{(o)}}{\partial \tau}. \tag{14.12}$$

Integration yields

$$\xi_0^{(o)}(\tau, r) = -c(r) \left[B_0^{(o)}(r) \cos \tau - A_0^{(o)}(r) \sin \tau \right] + C_0^{(o)}(r)\, \tau + D_0^{(o)}(r), \tag{14.13}$$

where $C_0^{(o)}(r)$ and $D_0^{(o)}(r)$ are yet undetermined functions of the slow variable. The order of the term $C_0^{(o)}(r)\, \tau$ in the small parameter is inconsistent, since the term can also be written as $C_0'(r)/\varepsilon$. Therefore, the term is removed from the solution (see also Kevorkian & Cole 1981).

At order ε^{-1}, it follows from Eq. (14.2) that

$$\frac{\partial^2 \alpha_1^{(o)}}{\partial \tau^2} + \alpha_1^{(o)} = -c \left[2 \frac{\partial^2 \alpha_0^{(o)}}{\partial \tau \, \partial r} - \left(\frac{1}{2} \frac{1}{c^2} \frac{dc^2}{dr} - K_2 \right) \frac{\partial \alpha_0^{(o)}}{\partial \tau} \right]. \tag{14.14}$$

Substitution of the solution for $\alpha_0^{(o)}(\tau, r)$ and removal of the resonant terms from the inhomogeneous part of the equation lead to

$$A_0^{(o)}(r) = A_0^* \, h(r), \quad B_0^{(o)}(r) = B_0^* \, h(r), \tag{14.15}$$

where A_0^* and B_0^* are arbitrary constants, and

$$h(r) = \left(\rho \, r^2 c^3\right)^{-1/2}. \tag{14.16}$$

The function $h(r)$ corresponds to Tassoul's function $h(r)$ (Tassoul 1990).

Hence, the lowest-order solutions $\alpha_0^{(o)}(\tau, r)$ and $\xi_0^{(o)}(\tau, r)$ can be rewritten as

$$\left.\begin{aligned} \alpha_0^{(o)}(\tau, r) &= h(r)\left(A_0^* \cos\tau + B_0^* \sin\tau\right), \\ \xi_0^{(o)}(\tau, r) &= -h(r)\, c(r)\left(B_0^* \cos\tau - A_0^* \sin\tau\right) + D_0^{(o)}(r). \end{aligned}\right\} \tag{14.17}$$

In these solutions, the function $D_0^{(o)}(r)$ is still undetermined. Its determination requires the derivation of higher-order asymptotic approximations.

By substitution of the solution $\alpha_0^{(o)}(\tau, r)$, Eq. (14.14) reduces to a homogeneous differential equation with general solution

$$\alpha_1^{(o)}(\tau, r) = A_1^{(o)}(r) \cos\tau + B_1^{(o)}(r) \sin\tau, \tag{14.18}$$

where $A_1^{(o)}(r)$ and $B_1^{(o)}(r)$ are yet undetermined functions of the slow variable.

At order ε^0, it follows from Eq. (14.3) that

$$\frac{\partial^2 \xi_1^{(o)}}{\partial \tau^2} = c\left[\frac{\partial \alpha_1^{(o)}}{\partial \tau} - 2\frac{\partial^2 \xi_0^{(o)}}{\partial \tau\, \partial r} + \left(\frac{1}{2}\frac{1}{c^2}\frac{dc^2}{dr} - \frac{4}{r}\right)\frac{\partial \xi_0^{(o)}}{\partial \tau}\right] + c^2\left(\frac{\partial \alpha_0^{(o)}}{\partial r} + \frac{2}{r}\alpha_0^{(o)}\right). \tag{14.19}$$

After substitution of the appropriate solutions and integration, one obtains

$$\begin{aligned} \xi_1^{(o)}(\tau, r) = &\left\{A_0^* h(r)\, c(r)\frac{d}{dr}\ln\left[h(r)\, c(r)\, r^2\right] - B_1^{(o)}(r)\right\} c(r) \cos\tau \\ &+ \left\{B_0^* h(r)\, c(r)\frac{d}{dr}\ln\left[h(r)\, c(r)\, r^2\right] + A_1^{(o)}(r)\right\} \\ &c(r) \sin\tau + D_1^{(o)}(r), \end{aligned} \tag{14.20}$$

where $D_1^{(o)}(r)$ is a yet undetermined function of the slow variable.

Also at order ε^0, it follows from Eq. (14.2) that

$$\begin{aligned} \frac{\partial^2 \alpha_2^{(o)}}{\partial \tau^2} + \alpha_2^{(o)} = &-c\left[2\frac{\partial^2 \alpha_1^{(o)}}{\partial \tau\, \partial r} - \left(\frac{1}{2}\frac{1}{c^2}\frac{dc^2}{dr} - K_2\right)\frac{\partial \alpha_1^{(o)}}{\partial \tau} + K_4\frac{\partial \xi_0^{(o)}}{\partial \tau}\right] \\ &- c^2\left(\frac{\partial^2 \alpha_0^{(o)}}{\partial r^2} + K_2\frac{\partial \alpha_0^{(o)}}{\partial r} + K_3\,\alpha_0^{(o)}\right). \end{aligned} \tag{14.21}$$

After substitution of the appropriate solutions and introduction of the function

$$W(r) = \frac{1}{h}\frac{d^2h}{dr^2} + K_2\,\frac{1}{h}\frac{dh}{dr} + (K_3 + K_4)\,, \tag{14.22}$$

the differential equation becomes

$$\begin{aligned}\frac{\partial^2\alpha_2^{(o)}}{\partial\tau^2} + \alpha_2^{(o)} = &-\left[2\,\frac{dB_1^{(o)}}{dr} + \left(K_2 - \frac{1}{c}\frac{dc}{dr}\right)B_1^{(o)} + A_0^*\,W(r)\,h(r)\,c(r)\right]c\,\cos\tau \\ &+\left[2\,\frac{dA_1^{(o)}}{dr} + \left(K_2 - \frac{1}{c}\frac{dc}{dr}\right)A_1^{(o)} - B_0^*\,W(r)\,h(r)\,c(r)\right]c\,\sin\tau.\end{aligned} \tag{14.23}$$

Removal of the resonant terms from the inhomogeneous part of the differential equation leads to

$$\left.\begin{aligned} A_1^{(o)}(r) &= h(r)\left[B_0^*\,F(r) + A_1^*\right], \\ B_1^{(o)}(r) &= -h(r)\left[A_0^*\,F(r) - B_1^*\right]. \end{aligned}\right\} \tag{14.24}$$

A_1^* and B_1^* are arbitrary constants, and the function $F(r)$ is defined as

$$F(r) = \frac{1}{2}\int_{r_0}^{r} c\left(r'\right) W\left(r'\right) dr', \tag{14.25}$$

where r_0 is the radial coordinate of an arbitrary point in the interval (0,1).

The function $W(r)$ is explicitly given by

$$\begin{aligned} W(r) = &-\frac{\ell(\ell+1)}{r^2} + \frac{4}{r}\frac{g}{c^2} + \frac{1}{2}\frac{1}{\rho}\frac{d^2\rho}{dr^2} - \frac{3}{4}\frac{1}{\rho^2}\left(\frac{d\rho}{dr}\right)^2 \\ &+ \frac{1}{r\,\rho}\frac{d\rho}{dr} - \frac{1}{4}\frac{1}{c^2}\left(\frac{dc}{dr}\right)^2 + \frac{1}{2}\frac{1}{c}\frac{d^2c}{dr^2} \end{aligned} \tag{14.26}$$

and corresponds to Tassoul's function $W_1(r)$ (Tassoul 1990). For $r \rightarrow 0$,

$$c(r)\,W(r) = -c_c\,\frac{\ell(\ell+1)}{r^2} + O\left(r^0\right) \tag{14.27}$$

and, for $r \rightarrow R$,

$$c(r)\,W(r) = -c_s\,\frac{(2n_e+1)\,(2n_e+3)}{16\,(R-r)^{3/2}} + O\left((R-r)^{-1/2}\right). \tag{14.28}$$

Hence, the integrand $c(r)\,W(r)$ in the definition of the function $F(r)$ becomes infinitely large as $r \to 0$ and as $r \to R$.

The solutions for the functions $\alpha_1^{(o)}(\tau, r)$ and $\xi_1^{(o)}(\tau, r)$ are then

$$\left.\begin{aligned} \alpha_1^{(o)}(\tau, r) &= h(r)\{\left[B_0^* \, F(r) + A_1^*\right] \cos\tau \\ &\quad - \left[A_0^* \, F(r) - B_1^*\right] \sin\tau\}, \\ \xi_1^{(o)}(\tau, r) &= c(r)\, h(r)\, \{\left[A_0^* \, G(r) - B_1^*\right] \cos\tau \\ &\quad + \left[B_0^* \, G(r) + A_1^*\right] \sin\tau\} + D_1^{(o)}(r), \end{aligned}\right\} \tag{14.29}$$

with

$$G(r) = c(r) \frac{d}{dr} \ln\left(r^2\, c(r)\, h(r)\right) + F(r). \tag{14.30}$$

For the determination of the function $D_1^{(o)}(r)$ too, higher-order asymptotic approximations must be derived.

Since Eq. (14.23) has been reduced to a homogeneous differential equation, a general solution for $\alpha_2^{(o)}(\tau, r)$ is given by

$$\alpha_2^{(o)}(\tau, r) = A_2^{(o)}(r)\, \cos\tau + B_2^{!o)}(r)\, \sin\tau, \tag{14.31}$$

where $A_2^{(o)}(r)$ and $B_2^{(o)}(r)$ are yet undetermined functions of the slow variable.

At order ε, Eq. (14.3) yields a differential equation for $\xi_2^{(o)}(\tau, r)$ of the form

$$\begin{aligned} \frac{\partial^2 \xi_2^{(o)}}{\partial \tau^2} &= H_1(r)\, \cos\tau + H_2(r)\, \sin\tau \\ &\quad - c^2 \left[\frac{d^2 D_0^{(o)}}{dr^2} + \frac{4}{r} \frac{d D_0^{(o)}}{dr} - \frac{\ell(\ell+1) - 2}{r^2} D_0^{(o)}\right]. \end{aligned} \tag{14.32}$$

The occurrence of a mixed secular term in the solution for $\xi_2^{(o)}(\tau, r)$ is prevented by the condition that

$$\frac{d^2 D_0^{(o)}}{dr^2} + \frac{4}{r} \frac{d D_0^{(o)}}{dr} - \frac{\ell(\ell+1) - 2}{r^2} D_0^{(o)} = 0. \tag{14.33}$$

It follows that a general solution for $D_0^{(o)}(r)$ is given by

$$D_0^{(o)}(r) = C_0^* \, r^{\ell - 1} + D_0^* \, r^{-(\ell + 2)}, \tag{14.34}$$

where C_0^* and D_0^* are arbitrary constants. The solution for $\xi_2^{(o)}(\tau, r)$ then takes the form

$$\xi_2^{(o)}(\tau, r) = -H_1(r) \cos\tau - H_2(r) \sin\tau + D_2^{(o)}(r), \tag{14.35}$$

where $D_2^{(o)}(r)$ is a yet undetermined function of the slow variable.

Also at order ε, Eq. (14.2) yields a differential equation for $\alpha_3^{(o)}(\tau, r)$ of the form

$$\frac{d^2\alpha_3^{(o)}}{d\tau^2} + \alpha_3^{(o)} = H_3(r) \cos\tau + H_4(r) \sin\tau - c^2 K_4 \frac{dD_0}{dr}. \tag{14.36}$$

The resonant terms are removed from the inhomogeneous part of the differential equation by the conditions that

$$H_3(r) = 0, \quad H_4(r) = 0. \tag{14.37}$$

A general solution for $\alpha_3^{(o)}(\tau, r)$ is then given by

$$\alpha_3(\tau, r) = A_3^{(o)}(r) \cos\tau + B_3^{(o)}(r) \sin\tau - c^2 K_4 \frac{dD_0}{dr}, \tag{14.38}$$

where $A_3^{(o)}(r)$ and $B_3^{(o)}(r)$ are yet undetermined functions of the slow variable.

Finally, at order ε^2, Eq. (14.3) leads to a differential equation for $\xi_3^{(o)}(\tau, r)$ of the form

$$\frac{\partial^2\xi_3^{(o)}}{\partial\tau^2} = H_5(r) \cos\tau + H_6(r) \sin\tau$$
$$-c^2 \left[\frac{d^2 D_1^{(o)}}{dr^2} + \frac{4}{r} \frac{dD_1^{(o)}}{dr} - \frac{\ell(\ell+1) - 2}{r^2} D_1^{(o)} \right]. \tag{14.39}$$

Proceeding as for Eq. (14.32), one obtains a general solution for $D_1^{(o)}(r)$ as

$$D_1^{(o)}(r) = C_1^* \, r^{\ell-1} + D_1^* \, r^{-(\ell+2)}, \tag{14.40}$$

where C_1^* and D_1^* are arbitrary constants.

At this stage, two-variable expansions of the divergence and the radial component of the Lagrangian displacement that are valid at larger distances from the singular boundary points $r = 0$ and $r = R$, have been determined up to the second order. When the sums of constants $A_0^* + \varepsilon A_1^*$, $B_0^* + \varepsilon B_1^*$, $C_0^* + \varepsilon C_1^*$, $D_0^* + \varepsilon D_1^*$ are renamed respectively as A_0^*, B_0^*, C_0^*, D_0^*, these expansions take the form

$$\left.\begin{aligned}\alpha^{(o)}(r;\varepsilon) &= h(r)\Big[A_0^* \cos\tau + B_0^* \sin\tau \\ &\quad +\varepsilon\, F(r)\left(B_0^* \cos\tau - A_0^* \sin\tau\right)\Big] + O\left(\varepsilon^2\right), \\ \xi^{(o)}(r;\varepsilon) &= -\varepsilon\, c(r)\, h(r)\Big[B_0^* \cos\tau - A_0^* \sin\tau \\ &\quad -\varepsilon\, G(r)\left(A_0^* \cos\tau + B_0^* \sin\tau\right)\Big] + \xi^{(o)\,(n.o.)}(r;\varepsilon) + O\left(\varepsilon^3\right),\end{aligned}\right\} \tag{14.41}$$

where $\xi^{(o)(n.o.)}(r;\varepsilon)$ is the non-oscillatory function of the slow variable

$$\xi^{(o)(n.o.)}(r;\varepsilon) = \varepsilon\left[C_0^*\, r^{\ell-1} + D_0^*\, r^{-(\ell+2)}\right]. \tag{14.42}$$

The two-variable expansion $\xi^{(o)}(r;\varepsilon)$ is of an order higher in ε than the two-variable expansion $\alpha^{(o)}(r;\varepsilon)$. The expansions contain the four arbitrary constants A_0^*, B_0^*, C_0^*, D_0^* and the arbitrarily chosen radial coordinate r_0. The four constants are fixed by the couplings with the asymptotic expansions valid near the singular boundary points. Furthermore, it will be shown that the choice of the radial coordinate r_0 has no bearing on the final results.

14.4 Boundary-Layer Expansions Near $r = 0$

Some coefficients of the terms in Eqs. (14.2) and (14.3) display a pole at the boundary point $r = 0$ or/and at the boundary point $r = R$. These terms must therefore be incorporated into the dominant asymptotic equations valid near the boundary point, besides the term involving the large parameter. For this purpose, boundary-layer theory is applied (Kevorkian & Cole 1981, 1996)[2]. A simple example of application is presented in Appendix E by way of illustration.

In this section, the construction of boundary-layer expansions near the singular boundary point $r = 0$ is described.

Because of the form of the first term in the two-variable expansion for $\xi^{(o)}(r;\varepsilon)$, it seems to be appropriate to pass on from the function $\xi(r)$ to the function $w(r)$ by means of the transformation

$$\xi(r) = c(r)\, w(r), \tag{14.43}$$

[2] The remaining part of this chapter is a reproduction of Smeyers, P.: The second-order asymptotic representation of higher-order non-radial p-modes in stars revisited. Astronomy & Astrophysics **451**, 237–249 (2006). With permission from Astronomy & Astrophysics, http://www.aanda.org.

so that Eqs. (14.2) and (14.3) become

$$\frac{d^2\alpha}{dr^2} + K_2(r)\,\frac{d\alpha}{dr} + \left[\frac{1}{\varepsilon^2}\,\frac{1}{c^2} + K_3(r) + \varepsilon^2\,K_1(r)\right]\alpha$$
$$= -K_4(r)\,c\left[\frac{dw}{dr} + \frac{1}{c}\,\frac{dc}{dr}\,w\right], \tag{14.44}$$

$$\frac{d^2w}{dr^2} + \left[\frac{4}{r} + \frac{2}{c}\,\frac{dc}{dr}\right]\frac{dw}{dr} + \left[-\frac{\ell(\ell+1)-2}{r^2} + \frac{4}{r}\,\frac{1}{c}\,\frac{dc}{dr} + \frac{1}{c}\,\frac{d^2c}{dr^2}\right]w$$
$$= \frac{1}{c}\,\frac{d\alpha}{dr} + \frac{1}{c}\left[\frac{2}{r} - \varepsilon^2\,\frac{c^2}{g}\,K_1(r)\right]\alpha. \tag{14.45}$$

Power series (6.37)–(6.40) lead to power series for the coefficients $K_1(r)$, $K_2(r)$, $K_3(r)$, $K_4(r)$ of the form

$$\left.\begin{aligned} K_1(r) &= \ell(\ell+1)\,N_c^2\left[1 + O\left(r^2\right)\right], \\ K_2(r) &= \frac{2}{r}\left[1 + O\left(r^2\right)\right], \\ K_3(r) &= -\frac{\ell(\ell+1)}{r^2}\left[1 + O\left(r^2\right)\right], \\ K_4(r) &= O\left(r^2\right). \end{aligned}\right\} \tag{14.46}$$

In the left-hand member of Eq. (14.44), the second derivative $d^2\alpha/dr^2$ and the singular terms are of the same order in ε as the term with the large parameter $\alpha/(\varepsilon^2\,c^2)$, when one introduces a boundary-layer coordinate r^* which is solution of the equation

$$\left(\frac{dr^*}{dr}\right)^2 = \frac{1}{\varepsilon^2}\,\frac{1}{c_c^2}. \tag{14.47}$$

If $r^*(0) = 0$, the positive boundary-layer coordinate is defined by

$$r^*(r) = \frac{1}{\varepsilon}\,\frac{r}{c_c}. \tag{14.48}$$

Correspondingly, boundary-layer expansions for the functions $\alpha(r)$ and $w(r)$ are introduced as

$$\left.\begin{aligned} \alpha^{(c)}(r;\varepsilon) &= \mu_0^{(c)}(\varepsilon)\,\alpha_0^{(c)}(r^*) + \mu_1^{(c)}(\varepsilon)\,\alpha_1^{(c)}(r^*) + \ldots, \\ w^{(c)}(r;\varepsilon) &= \nu_0^{(c)}(\varepsilon)\,w_0^{(c)}(r^*) + \nu_1^{(c)}(\varepsilon)\,w_1^{(c)}(r^*) + \ldots, \end{aligned}\right\} \tag{14.49}$$

where $\mu_0^{(c)}(\varepsilon)$, $\mu_1^{(c)}(\varepsilon)$, …and $\nu_0^{(c)}(\varepsilon)$, $\nu_1^{(c)}(\varepsilon)$, …are asymptotic series to be determined.

After transformation, Eq. (14.44) takes the form

$$\mu_0^{(c)}(\varepsilon)\left\{\frac{1}{\varepsilon^2}\left[\frac{d^2\alpha_0^{(c)}}{dr^{*2}}+\frac{2}{r^*}\frac{d\alpha_0^{(c)}}{dr^*}+\left(1-\frac{\ell(\ell+1)}{r^{*2}}\right)\alpha_0^{(c)}\right]+O\left(\varepsilon^0\right)\right\}$$
$$+\mu_1^{(c)}(\varepsilon)\left\{\frac{1}{\varepsilon^2}\left[\frac{d^2\alpha_1^{(c)}}{dr^{*2}}+\frac{2}{r^*}\frac{d\alpha_1^{(c)}}{dr^*}+\left(1-\frac{\ell(\ell+1)}{r^{*2}}\right)\alpha_1^{(c)}\right]+O\left(\varepsilon^0\right)\right\}$$
$$+\ldots=\nu_0^{(c)}(\varepsilon)\,O(\varepsilon)+\ldots \tag{14.50}$$

The first dominant boundary-layer equation is homogeneous when the function $\nu_0^{(c)}(\varepsilon)$ is of a higher order in ε than the function $\varepsilon^{-3}\,\mu_0^{(c)}(\varepsilon)$. It is then given by

$$\frac{d^2\alpha_0^{(c)}}{dr^{*2}}+\frac{2}{r^*}\frac{d\alpha_0^{(c)}}{dr^*}+\left[1-\frac{\ell(\ell+1)}{r^{*2}}\right]\alpha_0^{(c)}=0. \tag{14.51}$$

The solution that behaves as r^ℓ as $r\to 0$, is

$$\alpha_0^{(c)}\left(r^*\right)=A_{0,\mathrm{c}}\,r^{*-1/2}\,J_{\ell+1/2}\left(r^*\right), \tag{14.52}$$

where $J_{\ell+1/2}\left(r^*\right)$ is the Bessel function of the first kind of order $\ell+1/2$, and $A_{0,\mathrm{c}}$ an arbitrary constant (Abramowitz & Stegun 1965).

Next, Eq. (14.45) is transformed into

$$\nu_0(\varepsilon)\left\{\frac{1}{\varepsilon^2}\left[\frac{d^2w_0^{(c)}}{dr^{*2}}+\frac{4}{r^*}\frac{dw_0^{(c)}}{dr^*}-\frac{\ell(\ell+1)-2}{r^{*2}}w_0^{(c)}\right]+O\left(\varepsilon^0\right)\right\}$$
$$+\nu_1(\varepsilon)\left\{\frac{1}{\varepsilon^2}\left[\frac{d^2w_1^{(c)}}{dr^{*2}}+\frac{4}{r^*}\frac{dw_1^{(c)}}{dr^*}-\frac{\ell(\ell+1)-2}{r^{*2}}w_1^{(c)}\right]+O\left(\varepsilon^0\right)\right\}$$
$$+\ldots=\mu_0(\varepsilon)\left[\frac{1}{\varepsilon}\left(\frac{d\alpha_0^{(c)}}{dr^*}+\frac{2}{r^*}\alpha_0^{(c)}\right)+O(\varepsilon)\right]$$
$$+\mu_1(\varepsilon)\left[\frac{1}{\varepsilon}\left(\frac{d\alpha_1^{(c)}}{dr^*}+\frac{2}{r^*}\alpha_1^{(c)}\right)+O(\varepsilon)\right]+\ldots \tag{14.53}$$

The second dominant boundary-layer is inhomogeneous when $\nu_0^{(c)}(\varepsilon)=\varepsilon\,\mu_0^{(c)}(\varepsilon)$. It is then given by

$$\frac{d^2w_0^{(c)}}{dr^{*2}}+\frac{4}{r^*}\frac{dw_0^{(c)}}{dr^*}-\frac{\ell(\ell+1)-2}{r^{*2}}w_0^{(c)}=\frac{d\alpha_0^{(c)}}{dr^*}+\frac{2}{r^*}\alpha_0^{(c)}. \tag{14.54}$$

The solution that behaves as $r^{\ell-1}$ as $r \rightarrow 0$, is

$$w_0^{(c)}\left(r^*\right) = C_{0,c}\, r^{*(\ell-1)} + \frac{1}{2\ell+1}\left[r^{*(\ell-1)} \int_0^{r^*} r'^{-(\ell-2)} \left(\frac{d\alpha_0^{(c)}}{dr'} + \frac{2}{r'}\alpha_0^{(c)}\right) dr' \right.$$
$$\left. - r^{*-(\ell+2)} \int_0^{r^*} r'^{(\ell+3)} \left(\frac{d\alpha_0^{(c)}}{dr'} + \frac{2}{r'}\alpha_0^{(c)}\right) dr' \right]. \tag{14.55}$$

Here $C_{0,c}$ is an arbitrary constant, which must be set equal to zero when $\ell = 0$. The particular solution of the inhomogeneous differential equation can be transformed by a partial integration of the terms containing the first derivative $d\alpha_0^{(c)}\left(r'\right)/dr'$ and by use of the recurrence relations between Bessel functions

$$\left.\begin{aligned} z^{-(\nu-1)}\, J_\nu(z) &= -\frac{d}{dz}\left(z^{-(\nu-1)}\, J_{\nu-1}(z)\right), \\ z^{\nu+1}\, J_\nu(z) &= \frac{d}{dz}\left(z^{\nu+1}\, J_{\nu+1}(z)\right) \end{aligned}\right\} \tag{14.56}$$

(Abramowitz & Stegun 1965), so that it becomes

$$w_0^{(c)}\left(r^*\right) = C_{0,c}\, r^{*(\ell-1)} - \frac{A_{0,c}}{2\ell+1}\, r^{*-1/2}\left[\ell\, J_{\ell-1/2}\left(r^*\right) - (\ell+1)\, J_{\ell+3/2}\left(r^*\right)\right]. \tag{14.57}$$

For the determination of the next terms in the boundary-layer expansions, it is appropriate to set $\mu_1^{(c)}(\varepsilon) = \varepsilon\, \mu_0^{(c)}(\varepsilon)$ and $\nu_1^{(c)}(\varepsilon) = \varepsilon\, \nu_0^{(c)}(\varepsilon)$.

From Eq. (14.50), it then follows that

$$\frac{d^2\alpha_1^{(c)}}{dr^{*2}} + \frac{2}{r^*}\frac{d\alpha_1^{(c)}}{dr^*} + \left[1 - \frac{\ell(\ell+1)}{r^{*2}}\right]\alpha_1^{(c)} = 0, \tag{14.58}$$

so that

$$\alpha_1^{(c)}\left(r^*\right) = A_{1,c}\, r^{*-1/2}\, J_{\ell+1/2}\left(r^*\right), \tag{14.59}$$

where $A_{1,c}$ is an arbitrary constant.

Next, from Eq. (14.53), it follows that

$$\frac{d^2 w_1^{(c)}}{dr^{*2}} + \frac{4}{r^*}\frac{dw_1^{(c)}}{dr^*} - \frac{\ell(\ell+1)-2}{r^{*2}}\, w_1^{(c)} = \frac{d\alpha_1^{(c)}}{dr^*} + \frac{2}{r^*}\alpha_1^{(c)}, \tag{14.60}$$

so that

$$w_1^{(c)}\left(r^*\right) = C_{1,c}\, r^{*(\ell-1)}$$
$$- \frac{A_{1,c}}{2\ell+1}\, r^{*-1/2}\left[\ell\, J_{\ell-1/2}\left(r^*\right) - (\ell+1)\, J_{\ell+3/2}\left(r^*\right)\right], \tag{14.61}$$

where $C_{1,c}$ is an arbitrary constant.

When the sums of constants $A_{0,\mathrm{c}} + \varepsilon A_{1,\mathrm{c}}$ and $C_{0,\mathrm{c}} + \varepsilon C_{1,\mathrm{c}}$ are renamed respectively as $A_{0,\mathrm{c}}$ and $C_{0,\mathrm{c}}$, the boundary-layer expansions for the divergence and the radial component of the Lagrangian displacement near $r = 0$, which are valid up to the second order, become

$$\left.\begin{aligned} \alpha^{(\mathrm{c})}(r;\varepsilon) &= \mu_0^{(\mathrm{c})}(\varepsilon)\, A_{0,\mathrm{c}}\, r^{*-1/2}\, J_{\ell+1/2}\left(r^*\right), \\ \xi^{(\mathrm{c})}(r;\varepsilon) &= \varepsilon\, \mu_0^{(\mathrm{c})}(\varepsilon)\, c_{\mathrm{c}} \left\{ C_{0,\mathrm{c}}\, r^{*(\ell-1)} \right. \\ &\quad \left. - \frac{A_{0,\mathrm{c}}}{2\ell+1}\, r^{*-1/2} \left[\ell\, J_{\ell-1/2}\left(r^*\right) - (\ell+1)\, J_{\ell+3/2}\left(r^*\right)\right] \right\}. \end{aligned}\right\} \tag{14.62}$$

They involve the yet undetermined function $\mu_0^{(\mathrm{c})}(\varepsilon)$ and the two arbitrary constants $A_{0,\mathrm{c}}$ and $C_{0,\mathrm{c}}$, which are related to two independent particular solutions of the system of equations. With these two particular solutions, two functions $\mu_0^{(\mathrm{c})}(\varepsilon)$ can be associated. The introduction of distinct functions $\mu_0^{(\mathrm{c})}(\varepsilon)$ is seen to be required for the matching of the boundary-layer solutions with the two-variable expansions valid in the region far from the singular boundary points. The boundary-layer expansions then take the more general form

$$\left.\begin{aligned} \alpha^{(\mathrm{c})}(r;\varepsilon) &= \mu_0^{(\mathrm{c},1)}(\varepsilon)\, A_{0,\mathrm{c}}\, r^{*-1/2}\, J_{\ell+1/2}\left(r^*\right), \\ \xi^{(\mathrm{c})}(r;\varepsilon) &= \varepsilon\, \mu_0^{(\mathrm{c},2)}(\varepsilon)\, C_{0,\mathrm{c}}\, c_{\mathrm{c}}\, r^{*(\ell-1)} \\ &\quad - \varepsilon\, \mu_0^{(\mathrm{c},1)}(\varepsilon)\, \frac{A_{0,\mathrm{c}}}{2\ell+1}\, c_{\mathrm{c}}\, r^{*-1/2} \left[\ell\, J_{\ell-1/2}\left(r^*\right) - (\ell+1)\, J_{\ell+3/2}\left(r^*\right)\right]. \end{aligned}\right\} \tag{14.63}$$

Boundary-layer expansion $\alpha^{(\mathrm{c})}(r;\varepsilon)$ is purely oscillatory, while boundary-layer expansion $\xi^{(\mathrm{c})}(r;\varepsilon)$ contains a non-oscillatory part besides its oscillatory part.

14.5 Matching of the Boundary-Layer Expansions Valid Near $r = 0$

The boundary-layer expansions $\alpha^{(\mathrm{c})}(r;\varepsilon)$ and $\xi^{(\mathrm{c})}(r;\varepsilon)$, valid near the singular boundary point $r = 0$, must be matched with the two-variable expansions $\alpha^{(\mathrm{o})}(r;\varepsilon)$ and $\xi^{(\mathrm{o})}(r;\varepsilon)$, valid at larger distances from the singular boundary points $r = 0$ and $r = R$.

The matching condition relative to the divergence of the Lagrangian displacement is

$$\lim_{r\to\infty} \alpha^{(\mathrm{c})}(r;\varepsilon) = \lim_{r\to 0} \alpha^{(\mathrm{o})}(r;\varepsilon). \tag{14.64}$$

The condition requires that the behaviour of the two-variable expansion $\alpha^{(\mathrm{o})}(r;\varepsilon)$ be examined for small values of r, and the behaviour of the boundary-layer expansion $\alpha^{(\mathrm{c})}(r;\varepsilon)$, for large values of r.

For $r \to 0$, the fast independent variable $\tau(r)$ tends to the boundary-layer coordinate $r^*(r)$, so that

$$\tau(r) = \frac{1}{\varepsilon}\,\frac{r}{c_{\mathrm{c}}}. \tag{14.65}$$

The two-variable expansion $\alpha^{(\mathrm{o})}(r;\varepsilon)$ contains the function $F(r)$. The integrand in the function can be split up as

$$F(r) = \frac{1}{2}\int_{r_0}^{r}\left\{\left[c\left(r'\right) W\left(r'\right) + c_{\mathrm{c}}\,\frac{\ell(\ell+1)}{r'^2}\right] - c_{\mathrm{c}}\,\frac{\ell(\ell+1)}{r'^2}\right\} dr', \tag{14.66}$$

so that the function can be expressed as

$$F(r) = \frac{1}{2}\int_{r_0}^{r}\left[c\left(r'\right) W\left(r'\right) + c_{\mathrm{c}}\,\frac{\ell(\ell+1)}{r'^2}\right] dr' + c_{\mathrm{c}}\,\frac{\ell(\ell+1)}{2\,r} - c_{\mathrm{c}}\,\frac{\ell(\ell+1)}{2\,r_0} \tag{14.67}$$

and decomposed as

$$F(r) = F^{(\mathrm{c})}(r) - F^{(\mathrm{c})}\left(r_0\right) \tag{14.68}$$

with

$$F^{(\mathrm{c})}(r) = c_{\mathrm{c}}\,\frac{\ell(\ell+1)}{2\,r} + \frac{1}{2}\int_{0}^{r}\left[c\left(r'\right) W\left(r'\right) + c_{\mathrm{c}}\,\frac{\ell(\ell+1)}{r'^2}\right] dr'. \tag{14.69}$$

For $r \to 0$, it follows that

$$F(r) = c_{\mathrm{c}}\,\frac{\ell(\ell+1)}{2\,r} - F^{(\mathrm{c})}\left(r_0\right) + O(r). \tag{14.70}$$

Since furthermore

$$h(r) = \frac{h_{\mathrm{c}}}{r}\,[1 + O(r)], \tag{14.71}$$

one has

$$\lim_{r\to 0}\alpha^{(\mathrm{o})}(r;\varepsilon) = \frac{h_{\mathrm{c}}}{r}\left\{A_0^*\,\cos r^* + B_0^*\,\sin r^* \right. \\ \left. + \varepsilon\left[c_{\mathrm{c}}\,\frac{\ell(\ell+1)}{2\,r} - F^{(\mathrm{c})}\left(r_0\right)\right]\left(B_0^*\,\cos r^* - A_0^*\,\sin r^*\right)\right\}. \tag{14.72}$$

On the other hand, for $r \to \infty$, one obtains

$$\lim_{r\to\infty} \alpha^{(c)}(r;\varepsilon) = \mu_0^{(c,1)}(\varepsilon)\,\varepsilon\,A_{0,c}\left(\frac{2}{\pi}\right)^{1/2}\frac{c_c}{r}$$
$$\left[\sin\left(r^* - \frac{\ell\pi}{2}\right) + \varepsilon\,c_c\,\frac{\ell(\ell+1)}{2\,r}\cos\left(r^* - \frac{\ell\pi}{2}\right)\right]. \tag{14.73}$$

A matching of the boundary-layer expansion $\alpha^{(c)}(r;\varepsilon)$ with the two-variable expansion $\alpha^{(o)}(r;\varepsilon)$ is possible, when

$$\mu_0^{(c,1)}(\varepsilon) = \varepsilon^{-1} \tag{14.74}$$

and the coefficients of $\cos r^*$ and $\sin r^*$ are identically zero, i.e.,

$$\left.\begin{aligned}
&A_0^* + \varepsilon\left[c_c\,\frac{\ell(\ell+1)}{2\,r} - F^{(c)}(r_0)\right]B_0^* \\
&\quad = -A_{0,c}\left(\frac{2}{\pi}\right)^{1/2}\frac{c_c}{h_c}\left(\sin\frac{\ell\pi}{2} - \varepsilon\,c_c\,\frac{\ell(\ell+1)}{2\,r}\cos\frac{\ell\pi}{2}\right), \\
&B_0^* - \varepsilon\left[c_c\,\frac{\ell(\ell+1)}{2\,r} - F^{(c)}(r_0)\right]A_0^* \\
&\quad = A_{0,c}\left(\frac{2}{\pi}\right)^{1/2}\frac{c_c}{h_c}\left(\cos\frac{\ell\pi}{2} + \varepsilon\,c_c\,\frac{\ell(\ell+1)}{2\,r}\sin\frac{\ell\pi}{2}\right).
\end{aligned}\right\} \tag{14.75}$$

Since, at order ε^0,

$$\left.\begin{aligned}
A_0^* &= -A_{0,c}\left(\frac{2}{\pi}\right)^{1/2}\frac{c_c}{h_c}\sin\frac{\ell\pi}{2}, \\
B_0^* &= A_{0,c}\left(\frac{2}{\pi}\right)^{1/2}\frac{c_c}{h_c}\cos\frac{\ell\pi}{2},
\end{aligned}\right\} \tag{14.76}$$

the equations reduce to

$$\left.\begin{aligned}
A_0^* &= -A_{0,c}\left(\frac{2}{\pi}\right)^{1/2}\frac{c_c}{h_c}\left(\sin\frac{\ell\pi}{2} - \varepsilon\,F^{(c)}(r_0)\cos\frac{\ell\pi}{2}\right), \\
B_0^* &= A_{0,c}\left(\frac{2}{\pi}\right)^{1/2}\frac{c_c}{h_c}\left(\cos\frac{\ell\pi}{2} + \varepsilon\,F^{(c)}(r_0)\sin\frac{\ell\pi}{2}\right).
\end{aligned}\right\} \tag{14.77}$$

By these equations, the constants A_0^* and B_0^* are related to the constant $A_{0,c}$.

Due to the matching of the boundary-layer expansion $\alpha^{(c)}(r;\varepsilon)$ with the two-variable expansion $\alpha^{(o)}(r;\varepsilon)$, the oscillatory parts of the boundary-layer expansion $\xi^{(c)}(r;\varepsilon)$ and the two-variable expansion $\xi^{(o)}(r;\varepsilon)$ are automatically matched. The non-oscillatory part of the boundary-layer expansion $\xi^{(c)}(r;\varepsilon)$ corresponds to that of the two-variable expansion $\xi^{(o)}(r;\varepsilon)$ when

$$D_0^* = 0 \tag{14.78}$$

and

$$\mu_0^{(c,2)}(\varepsilon) = \varepsilon^{\ell-1}, \qquad C_0^* = c_c^{-(\ell-2)}\, C_{0,c}. \tag{14.79}$$

After this matching, the arbitrary constants $A_{0,c}$ and $C_{0,c}$ in the boundary-layer expansions valid near $r = 0$ remain to be determined.

14.6 Boundary-Layer Expansions Near $r = R$

In the boundary layer near $r = R$, it is appropriate to use the independent variable $z = R - r$ instead of r. After transformation, Eqs. (14.44) and (14.45) then take the form

$$\begin{aligned} &\frac{d^2\alpha}{dz^2} - K_2(r)\,\frac{d\alpha}{dz} + \left[\frac{1}{\varepsilon^2}\,\frac{1}{c^2} + K_3(r) + \varepsilon^2\,K_1(r)\right]\alpha \\ &\quad = K_4(r)\,c\left(\frac{dw}{dz} - \frac{1}{c}\,\frac{dc}{dr}\,w\right), \end{aligned} \tag{14.80}$$

$$\begin{aligned} &\frac{d^2w}{dz^2} - \left[\frac{4}{r} + \frac{2}{c}\,\frac{dc}{dr}\right]\frac{dw}{dz} + \left[-\frac{\ell(\ell+1)-2}{r^2} + \frac{4}{r}\,\frac{1}{c}\,\frac{dc}{dr} + \frac{1}{c}\,\frac{d^2c}{dr^2}\right] w \\ &\quad = -\frac{1}{c}\,\frac{d\alpha}{dz} + \frac{1}{c}\left[\frac{2}{r} - \varepsilon^2\,\frac{c^2}{g}\,K_1(r)\right]\alpha. \end{aligned} \tag{14.81}$$

In the supposition that the mass density is analytic at the boundary point $r = R$, power series (6.51) can be used. When one also makes the approximation that $m(r) \simeq M$, power series (6.53)–(6.57) hold. For the coefficients $K_1(r)$, $K_2(r)$, $K_3(r)$, $K_4(r)$, power series are then derived, which have the form

$$\left.\begin{aligned} K_1(r) &= \ell(\ell+1)\,\frac{N_s^2}{R^2}\,\frac{1}{z}\,[1 + O(z)] \\ &\equiv \frac{K_{1,s}}{z}\,[1 + O(z)], \\ K_2(r) &= -\,(n_e + 2)\,\frac{1}{z}\,[1 + O(z)] \\ &\equiv \frac{K_{2,s}}{z}\,[1 + O(z)], \\ K_3(r) &= \frac{K_{3,s}}{z}\,[1 + O(z)], \\ K_4(r) &= \frac{K_{4,s}}{z}\,[1 + O(z)]. \end{aligned}\right\} \tag{14.82}$$

In the left-hand member of Eq. (14.80), the second derivative and the singular terms are of the same order in ε as the term with the large parameter, when the boundary-layer coordinate $z^*(z)$ is introduced by means of the differential equation

$$\left(\frac{dz^*}{dz}\right)^2 = \frac{1}{\varepsilon^2}\frac{1}{c_s^2\,z}. \tag{14.83}$$

When one sets $z^*(0) = 0$, the positive boundary-layer coordinate is defined as

$$z^* = \frac{1}{\varepsilon}\frac{2}{c_s}\,z^{1/2}. \tag{14.84}$$

For the functions $\alpha(r)$ and $w(r)$, the boundary-layer expansions

$$\left.\begin{aligned} \alpha^{(s)}(r;\varepsilon) &= \mu_0^{(s)}(\varepsilon)\,\alpha_0^{(s)}(z^*) + \mu_1^{(s)}(\varepsilon)\,\alpha_1^{(s)}(z^*) + \ldots, \\ w^{(s)}(r;\varepsilon) &= \nu_0^{(s)}(\varepsilon)\,w_0^{(s)}(z^*) + \nu_1^{(s)}(\varepsilon)\,w_1^{(s)}(z^*) + \ldots \end{aligned}\right\} \tag{14.85}$$

are introduced, where $\mu_0^{(s)}(\varepsilon), \mu_1^{(s)}(\varepsilon), \ldots,$ and $\nu_0^{(s)}(\varepsilon), \nu_1^{(s)}(\varepsilon), \ldots$ are series of functions of ε to be determined.

Equation (14.80) then takes the form

$$\begin{aligned} &\mu_0^{(s)}(\varepsilon)\left[\frac{1}{\varepsilon^4}\left(\frac{d^2\alpha_0^{(s)}}{dz^{*2}} + \frac{2n_e+3}{z^*}\frac{d\alpha_0^{(s)}}{dz^*} + \alpha_0^{(s)}\right) + O\left(\varepsilon^{-2}\right)\right] \\ &+ \mu_1^{(s)}(\varepsilon)\left[\frac{1}{\varepsilon^4}\left(\frac{d^2\alpha_1^{(s)}}{dz^{*2}} + \frac{2n_e+3}{z^*}\frac{d\alpha_1^{(s)}}{dz^*} + \alpha_1^{(s)}\right) + O\left(\varepsilon^{-2}\right)\right] + \ldots \\ &= \nu_0^{(s)}(\varepsilon)\,O\left(\varepsilon^{-3}\right) + \ldots \end{aligned} \tag{14.86}$$

The first dominant boundary-layer equation is homogeneous when the function $\nu_0^{(s)}(\varepsilon)$ is at least of the same order in ε as the function $\mu_0^{(s)}(\varepsilon)$. It is then given by

$$\frac{d^2\alpha_0^{(s)}}{dz^{*2}} + \frac{2n_e+3}{z^*}\frac{d\alpha_0^{(s)}}{dz^*} + \alpha_0^{(s)} = 0. \tag{14.87}$$

The solution that satisfies the requirement that the divergence of the Lagrangian displacement be finite at $r = R$, is

$$\alpha_0^{(s)}\left(z^*\right) = A_{0,s}\,z^{*-(n_e+1)}\,J_{n_e+1}\left(z^*\right), \tag{14.88}$$

where $A_{0,s}$ is an arbitrary constant.

Next, Eq. (14.81) takes the form

$$\nu_0^{(s)}(\varepsilon)\left[\frac{1}{\varepsilon^4}\left(\frac{d^2w_0^{(s)}}{dz^{*2}}+\frac{1}{z^*}\frac{dw_0^{(s)}}{dz^*}-\frac{1}{z^{*2}}w_0^{(s)}\right)+O\left(\varepsilon^{-2}\right)\right]$$
$$+\nu_1^{(s)}(\varepsilon)\left[\frac{1}{\varepsilon^4}\left(\frac{d^2w_1^{(s)}}{dz^{*2}}+\frac{1}{z^*}\frac{dw_1^{(s)}}{dz^*}-\frac{1}{z^{*2}}w_1^{(s)}\right)+O\left(\varepsilon^{-2}\right)\right]+\ldots$$
$$=\mu_0^{(s)}(\varepsilon)\left[-\frac{1}{\varepsilon^3}\frac{d\alpha_0^{(s)}}{dz^*}+O\left(\varepsilon^{-1}\right)\right]+\mu_1^{(s)}(\varepsilon)\left[-\frac{1}{\varepsilon^3}\frac{d\alpha_1^{(s)}}{dz^*}+O\left(\varepsilon^{-1}\right)\right]+\ldots \tag{14.89}$$

The second dominant boundary-layer equation is inhomogeneous when $\nu_0^{(s)}(\varepsilon)=\varepsilon\,\mu_0^{(s)}(\varepsilon)$. It is then given by

$$\frac{d^2w_0^{(s)}}{dz^{*2}}+\frac{1}{z^*}\frac{dw_0^{(s)}}{dz^*}-\frac{1}{z^{*2}}w_0^{(s)}=-\frac{d\alpha_0^{(s)}}{dz^*}. \tag{14.90}$$

The solution that satisfies the requirement that the radial component of the Lagrangian displacement be finite at $r=R$, is

$$w_0^{(s)}\left(z^*\right)=-\frac{1}{2}z^*\,\alpha_0^{(s)}\left(z^*\right)+\frac{1}{2}\frac{1}{z^*}\int_0^{z^*}z'^2\,\frac{d\alpha_0^{(s)}}{dz'}\,dz'+C_{0,s}\,z^*+D_{0,s}\,z^{*-1}, \tag{14.91}$$

where $C_{0,s}$ and $D_{0,s}$ are arbitrary constants. The particular solution of the inhomogeneous differential equation can be transformed by partial integration and by use of the first recurrence relation (14.56) between Bessel functions. It results that

$$w_0^{(s)}\left(z^*\right)=A_{0,s}\,z^{*-(n_e+1)}\,J_{n_e}\left(z^*\right)+C_{0,s}\,z^*+D_{0,s}\,z^{*-1}. \tag{14.92}$$

When one sets $\mu_1^{(s)}(\varepsilon)=\varepsilon\,\mu_0^{(s)}(\varepsilon)$ and $\nu_1^{(s)}(\varepsilon)=\varepsilon\,\nu_0^{(s)}(\varepsilon)$ in the boundary-layer expansions, one obtains the boundary-layer equations of the next order

$$\left.\begin{aligned}&\frac{d^2\alpha_1^{(s)}}{dz^{*2}}+\frac{2n_e+3}{z^*}\frac{d\alpha_1^{(s)}}{dz^*}+\alpha_1^{(s)}=0,\\&\frac{d^2w_1^{(s)}}{dz^{*2}}+\frac{1}{z^*}\frac{dw_1^{(s)}}{dz^*}-\frac{1}{z^{*2}}w_1^{(s)}=-\frac{d\alpha_1^{(s)}}{dz^*}.\end{aligned}\right\} \tag{14.93}$$

Their solutions can be written as

$$\left.\begin{aligned}&\alpha_1^{(s)}\left(z^*\right)=A_{1,s}\,z^{*-(n_e+1)}\,J_{n_e+1}\left(z^*\right),\\&w_1^{(s)}\left(z^*\right)=A_{1,s}\,z^{*-(n_e+1)}\,J_{n_e}\left(z^*\right)+C_{1,s}\,z^*+D_{1,s}\,z^{*-1},\end{aligned}\right\} \tag{14.94}$$

where $A_{1,s}$, $C_{1,s}$, and $D_{1,s}$ are arbitrary constants.

By renaming the sums of constants $A_{0,s} + \varepsilon A_{1,s}$, $C_{0,s} + \varepsilon C_{1,s}$, $D_{0,s} + \varepsilon D_{1,s}$ respectively as $A_{0,s}$, $C_{0,s}$, $D_{0,s}$ and associating distinct functions $\mu_0^{(s)}(\varepsilon)$ with the three independent particular solutions, one obtains the boundary-layer expansions valid up to the second order near $r = R$

$$\left.\begin{aligned}
\alpha^{(s)}(r;\varepsilon) &= \mu_0^{(s,1)}(\varepsilon)\, A_{0,s}\, z^{*-(n_e+1)}\, J_{n_e+1}(z^*)\,, \\
w^{(s)}(r;\varepsilon) &= \varepsilon\, \mu_0^{(s,1)}(\varepsilon)\, A_{0,s}\, z^{*-(n_e+1)}\, J_{n_e}(z^*) \\
&\quad + \varepsilon\, \mu_0^{(s,2)}(\varepsilon)\, C_{0,s}\, z^* + \varepsilon\, \mu_0^{(s,3)}(\varepsilon)\, D_{0,s}\, z^{*-1}.
\end{aligned}\right\} \tag{14.95}$$

14.7 Matching of the Boundary-Layer Expansions Valid Near $r = R$

The boundary-layer expansions $\alpha^{(s)}(r;\varepsilon)$ and $\xi^{(s)}(r;\varepsilon)$ must also be matched with the two-variable expansions $\alpha^{(o)}(r;\varepsilon)$ and $\xi^{(o)}(r;\varepsilon)$ valid at larger distances from the singular boundary points $r = 0$ and $r = R$.

The matching condition relative to the divergence of the Lagrangian displacement

$$\lim_{z\to\infty} \alpha^{(s)}(r;\varepsilon) = \lim_{z\to 0} \alpha^{(o)}(r;\varepsilon) \tag{14.96}$$

requires that the behaviour of the two-variable expansion $\alpha^{(o)}(r;\varepsilon)$ be examined for small values of z, and the behaviour of the boundary-layer expansion $\alpha^{(s)}(r;\varepsilon)$, for large values of z.

It is convenient to decompose the fast independent variable $\tau(r)$ as

$$\tau(r) = \tau_R - \tau_s(r), \tag{14.97}$$

where $\tau_R = \tau(R)$ and

$$\tau_s(r) = \frac{1}{\varepsilon}\int_0^z \frac{dz'}{c(r')}. \tag{14.98}$$

For $z \to 0$, the fast independent variable $\tau_s(r)$ tends to

$$\tau_s(r) = \frac{1}{\varepsilon}\frac{2}{c_s} z^{1/2}, \tag{14.99}$$

so that it becomes equal to the boundary-layer coordinate $z^*(z)$.

The function $F(r)$ involved in the two-variable expansion $\alpha^{(o)}(r;\varepsilon)$ is now decomposed as

$$F(r) = F^{(s)}(z) - F^{(s)}(z_0) \tag{14.100}$$

with

$$F^{(s)}(z) = -c_s \frac{(2n_e + 1)(2n_e + 3)}{16 z^{1/2}} - \frac{1}{2} \int_0^z \left[c(r') W(r') + c_s \frac{(2n_e + 1)(2n_e + 3)}{16 z'^{3/2}} \right] dz', \tag{14.101}$$

so that, for $z \to 0$,

$$F(r) = -c_s \frac{(2n_e + 1)(2n_e + 3)}{16 z^{1/2}} - F^{(s)}(z_0) + O\left(z^{1/2}\right). \tag{14.102}$$

Since furthermore

$$h(r) = h_s z^{-(n_e+3/2)/2}, \tag{14.103}$$

it follows that

$$\lim_{z\to 0} \alpha^{(o)}(r;\varepsilon) = h_s z^{-(n_e+3/2)/2} \Bigg\{ A_0^* \cos\left(\tau_R - z^*\right) + B_0^* \sin\left(\tau_R - z^*\right) - \varepsilon \left[c_s \frac{(2n_e + 1)(2n_e + 3)}{16 z^{1/2}} + F^{(s)}(z_0) \right] \left[B_0^* \cos\left(\tau_R - z^*\right) - A_0^* \sin\left(\tau_R - z^*\right) \right] \Bigg\}. \tag{14.104}$$

On the other hand, for $z \to \infty$, one has that

$$\lim_{z\to\infty} \alpha^{(s)}(r;\varepsilon) = \mu_0^{(s,1)}(\varepsilon)\, \varepsilon^{n_e+3/2} A_{0,s} E_s z^{-(n_e+3/2)/2} \left[\sin\left(z^* - \chi\right) + \varepsilon c_s \frac{(2n_e + 1)(2n_e + 3)}{16 z^{1/2}} \cos\left(z^* - \chi\right) \right] \tag{14.105}$$

with $\chi = (2n_e + 1)\pi/4$ and $E_s = (2/\pi)^{1/2} (2/c_s)^{-(n_e+3/2)}$.

A matching of the boundary-layer expansion $\alpha^{(s)}(r;\varepsilon)$ with the two-variable expansion $\alpha^{(o)}(r;\varepsilon)$ is possible when

$$\mu_0^{(s,1)}(\varepsilon) = \varepsilon^{-(n_e+3/2)}, \tag{14.106}$$

and the coefficients of $\cos z^*$ and $\sin z^*$ are identically zero, i.e.,

$$\left.\begin{aligned}
&A_0^* \cos\tau_R + B_0^* \sin\tau_R \\
&\quad -\varepsilon\left[c_{\mathrm{s}}\,\frac{(2n_{\mathrm{e}}+1)(2n_{\mathrm{e}}+3)}{16\,z^{1/2}} + F^{(\mathrm{s})}(z_0)\right] \\
&\qquad \left(B_0^* \cos\tau_R - A_0^* \sin\tau_R\right) \\
&\quad = -A_{0,\mathrm{s}}\,\frac{E_{\mathrm{s}}}{h_{\mathrm{s}}}\left[\sin\chi - \varepsilon\,c_{\mathrm{s}}\,\frac{(2n_{\mathrm{e}}+1)(2n_{\mathrm{e}}+3)}{16\,z^{1/2}}\cos\chi\right], \\
&A_0^* \sin\tau_R - B_0^* \cos\tau_R \\
&\quad -\varepsilon\left[c_{\mathrm{s}}\,\frac{(2n_{\mathrm{e}}+1)(2n_{\mathrm{e}}+3)}{16\,z^{1/2}} + F^{(\mathrm{s})}(z_0)\right] \\
&\qquad \left(B_0^* \sin\tau_R + A_0^* \cos\tau_R\right) \\
&\quad = A_{0,\mathrm{s}}\,\frac{E_{\mathrm{s}}}{h_{\mathrm{s}}}\left[\cos\chi + \varepsilon\,c_{\mathrm{s}}\,\frac{(2n_{\mathrm{e}}+1)(2n_{\mathrm{e}}+3)}{16\,z^{1/2}}\sin\chi\right].
\end{aligned}\right\} \tag{14.107}$$

Since, at order ε^0,

$$\left.\begin{aligned}
A_0^* &= A_{0,\mathrm{s}}\,\frac{E_{\mathrm{s}}}{h_{\mathrm{s}}}\,\sin(\tau_R - \chi), \\
B_0^* &= -A_{0,\mathrm{s}}\,\frac{E_{\mathrm{s}}}{h_{\mathrm{s}}}\,\cos(\tau_R - \chi),
\end{aligned}\right\} \tag{14.108}$$

the equations reduce to

$$\left.\begin{aligned}
A_0^* &= A_{0,\mathrm{s}}\,\frac{E_{\mathrm{s}}}{h_{\mathrm{s}}}\left[\sin(\tau_R - \chi) - \varepsilon\,F^{(\mathrm{s})}(z_0)\cos(\tau_R - \chi)\right], \\
B_0^* &= -A_{0,\mathrm{s}}\,\frac{E_{\mathrm{s}}}{h_{\mathrm{s}}}\left[\cos(\tau_R - \chi) + \varepsilon\,F^{(\mathrm{s})}(z_0)\sin(\tau_R - \chi)\right].
\end{aligned}\right\} \tag{14.109}$$

The constants A_0^* and B_0^* are here related to the constant $A_{0,\mathrm{s}}$.

Due to the matching of the boundary-layer expansion $\alpha^{(\mathrm{s})}(r;\varepsilon)$ with the two-variable expansion $\alpha^{(\mathrm{o})}(r;\varepsilon)$, the oscillatory parts of the boundary-layer expansion $\xi^{(\mathrm{s})}(r;\varepsilon)$ and the two-variable expansion are automatically matched. For $z \to 0$, the non-oscillatory part of the two-variable expansion $\xi^{(\mathrm{o})}(r;\varepsilon)$ corresponds to that of boundary-layer expansion $\xi^{(\mathrm{s})}(r;\varepsilon)$, when

$$C_{0,\mathrm{s}} = 0 \tag{14.110}$$

and

$$\mu_0^{(\mathrm{s},3)}(\varepsilon) = \varepsilon^{-1}, \qquad C_0^* = D_{0,\mathrm{s}}\,R^{-(\ell-1)}\,\frac{c_{\mathrm{s}}^2}{2}. \tag{14.111}$$

Because of the second Eq. (14.79), it follows that

$$C_{0,\mathrm{c}} = \frac{1}{2}\, D_{0,\mathrm{s}}\, R^{-(\ell-1)}\, c_{\mathrm{c}}^{\ell-2}\, c_{\mathrm{s}}^{2}\,, \tag{14.112}$$

so that the constant $C_{0,\mathrm{c}}$ is related to the constant $D_{0,\mathrm{s}}$.

After this matching, three constants remain to be determined: $A_{0,\mathrm{c}}$, $A_{0,\mathrm{s}}$, and $D_{0,\mathrm{s}}$.

14.8 Eigenfrequency Equation

The elimination of the constants A_0^* and B_0^* from Eqs. (14.77) and (14.109) leads to a homogeneous system of linear, algebraic equations for the constants $A_{0,\mathrm{c}}$ and $A_{0,\mathrm{s}}$. A necessary and sufficient condition that the equations admit of a non-trivial solution is

$$\tan\left(\frac{\ell\pi}{2} + \chi - \tau_R\right) = -\varepsilon\, T_\ell\,, \tag{14.113}$$

where T_ℓ is a constant defined as

$$T_\ell = F^{(\mathrm{s})}\,(z_0) - F^{(\mathrm{c})}\,(r_0)\,. \tag{14.114}$$

Taking into account that $\tan\,(\varepsilon\, T_\ell) \simeq \varepsilon\, T_\ell$, one derives the quadratic equation for the absolute values of the eigenfrequencies

$$\tau_R \equiv |\sigma| \int_0^R \frac{dr}{c(r)} = \left(2n + \ell + n_{\mathrm{e}} + \frac{1}{2}\right)\frac{\pi}{2} + \frac{T_\ell}{|\sigma|} \tag{14.115}$$

with $n = 1, 2, 3, \ldots$ The relation between n and the radial order of the p-mode considered is established in Sect. 14.11.

The value of the constant T_ℓ is independent of the choice of the point with coordinates r_0 and z_0, which is situated in the domain at larger distances from the singular boundary points. Indeed, on the grounds of definitions (14.69) and (14.101), the constant can be expressed as

$$\begin{aligned} T_\ell = -\frac{1}{2}\Biggl\{ & c_{\mathrm{c}}\,\frac{\ell(\ell+1)}{r_0} + c_{\mathrm{s}}\,\frac{(2n_{\mathrm{e}}+1)\,(2n_{\mathrm{e}}+3)}{8\,(R-r_0)^{1/2}} \\ & + \int_0^{r_0}\left[c(r)\,W(r) + c_{\mathrm{c}}\,\frac{\ell(\ell+1)}{r^2}\right]dr \\ & + \int_{r_0}^{R}\left[c(r)\,W(r) + c_{\mathrm{s}}\,\frac{(2n_{\mathrm{e}}+1)\,(2n_{\mathrm{e}}+3)}{16\,(R-r)^{3/2}}\right]dr\Biggr\}, \end{aligned} \tag{14.116}$$

or, equivalently, as

$$T_\ell = -\frac{1}{2}\left[\int_0^R I(r)\,dr + c_c\,\frac{\ell(\ell+1)}{R} + c_s\,\frac{(2n_e+1)\,(2n_e+3)}{8\,R^{1/2}}\right]. \qquad (14.117)$$

The integrand $I(r)$ is defined as

$$I(r) = c(r)\,W(r) + c_c\,\frac{\ell(\ell+1)}{r^2} + c_s\,\frac{(2n_e+1)\,(2n_e+3)}{16\,(R-r)^{3/2}}. \qquad (14.118)$$

Equation (14.115) for the eigenfrequencies corresponds to Eq. (96) of Tassoul (1990). In the first-order approximation, the equation reduces to

$$|\sigma| \int_0^R \frac{dr}{c(r)} = \left(2n + \ell + n_e + \frac{1}{2}\right)\frac{\pi}{2}. \qquad (14.119)$$

With this approximation, Eq. (14.115) can also be expressed as

$$|\sigma| = \frac{\left(2n + \ell + n_e + \frac{1}{2}\right)\frac{\pi}{2}}{\int_0^R \frac{dr}{c(r)}} + \frac{T_\ell}{\left(2n + \ell + n_e + \frac{1}{2}\right)\frac{\pi}{2}}. \qquad (14.120)$$

For radial modes, Eq. (14.115) becomes

$$|\sigma| \int_0^R \frac{dr}{c(r)} = \left(2n + n_e + \frac{1}{2}\right)\frac{\pi}{2} + \frac{T_0}{|\sigma|}. \qquad (14.121)$$

In the first-order approximation, it agrees with the eigenfrequency equation derived by Ledoux (1962), and up to the second-order approximation, with the eigenfrequency equation derived by Tassoul & Tassoul (1968).

14.9 Condition on the Eulerian Perturbation of the Gravitational Potential at $r = R$

Relative to the boundary condition on the Eulerian perturbation of the gravitational potential at $r = R$, Smeyers et al. (1995) derived two equations that relate the Eulerian perturbation of the gravitational potential and its first derivative to the divergence and the radial component of the Lagrangian displacement at an arbitrary point. From Eqs. (5.89)–(5.91) and Eq. (5.93), it follows that

$$\Phi' = \left(c^2\,\alpha - g\,\xi\right) + \varepsilon^2\,\frac{r^2}{\ell(\ell+1)}\left[\frac{1}{r^2}\frac{d}{dr}\left(r^2\,\xi\right) - \alpha\right]. \qquad (14.122)$$

Next, differentiation and elimination of $d^2\xi/dr^2$ and dg/dr by means of Eq. (6.8) and Poisson's differential equation (2.14) yield

$$\frac{d\Phi'}{dr} = \frac{d\left(c^2\,\alpha\right)}{dr} - \frac{N^2}{g}\,c^2\,\alpha - g\,\frac{d\xi}{dr} + \left(2\,\frac{g}{r} - 4\pi G\rho\right)\xi + \varepsilon^2\,\xi. \tag{14.123}$$

Boundary condition (5.97) then takes the form

$$\frac{1}{\varepsilon^2}\,\frac{R}{\ell}\left[\left(\frac{d\xi}{dr}\right)_R + \frac{\ell+2}{R}\,\xi_R - \alpha_R\right] + O(1) = 0. \tag{14.124}$$

The oscillatory parts of boundary-layer expansions $\alpha^{(s)}(r;\varepsilon)$ and $\xi^{(s)}(r;\varepsilon)$ yield, at $r = R$,

$$\left.\begin{aligned}
\alpha_R^{(s)} &= \varepsilon^{-(n_e+3/2)}\,A_{0,s}\,\frac{2^{-(n_e+1)}}{\Gamma\left(n_e+2\right)},\\
\xi_R^{(s)} &= \varepsilon^{-(n_e-1/2)}\,A_{0,s}\,\frac{2^{-(n_e+1)}}{\Gamma\left(n_e+1\right)}\,c_s^2,\\
\left(\frac{d\xi^{(s)}}{dr}\right)_R &= \varepsilon^{-(n_e+3/2)}\,A_{0,s}\,\frac{2^{-(n_e+1)}}{\Gamma\left(n_e+2\right)}.
\end{aligned}\right\} \tag{14.125}$$

With these values, the lowest-order term of the oscillatory part in the boundary condition is identically zero.

The lowest-order term of the non-oscillatory part is also equal to zero, when

$$D_{0,s} = 0. \tag{14.126}$$

Because of the second equation (14.111), it follows that

$$C_0^* = 0. \tag{14.127}$$

Consequently, the asymptotic representation of the radial component of the Lagrangian displacement is purely oscillatory in the interval [0, 1], as well as that of the divergence.

14.10 Uniformly Valid Asymptotic Expansions

Since all constants involved in the various asymptotic expansions are now fixed, it is possible to construct uniformly valid asymptotic expansions for the divergence and the radial component of the Lagrangian displacement in a final form. A uniformly valid asymptotic expansion is constructed by the addition of the boundary-layer expansion and the two-variable expansion valid at larger distances from the singular boundary point considered, and the subtraction, from this sum, of the part common to both expansions. The constants A_0^* and B_0^* are here expressed in terms of the constants $A_{0,\mathrm{c}}$ and $A_{0,\mathrm{s}}$ by means of Eqs. (14.77) and (14.109).

The asymptotic expansions that are uniformly valid from the boundary point $r = 0$ to a distance sufficiently large from the boundary point $r = R$ can be expressed as

$$\alpha^{(\mathrm{c,u})}(r;\varepsilon) = A_{0,\mathrm{c}} \Bigg\{ \varepsilon^{-1}\, r^{*-1/2}\, J_{\ell+1/2}\left(r^*\right) + \left(\frac{2}{\pi}\right)^{1/2} \frac{c_\mathrm{c}}{h_\mathrm{c}}\, h(r) \left[\sin\left(\tau - \frac{\ell\pi}{2}\right) + \varepsilon \left(F(r) + F^{(\mathrm{c})}\left(r_0\right)\right) \cos\left(\tau - \frac{\ell\pi}{2}\right) \right] - \left(\frac{2}{\pi}\right)^{1/2} \frac{c_\mathrm{c}}{r} \left[\sin\left(r^* - \frac{\ell\pi}{2}\right) + \varepsilon\, c_\mathrm{c}\, \frac{\ell(\ell+1)}{2\,r} \cos\left(r^* - \frac{\ell\pi}{2}\right) \right] \Bigg\}, \tag{14.128}$$

$$\xi^{(\mathrm{c,u})}(r;\varepsilon) = A_{0,\mathrm{c}} \Bigg\{ -\frac{c_\mathrm{c}}{2\ell+1}\, r^{*-1/2} \left[\ell J_{\ell-1/2}\left(r^*\right) - (\ell+1)\, J_{\ell+3/2}\left(r^*\right) \right] - \varepsilon \left(\frac{2}{\pi}\right)^{1/2} \frac{c_\mathrm{c}}{h_\mathrm{c}}\, c(r)\, h(r) \left[\cos\left(\tau - \frac{\ell\pi}{2}\right) - \varepsilon \left(G(r) + F^{(\mathrm{c})}\left(r_0\right)\right) \sin\left(\tau - \frac{\ell\pi}{2}\right) \right] + \varepsilon \left(\frac{2}{\pi}\right)^{1/2} \frac{c_\mathrm{c}^2}{r} \left\{ \cos\left(r^* - \frac{\ell\pi}{2}\right) - \varepsilon \left[\frac{c_\mathrm{c}}{r} + c_\mathrm{c}\, \frac{\ell(\ell+1)}{2\,r} \right] \sin\left(r^* - \frac{\ell\pi}{2}\right) \right\} \Bigg\}. \tag{14.129}$$

Next, the asymptotic expansions that are uniformly valid from the boundary point $r = R$ to a distance sufficiently large from the boundary point $r = 0$ can be expressed as

$$\alpha^{(\mathrm{s,u})}(r;\varepsilon) = A_{0,\mathrm{s}} \Bigg\{ \varepsilon^{-(n_\mathrm{e}+3/2)}\, z^{*-(n_\mathrm{e}+1)}\, J_{n_\mathrm{e}+1}\left(z^*\right) + \frac{E_\mathrm{s}}{h_\mathrm{s}}\, h(r) \left[\sin\left(\tau_\mathrm{s} - \chi\right) - \varepsilon \left(F(r) + F^{(\mathrm{s})}\left(z_0\right)\right) \cos\left(\tau_\mathrm{s} - \chi\right) \right] - E_\mathrm{s}\, z^{-(n_\mathrm{e}+3/2)/2} \left[\sin\left(z^* - \chi\right) + \varepsilon\, c_\mathrm{s}\, \frac{(2n_\mathrm{e}+1)\,(2n_\mathrm{e}+3)}{16\, z^{1/2}} \cos\left(z^* - \chi\right) \right] \Bigg\}, \tag{14.130}$$

$$\xi^{(\mathrm{s,u})}(r;\varepsilon) = A_{0,\mathrm{s}} \Bigg\{ \varepsilon^{-(n_\mathrm{e}-1/2)} \frac{c_\mathrm{s}^2}{2} z^{*-n_\mathrm{e}} J_{n_\mathrm{e}}\left(z^*\right)$$

$$+ \varepsilon \frac{E_\mathrm{s}}{h_\mathrm{s}} c(r)\, h(r) \left[\cos\left(\tau_\mathrm{s} - \chi\right) + \varepsilon \left(G(r) + F^{(\mathrm{s})}\left(z_0\right)\right) \sin\left(\tau_\mathrm{s} - \chi\right)\right]$$

$$- \varepsilon E_\mathrm{s} c_\mathrm{s} z^{-(n_\mathrm{e}+1/2)/2} \left[\cos\left(z^* - \chi\right) - \varepsilon c_\mathrm{s} \frac{(2n_\mathrm{e} - 1)(2n_\mathrm{e} + 1)}{16 z^{1/2}} \sin\left(z^* - \chi\right)\right]\Bigg\}. \tag{14.131}$$

14.11 Identification of the Radial Order of a p-Mode with a Given Eigenfrequency

In the Cowling classification, as described in Chap. 10, the radial order of a p-mode belonging to a degree ℓ is determined by the number of nodes that the eigenfunction $\xi(r)$ displays in the interval $(0, R)$. These numbers are conveniently determined by means of the lowest-order uniformly valid asymptotic solutions for $\xi(r)$, which can be expressed in the compact forms

$$\xi^{(\mathrm{c,u})}(r;\varepsilon) = -\varepsilon \frac{A_{0,\mathrm{c}}}{2\ell + 1} \frac{c_\mathrm{c}}{h_\mathrm{c}} c(r)\, h(r) \tau^{1/2} \left[\ell J_{\ell-1/2}(\tau) - (\ell + 1) J_{\ell+3/2}(\tau)\right], \tag{14.132}$$

$$\xi^{(\mathrm{s,u})}(r;\varepsilon) = \varepsilon A_{0,\mathrm{s}} \left(\frac{2}{c_\mathrm{s}}\right)^{-(n_\mathrm{e}+3/2)} c(r)\, h(r) \tau_\mathrm{s}^{1/2} J_{n_\mathrm{e}}\left(\tau_\mathrm{s}\right). \tag{14.133}$$

14.11.1 Radial Modes

For a radial mode, the number of nodes in the asymptotic eigenfunction for $\xi(r)$ can be determined by means of the procedure adopted by Ledoux (1962). In the uniformly valid asymptotic solution $\xi^{(\mathrm{c,u})}(r;\varepsilon)$, only the Bessel function $J_{3/2}(\tau)$ is involved. According to McMahon's expansions for the zeros of Bessel functions, it has zeros nearly at

$$\tau^0 = (2j + 1) \frac{\pi}{2}, \quad j = 1, 2, 3, \ldots \tag{14.134}$$

The highest possible value of j is imposed by the condition that $\tau^0 \leq \tau_R$, or, more explicitly, that

$$2j + 1 \leq 2n + n_\mathrm{e} + \frac{1}{2}. \tag{14.135}$$

As $n_e < 1/2$, it directly follows that $j < n$, so that $n - 1$ is the highest possible value of j. As $n_e \geq 1/2$, it is appropriate to consider the first zero of the Bessel function $J_{n_e}(\tau_s)$, which is involved in the uniformly valid asymptotic solution $\xi^{(s,u)}(r;\varepsilon)$ and has zeros nearly at

$$\tau_s^0 = \left(2r + n_e - \frac{1}{2}\right)\frac{\pi}{2}, \quad r = 1, 2, 3, \ldots \tag{14.136}$$

The last zero of the Bessel function $J_{3/2}(\tau)$, counted from $\tau = 0$, must coincide with the first zero of the Bessel function $J_{n_e}(\tau_s)$, counted from $\tau_s = 0$. Therefore, the position of the last zero of the Bessel function $J_{3/2}(\tau)$ is given by

$$\tau^0(\text{last}) = \tau_R - \tau_s^0(\text{first}), \tag{14.137}$$

where τ_R is determined as

$$\tau_R = \left(2n + n_e + \frac{1}{2}\right)\frac{\pi}{2} \tag{14.138}$$

according to eigenfrequency equation (14.115). One then has

$$\tau^0(\text{last}) = (2n - 1)\frac{\pi}{2}. \tag{14.139}$$

By comparison with Eq. (14.134), it results that $n - 1$ is again the highest possible value of j.

Consequently, in both cases, the asymptotic eigenfunction $\xi(r)$ of a radial mode whose eigenfrequency is determined by the lowest-order approximation of Eq. (14.121) for a given n, displays $n - 1$ nodes. The mode thus corresponds to the $(n - 1)$th radial overtone.

14.11.2 Non-Radial p-Modes

For the determination of the number of nodes in the asymptotic eigenfunction $\xi(r)$ of a low-degree, higher-order p-mode, the case $\ell = 2$ is taken as an example.

With regard to the uniformly valid asymptotic solution $\xi^{(c,u)}(r;\varepsilon)$, it is appropriate to consider the function

$$H(\tau) = -\frac{\tau^{1/2}}{5}\left[2\,J_{3/2}(\tau) - 3\,J_{7/2}(\tau)\right]$$

and its asymptotic form for larger values of τ,

$$H_{\text{asymp}}(\tau) = \left(\frac{2}{\pi}\right)^{1/2}\cos\tau,$$

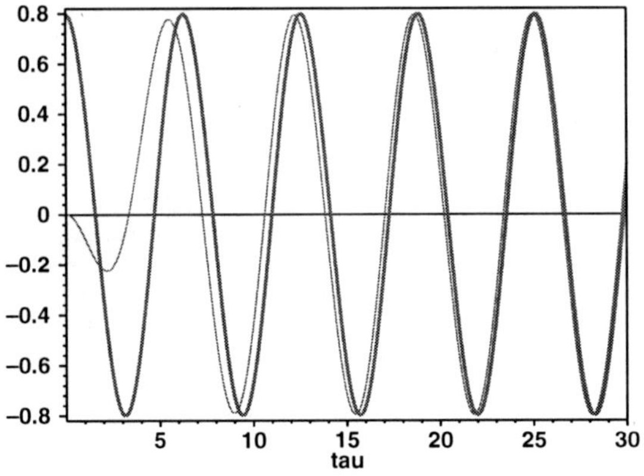

Fig. 14.1 The function $H(\tau)$ (*thinner line*) and its asymptotic form $H_{\text{asymp}}(\tau)$ for large values of τ (*thicker line*)

which displays zeros at

$$\tau^0 = (2j + 1)\,\frac{\pi}{2}, \qquad j = 0, 1, 2, \ldots \tag{14.140}$$

The function $H(\tau)$ and its asymptotic form $H_{\text{asymp}}(\tau)$ are represented in Fig. 14.1. From the figure, it appears that the zero of the asymptotic form $H_{\text{asymp}}(\tau)$ associated with $j = 0$ cannot be related to a zero of the function $H(\tau)$ with $\tau > 0$. Hence, the zeros of the asymptotic function $H_{\text{asymp}}(\tau)$ must be counted from $j = 1$.

In the asymptotic solution $\xi^{(\text{s,u})}(r;\varepsilon)$, the Bessel function $J_{n_\text{e}}(\tau_\text{s})$ has zeros nearly at

$$\tau_\text{s}^0 = \left(2k + n_\text{e} - \frac{1}{2}\right)\frac{\pi}{2}, \quad k = 1, 2, 3, \ldots \tag{14.141}$$

The last zero of the function $H_{\text{asymp}}(\tau)$, counted from $\tau = 0$, must coincide with the first zero of the Bessel function $J_{n_\text{e}}(\tau_\text{s})$, counted from $\tau_\text{s} = 0$. Therefore, its position is given by

$$\tau^0(\text{last}) = \tau_R - \tau_\text{s}^0(\text{first}), \tag{14.142}$$

where τ_R is determined by the lowest-order approximation of Eq. (14.115) or

$$\tau_R = \left(2n + 2 + n_\text{e} + \frac{1}{2}\right)\frac{\pi}{2}. \tag{14.143}$$

It results that

$$\tau^0(\text{last}) = (2n + 1)\,\frac{\pi}{2}. \tag{14.144}$$

By comparison with Eq. (14.140), one sees that n is the highest possible value of j. Consequently, the asymptotic eigenfunction $\xi(r)$ of a low-degree, higher-order

p-mode whose eigenfrequency is given by the first-order approximation of Eq. (14.119) displays n nodes, and the mode corresponds to the mode p_n.

The number of nodes in the asymptotic eigenfunction $\alpha(r)$ can be determined in a similar way. The lowest-order approximations of the uniformly valid asymptotic solutions for the divergence of the Lagrangian displacement can be expressed in the compact forms

$$\alpha^{(c,u)}(r;\varepsilon) = A_{0,c}\, c_c \,\frac{h(r)}{h_c}\, \tau^{1/2}\, J_{\ell+1/2}(\tau), \tag{14.145}$$

$$\alpha^{(s,u)}(r;\varepsilon) = A_{0,s} \left(\frac{2}{c_s}\right)^{-(n_e+3/2)} \frac{h(r)}{h_s}\, \tau_s^{1/2}\, J_{n_e+1}(\tau_s)\,. \tag{14.146}$$

Since only one Bessel function is involved in the asymptotic solution $\alpha^{(c,u)}(r;\varepsilon)$, it is possible to determine the number of nodes in a general way for any value of ℓ. The asymptotic solution $\alpha^{(c,u)}(r;\varepsilon)$ has zeros nearly at

$$\tau^0 = (2j+\ell)\,\frac{\pi}{2}, \quad j = 1,2,3,\ldots, \tag{14.147}$$

and the asymptotic solution $\alpha^{(s,u)}(r;\varepsilon)$, nearly at

$$\tau_s^0 = \left(2k + n_e + \frac{1}{2}\right)\frac{\pi}{2}, \qquad k = 1,2,3,\ldots \tag{14.148}$$

The last zero of the asymptotic solution $\alpha^{(c,u)}(r;\varepsilon)$, counted from $\tau = 0$, is situated at

$$\tau^0(\text{last}) = [2(n-1)+\ell]\,\frac{\pi}{2}. \tag{14.149}$$

By comparison with Eq. (14.147), it follows that $n-1$ is the highest possible value of j, and thus that the asymptotic solution for $\alpha(r)$ displays $n-1$ nodes for any value of ℓ.

14.12 Concluding Remarks

A main conclusion is that, in the three stellar regions distinguished above, namely the region at larger distances from the boundary points $r=0$ and $r=R$, the boundary layer near $r=0$, and the boundary layer near $r=R$, the lowest-order asymptotic approximation for the divergence of the Lagrangian displacement, $\alpha(r)$, can be constructed from the homogeneous second-order differential equation

$$\frac{d^2\alpha}{dr^2} + K_2(r)\,\frac{d\alpha}{dr} + \left[\frac{\sigma^2}{c^2} + K_3(r)\right]\alpha = 0. \tag{14.150}$$

By imposing that the asymptotic solutions satisfy the regularity conditions at the boundary points, and matching them appropriately, one derives the lowest-order asymptotic approximation of the eigenvalue equation, Eq. (14.119), without any foregoing knowledge of the asymptotic solutions for the radial component of the Lagrangian displacement. For the determination of the first asymptotic approximation of low-degree, higher-order p-modes, one can thus follow a procedure that is very similar to the procedure adopted for the resolution of the eigenvalue problem of the oscillations of the compressible equilibrium sphere of uniform mass density, as described in Sect. 10.4.1.3.

Another main conclusion is that the divergence of the Lagrangian displacement is an order ε larger than the radial component of the displacement and, according to Eq. (10.71), an order ε^2 larger than the horizontal component of the curl of the displacement. Low-degree p-modes of higher radial orders are thus divergence-dominated modes.

Table 14.1 First- and second-order asymptotic approximations and exact values for the eigenfrequencies of low-degree p-modes in the equilibrium sphere of uniform mass density, for $\Gamma_1 = 5/3$, according to Tassoul (1990)

ℓ	n	ω_1	ω_2	ω_{exact}
0	5	9.58514	9.36563	9.36305
	10	18.7139	18.6014	18.6011
	15	27.8426	27.7670	27.7669
	20	36.9713	36.9143	36.9143
1	5	10.4980	10.2182	10.2153
	10	19.6267	19.4771	19.4766
	15	28.7554	28.6533	28.6531
	20	37.8841	37.8066	37.8065
2	5	11.4109	11.0074	11.0023
	10	20.5396	20.3154	20.3146
	15	29.6683	29.5131	29.5128
	20	38.7970	38.6783	38.6782
3	5	12.3238	11.7473	11.7369
	10	21.4525	21.1213	21.1193
	15	30.5812	30.3489	30.3482
	20	39.7099	39.5309	39.5307
10	5	18.7139	16.1522	15.9614
	10	27.8426	26.1208	26.0671
	15	36.9713	35.6746	35.6523
	20	46.1000	45.0601	45.0487

Smeyers et al. (1996) verified that their asymptotic representation is equivalent with Tassoul's representation (1990). They noticed a formal difference relative to the first-order boundary-layer solution $\xi^{(c)}(r;\varepsilon)$ valid near the boundary point $r = 0$. The boundary-layer solution given by Eq. (14.61) involves a sum of two Bessel functions of the first kind. However, Tassoul's corresponding solution contains only one Bessel function, but, near the boundary point, an additional term comes in that is of an order higher at larger distances and involves a second Bessel function. The difference has its origin in the fact that, in Tassoul's procedure, asymptotic expansions for the divergence and the radial component of the Lagrangian displacement of a given form are introduced a priori and that these expansions are supposed to contain only one Bessel function at each order. It may be noted that asymptotic solution (14.61) is constructed in a natural way without any a priori assumption. The systematic construction of the asymptotic solutions at each level is an undeniable advantage of the combined use of two-variable expansions and boundary-layer expansions.

Tassoul (1990) examined the accuracy of the second-order asymptotic theory for eigenfrequencies of p-modes of the compressible equilibrium sphere of uniform mass density, since this model is the only one for which exact eigenfrequencies can be determined with a high degree of accuracy. In Table 14.1, Tassoul's results are presented for the exact eigenfrequencies ω_{exact} determined by means of Eq. (10.45), the first-order asymptotic approximations ω_1, and the second-order asymptotic approximations ω_2. These eigenfrequencies are determined for $\Gamma_1 = 5/3$. They were originally expressed in the unit $(\pi G\rho)^{1/2}$ but have been converted here to the unit $\left(GM/R^3\right)^{1/2}$.

From the table, it appears that the second-order asymptotic solution for the eigenfrequency, ω_2, appreciably increases the degree of approximation. The error on the eigenfrequency decreases as the radial order n of the mode increases, but, for a mode of a given radial order, it increases with the degree ℓ.

Chapter 15
Asymptotic Representation of Low-Degree and Intermediate-Degree p-Modes

15.1 Frequency Separations $D_{n,\ell}$ for Solar 5 Min-Oscillations

For low-degree p-modes in the solar 5 min-oscillations, the separations between the frequency of a p-mode of radial order n and degree ℓ, and the frequency of a p-mode of radial order $n - 1$ and degree $\ell + 2$ are seen to be small[1]. When the quantity

$$D_{n,\ell} = \frac{1}{2\ell + 3} \left(\nu_{n,\ell} - \nu_{n-1,\ell+2}\right) \tag{15.1}$$

is used as a measure, where ν is the cyclic frequency, the separations amount only to a few μHz. They are considered to be sensitive to physical conditions in the internal layers of the Sun (Gough 1983, Bahcall & Ulrich 1988, Gabriel 1989).

According to the first asymptotic approximation for eigenfrequencies of low-degree, higher-order p-modes, given by Eq. (14.119), the frequency separations $D_{n,\ell}$ should be strictly zero. Therefore, one has attempted to account for them by means of the second asymptotic approximation, given by Eq. (14.120). After transformation, one has

$$\nu_{n,\ell} = \nu_0 \left[\left(n + \frac{\ell}{2} + \frac{n_\mathrm{e}}{2} + \frac{1}{4}\right) + \frac{1}{2\pi^2 \nu_0} \frac{T_\ell}{n + \ell/2 + n_\mathrm{e}/2 + 1/4}\right], \tag{15.2}$$

where

$$\nu_0 = \left[2 \int_0^R \frac{dr}{c(r)}\right]^{-1}. \tag{15.3}$$

[1] This section is a reproduction of Van Hoolst, T., Smeyers, P.: The quantities $D_{n,\ell}$ as measures of small frequency separations in the Sun and their origin. Astronomy & Astrophysics **248**, 647–655 (1991). With permission from Astronomy & Astrophysics, http://www.aanda.org.

P. Smeyers and T. Van Hoolst, *Linear Isentropic Oscillations of Stars: Theoretical Foundations*, Astrophysics and Space Science Library 371,
DOI 10.1007/978-3-642-13030-4_16,

By the use of equalities (14.26) and (14.117), the constant T_ℓ can be split up into a part depending on the degree ℓ and a part independent of it:

$$T_\ell = \frac{1}{2}\,\ell(\ell+1)\left[\int_0^R \frac{c(r)-c_c}{r^2}\,dr - \frac{c_c}{R}\right] + \beta. \tag{15.4}$$

Partial integration in the first part yields

$$\int_0^R \frac{c(r)-c_c}{r^2}\,dr = \int_0^R \frac{1}{r}\frac{dc}{dr}\,dr - \frac{c(R)-c_c}{R}\,, \tag{15.5}$$

since, by the use of the second power series (6.40),

$$\lim_{r\to 0}\frac{c(r)-c_c}{r} = 0.$$

It then follows that

$$T_\ell = \frac{1}{2}\,\ell(\ell+1)\left[\int_0^R \frac{1}{r}\frac{dc}{dr}\,dr - \frac{c(R)}{R}\right] + \beta, \tag{15.6}$$

so that

$$D_{n,\ell} = \frac{1}{2\pi^2}\,\frac{1}{n+\ell/2+n_e/2+1/4}\left[\frac{c(R)}{R} - \int_0^R \frac{1}{r}\frac{dc}{dr}\,dr\right] \tag{15.7}$$

(see already Tassoul 1980). According to this equation, the frequency separations $D_{n,\ell}$ depend on the distribution of the first derivative of the isentropic sound velocity.

Nevertheless, the asymptotic theory for low-degree, higher-order p-modes does not give good results for the small frequency separations $D_{n,\ell}$ that are measured for low-degree p-modes in the solar 5 min-oscillations (see, e.g., Roxburgh & Vorontsov 1993, Roxburgh & Vorontsov 1994a, where reference is made to Tassoul 1990, Roxburgh & Vorontsov 1994b). In illustration, Table 15.1 contains the asymptotic approximations of a few frequency separations $D_{n,\ell}$ that were determined by Van Hoolst & Smeyers (1991), by means of Eq. (15.7), for a polytropic model with index $n=3$ and with a mass and a radius equal to the mass and the radius of the Sun. The frequency separations are expressed in μHz. The table also contains the exact values of the considered frequency separations and the relative differences of the asymptotic approximations.

Table 15.1 Asymptotic approximations, exact values, and relative differences for a few frequency separations $D_{n,\ell}$ in the case of the polytropic model used by Van Hoolst & Smeyers (1991)

	Asymptotic approximation	Exact value	Relative difference
$D_{11,0}$	9.35	4.66	1.01
$D_{21,0}$	5.24	3.85	0.36
$D_{31,0}$	3.64	3.08	0.18
$D_{41,0}$	2.8	2.6	0.08

In the same study, Van Hoolst & Smeyers made it their object to localise the origin of the frequency separations $D_{n,\ell}$ in their model. To this end, they transformed definition (14.148) into

$$D_{n,\ell} = \frac{1}{2\ell+3}\,\frac{1}{2\pi}\,\frac{\sigma_{n,\ell}^2 - \sigma_{n-1,\ell+2}^2}{\sigma_{n,\ell} + \sigma_{n-1,\ell+2}} \tag{15.8}$$

and, by means of equality (8.103), into

$$D_{n,\ell} = \int_0^1 G_{n,\ell}(x)\,dx, \tag{15.9}$$

where $x = r/R$, and

$$G_{n,\ell}(x) = \frac{1}{2\ell+3}\,\frac{R}{2\pi\left(\sigma_{n,\ell} + \sigma_{n-1,\ell+2}\right)}\left[I_{n,\ell}(x) - I_{n-1,\ell+2}(x)\right]. \tag{15.10}$$

Van Hoolst & Smeyers concentrated on the frequency separation $D_{31,0}$. Note that the "radial mode p_{31}" corresponds to the thirtieth radial overtone (see, for instance, Fig. 10.4 relative to the compressible equilibrium sphere of uniform mass density). The authors examined the variation of the integrand $G_{31,0}(x)$, which displays 62 zeros in the interval (0,1). By successively integrating from the zero x_{i-1} to the zero x_i, for $i = 1, \ldots, 62$ with $x_0 = 0$, they determined the integrals

$$S_i = \int_{x_{i-1}}^{x_i} G_{31,0}(x)\,dx. \tag{15.11}$$

The central region of the model, which extends from $x = 0$ to $x = 0.06833$, yields the positive contribution $S_1 = 17.0\,\mu$Hz. From $x = 0.06833$ to $x = 1$ the quantities S_i are alternatively negative and positive. Therefore, Van Hoolst and Smeyers considered the sequence of sums $\left(S_{2j} + S_{2j+1}\right)$ for $j = 1, \ldots, 30$. The distribution of these sums is represented in Fig. 15.1. It appears that the layers in the region extending from $x = 0.06833$ to $x = 0.8350$ yield negative contributions to $D_{31,0}$, whose

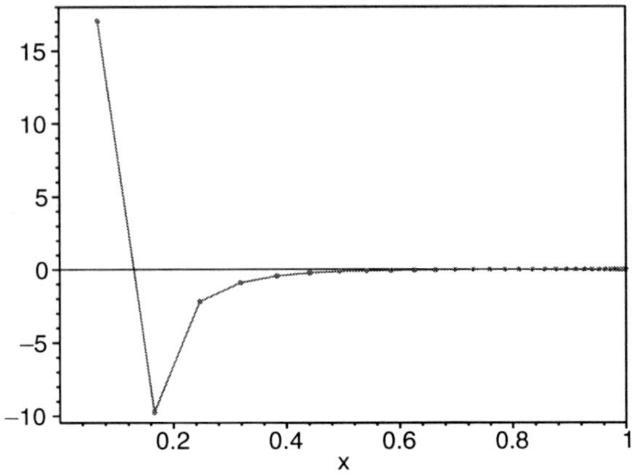

Fig. 15.1 Distribution of the sums $\left(S_{2j} + S_{2j+1}\right)$, $j = 1, \ldots, 30$, for $\ell = 0$ and $n = 31$, in the polytropic model considered by Van Hoolst & Smeyers (1991). The sums are expressed in μHz

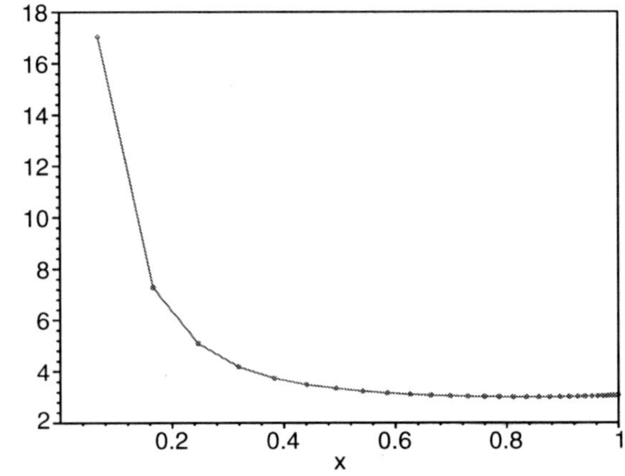

Fig. 15.2 Distribution of the partial sums T_{2k+1}, $k = 0, \ldots, 30$, for $\ell = 0$ and $n = 31$ in the polytropic model considered by Van Hoolst & Smeyers (1991). The partial sums are expressed in μHz

absolute values decrease as x increases. On the other hand, the contributions of the layers situated above $x = 0.8350$ are positive and small.

Next, Van Hoolst & Smeyers determined the partial sums

$$T_{2k+1} = S_1 + \sum_{j=1}^{k} \left(S_{2j} + S_{2j+1}\right) \tag{15.12}$$

for $k = 1, \ldots, 30$. The values of these sums and the value of $T_1 = S_1$ are represented in Fig. 15.2. The value 3.08 μHz of $D_{31,0}$ is nearly reached at the relative radial distance $x = 0.6632$.

By defining the inner boundary of the resonant acoustic cavity by means of the equation

$$r^2 = \frac{\ell(\ell + 1)\, c^2(r)}{\sigma^2}, \tag{15.13}$$

Van Hoolst & Smeyers observed that, for the mode p_{30} belonging to $\ell = 2$, with the cyclic frequency $\nu = 4018.56\,\mu$Hz, the inner boundary is situated at the relative radial distance from the centre $x = 0.0712$. From the comparison of this relative radial distance with the relative radial distance $x = 0.0683$ to which the central region extends, they inferred that the frequency separation $D_{31,0}$ is mainly formed in the region between the inner boundary of the resonant acoustic cavity for the radial mode p_{31}, which coincides with the centre of the model, and the inner boundary of the resonant acoustic cavity for the non-radial mode p_{30} belonging to $\ell = 2$.

The example of the second-degree mode p_{30} in the polytropic model of Van Hoolst & Smeyers illustrates that the inner boundaries of the resonant acoustic cavities for the low-degree p-modes involved in the solar 5 min-oscillations are not that close to the centre. However, in the asymptotic theory of low-degree, higher-order p-modes, the inner boundary of the acoustic cavity is supposed to be located close to the star's centre. A similar remark is made, e.g., in Lopes & Turck-Chièze (1994). For that reason, the asymptotic theory developed in the preceding chapter is inadequate to account for small frequency separations $D_{n,\ell}$ of low-degree p-modes in the solar 5 min-oscillations in a satisfactory way.

In order to meet the inadequacy, Smeyers (2003) developed a first-order asymptotic representation of the divergence of the Lagrangian displacement in p-modes for which the inner boundary of the resonant acoustic cavity is situated at larger distances from the star's centre. This representation applies to low-degree p-modes of less high radial orders as well as to intermediate-degree p-modes. It is presented hereafter.

15.2 Appropriate Equation

The approach rests on the use of Eq. (14.2), in which the term $K_4(r)\ d\xi/dr$ is neglected. The term is strictly equal to zero in the case of the equilibrium sphere of uniform mass density and is supposed to be negligible in other stellar models by virtue of one of the conclusions relative to low-degree, higher-order p-modes at the end of the preceding chapter[2].

Equation (14.2) is here written in the form

$$\frac{d^2\alpha}{dr^2} + K_2(r)\,\frac{d\alpha}{dr} + \left[\sigma^2\,\varphi(r) + K_3^*(r) + \frac{\ell(\ell+1)}{\sigma^2}\,\frac{N^2(r)}{r^2}\right]\alpha = 0, \qquad (15.14)$$

[2] The remaining part of this chapter is partially a reproduction of Smeyers, P.: Asymptotic representation of low- and intermediate-degree p-modes in stars. Astronomy & Astrophysics **407**, 643–653 (2003). With permission from Astronomy & Astrophysics, http://www.aanda.org.

where

$$\varphi(r) = \frac{1}{c^2(r)} - \frac{(\ell + 1/2)^2}{\sigma^2\, r^2}, \tag{15.15}$$

$$K_3^*(r) = K_3(r) + \frac{(\ell + 1/2)^2}{r^2}. \tag{15.16}$$

It differs from its usual form by the coefficient $\varphi(r)$ of the large parameter σ^2 and the coefficient $K_3^*(r)$. Moreover, it displays a turning point at the zero of the function $\varphi(r)$ that is solution of the equation

$$r_\mathrm{t} = \frac{(\ell + 1/2)\, c\,(r_\mathrm{t})}{\sigma} \equiv \frac{(\ell + 1/2)\, c_\mathrm{t}}{\sigma}. \tag{15.17}$$

The turning point is considered to be farther away from the star's centre and is identified with the inner boundary of the resonant acoustic cavity (see already Smeyers et al. 1988, Vorontsov 1991, Roxburgh & Vorontsov 1996).

The asymptotic solutions for the divergence of the Lagrangian displacement must satisfy the same boundary conditions as in the preceding chapter. In this connection, the parameter $(\ell + 1/2)^2$, and not the parameter $\ell(\ell + 1)$, appears in the definition of the function $\varphi(r)$. As it is shown below, the choice of the parameter is imposed by the requirement that the asymptotic solutions for $\alpha(r)$ should tend to zero as r^ℓ as $r \to 0$. The necessity of the use of the parameter $(\ell + 1/2)^2$ was emphasised before by Vorontsov (1991) and Roxburgh & Vorontsov (1996) on the ground of the argument that the second-order terms in their asymptotic expansions in terms of Airy functions must remain finite as $r \to 0$. Below their Eq. (15), Roxburgh & Vorontsov had the comment:

> It is well known that in quantum mechanics the replacement of $\ell(\ell + 1)$ by $(\ell + 1/2)^2$ improves the accuracy of JWKB formulae for the potentials of the type $\ell(\ell+1)/r^2$ (Kemble 1937; Langer 1937).

Equation (15.14) is made dimensionless as in the preceding chapter, so that it becomes

$$\frac{d^2\alpha}{dr^2} + K_2(r)\,\frac{d\alpha}{dr} + \left[\frac{1}{\varepsilon^2}\,\varphi(r) + K_3^*(r) + \varepsilon^2\,\ell(\ell + 1)\,\frac{N^2(r)}{r^2}\right]\alpha = 0 \tag{15.18}$$

with

$$\varphi(r) = \frac{1}{c^2(r)} - \frac{\varepsilon^2\,(\ell + 1/2)^2}{r^2}. \tag{15.19}$$

The factors $\varepsilon^2\,(\ell + 1/2)^2$ and $\varepsilon^2\,\ell(\ell + 1)$ are considered to be of the order of unity.

A lowest-order asymptotic approximation is determined by the use of two-variable expansions at larger distances from the turning point, and locally valid limit process expansions near the turning point (Kevorkian & Cole 1981, 1996).

15.3 Two-Variable Expansion at Larger Distances from the Turning Point and the Boundary Point $r = R$

In the region between the turning point $r = r_t$ and the boundary point $r = R$, the coefficient $\varphi(r)$ of the term with the large parameter in Eq. (15.18) is positive. At larger distances from the two points, a two-variable expansion procedure similar to that used in Sect. 14.3 is applied. The fast variable is defined as

$$\tau(r) = \frac{1}{\varepsilon} \int_{r_t}^{r} \sqrt{\varphi(r')}\, dr'. \tag{15.20}$$

The slow variable is still the radial coordinate r.

By introducing a two-variable expansion of the form

$$\alpha^{(o)}(r;\varepsilon) = \alpha_0^{(o)}(\tau, r) + \varepsilon\, \alpha_1^{(o)}(\tau, r) + \ldots \tag{15.21}$$

and solving the equations of the lowest orders in ε, one derives the lowest-order two-variable solution

$$\alpha^{(o)}(r;\varepsilon) = h(r)\, \left(A_0^* \cos\tau + B_0^* \sin\tau\right), \tag{15.22}$$

where A_0^* and B_0^* are arbitrary constants, and

$$h(r) = \left[r \sqrt{\rho(r)}\, c^2(r)\, \varphi^{1/4}(r)\right]^{-1}. \tag{15.23}$$

15.4 Boundary-Layer Expansion on the Outer Side of the Turning Point

In the boundary layer on the outer side of the turning point, it is convenient to use the positive radial coordinate

$$s(r) = r - r_t \tag{15.24}$$

and the Taylor expansions

$$\left.\begin{array}{l} \rho(r) = \rho_{\rm t}\,[1 + O(s)], \qquad c(r) = c_{\rm t}\left[1 + \dfrac{c_1}{c_{\rm t}}\,s + O\left(s^2\right)\right], \\ g(r) = g_{\rm t}\,[1 + O(s)], \qquad N^2(r) = N_{\rm t}^2\,[1 + O(s)], \\ \varphi(r) = \varphi_{\rm t}\,s\,\left[1 + O\left(s^2\right)\right], \; h(r) = h_{\rm t}\,s^{-1/4}\,[1 + O(s)], \\ K_2(r) = K_{2,{\rm t}}\,[1 + O\,(s)], \; K_3^*(r) = K_{3,{\rm t}}^*\,[1 + O\,(s)], \end{array}\right\} \tag{15.25}$$

where the coefficients $\varphi_{\rm t}$ and $h_{\rm t}$ are defined as

$$\varphi_{\rm t} = \frac{2}{c_{\rm t}^2}\left(\frac{1}{r_{\rm t}} - \frac{c_1}{c_{\rm t}}\right), \quad h_{\rm t} = \left(r_{\rm t}\,\sqrt{\rho_{\rm t}}\,c_{\rm t}^2\,\varphi_{\rm t}^{1/4}\right)^{-1}. \tag{15.26}$$

The coefficient $\varphi_{\rm t}$ is positive.

For the construction of a boundary-layer solution, the boundary-layer coordinate

$$s^*(r) = \frac{s(r)}{\delta(\varepsilon)}, \tag{15.27}$$

where the function $\delta(\varepsilon) \to 0$ as $\varepsilon \to 0$, and the boundary-layer expansion

$$\alpha^{(\rm t)}(r;\varepsilon) = \mu_0^{(\rm t)}(\varepsilon)\,\alpha_0^{(\rm t)}\left(s^*\right) + \mu_1^{(\rm t)}(\varepsilon)\,\alpha_1^{(\rm t)}\left(s^*\right) + \ldots \tag{15.28}$$

are introduced.

The dominant boundary-layer equation is given by

$$\frac{1}{\delta^2(\varepsilon)}\,\frac{d^2\alpha_0^{(\rm t)}}{ds^{*2}} + \frac{\delta(\varepsilon)}{\varepsilon^2}\,\varphi_{\rm t}\,s^*\,\alpha_0^{(\rm t)} = 0. \tag{15.29}$$

The term containing the second derivative is of the same order in ε as the term containing the large parameter, as

$$\delta(\varepsilon) = \varepsilon^{2/3}. \tag{15.30}$$

The dominant boundary-layer equation then becomes

$$\frac{d^2\alpha_0^{(\rm t)}}{ds^{*2}} + \varphi_{\rm t}\,s^*\,\alpha_0^{(\rm t)} = 0. \tag{15.31}$$

Its general solution can be expressed in terms of Airy functions as

$$\alpha_0^{(\rm t)}\left(s^*\right) = A_{0,{\rm t}}\,{\rm Ai}(-v) + B_{0,{\rm t}}\,{\rm Bi}(-v), \tag{15.32}$$

where $v = \varphi_{\rm t}^{1/3}\,s^*$, and $A_{0,{\rm t}}$ and $B_{0,{\rm t}}$ are arbitrary constants.

In view of the matching with the lowest-order two-variable solution that is valid at larger distances from the turning point, the lowest-order boundary-layer solution is written in the more general form

$$\alpha^{(t)}(r;\varepsilon) = \mu_0^{(t)}(\varepsilon)\,[A_{0,t}\,\mathrm{Ai}(-v) + B_{0,t}\,\mathrm{Bi}(-v)]\,, \tag{15.33}$$

where $\mu_0^{(t)}(\varepsilon)$ is a yet undetermined function of ε.

The matching condition is

$$\lim_{s\to\infty} \alpha^{(t)}(r;\varepsilon) = \lim_{s\to 0} \alpha^{(o)}(r;\varepsilon). \tag{15.34}$$

As $s \to 0$, $\tau(r) \to \zeta(r)$ with

$$\zeta(r) = \frac{2}{3}\, v^{3/2}, \tag{15.35}$$

so that

$$\lim_{s\to 0} \alpha^{(o)}(r;\varepsilon) = h_t\, s^{-1/4}\left(A_0^* \cos\zeta + B_0^* \sin\zeta\right). \tag{15.36}$$

On the other hand, as $s \to \infty$, the substitution of the asymptotic forms of the Airy functions in terms of harmonic functions yields

$$\lim_{s\to\infty} \alpha^{(t)}(r;\varepsilon) = \mu_0^{(t)}(\varepsilon)\,\varepsilon^{1/6}\,\pi^{-1/2}\,\varphi_t^{-1/12}\,s^{-1/4}$$
$$\left[A_{0,t}\,\sin\left(\zeta + \frac{\pi}{4}\right) + B_{0,t}\,\cos\left(\zeta + \frac{\pi}{4}\right)\right]. \tag{15.37}$$

Hence, the matching condition is satisfied when

$$\mu_0^{(t)}(\varepsilon) = \varepsilon^{-1/6} \tag{15.38}$$

and

$$\left.\begin{aligned} A_0^* &= \frac{1}{\sqrt{2\pi}}\, h_t^{-1}\, \varphi_t^{-1/12}\,(A_{0,t} + B_{0,t})\,, \\ B_0^* &= \frac{1}{\sqrt{2\pi}}\, h_t^{-1}\, \varphi_t^{-1/12}\,(A_{0,t} - B_{0,t})\,. \end{aligned}\right\} \tag{15.39}$$

By these equations, the constants A_0^* and B_0^* in the lowest-order two-variable solution $\alpha^{(o)}(r;\varepsilon)$ are related to the constants $A_{0,t}$ and $B_{0,t}$ in the lowest-order boundary-layer solution $\alpha^{(t)}(r;\varepsilon)$.

At the turning point $r = r_\mathrm{t}$, the lowest-order boundary-layer solution $\alpha^{(\mathrm{t})}(r;\varepsilon)$ and its first derivative are given by

$$\left.\begin{aligned}\alpha^{(\mathrm{t})}(r_\mathrm{t};\varepsilon) &= \frac{\varepsilon^{-1/6}}{3^{2/3}\,\Gamma(2/3)}\left(A_{0,\mathrm{t}} + \sqrt{3}\,B_{0,\mathrm{t}}\right),\\ \left(\frac{d\alpha^{(\mathrm{t})}(r;\varepsilon)}{dr}\right)_{r_\mathrm{t}} &= \varepsilon^{-5/6}\,\frac{3^{1/6}\,\Gamma(2/3)}{2\pi}\,\varphi_\mathrm{t}^{1/3}\left(A_{0,\mathrm{t}} - \sqrt{3}\,B_{0,\mathrm{t}}\right).\end{aligned}\right\} \tag{15.40}$$

15.5 Two-Variable Expansion at Larger Distances from the Boundary Point $r = 0$ and the Turning Point

In the region extending from the boundary point $r = 0$ to the turning point, $\varphi(r) \leq 0$. Therefore, it is appropriate to adopt there the function $-\varphi(r)$, instead of the function $\varphi(r)$, as coefficient of the term with the large parameter in Eq. (15.18).

At larger distances from the boundaries of the region considered, an asymptotic solution of Eq. (15.18) can again be constructed by means of a two-variable expansion procedure. The fast variable is now defined as

$$\tau_\mathrm{i}(r) = \frac{1}{\varepsilon}\int_r^{r_\mathrm{t}} \sqrt{-\varphi(r')}\,dr', \tag{15.41}$$

so that $\tau_\mathrm{i}(r) \geq 0$ and $d\tau_\mathrm{i}/dr \leq 0$. A two-variable expansion is introduced as

$$\alpha^{(\mathrm{i})}(r;\varepsilon) = \mu^{(\mathrm{i})}(\varepsilon)\left[\alpha_0^{(\mathrm{i})}(\tau_\mathrm{i},r) + \varepsilon\,\alpha_1^{(\mathrm{i})}(\tau_\mathrm{i},r) + \ldots\right]. \tag{15.42}$$

Proceeding as in Sect. 15.3, one derives the lowest-order two-variable solution

$$\alpha^{(\mathrm{i})}(r;\varepsilon) = \mu^{(\mathrm{i})}(\varepsilon)\,h_\mathrm{i}(r)\left[A_0^{**}\,\exp(\tau_\mathrm{i}) + B_0^{**}\,\exp(-\tau_\mathrm{i})\right], \tag{15.43}$$

where A_0^{**} and B_0^{**} are arbitrary constants, and

$$h_\mathrm{i}(r) = \left\{r\,\sqrt{\rho(r)}\,c^2(r)\,\left[-\varphi(r)\right]^{1/4}\right\}^{-1}. \tag{15.44}$$

As $r \to 0$, $\tau_\mathrm{i}(r)$ behaves as $\ln r^{-(\ell+1/2)}$, and $h_\mathrm{i}(r)$ as $r^{-1/2}$, so that

$$\alpha^{(\mathrm{i})}(r;\varepsilon) \propto \mu^{(\mathrm{i})}(\varepsilon)\,r^{-1/2}\left[A_0^{**}\,r^{-(\ell+1/2)} + B_0^{**}\,r^{\ell+1/2}\right]. \tag{15.45}$$

When $A_0^{**} = 0$, the lowest-order two-variable solution $\alpha^{(\mathrm{i})}(r;\varepsilon)$ remains finite for any ℓ and behaves as r^{ℓ} as $r \to 0$, in accordance with the analysis of the behaviours of the eigenfunctions near the singular point $r = 0$ made in Sect. 6.4.1. Notice that this behaviour is due to the presence of the factor $(\ell + 1/2)^2$ in definition (15.15) of the function $\varphi(r)$.

The admissible lowest-order two-variable solution thus takes the form

$$\alpha^{(\mathrm{i})}(r;\varepsilon) = \mu^{(\mathrm{i})}(\varepsilon)\, B_0^{**}\, h_{\mathrm{i}}(r)\, \exp\left(-\tau_{\mathrm{i}}\right). \tag{15.46}$$

15.6 Boundary-Layer Expansion on the Inner Side of the Turning Point

In the boundary layer on the inner side of the turning point $r = r_{\mathrm{t}}$, it is possible to proceed in a similar way as in the boundary layer on the outer side. One now introduces the positive radial coordinate

$$s_{\mathrm{i}}(r) = -s(r) = r_{\mathrm{t}} - r, \tag{15.47}$$

the associated boundary-layer coordinate

$$s_{\mathrm{i}}^{*}(r) = \frac{s_{\mathrm{i}}(r)}{\delta_{\mathrm{i}}(\varepsilon)}, \tag{15.48}$$

and the boundary-layer expansion

$$\alpha^{(\mathrm{t_i})}(r;\varepsilon) = \mu_0^{(\mathrm{t_i})}(\varepsilon)\, \alpha_0^{(\mathrm{t_i})}\left(s_{\mathrm{i}}^{*}\right) + \mu_1^{(\mathrm{t_i})}(\varepsilon)\, \alpha_1^{(\mathrm{t_i})}\left(s_{\mathrm{i}}^{*}\right) + \ldots \tag{15.49}$$

When one sets

$$\delta_{\mathrm{i}}(\varepsilon) = \varepsilon^{2/3}, \tag{15.50}$$

the dominant boundary-layer equation takes the form

$$\frac{d^2\alpha_0^{(\mathrm{t_i})}}{ds_i^{*2}} - \varphi_{\mathrm{t}}\, s_{\mathrm{i}}^{*}\, \alpha_0^{(\mathrm{t_i})} = 0. \tag{15.51}$$

It has the general solution

$$\alpha_0^{(\mathrm{t_i})}\left(s_{\mathrm{i}}^{*}\right) = A_{0,\mathrm{t_i}}\, \mathrm{Ai}\left(v_{\mathrm{i}}\right) + B_{0,\mathrm{t_i}}\, \mathrm{Bi}\left(v_{\mathrm{i}}\right), \tag{15.52}$$

where $v_{\mathrm{i}} = \varphi_{\mathrm{t}}^{1/3}\, s_{\mathrm{i}}^{*}$, and $A_{0,\mathrm{t_i}}$ and $B_{0,\mathrm{t_i}}$ are arbitrary constants.

The lowest-order boundary-layer solution can then be expressed in the more general form

$$\alpha^{(\mathrm{t_i})}(r;\varepsilon) = \mu_0^{(\mathrm{t_i})}(\varepsilon)\,[A_{0,\mathrm{t_i}}\,\mathrm{Ai}\,(v_\mathrm{i}) + B_{0,\mathrm{t_i}}\,\mathrm{Bi}\,(v_\mathrm{i})]\,, \tag{15.53}$$

where $\mu_0^{(\mathrm{t_i})}(\varepsilon)$ is a yet undetermined function of ε.

The condition for the matching with the lowest-order two-variable solution valid at larger distances from the turning point is

$$\lim_{s_\mathrm{i}\to\infty} \alpha^{(\mathrm{t_i})}(r;\varepsilon) = \lim_{s_\mathrm{i}\to 0} \alpha^{(\mathrm{i})}(r;\varepsilon). \tag{15.54}$$

As $s_\mathrm{i} \to 0$, $\tau_\mathrm{i}(r) \to \zeta_\mathrm{i}(r)$ with

$$\zeta_\mathrm{i}(r) = \frac{2}{3}\,v_\mathrm{i}^{3/2}, \tag{15.55}$$

so that

$$\lim_{s_\mathrm{i}\to 0} \alpha^{(\mathrm{i})}(r;\varepsilon) = \mu^{(\mathrm{i})}(\varepsilon)\,B_0^{**}\,h_\mathrm{t}\,s_\mathrm{i}^{-1/4}\,\exp\,(-\zeta_\mathrm{i})\,. \tag{15.56}$$

On the other hand, as $s_\mathrm{i} \to \infty$, the use of the asymptotic forms of the Airy functions leads to

$$\begin{aligned}\lim_{s_\mathrm{i}\to\infty} \alpha^{(\mathrm{t_i})}(r;\varepsilon) = {} & \mu_0^{(\mathrm{t_i})}(\varepsilon)\,\varepsilon^{1/6}\left(2\sqrt{\pi}\right)^{-1}\varphi_\mathrm{t}^{-1/12}\,s_\mathrm{i}^{-1/4} \\ & [A_{0,\mathrm{t_i}}\,\exp\,(-\zeta_\mathrm{i}) + 2\,B_{0,\mathrm{t_i}}\,\exp\,(\zeta_\mathrm{i})]\,.\end{aligned} \tag{15.57}$$

Hence, the matching condition is satisfied, when

$$\left.\begin{aligned}\mu^{(\mathrm{i})}(\varepsilon) &= \mu_0^{(\mathrm{t_i})}(\varepsilon)\,\varepsilon^{1/6}, \\ B_{0,\mathrm{t_i}} &= 0, \\ B_0^{**} &= \left(2\sqrt{\pi}\right)^{-1} h_\mathrm{t}^{-1}\,\varphi_\mathrm{t}^{-1/12}\,A_{0,\mathrm{t_i}}\,.\end{aligned}\right\} \tag{15.58}$$

At the turning point $r = r_\mathrm{t}$, the lowest-order boundary-layer solution $\alpha^{(\mathrm{t_i})}(r;\varepsilon)$ and its first derivative take the values

$$\left.\begin{aligned}\alpha^{(\mathrm{t_i})}(r_\mathrm{t};\varepsilon) &= \frac{\mu_0^{(\mathrm{t_i})}(\varepsilon)}{3^{2/3}\,\Gamma(2/3)}\,A_{0,\mathrm{t_i}}, \\ \left(\frac{d\alpha^{(\mathrm{t_i})}(r;\varepsilon)}{dr}\right)_{r_\mathrm{t}} &= \frac{\mu_0^{(\mathrm{t_i})}(\varepsilon)}{\varepsilon^{2/3}}\,\frac{3^{1/6}\,\Gamma(2/3)}{2\pi}\,\varphi_\mathrm{t}^{1/3}\,A_{0,\mathrm{t_i}}\,.\end{aligned}\right\} \tag{15.59}$$

From equalities (15.40) and (15.59), it results that the lowest-order boundary-layer solutions $\alpha^{(t)}(r;\varepsilon)$ and $\alpha^{(t_i)}(r;\varepsilon)$ and their first derivatives are continuous at the turning point $r = r_t$, when

$$\left.\begin{aligned} &\mu_0^{(t_i)}(\varepsilon) = \varepsilon^{-1/6}, \\ &A_{0,t} = A_{0,t_i}, \qquad B_{0,t} = 0. \end{aligned}\right\} \tag{15.60}$$

Equations (15.39) can then be rewritten as

$$\left.\begin{aligned} A_0^* &= \frac{1}{\sqrt{2\pi}}\, h_t^{-1}\, \varphi_t^{-1/12}\, A_{0,t}, \\ B_0^* &= \frac{1}{\sqrt{2\pi}}\, h_t^{-1}\, \varphi_t^{-1/12}\, A_{0,t}. \end{aligned}\right\} \tag{15.61}$$

The constants A_0^* and B_0^* in the lowest-order two-variable solution $\alpha^{(o)}(r;\varepsilon)$, valid at larger distances from the turning point and the boundary point $r = R$, are thus equal to each other.

15.7 Boundary-Layer Expansion Near the Boundary Point $r = R$

As in Sect. 14.6, the independent variable $z = R - r$ is adopted in the boundary layer near $r = R$. Equation (15.18) then becomes

$$\frac{d^2\alpha}{dz^2} - K_2(r)\,\frac{d\alpha}{dz} + \left[\frac{1}{\varepsilon^2}\,\varphi(r) + K_3^*(r) + \varepsilon^2\,\ell(\ell+1)\,\frac{N^2(r)}{(R-z)^2}\right]\alpha = 0. \tag{15.62}$$

Again in the supposition that the mass density is analytic at the boundary point $r = R$, power series (6.51) holds. When furthermore the approximation $m(r) \simeq M$ is valid, power series (6.53)–(6.57) and (14.82) can be used.

By the introduction of the positive boundary-layer coordinate

$$z^*(z) = \frac{1}{\varepsilon}\,\frac{2}{c_s}\,z^{1/2} \tag{15.63}$$

and the boundary-layer expansion

$$\alpha^{(s)}(r;\varepsilon) = \mu_0^{(s)}(\varepsilon)\,\alpha_0^{(s)}\left(z^*\right) + \mu_1^{(s)}(\varepsilon)\,\alpha_1^{(s)}\left(z^*\right) + \ldots, \tag{15.64}$$

the dominant boundary-layer equation is derived

$$\frac{d^2\alpha_0^{(s)}}{dz^{*2}} + \frac{2n_e + 3}{z^*}\frac{d\alpha_0^{(s)}}{dz^*} + \alpha_0^{(s)} = 0. \tag{15.65}$$

The solution that satisfies the requirement that the divergence of the Lagrangian displacement be finite at $r = R$, is given by

$$\alpha_0^{(s)}\left(z^*\right) = A_{0,s}\, z^{*-(n_e+1)}\, J_{n_e+1}\left(z^*\right), \tag{15.66}$$

where $A_{0,s}$ is an arbitrary constant.

The lowest-order boundary-layer solution admissible near the boundary point $r = R$ can be written in the more general form

$$\alpha^{(s)}(r;\varepsilon) = \mu_0^{(s)}(\varepsilon)\, A_{0,s}\, z^{*-(n_e+1)}\, J_{n_e+1}\left(z^*\right), \tag{15.67}$$

where the function $\mu_0^{(s)}(\varepsilon)$ is yet undetermined.

The condition for the matching with the lowest-order two-variable solution $\alpha^{(o)}(r;\varepsilon)$, valid at larger distances from the boundary point, is

$$\lim_{z\to\infty} \alpha^{(s)}(r;\varepsilon) = \lim_{z\to 0} \alpha^{(o)}(r;\varepsilon). \tag{15.68}$$

For $z \to 0$, the fast independent variable $\tau(r)$ is conveniently decomposed as

$$\tau(r) = \tau_R - \tau_s(r), \tag{15.69}$$

where $\tau_R = \tau(R)$, and $\tau_s(r)$ is defined as

$$\tau_s(r) = \frac{1}{\varepsilon}\int_0^z \sqrt{\varphi\left(r'\right)}\, dz' \tag{15.70}$$

and tends to the boundary-layer coordinate $z^*(z)$. Since

$$h(r) = h_s\, z^{-(n_e+3/2)/2}, \tag{15.71}$$

it follows that

$$\lim_{z\to 0} \alpha^{(o)}(r;\varepsilon) = h_s\, z^{-(n_e+3/2)/2}\left[A_0^* \cos\left(\tau_R - z^*\right) + B_0^* \sin\left(\tau_R - z^*\right)\right]. \tag{15.72}$$

On the other hand, for $z \to \infty$, use of the principal asymptotic form of the Bessel function $J_{n_e+1}\left(z^*\right)$ yields

$$\lim_{z\to\infty} \alpha^{(s)}(r;\varepsilon) = \mu_0^{(s)}(\varepsilon)\, \varepsilon^{n_e+3/2}\, A_{0,s}\, E_s\, z^{-(n_e+3/2)/2}\, \sin\left(z^* - \chi\right) \tag{15.73}$$

with $\chi = (2n_e + 1)\,\pi/4$, $E_s = (2/\pi)^{1/2}\,(2/c_s)^{-(n_e+3/2)}$. Hence, the matching condition is satisfied when

$$\mu_0^{(s)}(\varepsilon) = \varepsilon^{-(n_e+3/2)} \tag{15.74}$$

and

$$\left.\begin{aligned} A_0^* &= A_{0,s}\,\frac{E_s}{h_s}\,\sin(\tau_R - \chi)\,,\\ B_0^* &= -A_{0,s}\,\frac{E_s}{h_s}\,\cos(\tau_R - \chi)\,. \end{aligned}\right\} \tag{15.75}$$

15.8 Eigenfrequency Equation

Since the constants A_0^* and B_0^* are equal to each other, Eqs. (15.75) lead to the condition

$$\sin(\tau_R - \chi) + \cos(\tau_R - \chi) = 0 \tag{15.76}$$

or, equivalently, to the condition

$$2\,\sin\frac{\pi}{4}\,\cos\left[(\tau_R - \chi) - \frac{\pi}{4}\right] = 0. \tag{15.77}$$

The eigenfrequency equation is then given by

$$\tau_R \equiv |\sigma| \int_{r_t}^{R} \left[\frac{1}{c^2(r)} - \frac{(\ell + 1/2)^2}{\sigma^2 r^2}\right]^{1/2} dr = \left(2n + n_e\right)\frac{\pi}{2}, \quad n = 1, 2, 3, \ldots \tag{15.78}$$

In the particular case of the compressible equilibrium sphere of uniform mass density, it is shown that n corresponds to the radial order of the p-mode considered.

The integral is equal to the time needed for an acoustic wave to propagate in the radial direction from the turning point to the star's surface, with the local velocity of propagation

$$V(r) = \frac{c(r)}{\left[1 - \dfrac{(\ell + 1/2)^2\,c^2(r)}{\sigma^2\,r^2}\right]^{1/2}}. \tag{15.79}$$

This velocity is infinitely large at the turning point and decreases as the radial distance from the turning point increases. It is everywhere larger than the local isentropic sound velocity, except on the star's surface. With this definition, the eigenfrequency equation can be written as

$$|\sigma| = \frac{(2n + n_e)\,\dfrac{\pi}{2}}{\displaystyle\int_{r_t}^{R} \frac{dr}{V(r)}}. \tag{15.80}$$

15.9 Uniformly Valid Asymptotic Representation of the Divergence of the Lagrangian Displacement

When the constants A_0^* and B_0^* are expressed in terms of the constant $A_{0,t}$ by means of Eqs. (15.61), a lowest-order asymptotic representation of the divergence of the Lagrangian displacement that is uniformly valid from the turning point $r = r_t$ to a distance sufficiently large from the boundary point $r = R$ is given by

$$\alpha^{(t,u)}(r;\varepsilon) = A_{0,t} \left\{ \varepsilon^{-1/6}\,\mathrm{Ai}(-v) + \frac{1}{\sqrt{\pi}}\,\varphi_t^{-1/12} \left[h_t^{-1}\,h(r)\,\cos\left(\tau - \frac{\pi}{4}\right) - s^{-1/4}\,\cos\left(\zeta - \frac{\pi}{4}\right)\right]\right\}. \tag{15.81}$$

Similarly, a lowest-order asymptotic representation of the divergence of the Lagrangian displacement that is uniformly valid from the boundary point $r = R$ to a distance sufficiently large from the turning point $r = r_t$ is given by

$$\alpha^{(s,u)}(r;\varepsilon) = A_{0,s} \left\{ \varepsilon^{-(n_e+3/2)}\,z^{*-(n_e+1)}\,J_{n_e+1}\left(z^*\right) + E_s \left[h_s^{-1}\,h(r)\,\sin\left(\tau_s - \chi\right) - z^{-(n_e+3/2)/2}\,\sin\left(z^* - \chi\right)\right]\right\}. \tag{15.82}$$

Finally, by means of Eqs. (15.60), the lowest-order asymptotic representation of the divergence of the Lagrangian displacement that is uniformly valid from the turning point $r = r_t$ towards the star's centre, can be expressed as

$$\alpha^{(t_i,u)}(r;\varepsilon) = A_{0,t} \left\{ \varepsilon^{-1/6}\,\mathrm{Ai}\left(v_i\right) + \frac{1}{2\sqrt{\pi}}\,\varphi_t^{-1/12} \left[h_t^{-1}\,h_i(r)\,\exp\left(-\tau_i\right) - s_i^{-1/4}\,\exp\left(-\zeta_i\right)\right]\right\}. \tag{15.83}$$

15.10 Application to the Compressible Equilibrium Sphere of Uniform Mass Density

A convenient way to verify the validity of the asymptotic representation is to apply it to the simple case of the compressible equilibrium sphere of uniform mass density. It should be noted that, for this model, $K_4(r) = 0$ in the right-hand member of Eq. (6.10), so that the application does not test the validity of the supposition that the term $K_4(r)\, d\xi/dr$ is negligible for the asymptotic representation of p-modes in other models.

In the expression of the isentropic sound velocity, the value 5/3 has been adopted for the isentropic coefficient Γ_1. For a p-mode associated with an eigenfrequency σ and a degree ℓ, the turning point is then situated at the distance from the centre

$$r_{\rm t} = \left[1 + \frac{6\sigma^2}{5(\ell + 1/2)^2}\right]^{-1/2}. \tag{15.84}$$

According to Eq. (10.45), the exact eigenfrequency of a p-mode, expressed in the unit $\left(GM/R^3\right)^{1/2}$, is determined by means of the equation

$$\omega_{\rm exact} = \left\{ D_{\ell,k} + \left[D_{\ell,k}^2 + \ell(\ell + 1)\right]^{1/2} \right\}^{1/2}. \tag{15.85}$$

The parameter k is related to the radial order n of the mode by $k = n - 1$.

For a number of p-modes, Smeyers (2003) determined the asymptotic approximation of the eigenfrequency, $\omega_{\rm asymp}$, as well in the unit $\left(GM/R^3\right)^{1/2}$. To this end, he started from the first-order approximation ω_T, given by Eq. (14.119), and solved eigenfrequency equation (15.78) by means of the Newton-Raphson method.

The eigenfrequencies $\omega_{\rm exact}$, ω_T, and $\omega_{\rm asymp}$ are presented in Table 15.2 for the p-modes of the radial orders $n = 5, 10, 15, 20, 25, 30, 35, 40$ belonging to the lowest degrees $\ell = 1, 2, 3, 4$. In the same table, the relative radial distance of the turning point from the centre, $r_{\rm t}/R$, and the relative errors of the asymptotic approximations ω_T and $\omega_{\rm asymp}$ of the eigenfrequency are also presented.

In the case $\ell = 1$, the radial orders considered are limited to $n = 5, 10, 15, 20, 25$, since, for $n > 25$, the relative error of the asymptotic eigenfrequency $\omega_{\rm asym}$ appears to increase slightly. The increase is probably due to the fact that, for p-modes of radial orders larger than 25, the turning point is situated so close to the star's centre that it is located inside the boundary layer near $r = 0$. Under this circumstance, the asymptotic theory for low-degree p-modes of high-radial orders developed in the previous chapter should be applied rather than the asymptotic theory developed in the present chapter.

For the low-degree p-modes considered, the first-order asymptotic eigenfrequencies $\omega_{\rm asymp}$ are somewhat larger than the exact eigenfrequencies $\omega_{\rm exact}$. The largest

Table 15.2 Exact eigenfrequencies ω_{exact}, relative distances of the turning point from the centre, and first-order asymptotic approximations ω_T and ω_{asymp}, and their relative errors, for p-modes of the compressible equilibrium sphere of uniform mass density belonging to the degrees $\ell = 1, 2, 3, 4$

ℓ	n	ω_{exact}	r_t/R	ω_T	rel. error %	ω_{asymp}	rel. error %
1	5	10.2153	0.1329	10.4980	2.767	10.4083	1.890
	10	19.4766	0.07013	19.6267	0.7707	19.5789	0.5252
	15	28.6531	0.04773	28.7554	0.3570	28.7228	0.2432
	20	37.8065	0.03620	37.8841	0.2053	37.8594	0.1398
	25	46.9503	0.02915	47.0129	0.1332	46.9929	0.09068
2	5	11.0023	0.2031	11.4109	3.714	11.1803	1.619
	10	20.3146	0.1116	20.5396	1.108	20.4124	0.4817
	15	29.5128	0.07710	29.6683	0.5268	29.5804	0.2290
	20	38.6782	0.05890	38.7970	0.3072	38.7298	0.1335
	25	47.8296	0.04766	47.9257	0.2010	47.8714	0.08733
	30	56.9737	0.04002	57.0544	0.1417	57.0088	0.06157
	35	66.1135	0.03450	66.1831	0.1053	66.1438	0.04573
	40	75.2507	0.03031	75.3119	0.08127	75.2773	0.03530
3	5	11.7369	0.2627	12.3238	5.001	11.9024	1.410
	10	21.1193	0.1496	21.4525	1.577	21.2132	0.4444
	15	30.3482	0.1047	30.5812	0.7677	30.4138	0.2162
	20	39.5307	0.08056	39.7099	0.4533	39.5811	0.1277
	25	48.6930	0.06548	48.8386	0.2991	48.7340	0.08421
	30	57.8447	0.05515	57.9673	0.2120	57.8792	0.05970
	35	66.9901	0.04764	67.0960	0.1581	67.0199	0.04453
	40	76.1315	0.04193	76.2247	0.1225	76.1577	0.03448
4	5	12.4283	0.3138	13.2366	6.504	12.5831	1.245
	10	21.8946	0.1844	22.3653	2.150	21.9848	0.4120
	15	31.1612	0.1307	31.4940	1.068	31.2250	0.2047
	20	40.3652	0.1012	40.6228	0.6382	40.4145	0.1223
	25	49.5413	0.08264	49.7515	0.4242	49.5816	0.08129
	30	58.7027	0.06981	58.8802	0.3023	58.7367	0.05794
	35	67.8553	0.06043	68.0089	0.2264	67.8847	0.04338
	40	77.0022	0.05327	77.1376	0.1758	77.0281	0.03370

relative error found is that for $\ell = 1$ and $n = 5$ and amounts to 1.89%. Moreover, the relative errors of the first-order eigenfrequencies ω_{asymp} are all smaller than those of the first-order eigenfrequencies ω_T.

When p-modes of a given radial order are considered, and the degree ℓ or, equivalently, the distance of the turning point from the star's centre increases, the relative errors of the first-order asymptotic eigenfrequencies ω_{asymp} slowly decrease, while those of the eigenfrequencies ω_T increase. The ratios of the second relative errors to the first ones are approximately 1.46 for $\ell = 1$, approximately 2.29 for $\ell = 2$, approximately 3.55 for $\ell = 3$, and approximately 5.22 for $\ell = 4$.

Table 15.3 The exact frequency separations $\left(D_{n,\ell}\right)_{\text{exact}}$, and the asymptotic approximations $\left(D_{n,\ell}\right)_{\text{asymp}}$ and their relative errors, for the first- and second-degree p-modes considered in Table 15.2, relative to the compressible equilibrium sphere of uniform mass density

ℓ	n	$\left(D_{n,\ell}\right)_{\text{exact}}$	$\left(D_{n,\ell}\right)_{\text{asymp}}$	rel. error %
1	5	0.08220	0.08167	− 0.64
	10	0.04288	0.04280	− 0.20
	15	0.02911	0.02909	− 0.096
	20	0.02206	0.02204	− 0.055
	25	0.01776	0.01775	− 0.030
2	5	0.07675	0.07636	− 0.51
	10	0.04119	0.04111	− 0.18
	15	0.02829	0.02827	− 0.089
	20	0.02157	0.02156	− 0.053
	25	0.01744	0.01743	− 0.035
	30	0.01463	0.01463	− 0.024
	35	0.01261	0.01261	− 0.018
	40	0.01108	0.01108	− 0.013

The first-order asymptotic eigenfrequencies ω_{asymp} lead to small frequency separations for low-degree p-modes. The exact dimensionless frequency separations,

$$D_{n,\ell} = \frac{1}{2\ell + 3} \left(\omega_{n,\ell} - \omega_{n-1,\ell+2}\right), \tag{15.86}$$

and their asymptotic approximations are presented in Table 15.3 for the p-modes belonging to $\ell = 1$ and $\ell = 2$ that are considered in Table 15.2. Since the asymptotic approximations are somewhat smaller than the exact values, the relative errors are negative. In absolute value, they are smaller than 0.1% already from the radial order $n = 15$ on.

In order to gain insight into the origin of the small frequency separations between low-degree p-modes, Smeyers (2003) represented the variation of the velocity of acoustic propagation, defined by Eq. (15.79), for the second-degree mode p_{25}, with eigenfrequency $\omega_{\text{asymp}} = 47.8714$, and for the fourth-degree mode p_{24}, with eigenfrequency $\omega_{\text{asymp}} = 47.7493$. For the first mode, the turning point is located at $r_{\text{t}}/R = 0.04762$, for the second one, at $r_{\text{t}}/R = 0.08571$. The variations of the velocities of acoustic propagation as well as the variation of the isentropic sound velocity are represented in Fig. 15.3. The figure illustrates that the region situated in the vicinity of the two turning points largely contributes to the small frequency separation $D_{25,2}$. This conclusion agrees with the suggestion made earlier by Van Hoolst & Smeyers (1991) with regard to the origin of the small frequency separations in a polytropic model.

The asymptotic eigenfrequencies of the p-modes of the intermediate degrees $\ell =$ 10,20, 50 and the radial orders $n = 10, 20, 30, 40$ are contained in Table 15.4.

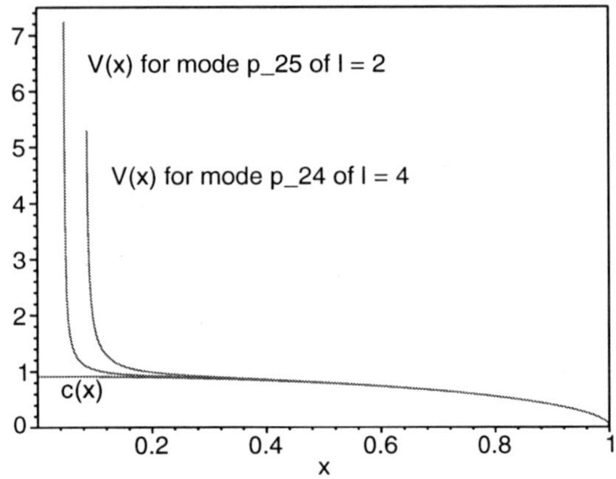

Fig. 15.3 The velocities of acoustic propagation for the modes p_{25} of $\ell = 2$ and p_{24} of $\ell = 4$, and the isentropic sound velocity, in the compressible equilibrium sphere of uniform mass density

Table 15.4 Exact eigenfrequencies ω_{exact}, relative distances of the turning point from the centre, and first-order asymptotic approximations ω_{asymp} and their relative errors, for p-modes belonging to the intermediate degrees $\ell = 10, 20, 50$, in the compressible equilibrium sphere of uniform mass density

ℓ	n	ω_{exact}	r_t/R	ω_{asymp}	rel. error %
10	10	26.0671	0.3451	26.1406	0.2820
	20	45.0487	0.2081	45.0925	0.09717
	30	63.6084	0.1490	63.6396	0.04908
	40	82.0326	0.1161	82.0569	0.02959
20	10	31.8289	0.5068	31.8852	0.1768
	20	51.9245	0.3391	51.9615	0.07127
	30	71.0358	0.2548	71.0634	0.03880
	40	89.7926	0.2040	89.8146	0.02448
50	10	44.8769	0.7165	44.9073	0.06785
	20	68.5313	0.5582	68.5565	0.03680
	30	89.7013	0.4571	89.7218	0.02284
	40	109.831	0.3870	109.848	0.01570

In all cases, the relative error of the asymptotic eigenfrequency ω_{asymp} with respect to the exact eigenfrequency ω_{exact} is smaller than 0.3%. The relative error decreases as the radial order n of the p-modes of a given degree ℓ increases as well as the degree ℓ of the p-modes of a given radial order n increases.

For the p-mode of degree $\ell = 10$ and radial order $n = 10$, the eigenfunction $\alpha(r)$ determined by means of the exact analytical solution, which is given by Eq. (10.46), is represented in Fig. 15.4. The eigenfunction is normalised by the condition that $\alpha(R) = 1$. The position of the turning point is indicated by the vertical line at $r_t/R = 0.3451$.

On the other hand, the asymptotic solution for the eigenfunction $\alpha(r)$ is represented in Fig. 15.5. With the eigenfrequency $\omega_{\text{asym}} = 26.1406$, the turning point is situated at $r_t/R = 0.3443$. From boundary-layer solution (15.67), it then results

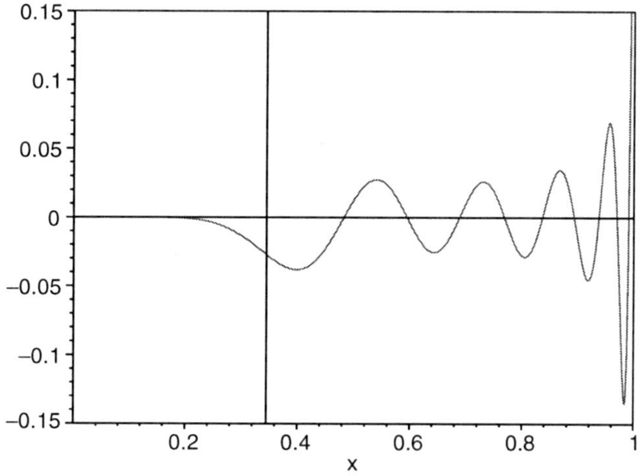

Fig. 15.4 The exact eigenfunction $\alpha(r)$ for the p-mode of degree $\ell = 10$ and radial order $n = 10$ in the compressible equilibrium sphere of uniform mass density

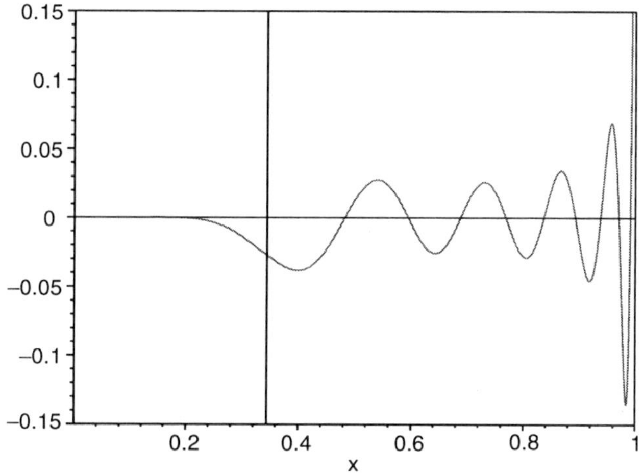

Fig. 15.5 Asymptotic representation of the eigenfunction $\alpha(r)$ for the p-mode of degree $\ell = 10$ and radial order $n = 10$ in the compressible equilibrium sphere of uniform mass density

that $A_{0,\mathrm{s}} = 0.01496$, and, from Eqs. (15.61) and (15.75), that $A_{0,\mathrm{t}} = -0.04416$. In the region from the turning point to the fitting point $r/R = 0.6721$, the asymptotic solution is determined by means of the uniformly valid solution $\alpha^{(\mathrm{t,u})}(r;\varepsilon)$, and, in the region from the boundary point $r/R = 1$ to the fitting point $r/R = 0.6721$, by means of the uniformly valid solution $\alpha^{(\mathrm{s,u})}(r;\varepsilon)$. On the inner side of the turning point, the uniformly valid solution $\alpha^{(\mathrm{t_i,u})}(r;\varepsilon)$ is used.

The asymptotic eigenfunction appears to be close to the exact analytical eigenfunction.

15.11 Eigenfrequency Equation with the Method of the Phase Functions

In a study devoted to the second-order asymptotic representation of solar p-modes of low and intermediate degrees, Vorontsov (1991) did no longer suppose that the mass density is analytic at the boundary point $r = R$, but constructed "exact", i.e. non-asymptotic, solutions for the surface layers within the framework of the Cowling approximation by applying the method of the phase functions known in quantum mechanics (Babikov 1976, see also Roxburgh & Vorontsov 1996). The connection between the phase shift and the structure of the envelope was described by Vorontsov & Zharkov (1989).

One derives the equation for the eigenfrequencies by joining the asymptotic solutions valid in the solar interior to the "exact" solutions valid in the envelope. The solutions are joined in the adiabatic region of the convective envelope below the ionisation layers of H and He, where the asymptotic solutions are supposed to be sufficiently accurate because of the smooth behaviour of the isentropic sound velocity in that region. Therefore, Vorontsov's representation is rather a composite than a purely asymptotic representation. Lopes & Turck-Chièze (1994) observed that it has some analogy with the wave ray theory applied in the seismology of the Earth (Woodhouse 1978).

Vorontsov started from two first-order differential equations which can be derived from Eqs. (6.15) and (6.16) by means of the transformation

$$\xi^*(r) = f_1(r)\,u(r), \qquad \eta^*(r) = f_2(r)\,y(r), \tag{15.87}$$

and a decomposition of the first derivative $d\Phi'/dr$ into a sum of two terms in accordance with the decomposition in the right-hand member of Eq. (7.5):

$$\frac{d\Phi'}{dr} = -\frac{4\pi G\rho}{r^2}\,u - \frac{\ell(\ell+1)}{r^2}\,S. \tag{15.88}$$

The functions $f_1(r)$ and $f_2(r)$ are given by definitions (7.35). The equations are

$$\frac{d\xi^*}{dr} = \left[\frac{\ell(\ell+1)}{\sigma^2} - \frac{r^2}{c^2}\right]\frac{f_1}{f_2}\,\eta^* + \frac{\ell(\ell+1)}{\sigma^2}\,f_1\,\Phi', \tag{15.89}$$

$$\frac{d\eta^*}{dr} = \left(1 - \frac{N^2}{\sigma^2} + \frac{4\pi G\rho}{\sigma^2}\right)\frac{\sigma^2}{r^2}\,\frac{f_2}{f_1}\,\xi^* + \frac{\ell(\ell+1)}{r^2}\,f_2\,S. \tag{15.90}$$

Near the boundary point $r = R$, it follows, in the Cowling approximation and in the approximation

$$\frac{4\pi G\rho}{\sigma^2} \ll 1,$$

that

$$\left.\begin{aligned}\frac{d\xi^*}{dr} &= -\frac{r^2 h^*}{c^2}\,\eta^*,\\ \frac{d\eta^*}{dr} &= \left(1-\frac{N^2}{\sigma^2}\right)\frac{\sigma^2}{r^2 h^*}\,\xi^*,\end{aligned}\right\} \tag{15.91}$$

with $h^*(r) = f_1(r)/f_2(r)$. By elimination of $\eta^*(r)$, one derives the second-order differential equation for $\xi^*(r)$

$$\frac{d^2\xi^*}{dr} + \frac{r^2 h^*}{c^2}\left(\frac{d}{dr}\frac{c^2}{r^2 h^*}\right)\frac{d\xi^*}{dr} + \left(1-\frac{N^2}{\sigma^2}\right)\frac{\sigma^2}{c^2}\,\xi^* = 0. \tag{15.92}$$

In addition, one can set

$$c(r) \simeq \varphi^{-1/2}(r), \tag{15.93}$$

where $\varphi(r)$ is defined by Eq. (15.19).

After introduction of the new independent variable

$$\tau^*(r) = \int_r^R \varphi^{1/2}\left(r'\right) dr', \tag{15.94}$$

and the function

$$\zeta\left(\tau^*\right) = B^{-1}(r)\,\xi^*(r) \tag{15.95}$$

with

$$B(r) = \left(r^2 h^* \varphi^{1/2}\right)^{1/2},$$

the equation reduces to the Schrödinger-type equation

$$\frac{d^2\zeta}{d\tau^{*2}} + \left[\sigma^2 - V\left(\tau^*\right)\right]\zeta = 0 \tag{15.96}$$

with the potential

$$V\left(\tau^*\right) = N^2 + \left(\frac{d}{d\tau^*}\ln B\right)^2 - \frac{d^2}{d\tau^{*2}}\ln B. \tag{15.97}$$

This equation corresponds to Vorontsov's equation (41).

Vorontsov sought exact solutions by using the method of the phase functions. He started from a solution of the form

$$\zeta\left(\tau^*\right) = A\left(\tau^*\right)\cos\left[\sigma\tau^* - \frac{\pi}{4} - \pi\,\widetilde{\alpha}\left(\tau^*\right)\right], \tag{15.98}$$

where $A\left(\tau^*\right)$ is the amplitude, and $\widetilde{\alpha}\left(\tau^*\right)$, the phase function.

A first differentiation yields

$$\frac{d\zeta}{d\tau^*} = -\sigma\, A\left(\tau^*\right)\, \sin\left[\sigma\tau^* - \frac{\pi}{4} - \pi\,\widetilde{\alpha}\left(\tau^*\right)\right]$$
$$+\pi\,\frac{d\widetilde{\alpha}}{d\tau^*}\, A\left(\tau^*\right)\, \sin\left[\sigma\tau^* - \frac{\pi}{4} - \pi\,\widetilde{\alpha}\left(\tau^*\right)\right]$$
$$+\frac{dA}{d\tau^*}\, \cos\left[\sigma\tau^* - \frac{\pi}{4} - \pi\,\widetilde{\alpha}\left(\tau^*\right)\right]. \tag{15.99}$$

By setting the sum of the last two terms equal to zero, one imposes the condition

$$\frac{1}{A}\frac{dA}{d\tau^*} = -\pi\,\frac{d\widetilde{\alpha}}{d\tau^*}\, \tan\left[\sigma\tau^* - \frac{\pi}{4} - \pi\widetilde{\alpha}\left(\tau^*\right)\right], \tag{15.100}$$

so that

$$\frac{d\zeta}{d\tau^*} = -\sigma\, A\left(\tau^*\right)\, \sin\left[\sigma\tau^* - \frac{\pi}{4} - \pi\,\widetilde{\alpha}\left(\tau^*\right)\right]. \tag{15.101}$$

A second differentiation of the solution yields

$$\frac{d^2\zeta}{d\tau^{*2}} = -\sigma^2\, A\left(\tau^*\right)\, \cos\left[\sigma\tau^* - \frac{\pi}{4} - \pi\,\widetilde{\alpha}\left(\tau^*\right)\right]$$
$$+\pi\,\sigma\, A\left(\tau^*\right)\,\frac{d\widetilde{\alpha}}{d\tau^*}\, \cos\left[\sigma\tau^* - \frac{\pi}{4} - \pi\,\widetilde{\alpha}\left(\tau^*\right)\right]$$
$$-\sigma\,\frac{dA}{d\tau^*}\, \sin\left[\sigma\tau^* - \frac{\pi}{4} - \pi\,\widetilde{\alpha}\left(\tau^*\right)\right]. \tag{15.102}$$

By elimination of $dA/d\tau^*$, one obtains

$$\frac{d^2\zeta}{d\tau^{*2}} = -\sigma^2\, A\left(\tau^*\right)\, \cos\left[\sigma\tau^* - \frac{\pi}{4} - \pi\,\widetilde{\alpha}\left(\tau^*\right)\right]$$
$$+\pi\,\sigma\, A\left(\tau^*\right)\,\frac{d\widetilde{\alpha}}{d\tau^*}\,\frac{1}{\cos\left[\sigma\tau^* - \frac{\pi}{4} - \pi\,\widetilde{\alpha}\left(\tau^*\right)\right]}. \tag{15.103}$$

Substitution into Eq. (15.96) leads to the non-linear differential equation for the phase function $\widetilde{\alpha}\left(\tau^*\right)$

$$\pi\,\frac{d\widetilde{\alpha}}{d\tau^*} = \frac{V\left(\tau^*\right)}{\sigma}\, \cos^2\left[\sigma\tau^* - \frac{\pi}{4} - \pi\,\widetilde{\alpha}\left(\tau^*\right)\right]. \tag{15.104}$$

To this equation, an initial condition for $\widetilde{\alpha}(0)$ is added which depends on the boundary condition adopted on the surface (Vorontsov & Zharkov 1989).

For the determination of the inner asymptotic solution of $\xi(r)$, we here observe that, for divergence-dominated p-modes, the approximation can be used

$$\alpha \simeq \frac{d\xi}{dr}. \tag{15.105}$$

When one adopts two-variable expansions similar to those introduced in Sect. 14.3,

$$\left.\begin{aligned} \alpha(r) &= \alpha_0(\tau, r) + \dots, \\ \xi(r) &= \nu(\varepsilon)\,[\xi_0(\tau, r) + \dots], \end{aligned}\right\} \tag{15.106}$$

where the fast variable $\tau(r)$ is defined by Eq. (14.5), it follows from Eq. (15.105), in the lowest-order asymptotic approximation, that

$$\alpha_0 = \nu(\varepsilon)\,\frac{\sqrt{\varphi(r)}}{\varepsilon}\,\frac{d\xi_0}{d\tau}. \tag{15.107}$$

The two members of the equation are of the same order in ε, when $\nu(\varepsilon) = \varepsilon$. It follows that

$$\frac{d\xi_0}{d\tau} = \frac{1}{\sqrt{\varphi(r)}}\,\alpha_0. \tag{15.108}$$

Use of solution (15.22) for α_0, in which $A_0^* = B_0^*$ according to Eqs. (15.61), and integration yield

$$\xi_0 = A_0^*\,\sqrt{2}\,\frac{h(r)}{\sqrt{\varphi(r)}}\,\sin\left(\tau - \frac{\pi}{4}\right), \tag{15.109}$$

so that the inner asymptotic solution for $\xi(r)$ is given by

$$\xi(r) = \varepsilon\,A_0^*\,\sqrt{2}\,\frac{h(r)}{\sqrt{\varphi(r)}}\,\sin\left(\tau - \frac{\pi}{4}\right). \tag{15.110}$$

By the use of definitions (14.5) and (15.94), the inner solution for $\zeta(r)$ then takes the form

$$\zeta_{\mathrm{i}} = \varepsilon\,F(r)\,\sin\left[\tau_R - \sigma\tau^*(r) - \frac{\pi}{4}\right] \tag{15.111}$$

with

$$F(r) = A_0^*\,\sqrt{2}\,f_1(r)\,r^2\,\left[r^2\,h^*(r)\,\varphi^{1/2}(r)\right]^{-1/2}\,\frac{h(r)}{\sqrt{\varphi(r)}}. \tag{15.112}$$

The first differentiation with respect to $\tau^*(r)$ yields

$$\frac{d\zeta_{\mathrm{i}}}{d\tau^*} = -\varepsilon\,\sigma\,F(r)\,\cos\left[\tau_R - \sigma\tau^*(r) - \frac{\pi}{4}\right]. \tag{15.113}$$

The necessary and sufficient condition for the function $\zeta(r)$ and its first derivative $d\zeta/d\tau^*$ to be continuous at the junction point $r = r_f$ leads to the eigenfrequency equation

$$\tau_R = \left[k + \widetilde{\alpha}\left(\tau_f^*\right)\right]\pi, \qquad k = 0, 1, 2, \ldots \tag{15.114}$$

This equation has the same form as eigenfrequency equation (15.78). The effective polytropic index n_{e} is here replaced by $2\left[\widetilde{\alpha}\left(\tau_f^*\right) + k - n\right]$.

Chapter 16
Asymptotic Representation of Low-Degree, Higher-Order g^+-Modes in Stars Containing a Convective Core

16.1 Introduction

A first-order asymptotic representation of low-degree, higher-order g^+-modes in quasi-static stars that contain a convective core has been developed by Smeyers & Moya (2007). The stars considered consist either of a convective core and a radiative envelope, or of a convective core, an intermediate radiative zone, and a convective envelope. The convective core is supposed to be in isentropic (adiabatic) equilibrium[1].

Smeyers & Moya used the full fourth-order system of differential equations (6.8) and (6.10) in the divergence and the radial component of the Lagrangian displacement. To this system, they applied two-variable expansion procedures at larger distances from the boundary and the turning points, and boundary-layer theory near these points (Kevorkian & Cole 1981, 1996). By their procedure, they improved the asymptotic representation of Willems et al. (1997) applying to stars that consist of a convective core and a radiative envelope. This asymptotic representation has partly been obscured by the fact that the authors adopted boundary-layer coordinates that are identical to the fast variables they used at larger distances from the boundary points.

16.2 Appropriate Equations

The small parameter is here

$$\varepsilon = |\sigma|, \tag{16.1}$$

[1] This chapter is a reproduction of Smeyers, P., Moya, A.: The asymptotic representation of higher-order g^+-modes in stars with a convective core. Astronomy & Astrophysics **465**, 509–524 (2007). With permission from Astronomy & Astrophysics, http://www.aanda.org.

P. Smeyers and T. Van Hoolst, *Linear Isentropic Oscillations of Stars: Theoretical Foundations*, Astrophysics and Space Science Library 371,
DOI 10.1007/978-3-642-13030-4_17,

so that Eqs. (6.8) and (6.10) take the form

$$\frac{d^2\alpha}{dr^2} + K_2(r)\,\frac{d\alpha}{dr} + \left[\frac{K_1(r)}{\varepsilon^2} + K_3(r) + \frac{\varepsilon^2}{c^2(r)}\right]\alpha = -K_4(r)\,\frac{d\xi}{dr}, \quad (16.2)$$

$$\frac{d^2\xi}{dr^2} + \frac{4}{r}\,\frac{d\xi}{dr} - \frac{\ell(\ell+1)-2}{r^2}\,\xi = \frac{d\alpha}{dr} - \left[\frac{K_1(r)}{\varepsilon^2}\,\frac{c^2(r)}{g(r)} - \frac{2}{r}\right]\alpha. \quad (16.3)$$

The asymptotic solutions for $\alpha(r)$ and $\xi(r)$ must satisfy the same boundary conditions as those imposed to the asymptotic solutions for low-degree, higher-order p-modes in Chap. 14. Furthermore, the variables in the equations are made dimensionless in the same way as in that chapter.

The asymptotic solutions are again constructed in a similar way in the various subintervals. From differential equation (16.2), a *homogeneous* second-order differential equation is derived for the lowest-order asymptotic approximation of the divergence $\alpha(r)$ of the Lagrangian displacement, while from differential equation (16.3), an *inhomogeneous* second-order differential equation is derived for the lowest-order asymptotic approximation of the radial component $\xi(r)$ of the Lagrangian displacement.

Smeyers & Moya made a distinction between stars consisting of a convective core and a radiative envelope, and stars consisting of a convective core, an intermediate radiative zone, and a convective envelope. The same distinction is made here.

16.3 Stars Consisting of a Convective Core and a Radiative Envelope

For stars consisting of a convective core and a radiative envelope, a turning point appears in the left-hand member of Eq. (16.2) at the boundary between the two regions. Be this turning point located at the radial distance $r = r_{\rm a}$ from the star's centre.

16.3.1 Two-Variable Solutions in the Radiative Envelope at Larger Distances from its Boundaries

In the region of the radiative envelope that is situated at larger distances from the boundary of the convective core at $r = r_{\rm a}$ and the star's surface at $r = R$, two-variable expansions similar to those developed by Smeyers et al. (1995) are derived[2].

[2] This section is a partial reproduction of Smeyers, P., De Boeck, I., Van Hoolst, T., Decock, L.: Asymptotic representation of linear, isentropic g-modes of stars. Astronomy & Astrophysics **301**, 105–122 (1995). With permission from Astronomy & Astrophysics, http://www.aanda.org.

Equation (16.2) is compared to a second-order differential equation that governs the motion of a linear oscillator with a small damping. Therefore, asymptotic solutions are constructed in terms of two independent variables. The requirement that the term involving the second derivative $d^2\alpha/dr^2$ be of the same order of magnitude as the term $\left[K_1(r)/\varepsilon^2\right]\alpha$, which contains the large parameter, leads to the introduction of a fast variable $\tau(r)$ obeying the differential equation

$$d\tau = \frac{1}{\varepsilon}\, K_1^{1/2}(r)\, dr. \tag{16.4}$$

If it is set equal to zero at $r = r_{\rm a}$, the fast variable is defined as

$$\tau(r) = \frac{1}{\varepsilon}\int_{r_{\rm a}}^{r} K_1^{1/2}\left(r'\right) dr'. \tag{16.5}$$

The slow variable is equal to the radial coordinate r.

For the functions $\alpha(r)$ and $\xi(r)$, asymptotic expansions in terms of the fast variable $\tau(r)$ and the slow variable r are introduced as

$$\left.\begin{aligned} \alpha^{(\rm o)}(r;\varepsilon) &= \alpha_0^{(\rm o)}(\tau,r) + \varepsilon\,\alpha_1^{(\rm o)}(\tau,r) + \varepsilon^2\,\alpha_2^{(\rm o)}(\tau,r) + O\left(\varepsilon^3\right), \\ \xi^{(\rm o)}(r;\varepsilon) &= \xi_0^{(\rm o)}(\tau,r) + \varepsilon\,\xi_1^{(\rm o)}(\tau,r) + \varepsilon^2\,\xi_2^{(\rm o)}(\tau,r) + O\left(\varepsilon^3\right). \end{aligned}\right\} \tag{16.6}$$

By transforming the differential operators in Eqs. (16.2) and (16.3), substituting the asymptotic expansions into the equations, and collecting the terms of the same orders in ε, one successively has

at order ε^{-2},

$$\frac{\partial^2\alpha_0^{(\rm o)}}{\partial\tau^2} + \alpha_0^{(\rm o)} = 0, \tag{16.7}$$

$$\frac{\partial^2\xi_0^{(\rm o)}}{\partial\tau^2} = -\frac{c^2}{g}\,\alpha_0^{(\rm o)}; \tag{16.8}$$

at order ε^{-1},

$$\frac{\partial^2\alpha_1^{(\rm o)}}{\partial\tau^2} + \alpha_1^{(\rm o)} = -\frac{1}{K_1^{1/2}}\left[2\,\frac{\partial^2\alpha_0^{(\rm o)}}{\partial\tau\,\partial r} + \left(\frac{1}{2\,K_1}\,\frac{dK_1}{dr} + K_2\right)\frac{\partial\alpha_0^{(\rm o)}}{\partial\tau} + K_4\,\frac{\partial\xi_0^{(\rm o)}}{\partial\tau}\right], \tag{16.9}$$

$$\frac{\partial^2\xi_1^{(\rm o)}}{\partial\tau^2} = -\frac{c^2}{g}\,\alpha_1^{(\rm o)} - \frac{1}{K_1^{1/2}}\left[2\,\frac{\partial^2\xi_0^{(\rm o)}}{\partial\tau\,\partial r} + \left(\frac{1}{2\,K_1}\,\frac{dK_1}{dr} + \frac{4}{r}\right)\frac{\partial\xi_0^{(\rm o)}}{\partial\tau} - \frac{\partial\alpha_0^{(\rm o)}}{\partial\tau}\right]; \tag{16.10}$$

and, at order ε^0,

$$\frac{\partial^2 \alpha_2^{(o)}}{\partial \tau^2} + \alpha_2^{(o)} = -\frac{1}{K_1^{1/2}} \left[2 \frac{\partial^2 \alpha_1^{(o)}}{\partial \tau\, \partial r} + \left(\frac{1}{2\, K_1} \frac{d K_1}{dr} + K_2 \right) \frac{\partial \alpha_1^{(o)}}{\partial \tau} + K_4 \frac{\partial \xi_1^{(o)}}{\partial \tau} \right] - \frac{1}{K_1} \left(\frac{\partial^2 \alpha_0^{(o)}}{\partial r^2} + K_2 \frac{\partial \alpha_0^{(o)}}{\partial r} + K_3\, \alpha_0^{(o)} + K_4 \frac{\partial \xi_0^{(o)}}{\partial r} \right), \tag{16.11}$$

$$\frac{\partial^2 \xi_2^{(o)}}{\partial \tau^2} = -\frac{c^2}{g}\, \alpha_2^{(o)} + \frac{1}{K_1^{1/2}} \left[\frac{\partial \alpha_1^{(o)}}{\partial \tau} - 2 \frac{\partial^2 \xi_1^{(o)}}{\partial \tau\, \partial r} - \left(\frac{1}{2\, K_1} \frac{d K_1}{dr} + \frac{4}{r} \right) \frac{\partial \xi_1^{(o)}}{\partial \tau} \right] + \frac{1}{K_1} \left(\frac{\partial \alpha_0^{(o)}}{\partial r} + \frac{2}{r}\, \alpha_0^{(o)} - \frac{\partial^2 \xi_0^{(o)}}{\partial r^2} - \frac{4}{r} \frac{\partial \xi_0^{(o)}}{\partial r} + \frac{\ell(\ell+1) - 2}{r^2}\, \xi_0^{(o)} \right). \tag{16.12}$$

Equations (16.7) and (16.8) form a fourth-order system of partial differential equations in the fast variable. A general solution of the homogeneous equation (16.7) is given by

$$\alpha_0^{(o)}(\tau, r) = A_0^{(o)}(r)\, \cos\tau + B_0^{(o)}(r)\, \sin\tau, \tag{16.13}$$

where $A_0^{(o)}(r)$ and $B_0^{(o)}(r)$ are yet unknown functions of the slow variable. Substitution into Eq. (16.8) and a twofold integration with respect to the fast variable yield the solution

$$\xi_0^{(o)}(\tau, r) = \frac{c^2}{g} \left[A_0^{(o)}(r)\, \cos\tau + B_0^{(o)}(r)\, \sin\tau \right] + F_0^{(o)}(r)\, \tau + G_0^{(o)}(r)\,, \tag{16.14}$$

where $F_0^{(o)}(r)$ and $G_0^{(o)}(r)$ are yet unknown functions of the slow variable. One removes the term $F_0^{(o)}(r)\, \tau$, which is secular in the fast variable, by setting

$$F_0^{(o)}(r) = 0. \tag{16.15}$$

The functions $A_0^{(o)}(r)$ and $B_0^{(o)}(r)$ are determined by use of Eq. (16.9). After substitution of the solutions for $\alpha_0^{(o)}(\tau, r)$ and $\xi_0^{(o)}(\tau, r)$, the equation becomes

$$\frac{\partial^2 \alpha_1^{(o)}}{\partial \tau^2} + \alpha_1^{(o)} = -K_1^{-1/2} \left[2 \frac{d B_0^{(o)}}{dr} + \left(\frac{c^2}{g}\, K_4 + \frac{1}{2\, K_1} \frac{d K_1}{dr} + K_2 \right) B_0^{(o)} \right] \cos\tau + K_1^{-1/2} \left[2 \frac{d A_0^{(o)}}{dr} + \left(\frac{c^2}{g}\, K_4 + \frac{1}{2\, K_1} \frac{d K_1}{dr} + K_2 \right) A_0^{(o)} \right] \sin\tau. \tag{16.16}$$

The resonant terms are removed from the inhomogeneous part of the equation by the requirement that their coefficients be equal to zero. Integration of the resulting differential equations with respect to the slow variable yields

$$A_0^{(o)}(r) = A_0^* \, K_5(r), \quad B_0^{(o)}(r) = B_0^* \, K_5(r), \tag{16.17}$$

where A_0^* and B_0^* are arbitrary constants, and

$$K_5(r) = \frac{g}{(N^2 \, r^6 \, c^8 \, \rho^2)^{1/4}}. \tag{16.18}$$

Substitution of the solutions for $A_0^{(o)}(r)$ and $B_0^{(o)}(r)$ into the solutions for $\alpha_0^{(o)}(\tau, r)$ and $\xi_0^{(o)}(\tau, r)$ leads to

$$\left.\begin{aligned} \alpha_0^{(o)}(\tau, r) &= K_5(r) \left(A_0^* \cos\tau + B_0^* \sin\tau \right), \\ \xi_0^{(o)}(\tau, r) &= K_6(r) \left(A_0^* \cos\tau + B_0^* \sin\tau \right) + G_0^{(o)}(r), \end{aligned}\right\} \tag{16.19}$$

where

$$K_6(r) = \frac{c^2}{g} \, K_5(r) = \left(N^2 \, r^6 \, \rho^2 \right)^{-1/4}. \tag{16.20}$$

Equation (16.16) reduces to a homogeneous equation, so that a general solution for the function $\alpha_1(\tau, r)$ is given by

$$\alpha_1^{(o)}(\tau, r) = A_1^{(o)}(r) \cos\tau + B_1^{(o)}(r) \sin\tau, \tag{16.21}$$

where $A_1^{(o)}(r)$ and $B_1^{(o)}(r)$ are yet undetermined functions of the slow variable.

The remaining equations are used for the derivation of a solution for the function $G_0^{(o)}(r)$.

Substitution of the solutions for $\alpha_0^{(o)}(\tau, r)$, $\xi_0^{(o)}(\tau, r)$, and $\alpha_1^{(o)}(\tau, r)$ into Eq. (16.10) and a twofold integration with respect to the fast variable lead to

$$\begin{aligned} \xi_1^{(o)}(\tau, r) = {} & K_1^{-1/2} \left[2 \frac{d}{dr} \left(\frac{c^2}{g} B_0^{(o)} \right) + \left(\frac{1}{2 K_1} \frac{dK_1}{dr} \frac{c^2}{g} + \frac{4}{r} \frac{c^2}{g} - 1 \right) B_0^{(o)} \right] \cos\tau \\ & - K_1^{-1/2} \left[2 \frac{d}{dr} \left(\frac{c^2}{g} A_0^{(o)} \right) + \left(\frac{1}{2 K_1} \frac{dK_1}{dr} \frac{c^2}{g} + \frac{4}{r} \frac{c^2}{g} - 1 \right) A_0^{(o)} \right] \sin\tau \\ & + \frac{c^2}{g} \left(A_1^{(o)} \cos\tau + B_1^{(o)} \sin\tau \right) + F_1^{(o)}(r) \, \tau + G_1^{(o)}(r). \end{aligned} \tag{16.22}$$

The term $F_1^{(o)}(r)\,\tau$, which is secular in the fast variable, is removed by the requirement that

$$F_1^{(o)}(r) = 0. \tag{16.23}$$

Substitution of the solutions for $\alpha_0^{(o)}(\tau, r)$, $\xi_0^{(o)}(\tau, r)$, $\alpha_1^{(o)}(\tau, r)$, and $\xi_1^{(o)}(\tau, r)$ into Eq. (16.11) and removal of the resonant terms from the inhomogeneous part of the equation lead to

$$\frac{\partial^2 \alpha_2^{(o)}}{\partial \tau^2} + \alpha_2^{(o)} = -\frac{K_4}{K_1}\,\frac{d G_0^{(o)}}{dr}, \tag{16.24}$$

so that a general solution for the function $\alpha_2^{(o)}(\tau, r)$ is given by

$$\alpha_2^{(o)}(\tau, r) = A_2^{(o)}(r)\,\cos\tau + B_2^{(o)}(r)\,\sin\tau - \frac{K_4}{K_1}\,\frac{d G_0^{(o)}}{dr}, \tag{16.25}$$

where $A_2^{(o)}(r)$ and $B_2^{(o)}(r)$ are yet unknown functions of the slow variable.

Finally, after substitution of the solutions for $\alpha_0^{(o)}(\tau, r)$, $\xi_0^{(o)}(\tau, r)$, $\alpha_1^{(o)}(\tau, r)$, $\xi_1^{(o)}(\tau, r)$, and $\alpha_2^{(o)}(\tau, r)$ into Eq. (16.12) and removal of the resonant terms from the inhomogeneous part of the equation, it results that

$$\frac{\partial^2 \xi_2^{(o)}}{\partial \tau^2} = -\frac{1}{K_1(r)}\left[\frac{d^2 G_0^{(o)}}{dr^2} + 2\left(\frac{1}{g}\,\frac{dg}{dr} + \frac{1}{r}\right)\frac{d G_0^{(o)}}{dr} - \frac{\ell(\ell+1)-2}{r^2}\,G_0^{(o)}\right]. \tag{16.26}$$

In order to exclude secular terms from the solution for $\xi_2^{(o)}(\tau, r)$, one must set the right-hand member of the equation equal to zero:

$$\frac{d^2 G_0^{(o)}}{dr^2} + 2\left(\frac{1}{g}\,\frac{dg}{dr} + \frac{1}{r}\right)\frac{d G_0^{(o)}}{dr} - \frac{\ell(\ell+1)-2}{r^2}\,G_0^{(o)} = 0. \tag{16.27}$$

This second-order differential equation is the required equation for the function $G_0^{(o)}(r)$. It has the same form as Clairaut's equation known in the theory of equilibrium tides. One verifies this by introducing the mean mass density inside the sphere with radius r, $\bar{\rho}(r)$, and making use of Eq. (2.20) for the gravity $g(r)$ and Eq. (2.22) for the mass $m(r)$ inside the sphere with radius r. The equation then becomes

$$r^2\,\frac{d^2}{dr^2}\,\frac{G_0^{(o)}}{r} + 6\,\frac{\rho(r)}{\bar{\rho}(r)}\left(r\,\frac{d}{dr}\,\frac{G_0^{(o)}}{r} + \frac{G_0^{(o)}}{r}\right) - \ell(\ell+1)\,\frac{G_0^{(o)}}{r} = 0 \tag{16.28}$$

(see, e.g., Kopal 1959).

A general solution of Eq. (16.27) consists of a linear combination of two particular solutions. Near the boundary point $r = 0$, which is a singular point of the equation, one can distinguish between a particular solution $y_1(r)$ that behaves as $r^{\ell-1}$ and a particular solution $y_2(r)$ that behaves as $r^{-(\ell+2)}$. In terms of these particular solutions, the general solution can be expressed as

$$G_0^{(\mathrm{o})}(r) = C_0^* \, y_1(r) + D_0^* \, y_2(r), \tag{16.29}$$

where C_0^* and D_0^* are arbitrary constants.

The lowest-order two-variable solutions for the divergence and the radial component of the Lagrangian displacement that are valid at larger distances from the boundary points $r = r_\mathrm{a}$ and $r = R$ of the radiative envelope are thus

$$\left.\begin{aligned} \alpha^{(\mathrm{o})}(r;\varepsilon) &= K_5(r)\left(A_0^* \cos\tau + B_0^* \sin\tau\right), \\ \xi^{(\mathrm{o})}(r;\varepsilon) &= K_6(r)\left[\left(A_0^* \cos\tau + B_0^* \sin\tau\right) + \frac{G_0^{(\mathrm{o})}(r)}{K_6(r)}\right]. \end{aligned}\right\} \tag{16.30}$$

The solution for the divergence of the Lagrangian displacement is purely oscillatory, while that for the radial component also contains a non-oscillatory part. These solutions involve the four arbitrary constants A_0^*, B_0^*, C_0^*, D_0^*.

16.3.2 Boundary-Layer Solutions on the Outer Side of the Boundary Between the Convective Core and the Radiative Envelope

Here boundary-layer solutions are constructed near the turning point $r = r_\mathrm{a}$ of Eq. (16.2), on the side of the radiative envelope.

Because of the form of the two-variable solutions $\alpha^{(\mathrm{o})}(r;\varepsilon)$ and $\xi^{(\mathrm{o})}(r;\varepsilon)$, it seems to be convenient to pass from the function $\xi(r)$ to the function $w(r)$ by means of the transformation

$$\xi(r) = \frac{c^2(r)}{g(r)}\, w(r). \tag{16.31}$$

Equations (16.2) and (16.3) then become

$$\frac{d^2\alpha}{dr^2} + K_2(r)\,\frac{d\alpha}{dr} + \left[\frac{K_1(r)}{\varepsilon^2} + K_3(r) + \frac{\varepsilon^2}{c^2(r)}\right]\alpha = -K_4(r)\,\frac{d}{dr}\left[\frac{c^2(r)}{g(r)}\, w\right], \tag{16.32}$$

$$\frac{d^2w}{dr^2} + \left[\frac{4}{r} + 2\frac{d}{dr}\ln\frac{c^2(r)}{g(r)}\right]\frac{dw}{dr}$$

$$+\left[-\frac{\ell(\ell+1)-2}{r^2} + \frac{4}{r}\frac{d}{dr}\ln\frac{c^2(r)}{g(r)} + \frac{g(r)}{c^2(r)}\frac{d^2}{dr^2}\frac{c^2(r)}{g(r)}\right]w$$

$$= \frac{g(r)}{c^2(r)}\frac{d\alpha}{dr} - \left[\frac{K_1(r)}{\varepsilon^2} - \frac{2}{r}\frac{g(r)}{c^2(r)}\right]\alpha. \tag{16.33}$$

An appropriate coordinate in the boundary layer is

$$s_{\rm a} = r - r_{\rm a}. \tag{16.34}$$

As $s_{\rm a} \to 0$, the functions involved in the coefficients of the equations behave as

$$\left.\begin{aligned}
&\rho(r) = \rho(r_{\rm a})\left[1 + O(s_{\rm a})\right], \quad g(r) = g(r_{\rm a})\left[1 + O(s_{\rm a})\right],\\
&c(r) = c(r_{\rm a})\left[1 + O(s_{\rm a})\right], \quad N^2(r) = N_{\rm a}^2\, s_{\rm a}\left[1 + O(s_{\rm a})\right],\\
&K_1(r) = \frac{\ell(\ell+1)}{r_{\rm a}^2} N_{\rm a}^2\, s_{\rm a}\left[1 + O(s_{\rm a})\right] \equiv K_{1,\rm a}\, s_{\rm a}\left[1 + O(s_{\rm a})\right],\\
&K_2(r) = K_{2,\rm a}\left[1 + O(s_{\rm a})\right], \quad K_3(r) = K_{3,\rm a}\left[1 + O(s_{\rm a})\right],\\
&K_4(r) = K_{4,\rm a}\left[1 + O(s_{\rm a})\right], \quad K_5(r) = K_{5,\rm a}\, s_{\rm a}^{-1/4}\left[1 + O(s_{\rm a})\right].
\end{aligned}\right\} \tag{16.35}$$

The boundary-layer coordinate

$$s_{\rm a}^*(r) = \frac{s_{\rm a}(r)}{\delta(\varepsilon)}, \tag{16.36}$$

where $\delta(\varepsilon) \to 0$ as $\varepsilon \to 0$, and boundary-layer expansions of the form

$$\left.\begin{aligned}
&\alpha^{(\rm a)}(r;\varepsilon) = \mu_0^{(\rm a)}(\varepsilon)\,\alpha_0^{(\rm a)}\left(s_{\rm a}^*\right) + \mu_1^{(\rm a)}(\varepsilon)\,\alpha_1^{(\rm a)}\left(s_{\rm a}^*\right) + \ldots,\\
&w^{(\rm a)}(r;\varepsilon) = \nu_0^{(\rm a)}(\varepsilon)\,w_0^{(\rm a)}\left(s_{\rm a}^*\right) + \nu_1^{(\rm a)}(\varepsilon)\,w_1^{(\rm a)}\left(s_{\rm a}^*\right) + \ldots
\end{aligned}\right\} \tag{16.37}$$

are introduced.

Equation (16.32) is transformed into

$$\mu_0^{(\rm a)}(\varepsilon)\left\{\frac{1}{\delta^2(\varepsilon)}\frac{d^2\alpha_0^{(\rm a)}}{ds_{\rm a}^{*2}} + \frac{K_{2,\rm a}}{\delta(\varepsilon)}\frac{d\alpha_0^{(\rm a)}}{ds_{\rm a}^*}\right.$$

$$\left.+\left[\frac{\delta(\varepsilon)}{\varepsilon^2}K_{1,\rm a}\,s_{\rm a}^* + K_{3,\rm a} + \frac{\varepsilon^2}{c_{\rm a}^2}\right]\alpha_0^{(\rm a)} + \ldots\right\} + \mu_1^{(\rm a)}(\varepsilon)\{\ldots\} + \ldots$$

$$= \nu_0^{(a)}(\varepsilon)\left(-\frac{K_{4,a}}{\delta(\varepsilon)}\frac{c_a^2}{g_a}\frac{dw_0^{(a)}}{ds_a^*}+\ldots\right)+\nu_1^{(a)}(\varepsilon)(\ldots)+\ldots \tag{16.38}$$

The first dominant boundary-layer equation is homogeneous, when $\nu_0^{(a)}(\varepsilon)$ is of a higher order in ε than $\mu_0^{(a)}(\varepsilon)/\delta(\varepsilon)$ and $\mu_0^{(a)}(\varepsilon)\,\delta^2(\varepsilon)/\varepsilon^2$. It then takes the form

$$\frac{1}{\delta^2(\varepsilon)}\frac{d^2\alpha_0^{(a)}}{ds_a^{*2}}+\frac{\delta(\varepsilon)}{\varepsilon^2}K_{1,a}\,s_a^*\,\alpha_0^{(a)}=0. \tag{16.39}$$

In the left-hand member, the term involving the second derivative is of the same order in ε as the term with the large parameter, when

$$\delta(\varepsilon)=\varepsilon^{2/3}. \tag{16.40}$$

A general solution of the boundary-layer equation in terms of Bessel functions of the first kind is given by

$$\alpha_0^{(a)}\left(s_a^*\right)=\sqrt{s_a^*}\left[A_{0,a}\,J_{1/3}\left(\frac{2}{3}\sqrt{K_{1,a}}\,s_a^{*3/2}\right)+B_{0,a}\,J_{-1/3}\left(\frac{2}{3}\sqrt{K_{1,a}}\,s_a^{*3/2}\right)\right], \tag{16.41}$$

where $A_{0,a}$ and $B_{0,a}$ are arbitrary constants.

Next, Eq. (16.33) is transformed into

$$\nu_0^{(a)}(\varepsilon)\left[\frac{1}{\varepsilon^{4/3}}\frac{d^2w_0^{(a)}}{ds_a^{*2}}+O\left(\varepsilon^{-2/3}\right)\right]+\nu_1^{(a)}(\varepsilon)(\ldots)+\ldots$$
$$=\mu_0^{(a)}(\varepsilon)\left[-\frac{K_{1,a}}{\varepsilon^{4/3}}\,s_a^*\,\alpha_0^{(a)}+O\left(\varepsilon^{-2/3}\right)\right]+\mu_1^{(a)}(\varepsilon)(\ldots)+\ldots \tag{16.42}$$

The second dominant boundary-layer equation is inhomogeneous when

$$\nu_0^{(a)}(\varepsilon)=\mu_0^{(a)}(\varepsilon). \tag{16.43}$$

This equality is compatible with the suppositions made above in the derivation of the first dominant boundary-layer equation. The second dominant boundary-layer equation then takes the form

$$\frac{d^2w_0^{(a)}}{ds_a^{*2}}=-K_{1,a}\,s_a^*\,\alpha_0^{(a)}. \tag{16.44}$$

By subtracting the first dominant boundary-layer equation, one obtains

$$\frac{d^2}{ds_a^{*2}}\left(w_0^{(a)}-\alpha_0^{(a)}\right)=0, \tag{16.45}$$

so that, after integration,

$$w_0^{(a)}\left(s_a^*\right) = \alpha_0^{(a)}\left(s_a^*\right) + C_{0,a}\, s_a^* + D_{0,a}, \tag{16.46}$$

where $C_{0,a}$ and $D_{0,a}$ are arbitrary constants.

Thus far, the function $\mu_0^{(a)}(\varepsilon)$ is undetermined. In view of the matching with the asymptotic solutions valid at larger distances from the turning point, it is appropriate to associate a different function of ε with each of the solutions $\alpha_0^{(a)}\left(s_a^*\right)$, $C_{0,a}\, s_a^*$, $D_{0,a}$ and to express the boundary-layer solutions in the more general form

$$\left.\begin{aligned}
\alpha^{(a)}(r;\varepsilon) &= \mu_0^{(a,1)}(\varepsilon)\,\sqrt{s_a^*}\Bigg[A_{0,a}\,J_{1/3}\left(\frac{2}{3}\,\sqrt{K_{1,a}}\,s_a^{*3/2}\right)\\
&\quad + B_{0,a}\,J_{-1/3}\left(\frac{2}{3}\,\sqrt{K_{1,a}}\,s_a^{*3/2}\right)\Bigg],\\
w^{(a)}(r;\varepsilon) &= \mu_0^{(a,1)}(\varepsilon)\,\sqrt{s_a^*}\Bigg[A_{0,a}\,J_{1/3}\left(\frac{2}{3}\,\sqrt{K_{1,a}}\,s_a^{*3/2}\right)\\
&\quad + B_{0,a}\,J_{-1/3}\left(\frac{2}{3}\,\sqrt{K_{1,a}}\,s_a^{*3/2}\right)\Bigg]\\
&\quad + \nu_0^{(a,2)}(\varepsilon)\,C_{0,a}\,s_a^* + \nu_0^{(a,3)}(\varepsilon)\,D_{0,a}.
\end{aligned}\right\} \tag{16.47}$$

The matching condition relative to the divergence of the Lagrangian displacement is

$$\lim_{s_a\to\infty} \alpha^{(a)}(r;\varepsilon) = \lim_{s_a\to 0} \alpha^{(o)}(r;\varepsilon). \tag{16.48}$$

The two-variable solution $\alpha^{(o)}(r;\varepsilon)$ must therefore be considered for small values of s_a, and the boundary-layer solution $\alpha^{(a)}(r;\varepsilon)$, for large values of s_a.

As $s_a \to 0$, the fast variable $\tau(r)$ tends to the first term in its Taylor series around $s_a = 0$, so that

$$\tau(r) = \frac{1}{\varepsilon}\frac{2}{3}\,\sqrt{K_{1,a}}\,s_a^{3/2}, \tag{16.49}$$

and, correspondingly,

$$\lim_{s_a\to 0} \alpha^{(o)}(r,\varepsilon) = \frac{K_{5,a}}{s_a^{1/4}}$$
$$\left[A_0^*\cos\left(\frac{1}{\varepsilon}\frac{2}{3}\,\sqrt{K_{1,a}}\,s_a^{3/2}\right) + B_0^*\sin\left(\frac{1}{\varepsilon}\frac{2}{3}\,\sqrt{K_{1,a}}\,s_a^{3/2}\right)\right]. \tag{16.50}$$

As $s_a \to \infty$, the use of the principal asymptotic form of the Bessel functions of the first kind for large arguments yields

$$\lim_{s_a\to\infty} \alpha^{(a)}(r;\varepsilon) = \mu_0^{(a,1)}(\varepsilon)\,\varepsilon^{1/6}\frac{\sqrt{3}}{\sqrt{\pi}K_{1,a}^{1/4}}\frac{1}{s_a^{1/4}}\left[A_{0,a}\sin\left(\frac{1}{\varepsilon}\frac{2}{3}\sqrt{K_{1,a}}\,s_a^{3/2}+\frac{\pi}{12}\right)\right.$$
$$\left.+B_{0,a}\sin\left(\frac{1}{\varepsilon}\frac{2}{3}\sqrt{K_{1,a}}\,s_a^{3/2}+\frac{5\pi}{12}\right)\right]. \tag{16.51}$$

The matching condition is satisfied, when

$$\mu_0^{(a,1)}(\varepsilon) = \varepsilon^{-1/6} \tag{16.52}$$

and

$$\left.\begin{aligned} A_0^* &= \frac{\sqrt{3}}{\sqrt{\pi}\,K_{1,a}^{1/4}\,K_{5,a}}\left(A_{0,a}\sin\frac{\pi}{12}+B_{0,a}\cos\frac{\pi}{12}\right),\\ B_0^* &= \frac{\sqrt{3}}{\sqrt{\pi}\,K_{1,a}^{1/4}\,K_{5,a}}\left(A_{0,a}\cos\frac{\pi}{12}+B_{0,a}\sin\frac{\pi}{12}\right). \end{aligned}\right\} \tag{16.53}$$

Next, the matching condition relative to the function $w(r)$ is

$$\lim_{s_a\to\infty} w^{(a)}(r;\varepsilon) = \lim_{s_a\to 0} w^{(o)}(r;\varepsilon). \tag{16.54}$$

For the oscillatory parts of the functions $w^{(a)}(r;\varepsilon)$ and $w^{(o)}(r;\varepsilon)$, the matching condition is automatically satisfied by conditions (16.52) and (16.53). For the non-oscillatory parts, it leads to

$$\left.\begin{aligned} &C_{0,a} = 0,\\ &\nu_0^{(a,3)}(\varepsilon) = \varepsilon^0, \quad D_{0,a} = \frac{g\,(r_a)}{c^2\,(r_a)}\,G_0^{(o)}\,(r_a). \end{aligned}\right\} \tag{16.55}$$

16.3.3 Junction with the Solutions Valid in the Convective Core

The convective core is considered to be in an isentropic (adiabatic) equilibrium, which is expressed by the equality

$$\frac{d\rho}{dr} = -\rho\,\frac{g}{c^2}, \tag{16.56}$$

so that $N^2(r) = 0$ and $K_1(r) = 0$. Consequently, the system of Eqs. (16.2) and (16.3) does contain no term of the order $1/\varepsilon^2$ in the core. At order ε^0, the system takes the form

$$\left.\begin{aligned} &\frac{d^2\alpha}{dr^2} + K_2(r)\,\frac{d\alpha}{dr} + K_3(r)\,\alpha = -K_4(r)\,\frac{d\xi}{dr}, \\ &\frac{d^2\xi}{dr^2} + \frac{4}{r}\,\frac{d\xi}{dr} - \frac{\ell(\ell+1)-2}{r^2}\,\xi = \frac{d\alpha}{dr} + \frac{2}{r}\,\alpha. \end{aligned}\right\} \tag{16.57}$$

The variations of the pressure and the mass density to which an isentropically moving mass element is subject in a region in isentropic equilibrium, result exclusively from the stratification of the equilibrium quantities in the radial direction. Therefore, the Lagrangian perturbations of the pressure and the mass density are given by

$$\delta P = \frac{dP}{dr}\,\xi, \quad \delta\rho = \frac{d\rho}{dr}\,\xi. \tag{16.58}$$

Since

$$\alpha = -\frac{\delta\rho}{\rho} = -\frac{\delta P}{\rho\, c^2}, \tag{16.59}$$

the divergence and the radial component of the Lagrangian displacement are related to each other as

$$\alpha = \frac{g}{c^2}\,\xi. \tag{16.60}$$

Equations (16.57) are satisfied by this relation, as can be verified by means of Poisson's equation (2.14) and Eq. (16.56).

By elimination of $\xi(r)$, the first Eq. (16.57) is transformed into a second-order differential equation for $\alpha(r)$:

$$\frac{d^2\alpha}{dr^2} + \left[K_2(r) + K_4(r)\,\frac{c^2}{g}\right]\frac{d\alpha}{dr} + \left[K_3(r) + K_4(r)\,\frac{d}{dr}\,\frac{c^2}{g}\right]\alpha = 0. \tag{16.61}$$

This equation admits of a particular solution that satisfies the requirement that the divergence of the Lagrangian displacement be finite at the boundary point $r = 0$. Be $\mu^{(c)}(\varepsilon)\,\alpha_0^{(c)}(r)$ this admissible solution, where $\mu^{(c)}(\varepsilon)$ is a yet undetermined function.

At the boundary between the convective core and the radiative envelope, the Lagrangian perturbation of the pressure, or, equivalently, the divergence of the Lagrangian displacement, and the Lagrangian displacement must be continuous. Because of the second condition, both the radial component, $\xi(r)$, and the transverse component, $\eta(r)$, of the Lagrangian displacement must be continuous. By virtue of

Eq. (5.93), these conditions imply that the first derivative of the radial component of the Lagrangian displacement, $d\xi/dr$, must also be continuous. The conditions imposed are therefore the continuity of the divergence of the Lagrangian displacement and that of the radial component of the Lagrangian displacement and its first derivative.

First, the continuity of the divergence of the Lagrangian displacement requires that

$$\mu^{(c)}(\varepsilon)\,\alpha_0^{(c)}(r_a) = \varepsilon^{-1/6}\,\frac{3^{1/3}}{\Gamma(2/3)\,K_{1,a}^{1/6}}\,B_{0,a}. \tag{16.62}$$

It follows that

$$\mu^{(c)}(\varepsilon) = \varepsilon^{-1/6}. \tag{16.63}$$

Furthermore, the arbitrary constant involved in the solution $\alpha_0^{(c)}(r)$ is fixed in terms of the constant $B_{0,a}$.

Next, the continuity of the radial component of the Lagrangian displacement requires that

$$\frac{c^2(r_a)}{g(r_a)}\,\alpha_0^{(c)}(r_a) = \frac{c^2(r_a)}{g(r_a)}\,\frac{3^{1/3}}{\Gamma(2/3)\,K_{1,a}^{1/6}}\,B_{0,a} + G_0^{(o)}(r_a)\,. \tag{16.64}$$

Because of the first condition, the function $G_0^{(o)}(r)$ must vanish at $r = r_a$, i.e.,

$$G_0^{(o)}(r_a) = 0, \tag{16.65}$$

so that, because of the third equality (16.55),

$$D_{0,a} = 0. \tag{16.66}$$

Thirdly, the continuity of the first derivative of the radial component of the Lagrangian displacement requires that

$$\frac{d\xi_0^{(a)}}{dr} = \varepsilon^{-5/6}\,\frac{3\,\Gamma(2/3)\,(3\,K_{1,a})^{1/6}}{2\,\pi}\,\frac{c^2(r_a)}{g(r_a)}\,A_{0,a}. \tag{16.67}$$

Since the left-hand member is of the order $\varepsilon^{-1/6}$, and the right-hand member, of the order $\varepsilon^{-5/6}$, one must set

$$A_{0,a} = 0. \tag{16.68}$$

Consequently, the outward boundary-layer solutions constructed from the lower boundary of the radiative envelope, given by Eqs. (16.47), involve only the arbitrary constant $B_{0,\mathrm{a}}$. Moreover, Eqs. (16.53) reduce to

$$\left.\begin{aligned} A_0^* &= B_{0,\mathrm{a}} \frac{\sqrt{3}}{\sqrt{\pi}\, K_{1,\mathrm{a}}^{1/4}\, K_{5,\mathrm{a}}} \cos\frac{\pi}{12}, \\ B_0^* &= B_{0,\mathrm{a}} \frac{\sqrt{3}}{\sqrt{\pi}\, K_{1,\mathrm{a}}^{1/4}\, K_{5,\mathrm{a}}} \sin\frac{\pi}{12}, \end{aligned}\right\} \tag{16.69}$$

so that the constants A_0^* and B_0^* are related to the constant $B_{0,\mathrm{a}}$.

16.3.4 Boundary-Layer Solutions Near the Boundary Point $r = R$

The region near the boundary point $r = R$ is treated as a small boundary layer, since the point is a singular point of Eqs. (16.32) and (16.33). As in Sect. 14.6, it is appropriate to introduce the coordinate

$$z = R - r \tag{16.70}$$

and to express the governing equations in terms of this coordinate. It is also appropriate to suppose that the mass density is analytical at the boundary point $r = R$ and tends to zero as z^{n_e}, with $n_\mathrm{e} > 0$, and to set $m(r) \simeq M$. Then power series (6.53)–(6.57) are valid, and the coefficients $K_1(r)$, $K_2(r)$, $K_3(r)$, $K_4(r)$ have power series of the form given by Eqs. (14.82). For the coefficient $K_5(r)$, one has

$$K_5(r) = K_{5,\mathrm{s}}\, z^{-(n_\mathrm{e}+3/2)/2}\, [1 + O(z)]. \tag{16.71}$$

The boundary-layer coordinate is now defined as

$$z^*(z) = \frac{1}{\varepsilon}\, 2\, K_{1,\mathrm{s}}^{1/2}\, z^{1/2}, \tag{16.72}$$

and asymptotic expansions of the form

$$\left.\begin{aligned} \alpha^{(\mathrm{s})}(r;\varepsilon) &= \mu_0^{(\mathrm{s})}(\varepsilon)\, \alpha_0^{(\mathrm{s})}(z^*) + \mu_1^{(\mathrm{s})}(\varepsilon)\, \alpha_1^{(\mathrm{s})}(z^*) + \dots, \\ w^{(\mathrm{s})}(r;\varepsilon) &= \nu_0^{(\mathrm{s})}(\varepsilon)\, w_0^{(\mathrm{s})}(z^*) + \nu_1^{(\mathrm{s})}(\varepsilon)\, w_1^{(\mathrm{s})}(z^*) + \dots \end{aligned}\right\} \tag{16.73}$$

are substituted into the equations.

Equation (16.32) can then be written in the form

$$\mu_0^{(s)}(\varepsilon)\left[\frac{1}{\varepsilon^4}\left(\frac{d^2\alpha_0^{(s)}}{dz^{*2}}+\frac{2n_e+3}{z^*}\frac{d\alpha_0^{(s)}}{dz^*}+\alpha_0^{(s)}\right)+O\left(\varepsilon^{-2}\right)\right]$$

$$+\mu_1^{(s)}(\varepsilon)\,[\ldots]+\ldots=\nu_0^{(s)}(\varepsilon)\left[O\left(\varepsilon^{-2}\right)\right]+\nu_1^{(s)}(\varepsilon)[\ldots]+\ldots \quad (16.74)$$

The first dominant boundary-layer equation is homogeneous when $\nu_0^{(s)}(\varepsilon)$ is of a higher order in ε than $\mu_0^{(s)}(\varepsilon)\,\varepsilon^{-2}$. It is given by

$$\frac{d^2\alpha_0^{(s)}}{dz^{*2}}+\frac{2n_e+3}{z^*}\frac{d\alpha_0^{(s)}}{dz^*}+\alpha_0^{(s)}=0 \quad (16.75)$$

and admits of the solution that satisfies the requirement that the divergence of the Lagrangian displacement be finite at $r=R$

$$\alpha_0^{(s)}\left(z^*\right)=A_{0,s}\,z^{*-(n_e+1)}\,J_{n_e+1}\left(z^*\right), \quad (16.76)$$

where $A_{0,s}$ is an arbitrary constant.

Next, Eq. (16.33) can be written in the form

$$\nu_0^{(s)}(\varepsilon)\left[\frac{1}{\varepsilon^4}\left(\frac{d^2w_0^{(s)}}{dz^{*2}}+\frac{3}{z^*}\frac{dw_0^{(s)}}{dz^*}\right)+O\left(\varepsilon^{-2}\right)\right]+\nu_1^{(s)}(\varepsilon)\,[\ldots]+\ldots$$

$$=\mu_0^{(s)}(\varepsilon)\left[-\frac{1}{\varepsilon^4}\left(2\,\frac{g_s}{c_s^2}\,\frac{1}{z^*}\,\frac{d\alpha_0^{(s)}}{dz^*}+\alpha_0^{(s)}\right)+O\left(\varepsilon^{-2}\right)\right]+\mu_1^{(s)}(\varepsilon)\,[\ldots]+\ldots \quad (16.77)$$

The second dominant boundary-layer equation is inhomogeneous when

$$\nu_0^{(s)}(\varepsilon)=\mu_0^{(s)}(\varepsilon). \quad (16.78)$$

It takes the form

$$\frac{d^2w_0^{(s)}}{dz^{*2}}+\frac{3}{z^*}\frac{dw_0^{(s)}}{dz^*}=-2\,\frac{g_s}{c_s^2}\,\frac{1}{z^*}\,\frac{d\alpha_0^{(s)}}{dz^*}-\alpha_0^{(s)}. \quad (16.79)$$

By subtracting the first dominant boundary-layer equation and introducing the function

$$w_0^*\left(z^*\right) = w_0^{(\mathrm{s})}\left(z^*\right) - \alpha_0^{(\mathrm{s})}\left(z^*\right), \tag{16.80}$$

one obtains the inhomogeneous differential equation

$$\frac{d^2 w_0^*}{d z^{*2}} + \frac{3}{z^*}\frac{d w_0^*}{d z^*} = 2\,\frac{N_\mathrm{s}^2}{g_\mathrm{s}}\,\frac{1}{z^*}\,\frac{d\alpha_0^{(\mathrm{s})}}{d z^*}. \tag{16.81}$$

A general solution of it is given by

$$\begin{aligned} w_0^*\left(z^*\right) &= C_{0,\mathrm{s}}\, z^{*-2} + D_{0,\mathrm{s}} \\ &\quad - \frac{N_\mathrm{s}^2}{g_\mathrm{s}}\left[z^{*-2}\int_0^{z^*} z'^2\,\frac{d\alpha_0^{(\mathrm{s})}\left(z'\right)}{d z'}\,d z' - \alpha_0^{(\mathrm{s})}\left(z^*\right)\right], \end{aligned} \tag{16.82}$$

where $C_{0,\mathrm{s}}$ and $D_{0,\mathrm{s}}$ are arbitrary constants.

After partial integration and use of the recurrence relation between Bessel functions

$$z'^{-n_\mathrm{e}}\, J_{n_\mathrm{e}+1}\left(z'\right) = -\frac{d}{d z'}\left[z'^{-n_\mathrm{e}}\, J_{n_\mathrm{e}}\left(z'\right)\right] \tag{16.83}$$

(Abramowitz & Stegun 1965), the solution becomes

$$w_0^*\left(z^*\right) = -2\,\frac{N_\mathrm{s}^2}{g_\mathrm{s}}\,A_{0,\mathrm{s}}\, z^{*-(n_\mathrm{e}+2)}\, J_{n_\mathrm{e}}\left(z^*\right) + C_{0,\mathrm{s}}\, z^{*-2} + D_{0,\mathrm{s}}. \tag{16.84}$$

By use of the second recurrence relation between Bessel functions

$$J_{n_\mathrm{e}}\left(z^*\right) = \frac{d J_{n_\mathrm{e}+1}\left(z^*\right)}{d z^*} + \frac{n_\mathrm{e}+1}{z^*}\, J_{n_\mathrm{e}+1}\left(z^*\right), \tag{16.85}$$

the solution can be rewritten as

$$\begin{aligned} w_0^*\left(z^*\right) &= C_{0,\mathrm{s}}\, z^{*-2} + D_{0,\mathrm{s}} \\ &\quad - \alpha_0^{(\mathrm{s})}\left(z^*\right) 2\,\frac{N_\mathrm{s}^2}{g_\mathrm{s}}\,\frac{1}{z^{*2}}\left[\frac{d \ln J_{n_\mathrm{e}+1}\left(z^*\right)}{d \ln z^*} + \left(n_\mathrm{e}+1\right)\right]. \end{aligned} \tag{16.86}$$

One then gets

$$w_0^{(s)}\left(z^*\right) = C_{0,s}\, z^{*-2} + D_{0,s} + \alpha_0^{(s)}\left(z^*\right)\left\{1 - 2\,\frac{N_s^2}{g_s}\,\frac{1}{z^{*2}}\left[\frac{d\ln J_{n_e+1}\left(z^*\right)}{d\ln z^*} + (n_e+1)\right]\right\}. \quad (16.87)$$

In view of the matching with the two-variable solutions valid at larger distances from the boundary point $r = R$, the boundary-layer solutions are written in the more general form

$$\left.\begin{aligned} \alpha^{(s)}(r;\varepsilon) &= \mu_0^{(s)}(\varepsilon)\, A_{0,s}\, z^{*-(n_e+1)}\, J_{n_e+1}\left(z^*\right), \\ \xi^{(s)}(r;\varepsilon) &= \frac{c_s^2}{g_s}\, z \left\{\alpha^{(s)}(r;\varepsilon)\right. \\ &\quad \left\{1 - 2\,\frac{N_s^2}{g_s}\,\frac{1}{z^{*2}}\left[\frac{d\ln J_{n_e+1}\left(z^*\right)}{d\ln z^*} + (n_e+1)\right]\right\} \\ &\quad \left. + \nu_0^{(s,2)}(\varepsilon)\, C_{0,s}\, z^{*-2} + \nu_0^{(s,3)}(\varepsilon)\, D_{0,s}\right\}. \end{aligned}\right\} \quad (16.88)$$

The matching condition relative to the divergence of the Lagrangian displacement is

$$\lim_{z\to\infty} \alpha^{(s)}(r;\varepsilon) = \lim_{z\to 0} \alpha^{(o)}(r;\varepsilon). \quad (16.89)$$

In the two-variable solution $\alpha^{(o)}(r;\varepsilon)$, the fast variable $\tau_s(r)$ is introduced as

$$\tau_s(r) \equiv \tau_R - \tau(r) = \frac{1}{\varepsilon}\int_0^z K_1^{1/2}\left(r'\right) dz' \quad (16.90)$$

with $\tau_R = \tau(R)$. As $z \to 0$, $\tau_s(r) = z^*$, so that

$$\lim_{z\to 0} \alpha^{(o)}(r;\varepsilon) = K_{5,s}\, z^{-(n_e+3/2)/2} \left[\left(A_0^* \cos\tau_R + B_0^* \sin\tau_R\right)\cos z^* + \left(A_0^* \sin\tau_R - B_0^* \cos\tau_R\right)\sin z^*\right]. \quad (16.91)$$

On the other hand, for the boundary-layer solution $\alpha^{(s)}(r;\varepsilon)$, one has that

$$\lim_{z\to\infty} \alpha^{(s)}(r;\varepsilon) = \mu_0^{(s)}(\varepsilon)\, \varepsilon^{n_e+3/2}\, A_{0,s}\left(\frac{2}{\pi}\right)^{1/2} \left(2\,K_{1,s}^{1/2}\right)^{-(n_e+3/2)} z^{-(n_e+3/2)/2} \sin\left[z^* - \left(n_e + \frac{1}{2}\right)\frac{\pi}{2}\right]. \quad (16.92)$$

The matching condition is satisfied when

$$\mu_0^{(s)}(\varepsilon) = \varepsilon^{-(n_e+3/2)} \tag{16.93}$$

and

$$\left.\begin{aligned} A_0^* &= A_{0,s}\, F \sqrt{\frac{2}{\pi}}\, \sin\left[\tau_R - \left(n_e + \frac{1}{2}\right)\frac{\pi}{2}\right], \\ B_0^* &= -A_{0,s}\, F \sqrt{\frac{2}{\pi}}\, \cos\left[\tau_R - \left(n_e + \frac{1}{2}\right)\frac{\pi}{2}\right], \end{aligned}\right\} \tag{16.94}$$

with

$$F = \left(2\, K_{1,s}^{1/2}\right)^{-(n_e+3/2)} K_{5,s}^{-1}. \tag{16.95}$$

Next, the matching condition relative to the function $w(r)$ is

$$\lim_{z\to\infty} w^{(s)}(r;\varepsilon) = \lim_{z\to 0} w^{(o)}(r;\varepsilon). \tag{16.96}$$

The oscillatory parts of the functions $w^{(s)}(r;\varepsilon)$ and $w^{(o)}(r;\varepsilon)$ are matched by the matching of the functions $\alpha^{(s)}(r;\varepsilon)$ and $\alpha^{(o)}(r;\varepsilon)$. The non-oscillatory parts are matched, when

$$\left.\begin{aligned} &D_{0,s} = 0, \\ &\nu_0^{(s,2)}(\varepsilon) = \varepsilon^{-2}, \quad C_{0,s} = \frac{g_s}{c_s^2}\, 4\, K_{1,s}\, G_0^{(o)}(R). \end{aligned}\right\} \tag{16.97}$$

16.3.5 Eigenfrequency Equation

The constants A_0^* and B_0^* are involved in Eqs. (16.69) and (16.94). These equations result from the matchings of the two-variable solution for $\alpha(r)$, valid in the radiative envelope at larger distances from its boundaries, with the boundary-layer solutions for that function, valid respectively near the boundary of the convective core and near the star's surface. Elimination of the constants leads to the system of two algebraic, linear, homogeneous equations for the constants $B_{0,a}$ and $A_{0,s}$

$$\left.\begin{aligned}
&B_{0,\mathrm{a}} \frac{\sqrt{3}}{\sqrt{\pi}\,K_{1,\mathrm{a}}^{1/4}\,K_{5,\mathrm{a}}} \cos\frac{\pi}{12} \\
&\qquad -A_{0,\mathrm{s}}\,F \sqrt{\frac{2}{\pi}} \sin\left[\tau_{\mathrm{R}} - \left(n_{\mathrm{e}} + \frac{1}{2}\right)\frac{\pi}{2}\right] = 0, \\
&B_{0,\mathrm{a}} \frac{\sqrt{3}}{\sqrt{\pi}\,K_{1,\mathrm{a}}^{1/4}\,K_{5,\mathrm{a}}} \sin\frac{\pi}{12} \\
&\qquad +A_{0,\mathrm{s}}\,F \sqrt{\frac{2}{\pi}} \cos\left[\tau_{\mathrm{R}} - \left(n_{\mathrm{e}} + \frac{1}{2}\right)\frac{\pi}{2}\right] = 0.
\end{aligned}\right\} \tag{16.98}$$

The condition for the system of equations to admit of a non-trivial solution is

$$\cos\left[\tau_R - \left(n_{\mathrm{e}} + \frac{1}{2}\right)\frac{\pi}{2} - \frac{\pi}{12}\right] = 0. \tag{16.99}$$

From this condition, the eigenfrequency equation follows

$$\tau_R - \left(n_{\mathrm{e}} + \frac{1}{2}\right)\frac{\pi}{2} - \frac{\pi}{12} = (2n-1)\,\frac{\pi}{2}, \quad n = 1, 2, 3, \ldots \tag{16.100}$$

or, more explicitly,

$$\begin{aligned}
\tau_R &\equiv \frac{[\ell(\ell+1)]^{1/2}}{|\sigma|} \int_{r_{\mathrm{a}}}^{R} \left(\frac{N^2(r)}{r^2}\right)^{1/2} dr \\
&= \left(2n + n_{\mathrm{e}} - \frac{1}{3}\right)\frac{\pi}{2}, \quad n = 1, 2, 3, \ldots
\end{aligned} \tag{16.101}$$

In Sect. 16.3.8, it is shown that the number n corresponds to the radial order of the g^+-mode considered.

Eigenfrequency equation (16.100) corresponds to the eigenfrequency equation established by Willems et al. (1997). It may be noted that eigenfrequency equation (A.12) of Tassoul (1980), which is derived in the Cowling approximation, agrees with eigenfrequency equation (16.100) derived from the full fourth-order system of equations.

Thus far, all constants involved in the asymptotic solutions have been determined, apart from the constant $C_{0,\mathrm{s}}$ and the two constants involved in the function $G_0^{(\mathrm{o})}(r)$. It is shown hereafter that the constant $C_{0,\mathrm{s}}$ is equal to zero. From Eq. (16.65) and the third Eq. (16.97), it then follows that the function $G_0^{(\mathrm{o})}(r)$ is identically zero.

16.3.6 The Condition on the Eulerian Perturbation of the Gravitational Potential at $r = R$

At the boundary point $r = R$, condition (5.97) about the Eulerian perturbation of the gravitational potential must be imposed. The Eulerian perturbation of the gravitational potential and its first derivative are related to the functions $\alpha(r)$ and $\xi(r)$ at an arbitrary point by Eqs. (14.122) and (14.123).

From boundary-layer solutions (16.88), it results that the divergence and the radial component of the Lagrangian displacement at $r = R$ are given by

$$\left.\begin{aligned} \alpha_R &= \varepsilon^{-(n_e+3/2)}\, A_{0,s}\, \frac{2^{-(n_e+1)}}{\Gamma\,(n_e+2)}, \\ \xi_R &= -\frac{c_s^2}{g_s}\,\frac{1}{4\,K_{1,s}} \left[\varepsilon^{-(n_e-1/2)}\, A_{0,s}\, \frac{2^{-(n_e-1)}}{\Gamma\,(n_e+1)}\,\frac{N_s^2}{g_s} - C_{0,s} \right], \end{aligned}\right\} \tag{16.102}$$

and their first derivatives at that point, by

$$\left.\begin{aligned} \left(\frac{d\alpha}{dr}\right)_R &= \varepsilon^{-(n_e+7/2)}\, A_{0,s}\, \frac{2^{-(n_e+1)}}{\Gamma\,(n_e+3)}\, K_{1,s}, \\ \left(\frac{d\xi}{dr}\right)_R &= -\varepsilon^{-(n_e+3/2)}\, A_{0,s}\, \frac{2^{-(n_e+1)}}{\Gamma\,(n_e+2)}\,\frac{c_s^2}{g_s} \left(\frac{N_s^2}{g_s}+1\right). \end{aligned}\right\} \tag{16.103}$$

In the boundary condition, terms involving the constant $A_{0,s}$ as well as terms involving the constant $C_{0,s}$ appear. The leading order in ε of the terms involving the constant $A_{0,s}$ is $-\,(n_e + 3/2)$, while the terms involving the constant $C_{0,s}$ are of the order ε^0. The sum of the terms involving the constant $A_{0,s}$ that are of the leading order in ε is equal to zero, so that the boundary condition is automatically satisfied by the boundary-layer solution $\alpha^{(s)}(r;\varepsilon)$ and the oscillatory part of the boundary-layer solution $\xi^{(s)}(r;\varepsilon)$. For the boundary condition to be also satisfied by the terms involving the constant $C_{0,s}$, one must set

$$C_{0,s} = 0. \tag{16.104}$$

From the third equation (16.97), it follows that

$$G_0^{(o)}(R) = 0. \tag{16.105}$$

In combination with Eq. (16.65), it results that

$$C_0^* = 0, \quad D_0^* = 0 \tag{16.106}$$

in solution (16.29) for the function $G_0^{(o)}(r)$, so that this function is identically zero at all points of the radiative envelope. Hence, the asymptotic solution for $\xi(r)$ is purely oscillatory in the whole radiative envelope, as well as the asymptotic solution for $\alpha(r)$.

16.3.7 Uniformly Valid Asymptotic Solutions

At this stage, all constants involved in the various asymptotic solutions are determined. Therefore, uniformly valid first-order asymptotic solutions for the divergence and the radial component of the Lagrangian displacement can be constructed in the radiative envelope. These uniformly valid asymptotic solutions are presented in terms of the constants $B_{0,\mathrm{a}}$ and $A_{0,\mathrm{s}}$ involved in Eqs. (16.98).

The first-order asymptotic solutions that are uniformly valid from the boundary between the convective core and the radiative envelope to a distance sufficiently large from the star's surface take the form

$$\left.\begin{aligned} \alpha^{(\mathrm{a,u})}(r;\varepsilon) = B_{0,\mathrm{a}} \Bigg[& \varepsilon^{-1/2}\sqrt{s_\mathrm{a}}\, J_{-1/3}\left(\frac{1}{\varepsilon}\frac{2}{3}\sqrt{K_{1,\mathrm{a}}}\, s_\mathrm{a}^{3/2}\right) \\ & + \frac{\sqrt{3}}{\sqrt{\pi}\, K_{1,\mathrm{a}}^{1/4}\, K_{5,\mathrm{a}}}\, K_5(r)\cos\left(\tau - \frac{\pi}{12}\right) \\ & - \frac{\sqrt{3}}{\sqrt{\pi}\, K_{1,\mathrm{a}}^{1/4}}\, \frac{1}{s_\mathrm{a}^{1/4}} \cos\left(\frac{1}{\varepsilon}\frac{2}{3}\sqrt{K_{1,\mathrm{a}}}\, s_\mathrm{a}^{3/2} - \frac{\pi}{12}\right)\Bigg], \\ \xi^{(\mathrm{a,u})}(r;\varepsilon) = {} & \frac{c^2(r)}{g(r)}\,\alpha^{(\mathrm{a,u})}(r;\varepsilon). \end{aligned}\right\} \tag{16.107}$$

The uniformly valid asymptotic solution $\alpha^{(\mathrm{a,u})}(r;\varepsilon)$ can be expressed in the compact form

$$\alpha^{(\mathrm{a,u})}(r;\varepsilon) = B_{0,\mathrm{a}}\, \frac{\sqrt{3}}{\sqrt{2}\, K_{1,\mathrm{a}}^{1/4}\, K_{5,\mathrm{a}}}\, K_5(r)\, \tau^{1/2}\, J_{-1/3}(\tau), \tag{16.108}$$

which corresponds to the first-order asymptotic solution given by Willems et al. (1997).

Next, the first-order asymptotic solutions for the divergence and the radial component of the Lagrangian displacement that are uniformly valid from the star's surface to a distance sufficiently large from the boundary between the convective core and the radiative envelope take the form

$$\left.\begin{aligned}
\alpha^{(s,u)}(r;\varepsilon) &= A_{0,s}\Bigg\{\varepsilon^{-(n_e+3/2)}\, z^{*-(n_e+1)}\, J_{n_e+1}\left(z^*\right) \\
&\quad + F\, K_5(r)\sqrt{\frac{2}{\pi}}\,\sin\left[\tau_s - \left(n_e+\frac{1}{2}\right)\frac{\pi}{2}\right] \\
&\quad - F\, K_{5,s}\, z^{-(n_e+3/2)/2} \\
&\qquad \sqrt{\frac{2}{\pi}}\,\sin\left[\frac{1}{\varepsilon}\,2\,K_{1,s}^{1/2}\, z^{1/2} - \left(n_e+\frac{1}{2}\right)\frac{\pi}{2}\right]\Bigg\}, \\
\xi^{(s,u)}(r;\varepsilon) &= A_{0,s}\Bigg\{\varepsilon^{-(n_e+3/2)}\,\frac{c_s^2}{g_s}\, z\, z^{*-(n_e+1)}\, J_{n_e+1}\left(z^*\right) \\
&\quad \left\{1 - 2\,\frac{N_s^2}{g_s}\,\frac{1}{z^{*2}}\left[\frac{d\ln J_{n_e+1}\left(z^*\right)}{d\ln z^*} + (n_e+1)\right]\right\} \\
&\quad + F\,\frac{c^2(r)}{g(r)}\, K_5(r)\sqrt{\frac{2}{\pi}}\,\sin\left[\tau_s - \left(n_e+\frac{1}{2}\right)\frac{\pi}{2}\right] \\
&\quad - F\,\frac{c_s^2}{g_s}\, K_{5,s}\, z^{-(n_e+1/2)/2} \\
&\qquad \sqrt{\frac{2}{\pi}}\,\sin\left[\frac{1}{\varepsilon}\,2\,K_{1,s}^{1/2}\, z^{1/2} - \left(n_e+\frac{1}{2}\right)\frac{\pi}{2}\right]\Bigg\}.
\end{aligned}\right\} \tag{16.109}$$

These uniformly valid asymptotic solutions can be expressed in the compact forms

$$\left.\begin{aligned}
\alpha^{(s,u)}(r;\varepsilon) &= A_{0,s}\, F\, K_5(r)\,\tau_s^{1/2}\, J_{n_e+1}\left(\tau_s\right), \\
\xi^{(s,u)}(r;\varepsilon) &= \frac{c^2(r)}{g(r)}\,\alpha^{(s,u)}(r;\varepsilon) \\
&\quad \left\{1 - 2\,\frac{N_s^2}{g_s}\,\frac{1}{\tau_s^2}\left[\frac{d\ln J_{n_e+1}\left(\tau_s\right)}{d\ln \tau_s} + (n_e+1)\right]\right\},
\end{aligned}\right\} \tag{16.110}$$

which correspond to the asymptotic solutions given by Willems et al. (1997).

16.3.8 *Identification of the Radial Order of a g^+-Mode with a given Eigenfrequency*

According to Cowling's classification, the radial order of a g^+-mode corresponds to the number of nodes the eigenfunction $\xi(r)$ displays between the boundary points $r = 0$ and $r = R$. In the case of a star composed of a convective core and a radiative envelope, the nodes of the asymptotic solution are situated in the radiative envelope.

For the determination of the number of nodes, it is convenient to start from the two-variable solution $\xi^{(o)}(r;\varepsilon)$, which is given by the second equation (16.30) and is valid between the lower boundary of the radiative envelope and the star's surface at distances sufficiently large from these boundaries. By means of Eqs. (16.69) for the constants A_0^* and B_0^*, this solution is transformed into

$$\xi^{(o)}(r;\varepsilon) = B_{0,a}\,\frac{\sqrt{3}}{\sqrt{\pi}\,K_{1,a}^{1/4}\,K_{5,a}}\,\frac{c^2(r)}{g(r)}\,K_5(r)\,\cos\left(\tau - \frac{\pi}{12}\right), \tag{16.111}$$

so that the positions of the nodes are given by

$$\tau^0 = \left(2j - \frac{5}{6}\right)\frac{\pi}{2}, \quad j = 1, 2, 3, \ldots \tag{16.112}$$

The node associated with $j = 1$ is located at $\tau^0 = 1.83$, while the first node of the asymptotic solution $\xi^{(a,u)}(r;\varepsilon)$ given by Eqs. (16.107) and (16.108), which is uniformly valid from the boundary between the convective core and the radiative envelope, is determined by the first zero of the Bessel function $J_{-1/3}(\tau)$. This zero is situated at $\tau = 1.87$. Hence, the node of the two-variable solution $\xi^{(o)}(r;\varepsilon)$ that is associated with $j = 1$ corresponds to the first node of the uniformly valid solution $\xi^{(a,u)}(r;\varepsilon)$ counted from the boundary between the convective core and the radiative envelope.

Next, the two-variable solution $\xi^{(o)}(r;\varepsilon)$ is transformed by means of Eqs. (16.94) for A_0^* and B_0^* into

$$\xi^{(o)}(r;\varepsilon) = A_{0,s}\,F\,\sqrt{\frac{2}{\pi}}\,\frac{c^2(r)}{g(r)}\,K_5(r)\,\sin\left[\tau_s - \left(n_e + \frac{1}{2}\right)\frac{\pi}{2}\right]. \tag{16.113}$$

For instance, for the case $n_e = 3$, the positions of the nodes of the two-variable solution $\xi^{(o)}(r;\varepsilon)$ in terms of τ_s are given by

$$\tau_s^0 = \left(2k - \frac{1}{2}\right)\frac{\pi}{2}, \quad k = 1, 2, 3, \ldots \tag{16.114}$$

Near $\tau_s = 0$, these positions must be compared with the positions of the nodes of the asymptotic solution $\xi^{(s,u)}(r;\varepsilon)$, which is uniformly valid from $r = R$ and is given by the second equation (16.110). To this end, the part of the two-variable solution $\xi^{(o)}(r;\varepsilon)$ that depends on the fast variable τ_s, i.e.,

$$H_2(\tau_s) \equiv \sqrt{\frac{2}{\pi}} \sin\left(\tau_s - \frac{7\pi}{4}\right),$$

is compared with that of the uniformly valid solution $\xi^{(s,u)}(r;\varepsilon)$, i.e.,

$$H_1(\tau_s) \equiv \tau_s^{1/2} J_4(\tau_s) \left\{1 - 2\frac{N_s^2}{g_s}\frac{1}{\tau_s^2}\left[\frac{d \ln J_4(\tau_s)}{d \ln \tau_s} + 4\right]\right\}.$$

Both parts are represented in Fig. 16.1 as functions of τ_s, for $\Gamma_1 = 5/3$. From the figure, it appears that the node of the two-variable solution $\xi^{(o)}(r;\varepsilon)$ that is associated with $k = 2$ and is located at $\tau_s = 7\pi/4$ must be related to the first node of the uniformly valid asymptotic solution $\xi^{(s,u)}(r;\varepsilon)$ counted from $\tau_s = 0$, i.e.,

$$\tau_s^0(\text{first}) = \frac{7\pi}{4}.$$

This node corresponds to the last node counted from the boundary between the convective core and the radiative envelope, so that it is equivalently located at

$$\tau^0(\text{last}) = \tau_R - \tau_s^0(\text{first}) = \left(2n - \frac{5}{6}\right)\frac{\pi}{2}.$$

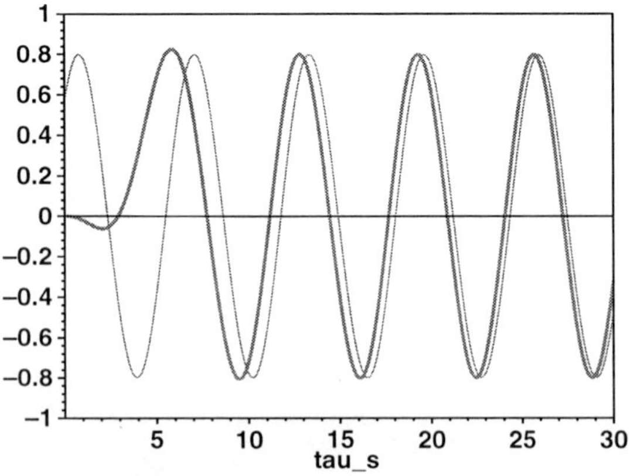

Fig. 16.1 The functions $H_1(\tau_s)$ (*thicker line*) and $H_2(\tau_s)$ (*thinner line*) for $n_e = 3$ and $\Gamma_1 = 5/3$

By comparison with Eq. (16.112), it follows that the last node of the uniformly valid asymptotic solution $\xi^{(s,u)}(r;\varepsilon)$, counted from the boundary between the convective core and the radiative envelope, has the number n. The asymptotic solution for the radial component of the Lagrangian displacement thus displays n nodes between the lower boundary of the radiative envelope and the star's surface, so that the g^+-mode considered has the radial order n.

The number of nodes of the asymptotic solution for the divergence of the Lagrangian displacement can be determined in a similar way. The nodes of the asymptotic solution $\alpha^{(a,u)}(r;\varepsilon)$, which is uniformly valid from $r = r_a$, coincide with those of the uniformly valid asymptotic solution $\xi^{(a,u)}(r;\varepsilon)$. Next, in order to relate the nodes of the two-variable solution $\alpha^{(o)}(r;\varepsilon)$ to those of the solution $\alpha^{(s,u)}(r;\varepsilon)$, which is given by the first Eq. (16.110) and is uniformly valid from $r = R$, one must compare the function $H_2(\tau_s)$ with the function

$$H_3(\tau_s) = \tau_s^{1/2} J_4(\tau_s).$$

From the representation of these functions in Fig. 16.2, it appears that the node of the two-variable solution $\alpha^{(o)}(r;\varepsilon)$ that is associated with $k = 3$ in Eq. (16.114) is related to the first node of the uniformly valid asymptotic solution $\alpha^{(s,u)}(r;\varepsilon)$ counted from $r = R$. Consequently, compared with that for the radial component of the Lagrangian displacement, the asymptotic solution for the divergence of the Lagrangian displacement displays one node less, i.e. $(n - 1)$ nodes, between the lower boundary of the radiative envelope and the star's surface. The additional node of the radial component is located close to the star's surface at the point whose coordinate τ_s is solution of the equation

$$1 - 2\,\frac{N_s^2}{g_s}\,\frac{1}{\tau_s^2}\left[\frac{d\ln J_4(\tau_s)}{d\ln\tau_s} + 4\right] = 0.$$

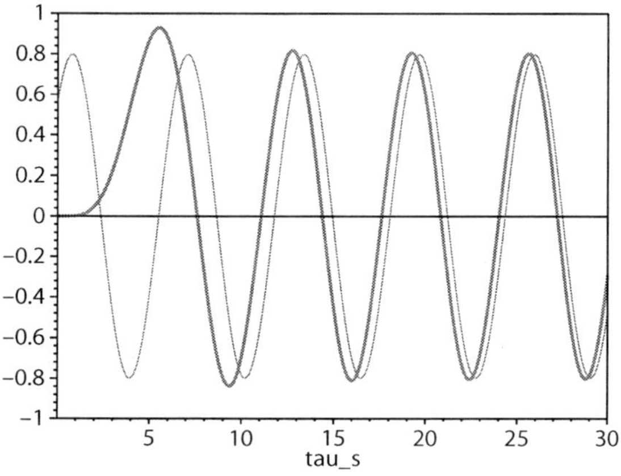

Fig. 16.2 The functions $H_3(\tau_s)$ (*thicker line*) and $H_2(\tau_s)$ (*thinner line*) for $n_e = 3$

16.4 Stars Consisting of a Convective Core, an Intermediate Radiative Zone, and a Convective Envelope

For stars consisting of a convective core, an intermediate radiative zone, and a convective envelope, a second turning point appears in Eq. (16.2) at the point $r = r_b$ corresponding to the radial distance of the boundary between the radiative zone and the convective envelope. The asymptotic solutions relative to the radiative envelope that are constructed in the previous section remain valid, except those constructed from the surface.

16.4.1 Boundary-Layer Solutions on the Inner Side of the Boundary Between the Intermediate Radiative Zone and the Convective Envelope

In the boundary-layer near the boundary between the intermediate radiative zone and the convective envelope, on the side of the radiative zone, it is appropriate to use the coordinate

$$s_b = r_b - r. \tag{16.115}$$

In the supposition that the coefficients of Eqs. (16.32) and (16.33) behave in a similar way as on the outer side of the turning point $r = r_a$, it suffices to replace the subscript "a" by the subscript "b" in Taylor series (16.35) to get the appropriate Taylor series. Proceeding as in Sect. 16.3.2, one introduces the boundary-layer coordinate

$$s_b^*(r) = \frac{s_b(r)}{\varepsilon^{2/3}} \tag{16.116}$$

and derives the boundary-layer solutions

$$\left.\begin{aligned} \alpha^{(b)}(r;\varepsilon) &= \mu_0^{(b)}(\varepsilon)\,\sqrt{s_b^*}\left[A_{0,b}\,J_{1/3}\left(\frac{2}{3}\,\sqrt{K_{1,b}}\,s_b^{*3/2}\right)\right.\\ &\quad\left.+B_{0,b}\,J_{-1/3}\left(\frac{2}{3}\,\sqrt{K_{1,b}}\,s_b^{*3/2}\right)\right],\\ \xi^{(b)}(r;\varepsilon) &= \frac{c^2\,(r_b)}{g\,(r_b)}\\ &\quad\left[\alpha^{(b)}(r;\varepsilon) + \nu_0^{(b,2)}(\varepsilon)\,C_{0,b}\,s_b^* + \nu_0^{(b,3)}(\varepsilon)\,D_{0,b}\right], \end{aligned}\right\} \tag{16.117}$$

where $A_{0,b}$, $B_{0,b}$, $C_{0,b}$, $D_{0,b}$ are arbitrary constants, and $\mu_0^{(b)}(\varepsilon)$, $\nu_0^{(b,2)}(\varepsilon)$, $\nu_0^{(b,3)}(\varepsilon)$, yet undetermined functions.

In view of the matching of the boundary-layer solution $\alpha^{(b)}(r;\varepsilon)$ with the two-variable solution $\alpha^{(o)}(r;\varepsilon)$, valid at larger distances from the turning point, it is useful to observe that, for $s_b \to 0$, the fast variable $\tau(r)$, defined by Eq. (16.5), behaves as

$$\tau(r) = \tau_{\text{Rad}} - \frac{1}{\varepsilon}\frac{2}{3}\sqrt{K_{1,b}}\, s_b^{3/2} \tag{16.118}$$

with

$$\tau_{\text{Rad}} = \frac{1}{\varepsilon}\int_{r_a}^{r_b} K_1^{1/2}(r')\, dr', \tag{16.119}$$

so that

$$\lim_{s_b \to 0} \alpha^{(o)}(r,\varepsilon) = \frac{K_{5,b}}{s_b^{1/4}}\left[A_0^* \cos\left(\tau_{\text{Rad}} - \frac{1}{\varepsilon}\frac{2}{3}\sqrt{K_{1,b}}\, s_b^{3/2}\right)\right.$$
$$\left. + B_0^* \sin\left(\tau_{\text{Rad}} - \frac{1}{\varepsilon}\frac{2}{3}\sqrt{K_{1,b}}\, s_b^{3/2}\right)\right]. \tag{16.120}$$

On the other hand, for $s_b \to \infty$,

$$\lim_{s_b \to \infty} \alpha^{(b)}(r;\varepsilon) = \mu_0^{(b)}(\varepsilon)\,\varepsilon^{1/6}\frac{\sqrt{3}}{\sqrt{\pi}\,K_{1,b}^{1/4}}\frac{1}{s_b^{1/4}}\left[A_{0,b} \sin\left(\frac{1}{\varepsilon}\frac{2}{3}\sqrt{K_{1,b}}\, s_b^{3/2} + \frac{\pi}{12}\right)\right.$$
$$\left. + B_{0,b} \cos\left(\frac{1}{\varepsilon}\frac{2}{3}\sqrt{K_{1,b}}\, s_b^{3/2} - \frac{\pi}{12}\right)\right]. \tag{16.121}$$

The matching relative to the divergence of the Lagrangian displacement then leads to

$$\mu_0^{(b)}(\varepsilon) = \varepsilon^{-1/6} \tag{16.122}$$

and

$$\left.\begin{aligned}
A_0^* &= \frac{\sqrt{3}}{\sqrt{\pi}\,K_{1,b}^{1/4}\,K_{5,b}}\left[A_{0,b} \sin\left(\tau_{\text{Rad}} + \frac{\pi}{12}\right)\right. \\
&\quad \left. + B_{0,b} \cos\left(\tau_{\text{Rad}} - \frac{\pi}{12}\right)\right], \\
B_0^* &= -\frac{\sqrt{3}}{\sqrt{\pi}\,K_{1,b}^{1/4}\,K_{5,b}}\left[A_{0,b} \cos\left(\tau_{\text{Rad}} + \frac{\pi}{12}\right)\right. \\
&\quad \left. - B_{0,b} \sin\left(\tau_{\text{Rad}} - \frac{\pi}{12}\right)\right].
\end{aligned}\right\} \tag{16.123}$$

Moreover, the matching relative to the non-oscillatory parts in the radial component of the Lagrangian displacement leads to

$$\left.\begin{aligned} &C_{0,\mathrm{b}} = 0, \\ &\nu_0^{(\mathrm{b},3)}(\varepsilon) = \varepsilon^0, \quad D_{0,\mathrm{b}} = \frac{g\,(r_\mathrm{b})}{c^2\,(r_\mathrm{b})}\,G_0^{(\mathrm{o})}\,(r_\mathrm{b})\,. \end{aligned}\right\} \tag{16.124}$$

At the turning point $r = r_\mathrm{b}$, the divergence of the Lagrangian displacement, and the radial component of the Lagrangian displacement and its first derivative have the values

$$\left.\begin{aligned} &\alpha^{(\mathrm{b})}\,(r_\mathrm{b};\varepsilon) = \varepsilon^{-1/6}\,\frac{3^{1/3}}{\Gamma(2/3)\,K_{1,\mathrm{b}}^{1/6}}\,B_{0,\mathrm{b}}, \\ &\xi^{(\mathrm{b})}\,(r_\mathrm{b};\varepsilon) = \frac{c^2\,(r_\mathrm{b})}{g\,(r_\mathrm{b})}\,\alpha^{(\mathrm{b})}\,(r_\mathrm{b};\varepsilon) + G_0^{(\mathrm{o})}\,(r_\mathrm{b}), \\ &\left(\frac{d\xi^{(\mathrm{b})}(r;\varepsilon)}{dr}\right)_{r=r_\mathrm{b}} = -\varepsilon^{-5/6}\,\frac{3^{7/6}\,\Gamma(2/3)\,K_{1,\mathrm{b}}^{1/6}}{2\pi}\,\frac{c^2\,(r_\mathrm{b})}{g\,(r_\mathrm{b})}\,A_{0,\mathrm{b}}. \end{aligned}\right\} \tag{16.125}$$

The two constants A_0^* and B_0^* appear in the two-variable solutions $\alpha^{(\mathrm{o})}(r,\varepsilon)$ and $\xi^{(\mathrm{o})}(r,\varepsilon)$, which are given by Eqs. (16.30) and are valid in the intermediate radiative zone at larger distances from the boundaries of the zone. So far, these constants are related to the constant $B_{0,\mathrm{a}}$ by Eqs. (16.69), and to the two constants $A_{0,\mathrm{b}}$ and $B_{0,\mathrm{b}}$, by Eqs. (16.123). By the asymptotic solutions in the convective envelope, which are constructed hereafter, the two constants $A_{0,\mathrm{b}}$, and $B_{0,\mathrm{b}}$ will be related to a single constant denoted as $B_{0,\mathrm{d}}$. In the subsequent derivations, Eqs. (16.123) will therefore be replaced by the two equations that relate the constants A_0^* and B_0^* to the constant $B_{0,\mathrm{d}}$.

16.4.2 Asymptotic Solutions in the Convective Envelope

Since $N^2 < 0$ in the convective envelope, it is convenient to replace the negative coefficient $K_1(r)$ of the term with the large parameter in Eq. (16.2) by the positive coefficient

$$K_1'(r) \equiv -K_1(r) = -\ell(\ell+1)\,\frac{N^2(r)}{r^2}. \tag{16.126}$$

With this modification, Eqs. (16.2) and (16.3) become

$$\left.\begin{aligned}&\frac{d^2\alpha}{dr^2} + K_2(r)\,\frac{d\alpha}{dr} + \left[-\frac{K_1'(r)}{\varepsilon^2} + K_3(r) + \frac{\varepsilon^2}{c^2(r)}\right]\alpha\\&\quad = -K_4(r)\,\frac{d\xi}{dr},\\&\frac{d^2\xi}{dr^2} + \frac{4}{r}\,\frac{d\xi}{dr} - \frac{\ell(\ell+1)-2}{r^2}\,\xi = \frac{d\alpha}{dr} + \left[\frac{K_1'(r)}{\varepsilon^2}\,\frac{c^2(r)}{g(r)} + \frac{2}{r}\right]\alpha.\end{aligned}\right\} \tag{16.127}$$

16.4.2.1 Two-Variable Solutions in the Convective Envelope at Larger Distances from its Boundaries

For the construction of asymptotic solutions valid in the convective envelope at larger distances from its boundaries, a two-variable expansion procedure similar to that applied in the case of a radiative envelope can be used. The fast variable is defined as

$$\tau_{\mathrm{e}}(r) = \frac{1}{\varepsilon}\int_{r_{\mathrm{b}}}^{r} K_1'^{1/2}\left(r'\right)dr', \tag{16.128}$$

and the slow variable is still the radial coordinate r.

The two-variable solutions for the divergence and the radial component of the Lagrangian displacement take the form

$$\left.\begin{aligned}\alpha^{(\mathrm{e})}(r;\varepsilon) &= \mu_0^{(\mathrm{e})}(\varepsilon)\,K_5'(r)\left[A_0^{**}\,\exp\tau_{\mathrm{e}} + B_0^{**}\,\exp\left(-\tau_{\mathrm{e}}\right)\right],\\ \xi^{(\mathrm{e})}(r;\varepsilon) &= \frac{c^2(r)}{g(r)}\,\alpha^{(\mathrm{e})}(r;\varepsilon) + \mu_0^{(\mathrm{e})}(\varepsilon)\,G_0^{(\mathrm{e})}(r),\end{aligned}\right\} \tag{16.129}$$

where the function $K_5'(r)$ is defined as

$$K_5'(r) = g(r)\left[-N^2(r)\,r^6\,c^8(r)\,\rho^2(r)\right]^{-1/4}. \tag{16.130}$$

The function $G_0^{(\mathrm{e})}(r)$ is the general solution of Clairaut's second-order differential equation

$$G_0^{(\mathrm{e})}(r) = C_0^{**}\,y_1(r) + D_0^{**}\,y_2(r). \tag{16.131}$$

The particular solutions $y_1(r)$ and $y_2(r)$ correspond to those appearing in solution (16.29) for $G_0^{(\mathrm{o})}(r)$. The function $\mu_0^{(\mathrm{e})}(\varepsilon)$ is yet undetermined, and A_0^{**}, B_0^{**}, C_0^{**}, D_0^{**} are arbitrary constants.

The main differences between the two-variable solutions (16.129) valid in a convective envelope and the two-variable solutions (16.30) valid in a radiative envelope are that the trigonometric functions of the fast variable $\tau(r)$ are replaced by exponential functions of the fast variable $\tau_e(r)$, and that the function $K_5(r)$ is replaced by the function $K_5'(r)$.

With the two-variable solutions (16.129), boundary-layer solutions valid respectively near the inner boundary of the convective envelope and near the star's surface must be matched.

16.4.2.2 Boundary-Layer Solutions on the Outer Side of the Boundary Between the Intermediate Radiative Zone and the Convective Envelope

The slope of the Brunt–Väisälä frequency is not necessarily continuous at the boundary between the intermediate radiative zone and the convective envelope. Therefore, on the outer side of the boundary, Taylor series are used of the form

$$\left.\begin{aligned}
&\rho(r) = \rho\left(r_b\right)\left[1 + O\left(s_d\right)\right], \quad g(r) = g\left(r_b\right)\left[1 + O\left(s_d\right)\right],\\
&c(r) = c\left(r_b\right)\left[1 + O\left(s_d\right)\right], \quad N^2(r) = N_d^2\, s_d\left[1 + O\left(s_d\right)\right],\\
&K_1'(r) = -\frac{\ell(\ell+1)}{r_b^2}\, N_d^2\, s_d\left[1 + O\left(s_d\right)\right] \equiv K_{1,d}'\, s_d\left[1 + O\left(s_d\right)\right],\\
&K_2(r) = K_{2,d}\left[1 + O\left(s_d\right)\right], \quad K_3(r) = K_{3,d}\left[1 + O\left(s_d\right)\right],\\
&K_4(r) = K_{4,d}\left[1 + O\left(s_d\right)\right], \quad K_5'(r) = K_{5,d}'\, s_d^{-1/4}\left[1 + O\left(s_d\right)\right],
\end{aligned}\right\} \tag{16.132}$$

with $s_d(r) = r - r_b$. The coefficient N_d^2 in the Taylor series of $N^2(r)$ differs in absolute value from the coefficient N_b^2 involved in the Taylor series of $N^2(r)$ that is valid on the inner side of the boundary between the intermediate radiative zone and the convective envelope, when the slope of $N^2(r)$ is not continuous at this boundary.

The boundary-layer solutions can be constructed in a way similar to that followed in Sect. 16.3.2, but involve modified Bessel functions I and K instead of Bessel functions of the first kind. In terms of the boundary-layer coordinate

$$s_d^*(r) = \frac{s_d(r)}{\varepsilon^{2/3}}, \tag{16.133}$$

they can be written as

$$\left.\begin{aligned}
\alpha^{(d)}(r;\varepsilon) &= \mu_0^{(d)}(\varepsilon)\,\sqrt{s_d^*}\left[A_{0,d}\,I_{1/3}\left(\frac{2}{3}\sqrt{K'_{1,d}}\,s_d^{*3/2}\right)\right.\\
&\qquad\left.+B_{0,d}\,K_{1/3}\left(\frac{2}{3}\sqrt{K'_{1,d}}\,s_d^{*3/2}\right)\right],\\
\xi^{(d)}(r;\varepsilon) &= \frac{c^2(r_b)}{g(r_b)}\\
&\left[\alpha^{(d)}(r;\varepsilon)+\nu_0^{(d,2)}(\varepsilon)\,C_{0,d}\,s_d^* + \nu_0^{(d,3)}(\varepsilon)\,D_{0,d}\right],
\end{aligned}\right\} \tag{16.134}$$

where $\mu_0^{(d)}(\varepsilon)$, $\nu_0^{(d,2)}(\varepsilon)$, $\nu_0^{(d,3)}(\varepsilon)$ are yet undetermined functions, and $A_{0,d}$, $B_{0,d}$, $C_{0,d}$, $D_{0,d}$ are arbitrary constants.

The boundary-layer solutions are matched with the two-variable solutions $\alpha^{(e)}(r;\varepsilon)$ and $\xi^{(e)}(r;\varepsilon)$ valid at larger distances from the boundaries of the convective envelope. The matching relative to the divergence of the Lagrangian displacement leads to

$$\mu_0^{(e)}(\varepsilon) = \varepsilon^{1/6}\,\mu_0^{(d)}(\varepsilon) \tag{16.135}$$

and

$$\left.\begin{aligned}
A_{0,d} &= \frac{2\sqrt{\pi}}{\sqrt{3}}\,K_{1,d}'^{1/4}\,K'_{5,d}\,A_0^{**},\\
B_{0,d} &= \frac{2}{\sqrt{3\pi}}\,K_{1,d}'^{1/4}\,K'_{5,d}\,B_0^{**},
\end{aligned}\right\} \tag{16.136}$$

and the matching relative to the non-oscillatory parts in the radial component of the Lagrangian displacement, to

$$\left.\begin{aligned}
&C_{0,d} = 0,\\
&\nu_0^{(d,3)}(\varepsilon) = \mu_0^{(e)}(\varepsilon), \quad D_{0,d} = \frac{g(r_b)}{c^2(r_b)}\,G_0^{(e)}(r_b)\,.
\end{aligned}\right\} \tag{16.137}$$

At the turning point $r = r_b$, the divergence of the Lagrangian displacement, and the radial component of the Lagrangian displacement and its first derivative have the values

$$\left.\begin{aligned}
&\alpha^{(\mathrm{d})}\left(r_{\mathrm{b}};\varepsilon\right)=\mu_0^{(\mathrm{d})}(\varepsilon)\,\frac{3^{-1/6}\,\pi}{\Gamma(2/3)\,K_{1,\mathrm{d}}^{\prime 1/6}}\,B_{0,\mathrm{d}},\\
&\xi^{(\mathrm{d})}\left(r_{\mathrm{b}};\varepsilon\right)=\frac{c^2\left(r_{\mathrm{b}}\right)}{g\left(r_{\mathrm{b}}\right)}\,\alpha^{(\mathrm{d})}\left(r_{\mathrm{b}};\varepsilon\right)+\nu_0^{(\mathrm{d},3)}(\varepsilon)\,G_0^{(\mathrm{e})}\left(r_{\mathrm{b}}\right),\\
&\left(\frac{d\,\xi^{(\mathrm{d})}(r;\varepsilon)}{dr}\right)_{r_{\mathrm{b}}}=\mu_0^{(\mathrm{d})}(\varepsilon)\,\varepsilon^{-2/3}\,\frac{3^{7/6}\,\Gamma(2/3)}{2\pi}\\
&\qquad K_{1,\mathrm{d}}^{\prime 1/6}\,\frac{c^2\left(r_{\mathrm{b}}\right)}{g\left(r_{\mathrm{b}}\right)}\left(A_{0,\mathrm{d}}-\frac{\pi}{\sqrt{3}}\,B_{0,\mathrm{d}}\right).
\end{aligned}\right\} \tag{16.138}$$

The continuity of the Lagrangian displacement and the Lagrangian perturbation of the pressure at the boundary between the intermediate radiative zone and the convective envelope implies that the divergence, the radial component, and the first derivative of the radial component of the Lagrangian displacement be continuous at $r = r_{\mathrm{b}}$. On the grounds of equalities (16.125) and (16.138), it results that

$$\left.\begin{aligned}
&\mu_0^{(\mathrm{d})}(\varepsilon)=\varepsilon^{-1/6},\\
&B_{0,\mathrm{b}}=\frac{\pi}{\sqrt{3}}\left(\frac{K_{1,\mathrm{b}}}{K_{1,\mathrm{d}}^{\prime}}\right)^{1/6}B_{0,\mathrm{d}},\\
&G_0^{(\mathrm{e})}\left(r_{\mathrm{b}}\right)=G_0^{(\mathrm{o})}\left(r_{\mathrm{b}}\right),\\
&A_{0,\mathrm{b}}=-\left(\frac{K_{1,\mathrm{d}}^{\prime}}{K_{1,\mathrm{b}}}\right)^{1/6}\left(A_{0,\mathrm{d}}-\frac{\pi}{\sqrt{3}}\,B_{0,\mathrm{d}}\right).
\end{aligned}\right\} \tag{16.139}$$

The constant $B_{0,\mathrm{b}}$ is thus related to the constant $B_{0,\mathrm{d}}$, and the constant $A_{0,\mathrm{b}}$, to the constants $A_{0,\mathrm{d}}$ and $B_{0,\mathrm{d}}$. Hereafter, it will result from the matching of the boundary-layer solutions valid near the boundary point $r = R$ that the constant $A_{0,\mathrm{d}}$ is equal to zero, so that, finally, the constants $A_{0,\mathrm{b}}$ and $B_{0,\mathrm{b}}$ are only related to the constant $B_{0,\mathrm{d}}$ [see Sect. 16.4.2.4].

From Eq. (16.135) and the first equation (16.139), it follows that

$$\mu_0^{(\mathrm{e})}(\varepsilon)=\varepsilon^0. \tag{16.140}$$

16.4.2.3 Boundary-Layer Solutions Near the Boundary Point $r = R$

For the construction of boundary-layer solutions near the boundary point $r = R$ of a star with a convective envelope, one can proceed in a similar way as for the construction of boundary-layer solutions near the boundary point $r = R$ of a star with a radiative envelope.

The boundary-layer coordinate is here

$$z_e^*(z) = \frac{1}{\varepsilon} 2 K_{1,s}^{\prime 1/2} z^{1/2}. \tag{16.141}$$

The first dominant boundary-layer equation is

$$\frac{d^2 \alpha_0^{(s)}}{d z_e^{*2}} + \frac{2n_e + 3}{z_e^*} \frac{d\alpha_0^{(s)}}{d z_e^*} - \alpha_0^{(s)} = 0 \tag{16.142}$$

and admits of the solution that remains finite at $r = R$

$$\alpha_0^{(s)}\left(z_e^*\right) = A_{0,s}\, z_e^{*-(n_e+1)}\, I_{n_e+1}\left(z_e^*\right), \tag{16.143}$$

where $I_{n_e+1}\left(z_e^*\right)$ is a modified Bessel function, and $A_{0,s}$ an arbitrary constant.

The second dominant boundary-layer equation is

$$\frac{d^2 w_0^{(s)}}{d z_e^{*2}} + \frac{3}{z_e^*} \frac{d w_0^{(s)}}{d z_e^*} = -2 \frac{g_s}{c_s^2} \frac{1}{z_e^*} \frac{d\alpha_0^{(s)}}{d z_e^*} + \alpha_0^{(s)}. \tag{16.144}$$

The subtraction of the first dominant boundary-layer equation and the introduction of the function

$$w_0^*\left(z_e^*\right) = w_0^{(s)}\left(z_e^*\right) - \alpha_0^{(s)}\left(z_e^*\right) \tag{16.145}$$

lead to an inhomogeneous differential equation of the same form as Eq. (16.81), with a solution similar to solution (16.82). The integral in the solution is transformed by partial integration and use of a recurrence relation between modified Bessel functions I, so that

$$w_0^*\left(z_e^*\right) = 2\, A_{0,s} \frac{N_s^2}{g_s} z_e^{*-(n_e+2)}\, I_{n_e}\left(z_e^*\right) + C_{0,s}\, z_e^{*-2} + D_{0,s}. \tag{16.146}$$

By the use of a second recurrence relation between modified Bessel functions I, the solution is transformed into

$$\begin{aligned} w_0^*\left(z_e^*\right) &= C_{0,s}\, z_e^{*-2} + D_{0,s} \\ &\quad + \alpha_0^{(s)}\left(z_e^*\right) 2 \frac{N_s^2}{g_s} \frac{1}{z_e^{*2}} \left[\frac{d \ln I_{n_e+1}\left(z_e^*\right)}{d \ln z_e^*} + (n_e + 1) \right]. \end{aligned} \tag{16.147}$$

Consequently, the boundary-layer solution for $w_0^{(s)}\left(z_e^*\right)$ is given by

$$w_0^{(s)}\left(z_e^*\right) = C_{0,s}\, z_e^{*-2} + D_{0,s} + \alpha_0^{(s)}\left(z_e^*\right)\left\{1 + 2\,\frac{N_s^2}{g_s}\,\frac{1}{z_e^{*2}}\left[\frac{d\ln I_{n_e+1}\left(z_e^*\right)}{d\ln z_e^*} + (n_e+1)\right]\right\}. \quad (16.148)$$

In view of the matching with the two-variable solutions $\alpha^{(e)}(r;\varepsilon)$ and $\xi^{(e)}(r;\varepsilon)$, valid at larger distances from the boundary point, the boundary-layer solutions for the divergence and the radial component of the Lagrangian displacement are written in the more general form

$$\left.\begin{aligned} \alpha^{(s)}(r;\varepsilon) &= \mu_0^{(s)}(\varepsilon)\, A_{0,s}\, z_e^{*-(n_e+1)}\, I_{n_e+1}\left(z_e^*\right), \\ \xi^{(s)}(r;\varepsilon) &= \frac{c_s^2}{g_s}\, z\left\{\nu_0^{(s,2)}(\varepsilon)\, C_{0,s}\, z_e^{*-2} + \nu_0^{(s,3)}(\varepsilon)\, D_{0,s}\right. \\ &\quad \left. + \alpha^{(s)}(r;\varepsilon)\left\{1 + 2\,\frac{N_s^2}{g_s}\,\frac{1}{z_e^{*2}}\left[\frac{d\ln I_{n_e+1}\left(z_e^*\right)}{d\ln z_e^*} + (n_e+1)\right]\right\}\right\}, \end{aligned}\right\} \quad (16.149)$$

where $\mu_0^{(s)}(\varepsilon)$, $\nu_0^{(s,2)}(\varepsilon)$, $\nu_0^{(s,3)}(\varepsilon)$ are yet undetermined functions.

The matching relative to the divergence of the Lagrangian displacement leads to

$$\mu_0^{(s)}(\varepsilon) = \varepsilon^{-(n_e+3/2)} \quad (16.150)$$

and

$$\left.\begin{aligned} A_0^{**} &= 0, \\ B_0^{**} &= A_{0,s}\,\frac{\exp\left[\tau_e(R)\right]}{\sqrt{2\pi}\left(2\,K_{1,s}^{\prime 1/2}\right)^{n_e+3/2}\, K_{5,s}'}, \end{aligned}\right\} \quad (16.151)$$

and the matching relative to the non-oscillatory parts in the radial component of the Lagrangian displacement, to

$$\left.\begin{aligned} D_{0,s} &= 0, \\ \nu_0^{(s,2)}(\varepsilon) &= \varepsilon^{-2}\,\mu_0^{(e)}(\varepsilon), \quad C_{0,s} = \frac{g_s}{c_s^2}\, 4\, K_{1,s}'\, G_0^{(e)}(R). \end{aligned}\right\} \quad (16.152)$$

16.4.2.4 Main Result of the Asymptotic Solutions in the Convective Envelope

From the first equation (16.136) and the first equation (16.151), it results that

$$A_{0,\mathrm{d}} = 0, \tag{16.153}$$

so that the last equation (16.139) reduces to

$$A_{0,\mathrm{b}} = \frac{\pi}{\sqrt{3}} \left(\frac{K'_{1,\mathrm{d}}}{K_{1,\mathrm{b}}} \right)^{1/6} B_{0,\mathrm{d}}. \tag{16.154}$$

By this equation and the second equation (16.139), the constants $A_{0,\mathrm{b}}$ and $B_{0,\mathrm{b}}$ are now related to the single constant $B_{0,\mathrm{d}}$.

For the sake of simplification of the notations, the quantity K is introduced as

$$K = \frac{K'_{1,\mathrm{d}}}{K_{1,\mathrm{b}}} \tag{16.155}$$

or, equivalently, as

$$K = \left[\frac{\left[dK'_1(r)/ds_\mathrm{d}\right]_\mathrm{rh}}{\left[dK_1(r)/ds_\mathrm{b}\right]_\mathrm{lh}} \right]_{r=r_\mathrm{b}} = \left[\frac{\left(dN^2/dr\right)_\mathrm{rh}}{\left(dN^2/dr\right)_\mathrm{lh}} \right]_{r=r_\mathrm{b}}. \tag{16.156}$$

Hence, K is equal to the ratio of the right-hand first derivative of $N^2(r)$ to the left-hand first derivative of $N^2(r)$, both taken with respect to r and considered at the turning point $r = r_\mathrm{b}$. The ratio K is a positive quantity.

Equations (16.123) can then be written in the form

$$\left.\begin{aligned} A_0^* &= B_{0,\mathrm{d}} \frac{\sqrt{\pi}}{K_{1,\mathrm{b}}^{1/4}\, K_{5,\mathrm{b}}} \left[K^{1/6} \sin\left(\tau_\mathrm{Rad} + \frac{\pi}{12}\right) \right. \\ &\quad \left. + K^{-1/6} \cos\left(\tau_\mathrm{Rad} - \frac{\pi}{12}\right) \right], \\ B_0^* &= -B_{0,\mathrm{d}} \frac{\sqrt{\pi}}{K_{1,\mathrm{b}}^{1/4}\, K_{5,\mathrm{b}}} \left[K^{1/6} \cos\left(\tau_\mathrm{Rad} + \frac{\pi}{12}\right) \right. \\ &\quad \left. - K^{-1/6} \sin\left(\tau_\mathrm{Rad} - \frac{\pi}{12}\right) \right]. \end{aligned}\right\} \tag{16.157}$$

16.4.3 Eigenfrequency Equation

Elimination of the constants A_0^* and B_0^* from the foregoing equations and Eqs. (16.69) leads to the system of two algebraic, linear, homogeneous equations

$$\left.\begin{aligned}
&B_{0,\mathrm{a}} \frac{\sqrt{3}}{\sqrt{\pi}\, K_{1,\mathrm{a}}^{1/4}\, K_{5,\mathrm{a}}} \cos\frac{\pi}{12} - B_{0,\mathrm{d}} \frac{\sqrt{\pi}}{K_{1,\mathrm{b}}^{1/4}\, K_{5,\mathrm{b}}} \\
&\quad \left[K^{1/6} \sin\left(\tau_{\mathrm{Rad}} + \frac{\pi}{12}\right) + K^{-1/6} \cos\left(\tau_{\mathrm{Rad}} - \frac{\pi}{12}\right)\right] = 0, \\
&B_{0,\mathrm{a}} \frac{\sqrt{3}}{\sqrt{\pi}\, K_{1,\mathrm{a}}^{1/4}\, K_{5,\mathrm{a}}} \sin\frac{\pi}{12} + B_{0,\mathrm{d}} \frac{\sqrt{\pi}}{K_{1,\mathrm{b}}^{1/4}\, K_{5,\mathrm{b}}} \\
&\quad \left[K^{1/6} \cos\left(\tau_{\mathrm{Rad}} + \frac{\pi}{12}\right) - K^{-1/6} \sin\left(\tau_{\mathrm{Rad}} - \frac{\pi}{12}\right)\right] = 0.
\end{aligned}\right\} \tag{16.158}$$

The necessary and sufficient condition for the existence a non-trivial solution for the constants $B_{0,\mathrm{a}}$ and $B_{0,\mathrm{d}}$ yields the eigenfrequency equation

$$\begin{aligned}
&\left[K^{1/6} \cos\left(\tau_{\mathrm{Rad}} + \frac{\pi}{12}\right) - K^{-1/6} \sin\left(\tau_{\mathrm{Rad}} - \frac{\pi}{12}\right)\right] \cos\frac{\pi}{12} \\
&+ \left[K^{1/6} \sin\left(\tau_{\mathrm{Rad}} + \frac{\pi}{12}\right) + K^{-1/6} \cos\left(\tau_{\mathrm{Rad}} - \frac{\pi}{12}\right)\right] \sin\frac{\pi}{12} = 0.
\end{aligned} \tag{16.159}$$

In the supposition that $K \neq 0$, the eigenfrequency equation can be transformed into

$$\cos\tau_{\mathrm{Rad}} - K^{-1/3} \sin\left(\tau_{\mathrm{Rad}} - \frac{\pi}{6}\right) = 0. \tag{16.160}$$

In imitation of Tassoul (1980), the angle θ_2 is introduced such that

$$K^{-1/3} = \frac{\sin(\pi/6 - \theta_2)}{\sin(\pi/6 + \theta_2)}, \tag{16.161}$$

or

$$\theta_2 = -\frac{\pi}{6} + \arctan\frac{2\cos(\pi/6)}{2\,K^{-1/3} + 1}. \tag{16.162}$$

The eigenfrequency equation then becomes

$$\tan(\tau_{\mathrm{Rad}} - \theta_2) = \tan\frac{\pi}{3}, \tag{16.163}$$

so that

$$\tau_{\mathrm{Rad}} \equiv \frac{[\ell(\ell+1)]^{1/2}}{|\sigma|} \int_{r_a}^{r_b} \left(\frac{N^2(r)}{r^2}\right)^{1/2} dr$$
$$= 2\left(n - \frac{2}{3}\right)\frac{\pi}{2} + \theta_2. \qquad (16.164)$$

In the particular case in which the slope of $N^2(r)$ is continuous at the boundary between the intermediate radiative zone and the convective envelope, $K'_{1,\mathrm{d}} = K_{1,\mathrm{b}}$, so that $K = 1$ and $\theta_2 = 0$. For this case, it is shown in Sect. 16.4.6 that the number n corresponds to the radial order of the g^+-mode considered.

Eigenfrequency equation (A.13) of Tassoul (1980), which is derived in the Cowling approximation, agrees with Eq. (16.164), when, in that equation, θ_1 is set equal to $\pi/6$ because of the isentropic (adiabatic) equilibrium adopted in the convective core [see Tassoul's comment above her Eq. (A.9)], and κ is replaced by $\kappa - 1$.

16.4.4 The Condition on the Eulerian Perturbation of the Gravitational Potential at $r = R$

For the imposition of boundary condition (5.97) on the Eulerian perturbation of the gravitational potential at $r = R$, Eqs. (14.122) and (14.123), which determine the Eulerian perturbation of the gravitational potential and its first derivative in terms of the divergence and the radial component of the Lagrangian displacement at an arbitrary point, are again used. The divergence and the radial component of the Lagrangian displacement at $r = R$ are now given by

$$\left.\begin{aligned} \alpha_R &= \varepsilon^{-(n_\mathrm{e}+3/2)}\, A_{0,\mathrm{s}}\, \frac{2^{-(n_\mathrm{e}+1)}}{\Gamma\,(n_\mathrm{e}+2)}, \\ \xi_R &= \frac{c_\mathrm{s}^2}{g_\mathrm{s}}\, \frac{1}{4\,K'_{1,s}} \\ &\left[\varepsilon^{-(n_\mathrm{e}-1/2)}\, A_{0,\mathrm{s}}\, \frac{2^{-(n_\mathrm{e}-1)}}{\Gamma\,(n_\mathrm{e}+1)}\, \frac{N_\mathrm{s}^2}{g_\mathrm{s}} + \mu_0^{(\mathrm{e})}(\varepsilon)\, C_{0,\mathrm{s}}\right], \end{aligned}\right\} \qquad (16.165)$$

and their first derivatives, by

$$\left.\begin{aligned} \left(\frac{d\alpha}{dr}\right)_R &= -\varepsilon^{-(n_\mathrm{e}+7/2)}\, A_{0,\mathrm{s}}\, \frac{2^{-(n_\mathrm{e}+1)}}{\Gamma\,(n_\mathrm{e}+3)}\, K'_{1,\mathrm{s}}, \\ \left(\frac{d\xi}{dr}\right)_R &= -\varepsilon^{-(n_\mathrm{e}+3/2)}\, A_{0,\mathrm{s}}\, \frac{2^{-(n_\mathrm{e}+1)}}{\Gamma\,(n_\mathrm{e}+2)}\, \frac{c_\mathrm{s}^2}{g_\mathrm{s}} \left(\frac{N_\mathrm{s}^2}{g_\mathrm{s}} + 1\right). \end{aligned}\right\} \qquad (16.166)$$

As for a star with a radiative envelope, the boundary condition is automatically satisfied by the boundary-layer solution $\alpha^{(s)}(r;\varepsilon)$ and the oscillatory part of the boundary-layer solution $\xi^{(s)}(r;\varepsilon)$. The terms involving the non-oscillatory part of the boundary-layer solution $\xi^{(s)}(r;\varepsilon)$ are of the order ε^0. For the boundary condition to be also satisfied by these terms, one must set

$$C_{0,\mathrm{s}} = 0. \tag{16.167}$$

From the third equation (16.152), it then follows that

$$G_0^{(\mathrm{e})}(R) = 0. \tag{16.168}$$

The third equation (16.139) can be written explicitly as

$$C_0^{**}\, y_1\,(r_\mathrm{b}) + D_0^{**}\, y_2\,(r_\mathrm{b}) = C_0^{*}\, y_1\,(r_\mathrm{b}) + D_0^{*}\, y_2\,(r_\mathrm{b})\,, \tag{16.169}$$

or, because of Eqs. (16.65) and (16.168), as

$$C_0^{**}\left[y_1\,(r_\mathrm{b}) - \frac{y_1(R)}{y_2(R)}\, y_2\,(r_\mathrm{b})\right] = C_0^{*}\left[y_1\,(r_\mathrm{b}) - \frac{y_1\,(r_\mathrm{a})}{y_2\,(r_\mathrm{a})}\, y_2\,(r_\mathrm{b})\right]. \tag{16.170}$$

The constants C_0^* and C_0^{**} are thus related to each other. The relation is satisfied when one sets

$$C_0^* = 0, \quad C_0^{**} = 0. \tag{16.171}$$

It then follows that

$$G_0^{(\mathrm{o})}(r) \equiv 0, \quad G_0^{(\mathrm{e})}(r) \equiv 0, \tag{16.172}$$

so that the radial component of the Lagrangian displacement is purely oscillatory in the intermediate radiative zone.

16.4.5 Uniformly Valid Asymptotic Solutions

Uniformly valid first-order asymptotic solutions for the divergence and the radial component of the Lagrangian displacement are here constructed in terms of the constants $B_{0,\mathrm{a}}$ and $B_{0,\mathrm{d}}$, which are involved in Eqs. (16.158) leading to the eigenfrequency equation:

1. The uniformly valid first-order asymptotic solutions $\alpha^{(\mathrm{a,u})}(r;\varepsilon)$ and $\xi^{(\mathrm{a,u})}(r;\varepsilon)$, given by Eqs. (16.107), remain valid in the intermediate radiative zone from the upper boundary of the convective core to a distance sufficiently large from the lower boundary of the convective envelope.

2. The first-order asymptotic solutions that are uniformly valid in the intermediate radiative zone from the lower boundary of the convective envelope to a distance sufficiently large from the upper boundary of the convective core, can be expressed in the compact form

$$\left.\begin{aligned}\alpha^{(\mathrm{b,u})}(r;\varepsilon) &= B_{0,\mathrm{d}}\,\frac{\pi}{\sqrt{2}\,K_{1,\mathrm{b}}^{1/4}\,K_{5,\mathrm{b}}}\,K_5(r)\\ &\quad\tau_{\mathrm{b}}^{1/2}\left[K^{1/6}\,J_{1/3}\left(\tau_{\mathrm{b}}\right)+K^{-1/6}\,J_{-1/3}\left(\tau_{\mathrm{b}}\right)\right],\\ \xi^{(\mathrm{b,u})}(r;\varepsilon) &= \frac{c^2(r)}{g(r)}\,\alpha^{(\mathrm{b,u})}(r;\varepsilon),\end{aligned}\right\} \tag{16.173}$$

with $\tau_{\mathrm{b}}(r) \equiv \dfrac{1}{\varepsilon}\displaystyle\int_r^{r_{\mathrm{b}}} K_1^{1/2}\left(r'\right)dr' = \tau_{\mathrm{Rad}} - \tau(r)\,.$

3. The first-order asymptotic solutions that are uniformly valid in the convective envelope from the lower boundary of the envelope to a distance sufficiently large from the star's surface, can be expressed in the compact form

$$\left.\begin{aligned}\alpha^{(\mathrm{d,u})}(r;\varepsilon) &= B_{0,\mathrm{d}}\,\frac{\sqrt{3}}{\sqrt{2}\,K_{1,\mathrm{d}}^{\prime 1/4}\,K_{5,\mathrm{d}}^{\prime}}\,K_5^{\prime}(r)\,\tau_{\mathrm{e}}^{1/2}\,K_{1/3}\left(\tau_{\mathrm{e}}\right),\\ \xi^{(\mathrm{d,u})}(r;\varepsilon) &= \frac{c^2(r)}{g(r)}\,\alpha^{(\mathrm{d,u})}(r;\varepsilon),\end{aligned}\right\} \tag{16.174}$$

where the fast variable τ_{e} is defined by Eq. (16.128).

4. The first-order asymptotic solutions that are uniformly valid in the convective envelope, from the star's surface to a distance sufficiently large from the lower boundary of the envelope, can be expressed in the compact form

$$\left.\begin{aligned}&\alpha^{(\mathrm{s,u})}(r;\varepsilon)\\ &\quad= B_{0,\mathrm{d}}\,\frac{\sqrt{3}\pi\,\exp\left[-\tau_{\mathrm{e}}(R)\right]}{\sqrt{2}\,K_{1,\mathrm{d}}^{\prime 1/4}\,K_{5,\mathrm{d}}^{\prime}}\,K_5^{\prime}(r)\,\tau_{\mathrm{s}}^{1/2}\,I_{n_{\mathrm{e}}+1}\left(\tau_{\mathrm{s}}\right),\\ &\xi^{(\mathrm{s,u})}(r;\varepsilon) = \frac{c^2(r)}{g(r)}\,\alpha^{(\mathrm{s,u})}(r;\varepsilon)\\ &\quad\left\{1+2\,\frac{N_{\mathrm{s}}^2}{g_{\mathrm{s}}}\,\frac{1}{\tau_{\mathrm{s}}^2}\left[\frac{d\ln I_{n_{\mathrm{e}}+1}\left(\tau_{\mathrm{s}}\right)}{d\ln\tau_{\mathrm{s}}}+\left(n_{\mathrm{e}}+1\right)\right]\right\},\end{aligned}\right\} \tag{16.175}$$

with $\tau_{\mathrm{s}}(r) \equiv \dfrac{1}{\varepsilon}\displaystyle\int_0^{z} K_1^{\prime 1/2}\left(r'\right)dz' = \tau_{\mathrm{e}}(R) - \tau_{\mathrm{e}}(r)\,.$

16.4.6 Identification of the Radial Order of a g^+-Mode with a Given Eigenfrequency

In the intermediate radiative zone, the nodes of the uniformly valid solutions $\alpha^{(\mathrm{b,u})}(r;\varepsilon)$ and $\xi^{(\mathrm{b,u})}(r;\varepsilon)$ coincide as well as those of the uniformly valid solutions $\alpha^{(\mathrm{a,u})}(r;\varepsilon)$ and $\xi^{(\mathrm{a,u})}(r;\varepsilon)$ do. Hence, the asymptotic solutions for $\alpha(r)$ and $\xi(r)$ have the same numbers of nodes in that zone.

Approximate positions of the nodes can be determined by means of the two-variable solutions $\alpha^{(\mathrm{o})}(r;\varepsilon)$ and $\xi^{(\mathrm{o})}(r;\varepsilon)$, which are valid in the zone, at larger distances from its boundaries. These solutions are given by Eqs. (16.30), in which $G_0^{(\mathrm{o})}(r) = 0$.

When the constants A_0^* and B_0^* are eliminated by means of Eqs. (16.69), the asymptotic solution $\xi^{(\mathrm{o})}(r;\varepsilon)$ takes the form given by Eq. (16.111), so that the nodes counted from the lower boundary of the intermediate radiative zone are located at

$$\tau^0 = \left(2j - \frac{5}{6}\right)\frac{\pi}{2}, \quad j = 1, 2, 3, \ldots \tag{16.176}$$

In order to determine the position of the last node counted from the lower boundary of the intermediate radiative zone, one can transform the two-variable solution $\xi^{(\mathrm{o})}(r;\varepsilon)$ by eliminating the constants A_0^* and B_0^* by means of Eqs. (16.157). The solution then becomes

$$\xi^{(\mathrm{o})}(r;\varepsilon) = B_{0,\mathrm{d}}\,\frac{\sqrt{\pi}}{K_{1,\mathrm{b}}^{1/4}\,K_{5,\mathrm{b}}}\,\frac{c^2(r)}{g(r)}\,K_5(r)$$
$$\left[K^{1/6}\sin\left(\tau_{\mathrm{b}} + \frac{\pi}{12}\right) + K^{-1/6}\cos\left(\tau_{\mathrm{b}} - \frac{\pi}{12}\right)\right]. \tag{16.177}$$

If the slope of N^2 is continuous at the boundary between the intermediate radiative zone and the convective envelope, $K = 1$, and the solution reduces to

$$\xi^{(\mathrm{o})}(r;\varepsilon) = B_{0,\mathrm{d}}\,\frac{\sqrt{\pi}}{K_{1,\mathrm{b}}^{1/4}\,K_{5,\mathrm{b}}}\,\frac{c^2(r)}{g(r)}\,K_5(r)\,2\cos\frac{\pi}{6}\,\cos\left(\tau_{\mathrm{b}} - \frac{\pi}{4}\right). \tag{16.178}$$

The positions of the nodes counted from the upper boundary of the intermediate radiative zone are given by

$$\tau_{\mathrm{b}}^0 = \left(2k - \frac{1}{2}\right)\frac{\pi}{2}, \quad k = 1, 2, 3, \ldots \tag{16.179}$$

The node associated with $k = 1$ is located at $\tau_{\mathrm{b}}^0(\text{first}) = 3\pi/4$.

Since the last node of the two-variable solution $\xi^{(\mathrm{o})}(r;\varepsilon)$ counted from the lower boundary of the intermediate radiative zone must coincide with the first node counted from the upper boundary of the intermediate radiative zone, the relation holds

$$\tau^0(\text{last}) = \tau_{\text{Rad}} - \tau_{\text{b}}^0(\text{first}) = \left[2(n-1) - \frac{5}{6}\right]\frac{\pi}{2}.$$

By comparison with Eq. (16.176), it follows that the last node counted from the lower boundary of the intermediate radiative zone is associated with $j = n - 1$. Consequently, both the divergence and the radial component of the Lagrangian displacement display $n - 1$ nodes in the intermediate radiative zone.

Since $N_{\text{s}}^2 < 0$, it results from Eqs. (16.165) that the asymptotic solutions for the divergence and the radial component of the Lagrangian displacement have opposite signs at the boundary point $r = R$. Therefore, one of these solutions must display one more node than the other solution in the convective envelope. From the asymptotic solutions $\alpha^{(\text{s,u})}(r;\varepsilon)$ and $\xi^{(\text{s,u})}(r;\varepsilon)$, which are uniformly valid from the surface to a distance sufficiently large from the boundary between the intermediate radiative zone and the convective envelope, one sees that the radial component of the Lagrangian displacement displays an additional node in the convective envelope at the point at which

$$1 + 2\,\frac{N_{\text{s}}^2}{g_{\text{s}}}\,\frac{1}{\tau_{\text{s}}^2}\left[\frac{d\ln I_{n_{\text{e}}+1}(\tau_{\text{s}})}{d\ln \tau_{\text{s}}} + (n_{\text{e}}+1)\right] = 0.$$

In illustration, the parts of the uniformly valid asymptotic solutions $\alpha^{(\text{s,u})}(r;\varepsilon)$ and $\xi^{(\text{s,u})}(r;\varepsilon)$ that depend on the fast variable τ_{s}, i.e., respectively

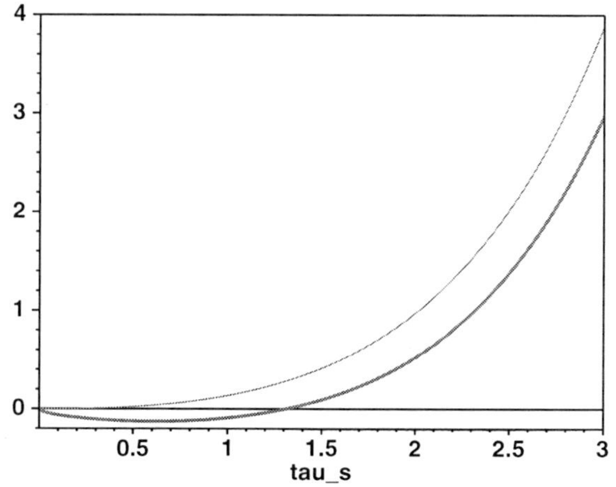

Fig. 16.3 The functions $H_4(\tau_{\text{s}})$ (*thinner line*) and $H_5(\tau_{\text{s}})$ (*thicker line*) for $n_{\text{e}} = 1$ and $\Gamma_{1,R} = 5/3$

$$H_4\left(\tau_{\mathrm{s}}\right) \equiv \tau_{\mathrm{s}}^{1/2}\, I_{n_{\mathrm{e}}+1}\left(\tau_{\mathrm{s}}\right)$$

and

$$H_5\left(\tau_{\mathrm{s}}\right) \equiv H_4\left(\tau_{\mathrm{s}}\right)\left\{1 + 2\,\frac{N_{\mathrm{s}}^2}{g_{\mathrm{s}}}\,\frac{1}{\tau_{\mathrm{s}}^2}\left[\frac{d\ln I_{n_{\mathrm{e}}+1}\left(\tau_{\mathrm{s}}\right)}{d\ln\tau_{\mathrm{s}}} + \left(n_{\mathrm{e}}+1\right)\right]\right\},$$

are represented in Fig. 16.3 for $n_{\mathrm{e}} = 1$ and $\Gamma_{1,R} = 5/3$.

When the slope of N^2 is continuous at the boundary between the convective core and the intermediate radiative zone, one arrives at the following conclusion. For the higher-order g^+-mode associated with the number n in eigenfrequency equation (16.164), the radial component of the Lagrangian displacement displays n nodes along the radius, so that the mode has the radial order n. Of these nodes, $n-1$ are situated in the intermediate radiative zone, and one is situated in the convective envelope. The divergence of the Lagrangian displacement displays $n-1$ nodes, which are all situated in the intermediate radiative zone and coincide there with the nodes of the radial component.

Chapter 17
Asymptotic Representation of Low-Degree, Higher-Order g^+-Modes in Stars Consisting of a Radiative Core and a Convective Envelope

17.1 Introduction

In this chapter, an asymptotic representation is developed for low-degree, higher-order g^+-modes in stars that consist of a radiative core and a convective envelope.

The appropriate equations are still Eqs. (16.2) and (16.3) for the functions $\alpha(r)$ and $\xi(r)$:

$$\frac{d^2\alpha}{dr^2} + K_2(r)\,\frac{d\alpha}{dr} + \left[\frac{K_1(r)}{\varepsilon^2} + K_3(r) + \frac{\varepsilon^2}{c^2(r)}\right]\alpha = -K_4(r)\,\frac{d\xi}{dr}, \quad (17.1)$$

$$\frac{d^2\xi}{dr^2} + \frac{4}{r}\frac{d\xi}{dr} - \frac{\ell(\ell+1)-2}{r^2}\,\xi = \frac{d\alpha}{dr} - \left[\frac{c^2(r)}{g(r)}\,\frac{K_1(r)}{\varepsilon^2} - \frac{2}{r}\right]\alpha. \quad (17.2)$$

Equation (17.1) displays a turning point at the radial distance $r = r_b$ of the boundary between the radiative core and the convective envelope. The boundary conditions are the same as those imposed in the previous chapter, and dimensionless quantities are introduced as before.

17.2 Asymptotic Solutions in the Radiative Core

17.2.1 Two-Variable Solutions at Larger Distances from the Boundaries of the Radiative Core

In the region of the radiative core that is situated at larger distances from the star's centre at $r = 0$ and the boundary between the radiative core and the convective envelope, two-variable solutions are constructed as in Sect. 16.3.1 for the radiative envelope of a star with a convective core, except that the fast variable is here defined as

$$\tau(r) = \frac{1}{\varepsilon}\int_0^r K_1^{1/2}\left(r'\right) dr'. \quad (17.3)$$

P. Smeyers and T. Van Hoolst, *Linear Isentropic Oscillations of Stars: Theoretical Foundations*, Astrophysics and Space Science Library 371,
DOI 10.1007/978-3-642-13030-4_18,

The lowest-order two-variable solutions for the divergence and the radial component of the Lagrangian displacement that are valid at larger distances from the boundary points $r = 0$ and $r = r_b$ of the radiative core are similar to solutions (16.30):

$$\left.\begin{aligned} \alpha^{(o)}(r;\varepsilon) &= K_5(r)\left(A_0^* \cos\tau + B_0^* \sin\tau\right), \\ \xi^{(o)}(r;\varepsilon) &= K_6(r)\left[\left(A_0^* \cos\tau + B_0^* \sin\tau\right) + \frac{G_0^{(o)}(r)}{K_6(r)}\right]. \end{aligned}\right\} \tag{17.4}$$

These solutions contain the four arbitrary constants A_0^*, B_0^*, C_0^*, D_0^*.

17.2.2 Boundary-Layer Solutions Near the Boundary Point $r = 0$

Some coefficients of Eqs. (17.1) and (17.2) display a pole at the boundary point $r = 0$. Therefore, boundary-layer theory is applied for the construction of asymptotic solutions near that point.

When one passes on from the functions $\alpha(r)$ and $\xi(r)$ to functions $v(r)$ and $w(r)$ by means of the transformations

$$\alpha(r) = K_5(r)\, v(r), \qquad \xi(r) = K_6(r)\, w(r), \tag{17.5}$$

Equations (17.1) and (17.2) become

$$\begin{aligned} &\frac{d^2 v}{dr^2} + \left(K_2 + \frac{2}{K_5}\frac{dK_5}{dr}\right)\frac{dv}{dr} + \left[\frac{K_1}{\varepsilon^2} + \left(K_3 + \frac{1}{K_5}\frac{d^2K_5}{dr^2} + \frac{K_2}{K_5}\frac{dK_5}{dr}\right) + \frac{\varepsilon^2}{c^2}\right] v \\ &= -K_4 \frac{K_6}{K_5}\left(\frac{dw}{dr} + \frac{1}{K_6}\frac{dK_6}{dr} w\right), \end{aligned} \tag{17.6}$$

$$\begin{aligned} &\frac{d^2 w}{dr^2} + \left(\frac{4}{r} + \frac{2}{K_6}\frac{dK_6}{dr}\right)\frac{dw}{dr} + \left[\frac{1}{K_6}\frac{d^2K_6}{dr^2} + \frac{4}{r}\frac{1}{K_6}\frac{dK_6}{dr} - \frac{\ell(\ell+1)-2}{r^2}\right] w \\ &= -\frac{K_1}{\varepsilon^2} v + \frac{K_5}{K_6}\left[\frac{dv}{dr} + \left(\frac{2}{r} + \frac{1}{K_5}\frac{dK_5}{dr}\right) v\right] \end{aligned} \tag{17.7}$$

(Smeyers et al. 1995).

Near $r = 0$, power series (6.37)–(6.40) and (14.46) are valid. Moreover, one has

$$K_5(r) = \frac{K_{5,c}}{r} + O(r), \qquad K_6(r) = \frac{K_{6,c}}{r^2} + O\left(r^0\right). \tag{17.8}$$

One can proceed as in Sect. 14.4. In the left-hand member of Eq. (17.6), the second derivative and the singular term

$$\left(K_3 + \frac{1}{K_5}\,\frac{d^2 K_5}{dr^2} + \frac{K_2}{K_5}\,\frac{dK_5}{dr} \right) v$$

are of the same order in ε as the term with the large parameter $(K_1/\varepsilon^2)\, v$, when one introduces a boundary-layer coordinate r^* satisfying the differential equation

$$\left(\frac{dr^*}{dr} \right)^2 = \frac{1}{\varepsilon^2}\, K_{1,c}. \tag{17.9}$$

If $r^*(0) = 0$, the positive boundary-layer coordinate is defined as

$$r^*(r) = \frac{1}{\varepsilon}\, K_{1,c}^{1/2}\, r. \tag{17.10}$$

Furthermore, boundary-layer expansions for the functions $v(r)$ and $w(r)$ are introduced as

$$\left. \begin{aligned} v^{(c)}(r;\varepsilon) &= \mu_0^{(c)}(\varepsilon)\, v_0^{(c)}(r^*) + \mu_1^{(c)}(\varepsilon)\, v_1^{(c)}(r^*) + \ldots, \\ w^{(c)}(r;\varepsilon) &= \nu_0^{(c)}(\varepsilon)\, w_0^{(c)}(r^*) + \nu_1^{(c)}(\varepsilon)\, w_1^{(c)}(r^*) + \ldots, \end{aligned} \right\} \tag{17.11}$$

where $\mu_0^{(c)}(\varepsilon)$, $\mu_1^{(c)}(\varepsilon)$, ... and $\nu_0^{(c)}(\varepsilon)$, $\nu_1^{(c)}(\varepsilon)$, ... are asymptotic series to be determined.

In Eq. (17.6), transformation of the differential operators into differential operators in terms of the boundary-layer coordinate r^*, substitution of the boundary-layer expansions, and expansion of the coefficients of the various terms for small values of r^* yield

$$\begin{aligned} &\mu_0^{(c)}(\varepsilon) \left\{ \frac{1}{\varepsilon^2} \left[\frac{d^2 v_0^{(c)}}{dr^{*2}} + \left(1 - \frac{\ell(\ell+1)}{r^{*2}} \right) v_0^{(c)} \right] + O\left(\varepsilon^0\right) \right\} \\ &\quad + \mu_1^{(c)}(\varepsilon)\, \{\ldots\} + \ldots = -\nu_0^{(c)}(\varepsilon)\, O\left(\varepsilon^0\right) + \nu_1^{(c)}(\varepsilon)\, \{\ldots\} + \ldots \end{aligned} \tag{17.12}$$

The first dominant boundary-layer equation is homogeneous when $\nu_0^{(c)}(\varepsilon)$ is of a higher order in ε than $\mu_0^{(c)}(\varepsilon)/\varepsilon^2$. It is then given by

$$\frac{d^2 v_0^{(c)}}{dr^{*2}} + \left[1 - \frac{\ell(\ell+1)}{r^{*2}} \right] v_0^{(c)} = 0. \tag{17.13}$$

The solution that satisfies the requirement that the divergence of the Lagrangian displacement be finite at $r = 0$ can be written as

$$v_0^{(c)}\left(r^*\right) = A_{0,c}\, r^{*1/2}\, J_{\ell+1/2}\left(r^*\right), \tag{17.14}$$

where $J_{\ell+1/2}\left(r^*\right)$ is a Bessel function of the first kind, and $A_{0,c}$, an arbitrary constant (Abramowitz & Stegun 1965).

Equation (17.7) is similarly transformed into

$$\nu_0^{(c)}(\varepsilon)\left\{\frac{1}{\varepsilon^2}\left[\frac{d^2 w_0^{(c)}}{dr^{*2}} - \frac{\ell(\ell+1)}{r^{*2}}\, w_0^{(c)}\right] + O\left(\varepsilon^0\right)\right\}$$

$$+\,\nu_1^{(c)}(\varepsilon)\,\{\ldots\} + \ldots = -\frac{\mu_0^{(c)}(\varepsilon)}{\varepsilon^2}\, v_0^{(c)} + \mu_1^{(c)}(\varepsilon)\,\{\ldots\} + \ldots \tag{17.15}$$

The second dominant boundary-layer equation is inhomogeneous when

$$\nu_0^{(c)}(\varepsilon) = \mu_0^{(c)}(\varepsilon). \tag{17.16}$$

It is then given by

$$\frac{d^2 w_0^{(c)}}{dr^{*2}} - \frac{\ell(\ell+1)}{r^{*2}}\, w_0^{(c)} = -v_0^{(c)}. \tag{17.17}$$

By subtraction of Eq. (17.13), it becomes

$$\frac{d^2\left(w_0^{(c)} - v_0^{(c)}\right)}{dr^{*2}} - \frac{\ell(\ell+1)}{r^{*2}}\left(w_0^{(c)} - v_0^{(c)}\right) = 0. \tag{17.18}$$

The solution that satisfies the requirement that the radial component of the Lagrangian displacement be finite at $r = 0$ is given by

$$w_0^{(c)}\left(r^*\right) = v_0^{(c)}\left(r^*\right) + C_{0,c}\, r^{*(\ell+1)}, \tag{17.19}$$

where $C_{0,c}$ is an arbitrary constant.

The lowest-order boundary-layer solutions for the divergence and the radial component of the Lagrangian displacement near $r = 0$ can be expressed in the more general form

$$\left.\begin{aligned}
\alpha^{(c)}(r;\varepsilon) &= \frac{\mu_0^{(c,1)}(\varepsilon)}{\varepsilon}\, A_{0,c}\, K_{1,c}^{1/2}\, K_{5,c}\, r^{*-1/2}\, J_{\ell+1/2}\left(r^*\right), \\
\xi^{(c)}(r;\varepsilon) &= \frac{\mu_0^{(c,1)}(\varepsilon)}{\varepsilon^2}\, A_{0,c}\, K_{1,c}\, K_{6,c}\, r^{*-3/2}\, J_{\ell+1/2}\left(r^*\right) \\
&+ \frac{\mu_0^{(c,2)}(\varepsilon)}{\varepsilon^2}\, C_{0,c}\, K_{1,c}\, K_{6,c}\, r^{*(\ell-1)}.
\end{aligned}\right\} \quad (17.20)$$

The boundary-layer solutions $\alpha^{(c)}(r;\varepsilon)$ and $\xi^{(c)}(r;\varepsilon)$ are matched with the two-variable solutions $\alpha^{(o)}(r;\varepsilon)$ and $\xi^{(o)}(r;\varepsilon)$ valid at larger distances from the boundaries of the radiative core.

The matching condition relative to the divergence of the Lagrangian displacement is

$$\lim_{r\to\infty} \alpha^{(c)}(r;\varepsilon) = \lim_{r\to 0} \alpha^{(o)}(r;\varepsilon). \quad (17.21)$$

For small values of r, the fast independent variable $\tau(r)$ tends to

$$\tau(r) = \frac{1}{\varepsilon}\, K_{1,c}^{1/2}\, r, \quad (17.22)$$

so that it becomes equal to the boundary-layer coordinate $r^*(r)$. One then has

$$\lim_{r\to 0} \alpha^{(o)}(r;\varepsilon) = \frac{1}{\varepsilon}\, K_{1,c}^{1/2}\, K_{5,c}\, \frac{1}{r^*}\left(A_0^* \cos r^* + B_0^* \sin r^*\right). \quad (17.23)$$

On the other hand, for large values of r, it follows that

$$\lim_{r\to\infty} \alpha^{(c)}(r;\varepsilon) = \frac{\mu_0^{(c,1)}(\varepsilon)}{\varepsilon}\, A_{0,c} \left(\frac{2}{\pi}\right)^{1/2} K_{1,c}^{1/2}\, K_{5,c}\, \frac{1}{r^*} \sin\left(r^* - \frac{\ell\pi}{2}\right). \quad (17.24)$$

A matching of the boundary-layer solution $\alpha^{(c)}(r;\varepsilon)$ with the two-variable solution $\alpha^{(o)}(r;\varepsilon)$ is possible, when

$$\mu_0^{(c,1)}(\varepsilon) = \varepsilon^0 \quad (17.25)$$

and

$$\left.\begin{aligned}
A_0^* &= -A_{0,c} \left(\frac{2}{\pi}\right)^{1/2} \sin\frac{\ell\pi}{2}, \\
B_0^* &= A_{0,c} \left(\frac{2}{\pi}\right)^{1/2} \cos\frac{\ell\pi}{2}.
\end{aligned}\right\} \quad (17.26)$$

Next, the matching condition relative to the radial component of the Lagrangian displacement is

$$\lim_{r\to\infty} \xi^{(c)}(r;\varepsilon) = \lim_{r\to 0} \xi^{(o)}(r;\varepsilon). \quad (17.27)$$

By the matching of the boundary-layer solution $\alpha^{(c)}(r;\varepsilon)$ and the two-variable solution $\alpha^{(o)}(r;\varepsilon)$, the oscillatory parts of the boundary-layer solution $\xi^{(c)}(r;\varepsilon)$ and the two-variable solution $\xi^{(o)}(r;\varepsilon)$ are automatically matched. The non-oscillatory parts of the two solutions are matched, when

$$D_0^* = 0 \tag{17.28}$$

and

$$\mu_0^{(c,2)}(\varepsilon) = \varepsilon^{\ell+1}, \qquad C_0^* = C_{0,c}\, K_{1,c}^{(\ell+1)/2}\, K_{6,c}. \tag{17.29}$$

17.2.3 *Boundary-Layer Solutions on the Inner Side of the Boundary Between the Radiative Core and the Convective Envelope*

The solutions in the boundary layer situated on the inner side of the boundary between the radiative core and the convective envelope are similar to those in the boundary layer situated on the inner side of the boundary between the intermediate radiative zone and the convective envelope in stars that consist of a convective core, a radiative zone, and a convective envelope. These boundary-layer solutions are given in Sect. 16.4.1. One then has

$$\left.\begin{aligned}
\alpha^{(b)}(r;\varepsilon) &= \mu_0^{(b)}(\varepsilon)\,\sqrt{s_b^*}\left[A_{0,b}\,J_{1/3}\left(\frac{2}{3}\,\sqrt{K_{1,b}}\,s_b^{*3/2}\right)\right.\\
&\quad\left.+B_{0,b}\,J_{-1/3}\left(\frac{2}{3}\,\sqrt{K_{1,b}}\,s_b^{*3/2}\right)\right],\\
\xi^{(b)}(r;\varepsilon) &= \frac{c^2(r_b)}{g(r_b)}\\
&\quad\left[\alpha^{(b)}(r;\varepsilon) + \nu_0^{(b,2)}(\varepsilon)\,C_{0,b}\,s_b^* + \nu_0^{(b,3)}(\varepsilon)\,D_{0,b}\right],
\end{aligned}\right\} \tag{17.30}$$

where

$$s_b^*(r) = \frac{r_b - r}{\varepsilon^{2/3}}. \tag{17.31}$$

$A_{0,b}$, $B_{0,b}$, $C_{0,b}$, $D_{0,b}$ are arbitrary constants, and $\mu_0^{(b)}(\varepsilon)$, $\nu_0^{(b,2)}(\varepsilon)$, $\nu_0^{(b,3)}(\varepsilon)$, yet undetermined functions.

The boundary-layer solution $\alpha^{(b)}(r;\varepsilon)$ can be matched with the two-variable solution $\alpha^{(o)}(r;\varepsilon)$, valid at larger distances from the turning point $r = r_b$, also as in Sect. 16.4.1, except that here

$$\tau_{\mathrm{Rad}} = \frac{1}{\varepsilon} \int_0^{r_\mathrm{b}} K_1^{1/2}\left(r'\right) dr'. \tag{17.32}$$

The matching leads to

$$\mu_0^{(\mathrm{b})}(\varepsilon) = \varepsilon^{-1/6} \tag{17.33}$$

and

$$\left.\begin{aligned} A_0^* &= \frac{\sqrt{3}}{\sqrt{\pi}\, K_{1,\mathrm{b}}^{1/4}\, K_{5,\mathrm{b}}} \left[A_{0,\mathrm{b}} \sin\left(\tau_{\mathrm{Rad}} + \frac{\pi}{12}\right) \right. \\ &\quad \left. + B_{0,\mathrm{b}} \cos\left(\tau_{\mathrm{Rad}} - \frac{\pi}{12}\right) \right], \\ B_0^* &= -\frac{\sqrt{3}}{\sqrt{\pi}\, K_{1,\mathrm{b}}^{1/4}\, K_{5,\mathrm{b}}} \left[A_{0,\mathrm{b}} \cos\left(\tau_{\mathrm{Rad}} + \frac{\pi}{12}\right) \right. \\ &\quad \left. - B_{0,\mathrm{b}} \sin\left(\tau_{\mathrm{Rad}} - \frac{\pi}{12}\right) \right]. \end{aligned}\right\} \tag{17.34}$$

Moreover, the matching of the non-oscillatory parts of the boundary-layer solution $\xi^{(\mathrm{b})}(r;\varepsilon)$ and the two-variable solution $\xi^{(\mathrm{o})}(r;\varepsilon)$ leads to

$$\left.\begin{aligned} & C_{0,\mathrm{b}} = 0, \\ & \nu_0^{(\mathrm{b},3)}(\varepsilon) = \varepsilon^0, \qquad D_{0,\mathrm{b}} = \frac{g\left(r_\mathrm{b}\right)}{c^2\left(r_\mathrm{b}\right)}\, G_0^{(\mathrm{o})}\left(r_\mathrm{b}\right). \end{aligned}\right\} \tag{17.35}$$

At the turning point $r = r_\mathrm{b}$, the divergence of the Lagrangian displacement, and the radial component of the Lagrangian displacement and its first derivative have the values

$$\left.\begin{aligned} \alpha^{(\mathrm{b})}\left(r_\mathrm{b};\varepsilon\right) &= \varepsilon^{-1/6}\, \frac{3^{1/3}}{\Gamma(2/3)\, K_{1,\mathrm{b}}^{1/6}}\, B_{0,\mathrm{b}}, \\ \xi^{(\mathrm{b})}\left(r_\mathrm{b};\varepsilon\right) &= \frac{c^2\left(r_\mathrm{b}\right)}{g\left(r_\mathrm{b}\right)}\, \alpha^{(\mathrm{b})}\left(r_\mathrm{b};\varepsilon\right) + G_0^{(\mathrm{o})}\left(r_\mathrm{b}\right), \\ \left(\frac{d\xi^{(\mathrm{b})}(r;\varepsilon)}{dr}\right)_{r=r_\mathrm{b}} &= -\varepsilon^{-5/6}\, \frac{3^{7/6}\, \Gamma(2/3)\, K_{1,\mathrm{b}}^{1/6}}{2\pi}\, \frac{c^2\left(r_\mathrm{b}\right)}{g\left(r_\mathrm{b}\right)}\, A_{0,\mathrm{b}}. \end{aligned}\right\} \tag{17.36}$$

These equations are similar to Eqs. (16.125).

Boundary-layer solutions $\alpha^{(\mathrm{b})}\left(r_\mathrm{b};\varepsilon\right)$ and $\xi^{(\mathrm{b})}\left(r_\mathrm{b};\varepsilon\right)$ must be joined to the asymptotic solutions valid in the convective envelope. The latter solutions are similar to those developed in Sect. 16.4.2 for the convective envelope of a star containing a convective core and an intermediate radiative region, besides its convective envelope.

17.3 Eigenfrequency Equation

The eigenfrequency equation is derived by the elimination of the constants A_0^* and B_0^* from Eqs. (17.26) and Eqs. (16.157). The elimination leads to a system of two, algebraic, linear, homogeneous equations for the constants $A_{0,\mathrm{c}}$ and $B_{0,\mathrm{d}}$:

$$\left.\begin{aligned} &-A_{0,\mathrm{c}}\left(\frac{2}{\pi}\right)^{1/2}\sin\frac{\ell\pi}{2} - B_{0,\mathrm{d}}\frac{\sqrt{\pi}}{K_{1,\mathrm{b}}^{1/4}\,K_{5,\mathrm{b}}} \\ &\quad\left[K^{1/6}\sin\left(\tau_{\mathrm{Rad}}+\frac{\pi}{12}\right)+K^{-1/6}\cos\left(\tau_{\mathrm{Rad}}-\frac{\pi}{12}\right)\right]=0, \\ &A_{0,\mathrm{c}}\left(\frac{2}{\pi}\right)^{1/2}\cos\frac{\ell\pi}{2} + B_{0,\mathrm{d}}\frac{\sqrt{\pi}}{K_{1,\mathrm{b}}^{1/4}\,K_{5,\mathrm{b}}} \\ &\quad\left[K^{1/6}\cos\left(\tau_{\mathrm{Rad}}+\frac{\pi}{12}\right)-K^{-1/6}\sin\left(\tau_{\mathrm{Rad}}-\frac{\pi}{12}\right)\right]=0. \end{aligned}\right\} \tag{17.37}$$

The necessary and sufficient condition for the existence a non-trivial solution for the constants $A_{0,\mathrm{c}}$ and $B_{0,\mathrm{d}}$ leads to the eigenfrequency equation

$$\begin{aligned} &\left[K^{1/6}\cos\left(\tau_{\mathrm{Rad}}+\frac{\pi}{12}\right)-K^{-1/6}\sin\left(\tau_{\mathrm{Rad}}-\frac{\pi}{12}\right)\right]\sin\frac{\ell\pi}{2} \\ &\quad-\left[K^{1/6}\sin\left(\tau_{\mathrm{Rad}}+\frac{\pi}{12}\right)+K^{-1/6}\cos\left(\tau_{\mathrm{Rad}}-\frac{\pi}{12}\right)\right]\cos\frac{\ell\pi}{2}=0. \end{aligned} \tag{17.38}$$

In the supposition that $K \neq 0$, the eigenfrequency equation can be transformed into

$$\sin\left[\frac{\ell\pi}{2}-\left(\tau_{\mathrm{Rad}}+\frac{\pi}{12}\right)\right]-K^{-1/3}\cos\left[\frac{\ell\pi}{2}-\left(\tau_{\mathrm{Rad}}-\frac{\pi}{12}\right)\right]=0, \tag{17.39}$$

or, equivalently, into

$$\cos\psi - K^{-1/3}\sin\left(\psi-\frac{\pi}{6}\right)=0 \tag{17.40}$$

with

$$\psi=\frac{\pi}{2}-\frac{\ell\pi}{2}+\tau_{\mathrm{Rad}}+\frac{\pi}{12}.$$

Since it is of the same form as Eq. (16.160), the eigenfrequency equation can be written as

$$\tan\left(\psi-\theta_2\right)=\tan\frac{\pi}{3}, \tag{17.41}$$

where θ_2 is defined by Eq. (16.162). Hence, it is given by

$$\tau_{\mathrm{Rad}} \equiv \frac{[\ell(\ell+1)]^{1/2}}{|\sigma|} \int_0^{r_b} \left(\frac{N^2(r)}{r^2}\right)^{1/2} dr = \left(2n + \ell - \frac{1}{2}\right)\frac{\pi}{2} + \theta_2. \tag{17.42}$$

In the particular case in which the slope of N^2 is continuous at the boundary between the radiative core and the convective envelope, $\theta_2 = 0$. For this case, it is shown in Sect. 17.6 that the number n corresponds to the radial order of the g^+-mode considered.

Eigenfrequency equation (A.7) of Tassoul (1980), which is derived in the Cowling approximation, agrees with Eq. (17.42).

17.4 The Condition on the Eulerian Perturbation of the Gravitational Potential at $r = R$

Boundary condition (5.97) on the Eulerian perturbation of the gravitational potential at $r = R$ is imposed as in Sect. 16.4.4. It results that

$$C_{0,s} = 0 \tag{17.43}$$

and, because of the third equation (16.152), that

$$G_0^{(e)}(R) = 0. \tag{17.44}$$

The third equation (16.139) can be written explicitly as

$$C_0^{**}\, y_1(r_b) + D_0^{**}\, y_2(r_b) = C_0^{*}\, y_1(r_b), \tag{17.45}$$

or, because of Eq. (17.44), as

$$C_0^{**}\left[y_1(r_b) - \frac{y_1(R)}{y_2(R)}\, y_2(r_b)\right] = C_0^{*}\, y_1(r_b). \tag{17.46}$$

The constants C_0^* and C_0^{**} are thus related to each other. When they are set equal to zero, it follows that, at all points,

$$G_0^{(o)}(r) = 0, \qquad G_0^{(e)}(r) = 0, \tag{17.47}$$

so that the radial component of the Lagrangian displacement, as well as the divergence, is purely oscillatory in the radiative core.

17.5 Uniformly Valid Asymptotic Solutions

Uniformly valid first-order asymptotic solutions for the divergence and the radial component of the Lagrangian displacement can be constructed, in terms of the constants $A_{0,c}$ and $B_{0,d}$, in the radiative core and the convective envelope:

1. The first-order asymptotic solutions that are uniformly valid in the radiative core, from the star's centre to a distance sufficiently large from the boundary between the core and the convective envelope, can be expressed in the compact form

$$\left.\begin{aligned} \alpha^{(o,u)}(r;\varepsilon) &= A_{0,c}\, K_5(r)\, \tau^{1/2}\, J_{\ell+1/2}(\tau), \\ \xi^{(o,u)}(r;\varepsilon) &= \frac{c^2(r)}{g(r)}\, \alpha^{(o,u)}(r;\varepsilon). \end{aligned}\right\} \tag{17.48}$$

2. The first-order asymptotic solutions that are uniformly valid in the radiative core, from the lower boundary of the convective envelope to a distance sufficiently large from the star's centre, are given in a compact form by Eqs. (16.173).
3. The first-order asymptotic solutions that are uniformly valid in the convective envelope, from the lower boundary of the envelope to a distance sufficiently large from the star's surface, are given in a compact form by Eqs. (16.174).
4. The first-order asymptotic solutions that are uniformly valid in the convective envelope, from the star's surface to a distance sufficiently large from the lower boundary of the envelope, are given in a compact form by Eqs. (16.175).

17.6 Identification of the Radial Order of a g^+-Mode

In the radiative core, the nodes of the uniformly valid asymptotic solutions $\alpha^{(o,u)}(r;\varepsilon)$ and $\xi^{(o,u)}(r;\varepsilon)$, and $\alpha^{(b,u)}(r;\varepsilon)$ and $\xi^{(b,u)}(r;\varepsilon)$ coincide, so that the asymptotic representations of the functions $\alpha(r)$ and $\xi(r)$ there have the same number of nodes. Approximate positions of the nodes can be determined by means of the two-variable solutions $\alpha^{(o)}(r;\varepsilon)$ and $\xi^{(o)}(r;\varepsilon)$, which are given by Eqs. (17.4) and are valid in the radiative core at larger distances from the boundaries of the core.

By elimination of the constants A_0^* and B_0^* by means of Eqs. (17.26), the two-variable solution $\xi^{(o)}(r;\varepsilon)$ becomes

$$\xi^{(o)}(r;\varepsilon) = A_{0,c} \left(\frac{2}{\pi}\right)^{1/2} K_6(r)\, \sin\left(\tau - \frac{\ell\pi}{2}\right), \tag{17.49}$$

so that the positions of the nodes counted from the star's centre are given by

$$\tau^0 = (2j + \ell)\,\frac{\pi}{2}, \qquad j = 1, 2, 3, \ldots \tag{17.50}$$

These positions correspond to the positions determined by McMahon's expansion for large zeros of the Bessel function of the first kind $J_{\ell+1/2}(\tau)$, which is involved in the uniformly valid asymptotic solution $\xi^{(\mathrm{o,u})}(r;\varepsilon)$ given by the second equation (17.48) (Abramowitz & Stegun 1965).

On the other hand, by the elimination of the constants A_0^* and B_0^* by means of Eqs. (16.157), the two-variable solution $\xi^{(\mathrm{o})}(r;\varepsilon)$ can be written in the form given by Eq. (16.177). If the slope of N^2 is continuous at the boundary between the radiative core and the convective envelope, $K = 1$, and the positions of the nodes are given by Eq. (16.179):

$$\tau_b^0 = \left(2k - \frac{1}{2}\right)\frac{\pi}{2}, \qquad k = 1, 2, 3, \ldots \tag{17.51}$$

The first node counted form the boundary between the radiative core and the convective envelope is then located at $\tau_b^0(\text{first}) = 3\pi/4$. Since the last node counted from the star's centre must coincide with the first node counted from the boundary between the radiative core and the convective envelope, the relation holds

$$\tau^0(\text{last}) = \tau_{\text{Rad}} - \tau_b^0(\text{first}) = [2(n-1) + \ell]\,\frac{\pi}{2}.$$

By comparison with Eq. (17.50), it follows that the last node counted from the star's centre is associated with $j = n - 1$. Consequently, the radial component of the Lagrangian displacement displays $n - 1$ nodes in the radiative core. The same holds true for the divergence of the Lagrangian displacement.

The radial displacement displays an additional node in the convective envelope, as in a star that consists of a convective core, an intermediate radiative zone, and a convective envelope. Hence, the radial displacement of the higher-order g^+-mode whose eigenfrequency is associated with a number n in Eq. (17.42) displays n nodes along the radius, so that the mode has the radial order n.

17.7 Global Conclusion from the Asymptotic Theory of Low-Degree, Higher-Order p- and g^+-Modes

In the last two chapters, first-order asymptotic representations of low-degree, higher-order g^+-modes are developed for stars containing a radiative region and one or two convective regions. The radiative region may be a radiative envelope, a region intermediate between a convective core and a convective envelope, or a radiative core. When the core is convective, it is supposed to be in isentropic equilibrium.

A first conclusion of these asymptotic representations is that the radial component of the Lagrangian displacement is related to the divergence of the displacement as

$$\xi(r) = \frac{c^2(r)}{g(r)}\,\alpha(r), \tag{17.52}$$

except in a small region near $r = R$. Because of the relation existing between the Lagrangian and the Eulerian perturbation of any physical quantity, this result implies that the Eulerian perturbation of the pressure is everywhere zero, except in a small region near $r = R$. In the first-order asymptotic approximation, the Lagrangian perturbation of the pressure is thus entirely determined by the effect of the transport along the equilibrium pressure gradient. This approximation is commonly adopted in the theory of stellar structure, both in the derivation of the local criterion for stability against convection (see Sect. 3.5) and in the mixing length theory of convection. In terrestrial seismology, it is known as the subseismic approximation for low-frequency gravity modes (Smylie & Rochester 1981, Smylie et al. 1984). The use of the subseismic approximation in the whole star leads to an inadequate asymptotic representation of the radial component of the Lagrangian displacement near the star's surface for low-degree, higher-order g^+-modes (De Boeck et al. 1992).

Another conclusion of the asymptotic representations is that, in the various regions of the stellar models considered, the lowest-order asymptotic approximation of the Lagrangian displacement of a low-degree, higher-order g^+-mode is solution of a homogeneous second-order differential equation that stems from the equation

$$\frac{d^2\alpha}{dr^2} + K_2(r)\,\frac{d\alpha}{dr} + \left[\frac{K_1(r)}{\sigma^2} + K_3(r)\right]\alpha = 0. \tag{17.53}$$

The asymptotic solutions of neighbouring regions can adequately be matched with each other. At the turning points, situated at the boundaries between a radiative region and a convective region, the continuity of $d\xi(r)/dr$ is imposed in the foregoing asymptotic treatments, besides the continuity of $\alpha(r)$ and $\xi(r)$. The continuity of $d\xi(r)/dr$ can be replaced by the continuity of $d\alpha(r)/dr$ because of Eq. (17.52). Therefore, in the case of the exclusive use of Eq. (17.53), it suffices to impose the continuity of $\alpha(r)$ and $d\alpha(r)/dr$ at the turning points. These procedures make it possible to construct asymptotic solutions for the divergence of the Lagrangian displacement that are valid over the entire domain $[0, R]$, and even to derive the eigenfrequency equation without any preceding construction of lowest-order asymptotic approximations for the radial component of the Lagrangian displacement. This conclusion is similar to the conclusion drawn in Sect. 14.12 relative to low-degree, higher-order p-modes.

Equations (14.150) and (17.53) can be combined into a single second-order, homogeneous differential equation:

$$\frac{d^2\alpha}{dr^2} + K_2(r)\,\frac{d\alpha}{dr} + \left[\frac{\sigma^2}{c^2} + K_3(r) + \frac{K_1(r)}{\sigma^2}\right]\alpha = 0. \tag{17.54}$$

The conclusion is then that, from this equation, first asymptotic approximations of the divergence of the Lagrangian displacement can be constructed, and the eigenfrequency equation can be derived for low-degree, higher-order p- and g^+-modes of a quasi-static star.

Equation (17.54) is strictly valid in the case of an equilibrium sphere with uniform mass density (see Sect. 6.2.1). It can be brought into its self-adjoint form

$$\frac{d}{dr}\left(r^2\rho c^4 \frac{d\alpha}{dr}\right) + r^2\rho c^4\left[\frac{\sigma^2}{c^2} + K_3(r) + \frac{K_1(r)}{\sigma^2}\right]\alpha = 0. \tag{17.55}$$

On the basis of this equation, Cowling's reasoning relative to the existence of p- and g-modes can be started over again. Equation (17.55) tends to a Sturm–Liouville equation with singular end points, for both large and small values of $|\sigma^2|$. In the first limit, the equation suggests the existence of an infinite spectrum of increasing, discrete eigenvalues σ^2 for each degree ℓ, and in the second limit, the equation suggests the existence of an infinite spectrum of increasing, discrete eigenvalues $1/|\sigma^2|$ for each degree ℓ different from zero. Equation (17.55) has the advantage over Eqs. (10.13) and (10.14) to contain no terms that become increasingly large in some part of the interval $(0, R)$ and therefore must be excluded from that part. The eigenfunctions $\alpha(r)$ of the p- and g-modes all have the same numbers of nodes as the corresponding modes in the Sturm–Liouville eigenvalue problems. An obvious inference is that it is preferable to determine the radial orders of the p- and g-modes on the basis of the numbers of nodes displayed by their eigenfunctions $\alpha(r)$ rather than on the basis of those displayed by their eigenfunctions $\xi(r)$.

Chapter 18
High-Degree, Low-Order Modes

18.1 Introduction

As a preliminary to the study of high-degree, low-order modes in stars, the propagation diagram of the polytropic model with index $n = 3$ for the degree $\ell = 100$ and $\Gamma_1 = 5/3$ is represented in Fig. 18.1. One observes that the lower part of the A-region is reduced to a narrow strip near the surface, which terminates in a point, and that the G-region displays a peak in the surface layers, which exceeds the top of this region in the main body of the model. Hence, for high degrees ℓ, low-order p-, f-, and g^+-modes are possible that are trapped near the surface. In addition, low-order g^+-modes may be expected that are trapped near a local maximum of the upper boundary of the G-region. An asymptotic study of these two types of modes was made by Christensen-Dalsgaard (1980) and is presented in this chapter.

18.2 High-Degree, Low-Order p- and g^+-Modes Trapped Near the Surface

For high-degree, low-order modes trapped near the surface, convenient starting equations are the radial component of Eq. (5.4) considered after separation of the angular variables by means a spherical harmonic $Y_\ell^m(\theta, \phi)$, and Eq. (5.93) for the divergence of the Lagrangian displacement:

$$\sigma^2 \xi = \frac{d}{dr}\left(\Phi' + \frac{P'}{\rho}\right) + \frac{N^2}{g}\, c^2\, \alpha, \tag{18.1}$$

$$\alpha = \frac{1}{r^2}\frac{d}{dr}\left(r^2\, \xi\right) - \frac{\ell(\ell+1)}{r^2}\,\eta. \tag{18.2}$$

When $\ell > 0$ and $\sigma^2 \neq 0$, the function $\eta(r)$ can be expressed, according to Eq. (5.89), as

P. Smeyers and T. Van Hoolst, *Linear Isentropic Oscillations of Stars: Theoretical Foundations*, Astrophysics and Space Science Library 371,
DOI 10.1007/978-3-642-13030-4_19, © Springer-Verlag Berlin Heidelberg 2010

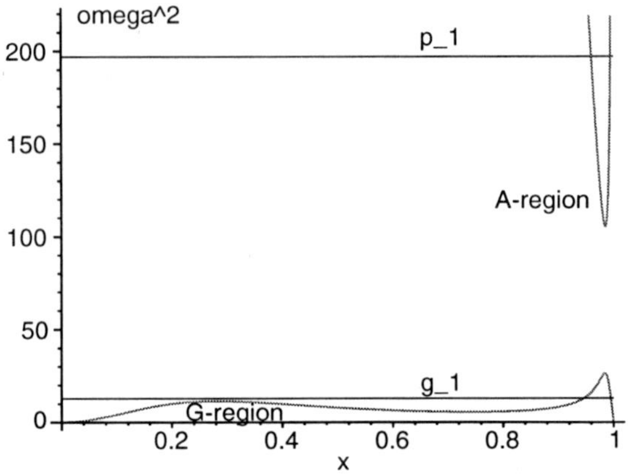

Fig. 18.1 Propagation diagram of the polytropic model with index $n = 3$, for the degree $\ell = 100$ and $\Gamma_1 = 5/3$. The square of the dimensionless frequency ω is plotted versus the relative radial distance $x = r/R$. The positions of the horizontal lines correspond to the values of ω^2 for the envelope modes p_1 and g_1^+

$$\eta = \frac{1}{\sigma^2}\left(\Phi' + \frac{P'}{\rho}\right). \tag{18.3}$$

In the case of envelope modes, the Eulerian perturbation of the gravitational potential may be neglected. By elimination of $\eta(r)$, and elimination of $P'(r)$ by means of the equation

$$\frac{P'}{\rho} = g\,\xi - c^2\,\alpha, \tag{18.4}$$

Equations (18.1) and (18.2) are transformed into

$$\left.\begin{aligned} \sigma^2\,\xi &= \frac{d}{dr}\left(g\,\xi - c^2\,\alpha\right) + \frac{N^2}{g}\,c^2\,\alpha, \\ \alpha &= \frac{1}{r^2}\,\frac{d}{dr}\left(r^2\,\xi\right) - \frac{\ell(\ell+1)}{r^2}\,\frac{1}{\sigma^2}\left(g\,\xi - c^2\,\alpha\right). \end{aligned}\right\} \tag{18.5}$$

It is convenient to introduce the downwards vertical coordinate $z = R - r$ and the associated vertical component of the Lagrangian displacement $\xi_z = -\,\xi$. When the gravity is considered to be constant, the equations become

$$\left.\begin{aligned} g\,\frac{d\xi_z}{dz} + \sigma^2\,\xi_z &= -\frac{d}{dz}\left(c^2\,\alpha\right) - \frac{N^2}{g}\,c^2\,\alpha, \\ \alpha &= \frac{d\xi_z}{dz} - \frac{2}{R-z}\,\xi_z + \frac{\ell(\ell+1)}{r^2}\,\frac{1}{\sigma^2}\left(g\,\xi_z + c^2\,\alpha\right). \end{aligned}\right\} \tag{18.6}$$

For $R \to \infty$, the modes can be regarded as modes of a plane-parallel layer with a constant gravity that depend on the horizontal coordinates x and y by $\exp\left[\mathrm{i}\left(k_x x + k_y y\right)\right]$. The coefficient of the second term in the right-hand member of the second Eq. (18.6) tends to zero, and the coefficient $\ell(\ell + 1)/r^2$ of the third term can be replaced by the square of the horizontal wave number

$$k_h = \sqrt{k_x^2 + k_y^2}. \tag{18.7}$$

By elimination of $d\xi_z/dz$ from Eqs. (18.6), it results that

$$\left(\sigma^4 - g^2 k_h^2\right)\xi_z = -c^2\left\{\sigma^2 \frac{d\alpha}{dz} + \left[\sigma^2 \frac{d}{dz}\ln\left(\rho c^2\right) - g\,k_h^2\right]\alpha\right\}. \tag{18.8}$$

The oscillation modes of a plane-parallel layer with a constant gravity and a uniform temperature gradient were studied by Lamb (1932). Like Lamb, Christensen-Dalsgaard used a polytropic equilibrium model with index n, in which he introduced a uniform temperature by adopting the perfect gas law with constant molecular weight. When γ is the ratio of the specific heats, the square of the sound speed can be expressed as $c^2 = g\gamma z/(n + 1)$, in accordance with the first term in the right-hand member of Eq. (6.56). Equation (18.8) and the first equation (18.6) can then be written as

$$\left.\begin{aligned}
\left(\sigma^4 - g^2 k_h^2\right)\xi_z &= -\sigma^2 c^2 \frac{d\alpha}{dz} - g\left(\sigma^2\gamma - k_h^2 c^2\right)\alpha,\\
g\frac{d\xi_z}{dz} + \sigma^2\xi_z &= -c^2\frac{d\alpha}{dz} - (\gamma - 1)\,g\,\alpha.
\end{aligned}\right\} \tag{18.9}$$

These equations correspond to Lamb's equations (312.7) and (312.8). From them, Christensen–Dalsgaard derived the second-order differential equation for $\alpha(z)$, equivalent to Lamb's equation (312.13),

$$z\frac{d^2\alpha}{dz^2} + (n + 2)\frac{d\alpha}{dz} + (C - k_h z)\,k_h\,\alpha = 0, \tag{18.10}$$

where

$$\left.\begin{aligned}
C &= \frac{n+1}{\gamma}\tau^2 + \left(\frac{\nabla_{\mathrm{ad}}}{\nabla} - 1\right)\tau^{-2}, \qquad \tau^2 = \frac{\sigma^2}{g\,k_h},\\
\nabla_{\mathrm{ad}} &\equiv \left(\frac{\partial \ln T}{\partial \ln P}\right)_S = \frac{\gamma - 1}{\gamma}, \qquad \nabla \equiv \frac{d\ln T}{d\ln P} = \frac{1}{n+1}.
\end{aligned}\right\} \tag{18.11}$$

The point $z = 0$ is a singular point of the differential equation.

As given by Lamb, the solution that is regular at $z = 0$, is

$$\alpha(z) = {}_1F_1\left(-B; n+2; 2k_h z\right), \tag{18.12}$$

where ${}_1F_1$ is the confluent hypergeometric function, and

$$B \equiv \frac{1}{2}\left[C - (n+2)\right]. \tag{18.13}$$

For the solution to remain bounded as $z \to \infty$, Christensen–Dalsgaard connected the confluent hypergeometric function with a Laguerre polynomial by imposing that B is a non-negative integer:

$$B = k - 1, \qquad k = 1, 2, 3, \ldots \tag{18.14}$$

(Abramowitz & Stegun 1965). One then derives the quadratic eigenvalue equation for τ^2

$$\frac{n+1}{\gamma}\tau^4 - (n+2k)\,\tau^2 + \left(\frac{\nabla_{\rm ad}}{\nabla} - 1\right) = 0. \tag{18.15}$$

Its discriminant can be written as

$$D = (n+2k)^2\left[1 - \frac{(n+1)^2\,(\nabla_{\rm ad} - \nabla)}{\gamma\left(\frac{n}{2} + k\right)^2}\right]. \tag{18.16}$$

It is manifestly positive when $\nabla > \nabla_{\rm ad}$. In the case $\nabla < \nabla_{\rm ad}$, one observes that

$$\frac{(n+1)^2\,(\nabla_{\rm ad} - \nabla)}{\gamma\left(\frac{n}{2} + k\right)^2} < \frac{(n+1)^2\,\nabla_{\rm ad}}{\gamma\left(\frac{n}{2} + \frac{1}{2}\right)^2} = \frac{4(\gamma - 1)}{\gamma^2} < 1, \tag{18.17}$$

since $(\gamma - 2)^2 > 0$. Hence, the quadratic eigenvalue equation admits of two distinct real roots for each value of k:

$$\tau_k^2 = \frac{\gamma}{n+1}\left(\frac{n}{2} + k\right)\left\{1 \pm \left[1 - \frac{(n+1)^2\,(\nabla_{\rm ad} - \nabla)}{\gamma\left(\frac{n}{2} + k\right)^2}\right]^{1/2}\right\}. \tag{18.18}$$

Two infinite spectra of discrete eigenvalues τ^2 thus exist: a first spectrum consists of the eigenvalues τ^2 associated with the plus sign, which are positive and increase indefinitely with k; a second spectrum consists of the eigenvalues τ^2 associated with the minus sign, which decrease in absolute value towards zero for $k \to \infty$ and are positive as $\nabla < \nabla_{\rm ad}$, equal to zero as $\nabla = \nabla_{\rm ad}$, and negative as $\nabla > \nabla_{\rm ad}$.

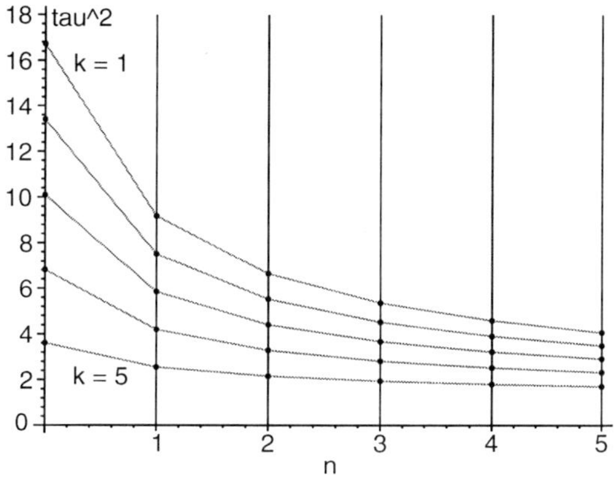

Fig. 18.2 Eigenvalues τ^2 determined by the use of the plus sign in Eq. (18.18), as functions of the polytropic index n, for $k = 1, \ldots, 5$ and $\gamma = 5/3$

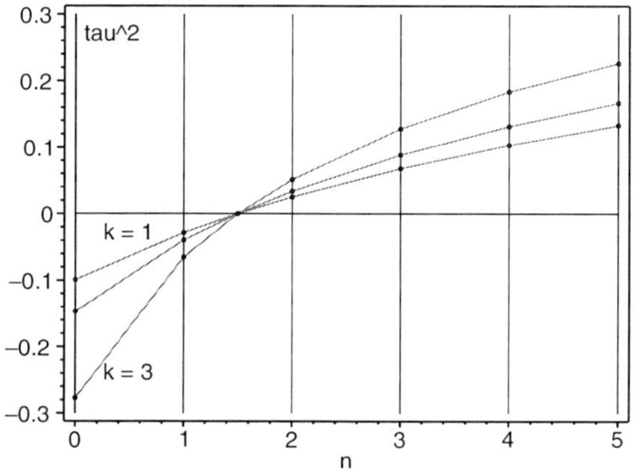

Fig. 18.3 Eigenvalues τ^2 determined by the use of the minus sign in Eq. (18.18), as functions of the polytropic index n, for $k = 1, 2, 3$ and $\gamma = 5/3$

In Figs. 18.2 and 18.3, eigenvalues τ^2 are represented as functions of the polytropic index n for a few values of k and for $\gamma = 5/3$. The eigenvalues represented in Fig. 18.2 result from the use of the plus sign in Eq. (18.18), and those represented in Fig. 18.3, from the use of the minus sign.

From an eigenvalue τ^2, one derives the square of the dimensionless eigenfrequency $\omega = \left[R^3/(GM)\right]^{1/2}\sigma$ associated with a given degree ℓ by using the relations $\tau^2 = \sigma^2/\,(g\,k_h)$ and $k_h = \sqrt{\ell(\ell+1)}/R$. It then follows that

$$\omega^2 = \sqrt{\ell(\ell+1)}\,\tau^2 \simeq \ell\,\tau^2. \tag{18.19}$$

For example, for the polytropic model $n = 3$ and the degree $\ell = 100$, one gets $\omega^2_{k=1} = 197$ by the use of the plus sign in Eq. (18.18), and $\omega^2_{k=1} = 12.9$ by the use of the minus sign. These eigenfrequencies are represented by horizontal lines in the propagation diagram displayed in Fig. 18.1. As shown hereafter, the first eigenfrequency corresponds to that of the mode p_1, and the second eigenfrequency, to that of the mode g_1^+. The horizontal lines cross the A- and G-region, respectively, near the surface.

With the two eigenvalues τ_k^2 determined by means of Eq. (18.18) only one eigenfunction $\alpha(z)$ is associated. In terms of the relative radial distance from the surface, $z' = z/R$, the eigenfunctions of the lowest-order modes associated with the degree $\ell = 100$ in the polytropic model with index $n = 3$ are given by

$$\left.\begin{aligned}
\alpha_{k=1}(z) &= \exp\left(-10.49\,z'\right),\\
\alpha_{k=2}(z) &= \left(1 - 4.195\,z'\right)\exp\left(-10.49\,z'\right),\\
\alpha_{k=3}(z) &= \left(1 - 8.390\,z' + 14.67\,z'^2\right)\exp\left(-10.49\,z'\right),\\
\alpha_{k=4}(z) &= \left(1 - 12.59\,z' + 44.00\,z'^2 - 43.95\,z'^3\right)\exp\left(-10.49\,z'\right).
\end{aligned}\right\} \tag{18.20}$$

The first three eigenfunctions are represented in Fig. 18.4. They successively display no node, one node, and two nodes. The associated low-order modes are thus respectively a p_1- or a g_1^+-mode, a p_2- or a g_2^+-mode, and a p_3- or a g_3^+-mode.

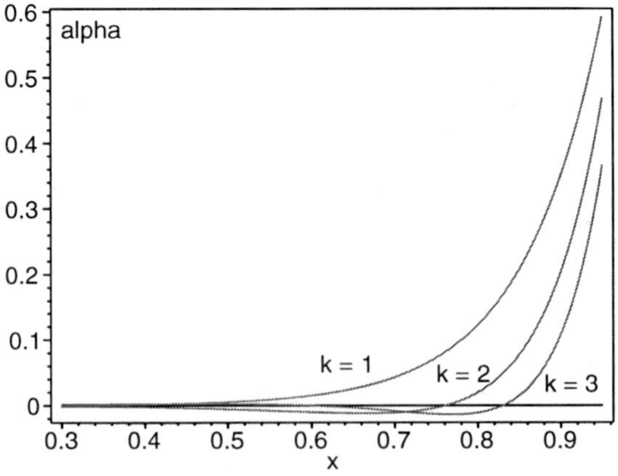

Fig. 18.4 Eigenfunctions $\alpha(x)$ associated with $\ell = 100$ and $k = 1, 2, 3$ in the polytropic model with index $n = 3$

In addition to the modes found up to here, a divergence-free envelope mode with a frequency different from zero is possible. For any value of the polytropic index, it follows from the first Eq. (18.9) that

$$\sigma^2 = g\, k_h, \tag{18.21}$$

or, equivalently, that

$$\tau^2 = 1, \tag{18.22}$$

and from the second Eq. (18.9), that

$$\xi_z = A\, \exp\left(-k_h z\right), \tag{18.23}$$

where A is an arbitrary constant. Equation (18.21) is similar to Eq. (10.18). The divergence-free mode is as an envelope f-mode.

18.3 High-Degree, Low-Order g^+-Modes Trapped Near a Maximum of the Boundary of the G-Region

For a polytropic model with index $n = 3$, the relative difference between N^2 and the upper boundary of the G-region increases from nearly 1.2×10^{-5} to 0.0032 in the region extending from the relative radial distance $x = 0.17$ to the relative radial distance $x = 0.84$, so that the upper boundary of the G-region is there almost determined by N^2. Be r_m the radial distance of a local maximum of N^2, and suppose that N^2 has a non-vanishing second derivative at that distance, so that it can be approximated as

$$N^2 = N_\mathrm{m}^2 \left[1 - \beta^2 \left(\frac{r}{r_\mathrm{m}} - 1\right)^2\right], \tag{18.24}$$

where $N_\mathrm{m}^2 = N^2\left(r_\mathrm{m}\right)$, and

$$\beta^2 = -\frac{1}{2}\, \frac{r_\mathrm{m}^2}{N_\mathrm{m}^2} \left(\frac{d^2 N^2}{d r^2}\right)_{r\, =\, r_\mathrm{m}}. \tag{18.25}$$

In this context, Christensen–Dalsgaard used Eqs. (7.36), in which he neglected the variations of $h(r)\left(\sigma^{-2} - S_\ell^{-2}\right)$ and $h^{-1}(r)\, r^{-2}$ around r_m. The equations then are

$$\left.\begin{aligned} \frac{dv}{dr} &= \ell(\ell+1)\left[\sigma^{-2} - S_\ell^{-2}\left(r_\mathrm{m}\right)\right] h\left(r_\mathrm{m}\right) w, \\ \frac{dw}{dr} &= \left\{\sigma^2 - N_\mathrm{m}^2\left[1 - \beta^2\left(\frac{r}{r_\mathrm{m}} - 1\right)^2\right]\right\} \frac{v}{r_\mathrm{m}^2\, h\left(r_\mathrm{m}\right)}. \end{aligned}\right\} \tag{18.26}$$

By elimination of the function $w(r)$ and introduction of the new independent variable

$$\zeta = \mu\left(\frac{r}{r_m} - 1\right), \tag{18.27}$$

one derives the second-order differential equation

$$\frac{d^2 v}{d\zeta^2} - \left(a + \frac{\zeta^2}{4}\right) v = 0, \tag{18.28}$$

where

$$\mu^2 = 2\left[\beta^2 \frac{N_{\rm m}^2}{\sigma^2}\, \ell(\ell+1)\left(1 - \frac{\sigma^2}{S_\ell^2\,(r_{\rm m})}\right)\right]^{1/2}, \tag{18.29}$$

and

$$a = -\frac{\ell(\ell+1)}{\mu^2}\left(1 - \frac{\sigma^2}{S_\ell^2\,(r_{\rm m})}\right)\left(\frac{N_{\rm m}^2}{\sigma^2} - 1\right) \tag{18.30}$$

or, equivalently,

$$a = -\frac{1}{2}\left(\frac{N_{\rm m}^2}{\sigma^2} - 1\right)\left[\frac{\sigma^2}{N_{\rm m}^2}\,\frac{\ell(\ell+1)}{\beta^2}\left(1 - \frac{\sigma^2}{S_\ell^2\,(r_{\rm m})}\right)\right]^{1/2}. \tag{18.31}$$

It is supposed that $\sigma^2 < N_{\rm m}^2$ and $N_{\rm m}^2 < S_\ell^2\,(r_{\rm m})$, so that μ^2 is real, and $a < 0$.

The solutions are parabolic cylinder functions. A standard solution is given by

$$v = C_1\, U(a,\zeta) + C_2\, V(a,\zeta), \tag{18.32}$$

where

$$\left.\begin{aligned} U(a,\zeta) &= \cos\left[\pi\left(\frac{1}{4} + \frac{1}{2}a\right)\right] Y_1 - \sin\left[\pi\left(\frac{1}{4} + \frac{1}{2}a\right)\right] Y_2, \\ V(a,\zeta) &= \frac{1}{\Gamma\left(\frac{1}{2} - a\right)} \\ &\quad \left\{\sin\left[\pi\left(\frac{1}{4} + \frac{1}{2}a\right)\right] Y_1 + \cos\left[\pi\left(\frac{1}{4} + \frac{1}{2}a\right)\right] Y_2\right\}, \end{aligned}\right\} \tag{18.33}$$

with

$$Y_1 = \frac{1}{\sqrt{\pi}}\,\frac{\Gamma\left(\frac{1}{4} - \frac{1}{2}a\right)}{2^{a/2+1/4}}\, y_1, \qquad Y_2 = \frac{1}{\sqrt{\pi}}\,\frac{\Gamma\left(\frac{3}{4} - \frac{1}{2}a\right)}{2^{a/2-1/4}}\, y_2, \tag{18.34}$$

and

$$\left.\begin{aligned} y_1 &= \exp\left(-\zeta^2/4\right) M\left(\frac{1}{2}a + \frac{1}{4}, \frac{1}{2}, \frac{1}{2}\zeta^2\right), \\ y_2 &= \zeta \exp\left(-\zeta^2/4\right) M\left(\frac{1}{2}a + \frac{3}{4}, \frac{3}{2}, \frac{1}{2}\zeta^2\right). \end{aligned}\right\} \tag{18.35}$$

M is a confluent hypergeometric function of Kummer, and C_1 and C_2 are arbitrary constants (Abramowitz & Stegun 1965).

When $\zeta >> |a|$ and $\zeta \to \infty$, the functions (a, ζ) and $V(a, \zeta)$ have the asymptotic behaviour

$$\left.\begin{aligned} U(a,\zeta) &\sim \exp\left(-\zeta^2/4\right) \zeta^{-a-1/2}, \\ V(a,\zeta) &\sim \sqrt{\frac{2}{\pi}} \exp\left(\zeta^2/4\right) \zeta^{a-1/2}. \end{aligned}\right\} \tag{18.36}$$

Furthermore, the relation holds

$$U(a,-x) = \frac{\pi}{\Gamma(1/2+a)} V(a,x) - \sin(\pi a)\, U(a,x). \tag{18.37}$$

Hence, for v to remain finite as $\zeta \to \infty$, one must set $C_2 = 0$ in solution (18.32), and for v to remain finite as $\zeta \to -\infty$, it is required that, in relation (18.37),

$$\frac{1}{\Gamma(1/2+a)} = 0, \tag{18.38}$$

or, equivalently, that

$$a = -\frac{1}{2} - k, \qquad k = 0, 1, 2, \ldots \tag{18.39}$$

The conditions lead to the eigenvalue equation

$$K = \left(\frac{N_{\mathrm{m}}^2}{\sigma^2} - 1\right) \left[\frac{\sigma^2}{N_{\mathrm{m}}^2} \frac{\ell(\ell+1)}{\beta^2} \left(1 - \frac{\sigma^2}{S_\ell^2(r_{\mathrm{m}})}\right)\right]^{1/2} \tag{18.40}$$

with $K = 2k + 1$. In this cubic equation for σ^2, the term $\sigma^2/S_\ell^2(r_{\mathrm{m}})$ may be neglected compared with 1 for sufficiently high degrees ℓ. The resulting quadratic equation yields the positive root

$$1 - \frac{\sigma^2}{N_{\mathrm{m}}^2} = \frac{K\beta}{2\ell(\ell+1)} \left(-K\beta + \sqrt{K^2\beta^2 + 4\ell(\ell+1)}\right). \tag{18.41}$$

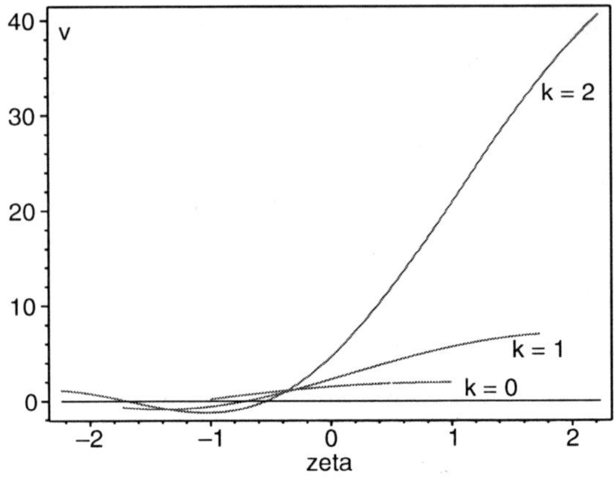

Fig. 18.5 The eigenfunctions $v(\zeta)$ associated with $k = 0, 1, 2$ in their oscillatory region near the local maximum $r = r_{\mathrm{m}}$ of the upper boundary of the G-region

Hence, for sufficiently high degrees ℓ,

$$1 - \frac{\sigma^2}{N_{\mathrm{m}}^2} \simeq \beta \, \frac{2k+1}{\ell + 1/2}, \qquad k = 0, 1, 2, \ldots \tag{18.42}$$

so that the relative difference $\left(N_{\mathrm{m}}^2 - \sigma^2\right) / N_{\mathrm{m}}^2$ tends to zero approximately as ℓ^{-1}.

From Eq. (18.28), it follows that the eigenfunctions $v(\zeta)$ are oscillatory in the region near $r = r_{\mathrm{m}}$ where

$$a + \frac{\zeta^2}{4} < 0. \tag{18.43}$$

Christensen–Dalsgaard noted that the eigenfunctions have successively no zero for $k = 0$, one zero for $k = 1$, two zeros for $k = 2, \ldots$ In illustration, the eigenfunctions $v(\zeta)$ for the modes associated with $k = 0, 1, 2$ are represented in Fig. 18.5. In accordance with Christensen–Dalsgaard's suggestion, the mode associated with $k = 0$ can be classified as a local f-mode, and the other modes, as local g^+-modes.

Chapter 19
Period Changes in a Rapidly Evolving Pulsating Star

19.1 Introduction

In a pulsating star that evolves on its Helmholtz–Kelvin time-scale, the rapid evolution leads to a slow change of the single or multiple pulsation periods. A theoretical expression for this change has been derived by Smeyers & Bruggen (1984) in the case of a radial pulsation mode, and by Bruggen & Smeyers (1987) in the case of a non-radial pulsation mode. The star's pulsation is here supposed to remain isentropic during the evolution[1].

19.2 Appropriate Equations

Consider a pulsating spherically symmetric star with a mass M and a varying radius R that is evolving on its Helmholtz–Kelvin time-scale. The star is in thermal imbalance. Radial motions of contraction and dilation take place in it due to the evolution. The rapidly evolving star is nevertheless regarded to be in an instantaneous hydrostatic equilibrium. Furthermore, the period of the pulsation, radial or non-radial, is assumed to be of the order of the star's dynamic time-scale and to be much shorter than the star's Helmholz–Kelvin time scale.

Dimensionless physical quantities in terms of the star's mass, and the star's radius at the time considered are used. The time t, the pressure P, the mass density ρ, the gravity g, the isentropic sound velocity c are expressed respectively in the unit $\left[R^3/(GM)\right]^{1/2}$, $GM^2/\left(4\pi R^4\right)$, $M/\left(4\pi R^3\right)$, GM/R^2, $(GM/R)^{1/2}$.

It is appropriate to adopt a Lagrangian description for both the evolutionary and the pulsational motions of the mass elements. The elements are characterised by the following three parameters: the mass m contained inside the sphere with a radius

[1] This chapter is partially a reproduction of Bruggen, P., Smeyers, P.: Theoretical expressions for evolutionary period changes in non-radially pulsating stars. Astronomy & Astrophysics **186**, 170–176 (1987). With permission from Astronomy & Astrophysics, http://www.aanda.org.

P. Smeyers and T. Van Hoolst, *Linear Isentropic Oscillations of Stars: Theoretical Foundations*, Astrophysics and Space Science Library 371,
DOI 10.1007/978-3-642-13030-4_20,

equal to the distance of the mass element from the star's centre, and the angular variables θ and ϕ. Hence, the spherical coordinates of a mass element in the evolving star are represented as

$$r = r(t, m), \qquad \theta = \theta, \qquad \phi = \phi. \tag{19.1}$$

Correspondingly, the physical quantities, as the mass density, the pressure, the gravitational potential, are expressed as

$$\rho = \rho(t, m), \qquad P = P(t, m), \qquad \Phi = \Phi(t, m). \tag{19.2}$$

The instantaneous hydrostatic equilibrium in the layer characterised by the mass m is given by

$$\frac{\partial P}{\partial m} = -\rho \, \frac{\partial \Phi}{\partial m}. \tag{19.3}$$

The condition can also be written as

$$\frac{1}{\frac{\partial r}{\partial m}} \, \frac{\partial P}{\partial m} = -\rho \, g \tag{19.4}$$

with

$$g = \frac{m}{r^2}. \tag{19.5}$$

The infinitesimal mass dm contained in the layer between m and $m + dm$ obeys the equation

$$\rho \, r^2 \, \frac{\partial r}{\partial m} = 1. \tag{19.6}$$

For the description of the oscillatory motions of the mass elements in the evolving star, Bruggen & Smeyers (1987) started from the covariant form of the equations of motion in terms of generalised coordinates q^1, q^2, q^3:

$$\frac{\partial}{\partial t}\left(g_{jk}\, \dot{q}^k\right) + \dot{q}^\ell \, \nabla_\ell \left(g_{jk}\, \dot{q}^k\right) = -\frac{\partial \Phi}{\partial q^j} - \frac{1}{\rho}\, \frac{\partial P}{\partial q^j}, \qquad j = 1, 2, 3 \tag{19.7}$$

or, equivalently,

$$\frac{d}{dt}\left(g_{jk}\, \dot{q}^k\right) = -\frac{\partial \Phi}{\partial q^j} - \frac{1}{\rho}\, \frac{\partial P}{\partial q^j}, \qquad j = 1, 2, 3 \tag{19.8}$$

[see Eq. (Appendix C.26)]. With $q^1 = r$, $q^2 = \theta$, $q^3 = \phi$, these equations can be written in the Lagrangian form

$$
\left.\begin{aligned}
\left[\frac{\partial}{\partial t}\left(g_{1k}\,\dot{q}^k\right)\right]_{m,\theta,\phi} &= -\frac{1}{\frac{\partial r}{\partial m}}\left(\frac{\partial \Phi}{\partial m}+\frac{1}{\rho}\frac{\partial P}{\partial m}\right),\\
\left[\frac{\partial}{\partial t}\left(g_{2k}\,\dot{q}^k\right)\right]_{m,\theta,\phi} &= -\left(\frac{\partial \Phi}{\partial \theta}+\frac{1}{\rho}\frac{\partial P}{\partial \theta}\right),\\
\left[\frac{\partial}{\partial t}\left(g_{3k}\,\dot{q}^k\right)\right]_{m,\theta,\phi} &= -\left(\frac{\partial \Phi}{\partial \phi}+\frac{1}{\rho}\frac{\partial P}{\partial \phi}\right).
\end{aligned}\right\} \tag{19.9}
$$

By taking the Lagrangian perturbation of both members, considering that $g_{11} = 1$, $\dot{\theta} = 0$, $\dot{\phi} = 0$, and taking into account the instantaneous hydrostatic equilibrium in the evolving star, one gets

$$
\left.\begin{aligned}
&\left[\frac{\partial}{\partial t}\left(g_{1k}\,\frac{\partial\left(\delta q^k\right)}{\partial t}\right)\right]_{m,\theta,\phi}\\
&\quad = -\frac{1}{\frac{\partial r}{\partial m}}\left(\frac{\partial(\delta\Phi)}{\partial m}-\frac{\delta\rho}{\rho^2}\frac{\partial P}{\partial m}+\frac{1}{\rho}\frac{\partial(\delta P)}{\partial m}\right),\\
&\left[\frac{\partial}{\partial t}\left(g_{2k}\,\frac{\partial\left(\delta q^k\right)}{\partial t}\right)\right]_{m,\theta,\phi} = -\left(\frac{\partial(\delta\Phi)}{\partial \theta}+\frac{1}{\rho}\frac{\partial(\delta P)}{\partial \theta}\right),\\
&\left[\frac{\partial}{\partial t}\left(g_{3k}\,\frac{\partial\left(\delta q^k\right)}{\partial t}\right)\right]_{m,\theta,\phi} = -\left(\frac{\partial(\delta\Phi)}{\partial \phi}+\frac{1}{\rho}\frac{\partial(\delta P)}{\partial \phi}\right),
\end{aligned}\right\} \tag{19.10}
$$

or, in terms of the operators U_{ik} defined by Eqs. (4.14),

$$
\left.\begin{aligned}
\left[\frac{\partial}{\partial t}\left(g_{1k}\,\frac{\partial\left(\delta q^k\right)}{\partial t}\right)\right]_{m,\theta,\phi} &= -U_{1k}\,\delta q^k,\\
\left[\frac{\partial}{\partial t}\left(g_{2k}\,\frac{\partial\left(\delta q^k\right)}{\partial t}\right)\right]_{m,\theta,\phi} &= -U_{2k}\,\delta q^k,\\
\left[\frac{\partial}{\partial t}\left(g_{3k}\,\frac{\partial\left(\delta q^k\right)}{\partial t}\right)\right]_{m,\theta,\phi} &= -U_{3k}\,\delta q^k.
\end{aligned}\right\} \tag{19.11}
$$

To these equations, the following equations are added: continuity equation (2.53)

$$
\delta\rho = -\rho\,\alpha, \tag{19.12}
$$

with Eq. (2.35), which relates the divergence of the Lagrangian displacement to the components of the displacement,

$$\alpha = \frac{1}{r^2 \dfrac{\partial r}{\partial m}} \frac{\partial}{\partial m}\left(r^2 \, \delta r\right) + \frac{1}{\sin\theta}\frac{\partial}{\partial \theta}\left(\sin\theta \, \delta\theta\right) + \frac{\partial(\delta\phi)}{\partial\phi}; \tag{19.13}$$

Equation (3.24), which expresses that the star's pulsation is isentropic,

$$\delta P = \frac{\Gamma_1 P}{\rho}\, \delta\rho; \tag{19.14}$$

and an equation for the Lagrangian perturbation of the gravitational potential.

19.3 Two-Time-Variable Expansion Procedure

The motions of the mass elements in the evolving pulsating star consist of a superimposition of a more rapid oscillatory motion, with a time scale of the order of the star's dynamic time-scale, on a slower evolutionary motion, with the star's Helmholtz–Kelvin time-scale. Hence, two largely different time scales are involved, and the motions of the mass elements related to the shortest time scale are periodic. The use of an expansion procedure with two time variables is therefore appropriate. The slow time variable is defined as

$$\tilde{t}(t) = C\, t, \tag{19.15}$$

where C is the ratio of the star's Helmholtz–Kelvin time-scale to the star's dynamic time-scale at the time t considered, as defined by Eq. (3.21). The ratio C is supposed to be small. Moreover, the fast time variable is defined as

$$dt^{+}(t) = \mu\left(\tilde{t}\,\right) dt, \tag{19.16}$$

where the function $\mu\left(\tilde{t}\,\right)$ is specified subsequently (Kevorkian & Cole 1981).

The evolutionary variations in the star depend on the slow time variable. Therefore, the radial coordinate of a mass element, the mass density, the pressure, and the gravitational potential are expressed as

$$\left.\begin{aligned} r &= r\left[\tilde{t}(t), m\right], & \rho &= \rho\left[\tilde{t}(t), m\right], \\ P &= P\left[\tilde{t}(t), m\right], & \Phi &= \Phi\left[\tilde{t}(t), m\right]. \end{aligned}\right\} \tag{19.17}$$

On their side, the components of the Lagrangian displacement of a mass element related to the pulsation depend on both time variables t^+ and $\tilde{t}$, and also on the Lagrangian parameters m, θ, ϕ, which characterise the mass element:

$$\delta q^k = \delta q^k\left[t^{+}(t), \tilde{t}(t); m, \theta, \phi\right]. \tag{19.18}$$

To the first order in the small parameter C, the left-hand members of Eqs. (19.11) then become

$$\left[\frac{\partial}{\partial t}\left(g_{jk}\,\frac{\partial\left(\delta q^{k}\right)}{\partial t}\right)\right]_{m,\theta,\phi}$$
$$= g_{jk}\left[\mu^{2}\left(\tilde{t}\right)\frac{\partial^{2}\left(\delta q^{k}\right)}{\partial t^{+2}} + 2\,C\,\mu\left(\tilde{t}\right)\frac{\partial^{2}\left(\delta q^{k}\right)}{\partial t^{+}\,\partial\tilde{t}} + C\,\frac{d\mu}{d\tilde{t}}\,\frac{\partial\left(\delta q^{k}\right)}{\partial t^{+}}\right]$$
$$+ C\,\mu\left(\tilde{t}\right)\frac{\partial g_{jk}}{\partial\tilde{t}}\,\frac{\partial\left(\delta q^{k}\right)}{\partial t^{+}}, \qquad j = 1, 2, 3. \tag{19.19}$$

The eigenvalue problem of the linear oscillations of the evolving spherically symmetric star is supposed to have been solved in the isentropic approximation for the star "frozen" at the slow time $\tilde{t}$ considered. For a given mode n in the evolving star, the function $\mu\left(\tilde{t}\right)$ and the components of the Lagrangian displacement $\left(\delta q^{k}\right)_{n}\left(t^{+}, \tilde{t}; m, \theta, \phi\right), k = 1, 2, 3$, are expanded, respectively, from the isentropic approximation of the eigenfrequency, $\omega_{n,0}\left(\tilde{t}\right)$, and the isentropic approximations of the components of the Lagrangian displacement, $\left(\delta q^{k}\right)_{n,0}\left(\tilde{t}; m, \theta, \phi\right), k = 1, 2, 3$:

$$\left.\begin{aligned}
&\mu\left(\tilde{t}\right) = \omega_{n,0}\left(\tilde{t}\right) + C\,\omega_{n,1}\left(\tilde{t}\right) + \ldots,\\
&\left(\delta q^{k}\right)_{n}\left(t^{+}, \tilde{t}; m, \theta, \phi\right) = F_{n,0}\left(t^{+}\right)\left(\delta q^{k}\right)_{n,0}\left(\tilde{t}; m, \theta, \phi\right)\\
&\quad + C\,\sum_{s} F_{s,1}\left(t^{+}, \tilde{t}\right)\left(\delta q^{k}\right)_{s,0}\left(\tilde{t}; m, \theta, \phi\right) + \ldots, \quad k = 1, 2, 3.
\end{aligned}\right\} \tag{19.20}$$

The summation over s represents a summation over all isentropic modes of the evolving star "frozen" at time $\tilde{t}$.

After substitution into Eqs. (19.11) and (19.19), it follows, at order C^{0},

$$\omega_{n,0}^{2}\,\frac{d^{2}F_{n,0}}{dt^{+2}}\,g_{jk}\left(\delta q^{k}\right)_{n,0} + F_{n,0}\,U_{jk}\left(\delta q^{k}\right)_{n,0} = 0, \qquad j = 1, 2, 3\,, \tag{19.21}$$

and, at order C,

$$\omega_{n,0}^{2}\,g_{jk}\sum_{s}\frac{\partial^{2}F_{s,1}}{\partial t^{+2}}\left(\delta q^{k}\right)_{s,0} + \sum_{s} F_{s,1}\,U_{jk}\left(\delta q^{k}\right)_{s,0} = -\frac{dF_{n,0}}{dt^{+}}$$
$$\left[\frac{d\omega_{n,0}}{d\tilde{t}}\,g_{jk}\left(\delta q^{k}\right)_{n,0} + \omega_{n,0}\,\frac{\partial g_{jk}}{\partial\tilde{t}}\left(\delta q^{k}\right)_{n,0} + 2\,\omega_{n,0}\,g_{jk}\,\frac{\partial\left(\delta q^{k}\right)_{n,0}}{\partial\tilde{t}}\right]$$
$$-2\,\omega_{n,0}\,\omega_{n,1}\,\frac{d^{2}F_{n,0}}{dt^{+2}}\,g_{jk}\left(\delta q^{k}\right)_{n,0}, \qquad j = 1, 2, 3. \tag{19.22}$$

For any isentropic mode s of the evolving star "frozen" at time $\tilde{t}$, the components of the Lagrangian displacement satisfy the wave equation

$$\omega_{s,0}^2 \, g_{jk} \left(\delta q^k\right)_{s,0} - U_{jk} \left(\delta q^k\right)_{s,0} = 0, \qquad j = 1, 2, 3, \tag{19.23}$$

in accordance with Eq. (4.23). Furthermore, the orthogonnality relation between the isentropic mode s and an isentropic mode i holds

$$\begin{aligned} &\int_M g_{jk} \left(\delta q^k\right)_{s,0} \overline{\left(\delta q^j\right)_{i,0}} \, dm^* \\ &= \delta_{s,i} \int_M g_{jk} \left(\delta q^k\right)_{i,0} \overline{\left(\delta q^j\right)_{i,0}} \, dm^* \equiv \delta_{s,i} \, N_{i,0}. \end{aligned} \tag{19.24}$$

Here dm^* is the infinitesimal mass contained between r and $r + dr$, θ and $\theta + d\theta$, ϕ and $\phi + d\phi$. It is related to the infinitesimal mass dm contained inside the layer between r and $r + dr$ as

$$dm^* = \frac{dm}{4\pi} \sin\theta \, d\theta \, d\phi = \frac{1}{4\pi} \, d\omega \, dm. \tag{19.25}$$

By using the wave equation for $s = n$, multiplying Eq. (19.21) by $\overline{\left(\delta q^j\right)_{n,0}}$, and integrating over the whole mass, one derives the equation

$$\frac{d^2 F_{n,0}}{dt^{+2}} + F_{n,0} = 0. \tag{19.26}$$

A general solution is given by

$$F_{n,0}\left(t^+\right) = A_{n,0} \cos t^+ + B_{n,0} \sin t^+, \tag{19.27}$$

where $A_{n,0}$ and $B_{n,0}$ are arbitrary constants. Hence, the function $F_{n,0}\left(t^+\right)$ consists of a linear combination of harmonic functions of the fast time variable t^+ with period 2π.

Next, Eq. (19.22) is transformed into

$$\begin{aligned} &\omega_{n,0}^2 \sum_s \left(\frac{\partial^2 F_{s,1}}{\partial t^{+2}} + \frac{\omega_{s,0}^2}{\omega_{n,0}^2} F_{s,1} \right) g_{jk} \left(\delta q^k\right)_{s,0} \\ &= \left[\frac{d\omega_{n,0}}{d\tilde{t}} \, g_{jk} \left(\delta q^k\right)_{n,0} + \omega_{n,0} \frac{\partial g_{jk}}{\partial \tilde{t}} \left(\delta q^k\right)_{n,0} + 2\, \omega_{n,0} \, g_{jk} \frac{\partial \left(\delta q^k\right)_{n,0}}{\partial \tilde{t}} \right] \\ &\quad \left(A_{n,0} \sin t^+ - B_{n,0} \cos t^+\right) \\ &+ 2\, \omega_{n,0} \, \omega_{n,1} \, g_{jk} \left(\delta q^k\right)_{n,0} \left(A_{n,0} \cos t^+ + B_{n,0} \sin t^+\right), \quad j = 1, 2, 3. \end{aligned} \tag{19.28}$$

By multiplying by $\overline{\left(\delta q^j\right)_{i,0}}$ and integrating over the whole mass, one obtains an equation for the functions $F_{i,1}\left(t^+,\tilde{t}\right)$:

$$\frac{\partial^2 F_{i,1}}{\partial t^{+2}}+\frac{\omega_{i,0}^2}{\omega_{n,0}^2}F_{i,1}=V_{i,n}\left(A_{n,0}\sin t^+-B_{n,0}\cos t^+\right)$$
$$+2\,\delta_{i,n}\frac{\omega_{n,1}}{\omega_{n,0}}\left(A_{n,0}\cos t^+ + B_{n,0}\sin t^+\right), \tag{19.29}$$

where

$$V_{i,n}\left(\tilde{t}\right)=\frac{1}{\omega_{n,0}^2 N_{i,0}}\left[\delta_{i,n}\frac{d\omega_{n,0}}{d\tilde{t}}N_{n,0}\right.$$
$$+\omega_{n,0}\int_M \frac{\partial g_{jk}}{\partial \tilde{t}}\overline{\left(\delta q^j\right)_{i,0}}\left(\delta q^k\right)_{n,0}dm^*$$
$$\left.+2\,\omega_{n,0}\int_M g_{jk}\overline{\left(\delta q^j\right)_{i,0}}\frac{\partial\left(\delta q^k\right)_{n,0}}{\partial\tilde{t}}dm^*\right]. \tag{19.30}$$

For $i=n$, the inhomogeneous part of Eq. (19.29) contains resonant terms. They are removed under the conditions that

$$V_{n,n}\left(\tilde{t}\right)=0, \tag{19.31}$$

and

$$\omega_{n,1}=0. \tag{19.32}$$

The function $F_{n,1}\left(t^+,\tilde{t}\right)$ is then given by

$$F_{n,1}\left(t^+,\tilde{t}\right)=A_{n,1}\left(\tilde{t}\right)\cos t^+ + B_{n,1}\left(\tilde{t}\right)\sin t^+, \tag{19.33}$$

where $A_{n,1}\left(\tilde{t}\right)$ and $B_{n,1}\left(\tilde{t}\right)$ are yet undetermined functions of the slow time variable.

For $i\neq n$, the functions $F_{i,1}\left(t^+,\tilde{t}\right)$ are given by

$$F_{i,1}\left(t^+,\tilde{t}\right)=A_{i,1}\left(\tilde{t}\right)\cos\left(\frac{\omega_{i,0}}{\omega_{n,0}}t^+\right)+B_{i,1}\left(\tilde{t}\right)\sin\left(\frac{\omega_{i,0}}{\omega_{n,0}}t^+\right)$$
$$-\frac{V_{i,n}\left(\tilde{t}\right)}{1-\dfrac{\omega_{i,0}^2}{\omega_{n,0}^2}}\left(A_{n,0}\sin t^+-B_{n,0}\cos t^+\right), \tag{19.34}$$

where $A_{i,1}\left(\tilde{t}\right)$ and $B_{i,1}\left(\tilde{t}\right)$ are also yet undetermined functions of the slow time variable. When the mode i is supposed to be non excited, one can set $A_{i,1}=0$ and $B_{i,1}=0$, so that the functions $F_{i,1}\left(t^+,\tilde{t}\right)$ reduce to

$$F_{i,1}\left(t^+,\tilde{t}\right)=-\frac{V_{i,n}\left(\tilde{t}\right)}{1-\dfrac{\omega_{i,0}^2}{\omega_{n,0}^2}}\left(A_{n,0}\,\sin t^+ - B_{n,0}\,\cos t^+\right). \tag{19.35}$$

19.4 Rate of Change of an Isentropic Pulsation Period

For the rate of change of an isentropic oscillation period due to the star's evolution, an appropriate equation is obtained by the differentiation of wave equation (19.23) with respect to the slow time variable:

$$\begin{aligned}2\,\omega_{n,0}\,\frac{d\omega_{n,0}}{d\tilde{t}}\,g_{jk}\left(\delta q^k\right)_{n,0}+\omega_{n,0}^2\,\frac{\partial g_{jk}}{\partial\tilde{t}}\left(\delta q^k\right)_{n,0}+\omega_{n,0}^2\,g_{jk}\,\frac{\partial\left(\delta q^k\right)_{n,0}}{\partial\tilde{t}}\\-\frac{\partial U_{jk}}{\partial\tilde{t}}\left(\delta q^k\right)_{n,0}-U_{jk}\,\frac{\partial\left(\delta q^k\right)_{n,0}}{\partial\tilde{t}}=0,\qquad j=1,2,3.\end{aligned} \tag{19.36}$$

The partial derivative $\partial\left(\delta q^k\right)_{n,0}/\partial\tilde{t}$ can be eliminated from this equation as follows. Substitution of solutions (19.33) and (19.35) into Eq. (19.28), multiplication by $\cos t^+$ or $\sin t^+$, and integration over $2\,\pi$ in the fast time variable t^+ lead to

$$\begin{aligned}g_{jk}\,\frac{\partial\left(\delta q^k\right)_{n,0}}{\partial\tilde{t}}&=-\frac{1}{2\,\omega_{n,0}}\,\frac{d\omega_{n,0}}{d\tilde{t}}\,g_{jk}\left(\delta q^k\right)_{n,0}-\frac{1}{2}\,\frac{\partial g_{jk}}{\partial\tilde{t}}\left(\delta q^k\right)_{n,0}\\&+\frac{1}{2}\,\omega_{n,0}\,g_{jk}\sum_{s\neq n}V_{s,n}\left(\delta q^k\right)_{s,0},\qquad j=1,2,3.\end{aligned} \tag{19.37}$$

Multiplication by $g^{j\ell}$ then yields

$$\begin{aligned}\frac{\partial\left(\delta q^\ell\right)_{n,0}}{\partial\tilde{t}}&=-\frac{1}{2\,\omega_{n,0}}\,\frac{d\omega_{n,0}}{d\tilde{t}}\left(\delta q^\ell\right)_{n,0}-\frac{1}{2}\,g^{i\ell}\,\frac{\partial g_{ir}}{\partial\tilde{t}}\left(\delta q^r\right)_{n,0}\\&+\frac{1}{2}\,\omega_{n,0}\sum_{s\neq n}V_{s,n}\left(\delta q^\ell\right)_{s,0},\qquad \ell=1,2,3.\end{aligned} \tag{19.38}$$

After elimination of the partial derivative $\partial\left(\delta q^k\right)_{n,0}/\partial\tilde{t}$, multiplication by $\overline{\left(\delta q^j\right)_{n,0}}$, and integration over the total mass, Eq. (19.36) becomes

$$\begin{aligned}
\frac{1}{\omega_{n,0}} \frac{d\omega_{n,0}}{d\tilde{t}} = -\frac{1}{N_{n,0}} \Bigg\{ & -\frac{1}{2\,\omega_{n,0}^2} \int_M \overline{\left(\delta q^j\right)_{n,0}} \frac{\partial U_{jk}}{\partial \tilde{t}} \left(\delta q^k\right)_{n,0} dm^* \\
& + \frac{1}{4} \int_M \frac{\partial g_{jk}}{\partial \tilde{t}} \overline{\left(\delta q^j\right)_{n,0}} \left(\delta q^k\right)_{n,0} dm^* \\
& + \frac{1}{4}\,\omega_{n,0} \int_M g_{jk} \overline{\left(\delta q^j\right)_{n,0}} \left[\sum_{s \neq n} V_{s,n} \left(\delta q^k\right)_{s,0} \right] dm^* \\
& + \frac{1}{4\,\omega_{n,0}^2} \int_M \overline{\left(\delta q^j\right)_{n,0}} U_{jk} \left[g^{ik} \frac{\partial g_{ir}}{\partial \tilde{t}} \left(\delta q^r\right)_{n,0} \right] dm^* \\
& - \frac{1}{4\,\omega_{n,0}} \int_M \overline{\left(\delta q^j\right)_{n,0}} U_{jk} \left[\sum_{s \neq n} V_{s,n} \left(\delta q^k\right)_{s,0} \right] dm^* \Bigg\}. \qquad (19.39)
\end{aligned}$$

The third and the last term inside the braces in the right-hand member are equal to zero because of the orthogonality relation between the isentropic modes.

The relative rate of change of the isentropic pulsation period, $\Pi_{n,0}$, is then given by

$$\begin{aligned}
\frac{1}{\Pi_{n,0}} \frac{d\Pi_{n,0}}{d\tilde{t}} &= -\frac{1}{\omega_{n,0}} \frac{d\omega_{n,0}}{d\tilde{t}} \\
&= \frac{1}{\omega_{n,0}^2 N_{n,0}} \Bigg\{ -\frac{1}{2} \int_M \overline{\left(\delta q^j\right)_{n,0}} \frac{\partial U_{jk}}{\partial \tilde{t}} \left(\delta q^k\right)_{n,0} dm^* \\
&\quad + \frac{1}{4}\,\omega_{n,0}^2 \int_M \frac{\partial g_{jk}}{\partial \tilde{t}} \overline{\left(\delta q^j\right)_{n,0}} \left(\delta q^k\right)_{n,0} dm^* \\
&\quad + \frac{1}{4} \int_M \overline{\left(\delta q^j\right)_{n,0}} U_{jk} \left[g^{ik} \frac{\partial g_{ir}}{\partial \tilde{t}} \left(\delta q^r\right)_{n,0} \right] dm^* \Bigg\}. \qquad (19.40)
\end{aligned}$$

The relative rate of change of an isentropic oscillation period is thus determined by the rate of change of physical quantities in the rapidly evolving star and the isentropic approximation of the pulsation mode, all considered at the slow time $\tilde{t}$.

Equation (19.40) is valid for periods of both radial and non-radial pulsation modes. For radial pulsation modes, the equation reduces to

$$\frac{1}{\Pi_{n,0}} \frac{d\Pi_{n,0}}{d\tilde{t}} = -\frac{1}{2\,\omega_{n,0}^2 N_{n,0}} \int_M \overline{\left(\delta q^1\right)_{n,0}} \frac{\partial U_{11}}{\partial \tilde{t}} \left(\delta q^1\right)_{n,0} dm^*. \qquad (19.41)$$

Once the rate of change of the isentropic pulsation period is determined, the rates of change of the components $\left(\delta q^\ell\right)_{n,0}$ of the Lagrangian displacement, $\ell = 1, 2, 3$, can be derived by means of Eq. (19.38).

19.5 Use of the Equality of the Mean Kinetic and the Mean Potential Energy of Pulsation

Equation (19.36) integrated over the whole mass of the star also follows from the equality of the mean kinetic and the mean potential energy of oscillation over a period.

The equality of the mean energies is expressed by Eq. (8.47) for a normal isentropic mode. In terms of the two-time-variable formalism and the dimensionless quantities adopted here, the lowest-order asymptotic approximation of the normal isentropic displacement field is represented as

$$\left(\delta q^k\right)_n = F_{n,0}\left(t^+\right)\left(\delta q^k\right)_{n,0}\left(\tilde{t}; m, \theta, \phi\right), \qquad k = 1, 2, 3. \tag{19.42}$$

Hence, the lowest-order asymptotic approximations of the star's kinetic energy and potential energy of pulsation are given by

$$\delta_2 T_{\text{tot}} = \frac{1}{2}\,\omega_{n,0}^2 \left(\frac{dF_{n,0}}{dt^+}\right)^2 \int_M g_{ij}\,\overline{\left(\delta q^i\right)_{n,0}}\,\left(\delta q^j\right)_{n,0}\,dm^*, \tag{19.43}$$

$$\delta_2 V_{\text{tot}} = \frac{1}{2}\,F_{n,0}^2 \int_M \overline{\left(\delta q^i\right)_{n,0}}\,U_{ij}\,\left(\delta q^i\right)_{n,0}\,dm^*. \tag{19.44}$$

After substitution of solution (19.27) for $F_{n,0}\left(t^+\right)$, one obtains for the mean values of the kinetic and the potential energy over a period 2π in the fast time variable

$$\left.\begin{aligned}
\frac{1}{2\pi}\int_{-\pi}^{\pi} \delta_2 T_{\text{tot}}\,dt^+ &= \frac{1}{4}\,\omega_{n,0}^2\left(A_{n,0}^2 + B_{n,0}^2\right)\int_M g_{ij}\,\overline{\left(\delta q^i\right)_{n,0}}\,\left(\delta q^j\right)_{n,0}\,dm^*,\\
\frac{1}{2\pi}\int_{-\pi}^{\pi} \delta_2 V_{\text{tot}}\,dt^+ &= \frac{1}{4}\left(A_{n,0}^2 + B_{n,0}^2\right)\int_M \overline{\left(\delta q^i\right)_{n,0}}\,U_{ij}\,\left(\delta q^j\right)_{n,0}\,dm^*.
\end{aligned}\right\} \tag{19.45}$$

Partial differentiation with respect to the slow time variable yields the rates of change of the mean energies

$$\begin{aligned}
&\frac{\partial}{\partial \tilde{t}}\left(\frac{1}{2\pi}\int_{-\pi}^{\pi} \delta_2 T_{\text{tot}}\,dt^+\right) = \left(A_{n,0}^2 + B_{n,0}^2\right)\\
&\left\{\frac{1}{2}\,\omega_{n,0}\,\frac{d\omega_{n,0}}{d\tilde{t}}\,N_{n,0} + \frac{1}{4}\,\omega_{n,0}^2\left[\int_M \frac{\partial g_{ij}}{\partial \tilde{t}}\,\overline{\left(\delta q^i\right)_{n,0}}\,\left(\delta q^j\right)_{n,0}\,dm^*\right.\right.\\
&\quad + \frac{1}{4}\,\omega_{n,0}^2 \int_M g_{ij}\,\frac{\partial \overline{\left(\delta q^i\right)_{n,0}}}{\partial \tilde{t}}\,\left(\delta q^j\right)_{n,0}\,dm^*\\
&\quad \left.\left. + \frac{1}{4}\,\omega_{n,0}^2 \int_M g_{ij}\,\overline{\left(\delta q^i\right)_{n,0}}\,\frac{\partial\left(\delta q^j\right)_{n,0}}{\partial \tilde{t}}\,dm^*\right]\right\},
\end{aligned} \tag{19.46}$$

$$\frac{\partial}{\partial \tilde{t}} \left(\frac{1}{2\pi} \int_{-\pi}^{\pi} \delta_2 V_{\text{tot}} \, dt^+ \right) = \left(A_{n,0}^2 + B_{n,0}^2 \right)$$

$$\frac{1}{4} \left[\int_M \overline{\frac{\partial (\delta q^i)_{n,0}}{\partial \tilde{t}}} \, U_{ij} \left(\delta q^j\right)_{n,0} dm^* + \int_M \overline{\left(\delta q^i\right)_{n,0}} \frac{\partial U_{ij}}{\partial \tilde{t}} \left(\delta q^j\right)_{n,0} dm^* \right.$$

$$\left. + \int_M \overline{\left(\delta q^i\right)_{n,0}} \, U_{ij} \frac{\partial \left(\delta q^j\right)_{n,0}}{\partial \tilde{t}} dm^* \right]. \tag{19.47}$$

By using wave equation (19.23) in the first term of the right-hand member of Eq. (19.47), equating the expressions for the rates of change of the mean kinetic and the mean potential energy, multiplying all terms by $2/\left[\omega_{n,0}^2 \left(A_{n,0}^2 + B_{n,0}^2\right) N_{n,0}\right]$, one obtains an equation which corresponds to Eq. (19.36) integrated over the star's mass.

19.6 Explicit Expression for the Rate of Change of a Period in a Radially Pulsating Star

Be $(\delta r)_{n,0}\left(\tilde{t}; m\right)$ an isentropic radial pulsation whose period changes due to the evolution of the star on its Helmholtz–Kelvin time-scale. The divergence of the radial displacement can be expressed as

$$\alpha_{n,0}\left(\tilde{t}; m\right) = \frac{1}{r^2} \frac{1}{\frac{\partial r}{\partial m}} \frac{\partial \left[r^2 (\delta r)_{n,0}\right]}{\partial m}, \tag{19.48}$$

and the Eulerian perturbation of the mass density, as

$$\rho'_{n,0}\left(\tilde{t}; m\right) = -\rho \left(\alpha_{n,0} + \frac{1}{\rho} \frac{1}{\frac{\partial r}{\partial m}} \frac{\partial \rho}{\partial m} (\delta r)_{n,0} \right). \tag{19.49}$$

For the derivation of the rate of change of the period, it is convenient to transform the sum of terms $U_{11} \left(\delta q^1\right)_{n,0}$ into

$$U_{11} \left(\delta q^1\right)_{n,0} = \frac{1}{\frac{\partial r}{\partial m}} \left[\frac{\partial \Phi'_{n,0}}{\partial m} - \frac{1}{\rho} \frac{\partial \left(\rho c^2 \alpha_{n,0}\right)}{\partial m} + \frac{1}{\rho} \frac{\partial \left[\rho g (\delta r)_{n,0}\right]}{\partial m} - \frac{\rho'_{n,0}}{\rho^2} \frac{\partial P}{\partial m} \right] \tag{19.50}$$

and to decompose this sum into the three parts

$$\left.\begin{aligned} G_{11}\left(\delta q^1\right)_{n,0} &= \frac{1}{\rho}\,\frac{1}{\frac{\partial r}{\partial m}}\left[\frac{\partial\left[\rho\, g\,(\delta r)_{n,0}\right]}{\partial m} - \frac{\rho'_{n,0}}{\rho}\,\frac{\partial P}{\partial m}\right], \\ C_{11}\left(\delta q^1\right) &= -\frac{1}{\rho}\,\frac{1}{\frac{\partial r}{\partial m}}\,\frac{\partial\left(\rho\, c^2\,\alpha_{n,0}\right)}{\partial m}, \\ F_{11}\left(\delta q^1\right)_{n,0} &= \frac{1}{\frac{\partial r}{\partial m}}\,\frac{\partial \Phi'_{n,0}}{\partial m}. \end{aligned}\right\} \tag{19.51}$$

These three parts are treated separately hereafter.

19.6.1 The First Part

The term $G_{11}\left(\delta q^1\right)_{n,0}$ can be simplified by means of Eq. (19.5) and the condition of instantaneous hydrostatic equilibrium, expressed by Eq. (19.3), so that

$$G_{11}\left(\delta q^1\right)_{n,0} = \left(\rho - 4\,\frac{g}{r}\right)(\delta r)_{n,0}. \tag{19.52}$$

Partial differentiation of the operator G_{11} with respect to the slow time variable yields

$$\frac{\partial G_{11}}{\partial \tilde{t}}\left(\delta q^1\right)_{n,0} = \left(12\,\frac{g}{r^2}\,\frac{\partial r}{\partial \tilde{t}} + \frac{\partial \rho}{\partial \tilde{t}}\right)(\delta r)_{n,0}. \tag{19.53}$$

It follows that

$$\begin{aligned} S_1\left(\tilde{t}\right) &\equiv -\frac{1}{2\,\omega_{n,0}^2\,N_{n,0}}\int_M \left(\delta q^1\right)_{n,0}\,\frac{\partial G_{11}}{\partial \tilde{t}}\left(\delta q^1\right)_{n,0}\,dm^* \\ &= -\frac{1}{2\,\omega_{n,0}^2\,N_{n,0}}\int_0^1 \left(12\,\frac{g}{r^2}\,\frac{\partial r}{\partial \tilde{t}} + \frac{\partial \rho}{\partial \tilde{t}}\right)(\delta r)_{n,0}^2\,dm \end{aligned} \tag{19.54}$$

with

$$N_{n,0} = \int_0^1 (\delta r)_{n,0}^2\,dm.$$

19.6.2 The Second Part

Partial differentiation of the operator C_{11} with respect to the slow time variable yields

$$\frac{\partial C_{11}}{\partial \tilde{t}} \left(\delta q^1\right)_{n,0} = -2\, r \,\frac{\partial r}{\partial \tilde{t}} \frac{\partial \left(\rho\, c^2\, \alpha_{n,0}\right)}{\partial m}$$
$$-r^2 \frac{\partial}{\partial m} \left[\rho\, c^2 \left(\frac{2}{\rho} \frac{\partial \rho}{\partial \tilde{t}} + \frac{1}{c^2} \frac{\partial c^2}{\partial \tilde{t}} + \frac{2}{r} \frac{\partial r}{\partial \tilde{t}} \right) \alpha_{n,0} \right.$$
$$\left. -2 \frac{\rho\, c^2}{r} \left(\frac{1}{\rho} \frac{\partial \rho}{\partial \tilde{t}} + \frac{3}{r} \frac{\partial r}{\partial \tilde{t}} \right) (\delta r)_{n,0} \right]. \qquad (19.55)$$

After multiplication by $(\delta r)_{n,0}$ and integration over the total mass of the star, the first term of the right-hand member yields the contribution to the relative rate of period change

$$S_{2,a}\left(\tilde{t}\right) \equiv \frac{1}{2\, \omega_{n,0}^2\, N_{n,0}} \int_M 2\, (\delta r)_{n,0}\, r\, \frac{\partial r}{\partial \tilde{t}} \frac{\partial \left(\rho\, c^2\, \alpha_{n,0}\right)}{\partial m}\, dm^*$$
$$= \frac{1}{2\, \omega_{n,0}^2\, N_{n,0}} \int_0^1 2\, (\delta r)_{n,0}\, r\, \frac{\partial r}{\partial \tilde{t}} \frac{\partial \left(\rho\, c^2\, \alpha_{n,0}\right)}{\partial m}\, dm. \qquad (19.56)$$

By partial integration and use of the boundary conditions $(\delta r)_{n,0}\left(\tilde{t}, 0\right) = 0$ and $\rho\left(\tilde{t}, 1\right)\, c^2\left(\tilde{t}, 1\right) = 0$, it follows that

$$S_{2,a}\left(\tilde{t}\right) = -\frac{1}{2\, \omega_{n,0}^2\, N_{n,0}} \int_0^1 2\, \rho\, c^2\, \alpha_{n,0}$$
$$\left[\frac{\partial r}{\partial m} \frac{\partial r}{\partial \tilde{t}} (\delta r)_{n,0} + r\, \frac{\partial^2 r}{\partial m\, \partial \tilde{t}} (\delta r)_{n,0} + r\, \frac{\partial r}{\partial \tilde{t}} \frac{\partial (\delta r)_{n,0}}{\partial m} \right] dm. \qquad (19.57)$$

Elimination of $\partial r/\partial m$, $\partial^2 r/\partial m\, \partial \tilde{t}$, and $\partial (\delta r)_{n,0}/\partial m$ by means of Eqs. (19.6) and (19.48) leads to

$$S_{2,a}\left(\tilde{t}\right) = -\frac{1}{2\, \omega_{n,0}^2\, N_{n,0}} \int_0^1 2\, c^2\, \alpha_{n,0}$$
$$\left[\frac{1}{r} \frac{\partial r}{\partial \tilde{t}} \alpha_{n,0} - \left(\frac{1}{\rho} \frac{\partial \rho}{\partial \tilde{t}} + \frac{3}{r} \frac{\partial r}{\partial \tilde{t}} \right) \frac{1}{r} (\delta r)_{n,0} \right] dm. \qquad (19.58)$$

Similarly, after multiplication by $(\delta r)_{n,0}$ and integration over the total mass of the star, the second term in the right-hand member of Eq. (19.55) yields the contribution to the relative rate of period change

$$S_{2,b}\left(\tilde{t}\right) \equiv \frac{1}{2\, \omega_{n,0}^2\, N_{n,0}} \int_M (\delta r)_{n,0}\, r^2$$
$$\frac{\partial}{\partial m} \left[\rho\, c^2 \left(\frac{2}{\rho} \frac{\partial \rho}{\partial \tilde{t}} + \frac{1}{c^2} \frac{\partial c^2}{\partial \tilde{t}} + \frac{2}{r} \frac{\partial r}{\partial \tilde{t}} \right) \alpha_{n,0} \right.$$

$$
\left. -2\,\frac{\rho\,c^2}{r}\left(\frac{1}{\rho}\,\frac{\partial\rho}{\partial\tilde{t}}+\frac{3}{r}\,\frac{\partial r}{\partial\tilde{t}}\right)(\delta r)_{n,0}\,\right]dm^*
$$

$$
=\frac{1}{2\,\omega_{n,0}^2\,N_{n,0}}\int_0^1(\delta r)_{n,0}\,r^2
$$

$$
\frac{\partial}{\partial m}\left[\rho\,c^2\left(\frac{2}{\rho}\,\frac{\partial\rho}{\partial\tilde{t}}+\frac{1}{c^2}\,\frac{\partial c^2}{\partial\tilde{t}}+\frac{2}{r}\,\frac{\partial r}{\partial\tilde{t}}\right)\alpha_{n,0}\right.
$$

$$
\left. -2\,\frac{\rho\,c^2}{r}\left(\frac{1}{\rho}\,\frac{\partial\rho}{\partial\tilde{t}}+\frac{3}{r}\,\frac{\partial r}{\partial\tilde{t}}\right)(\delta r)_{n,0}\,\right]dm. \tag{19.59}
$$

By partial integration and use of the boundary conditions $r\left(\tilde{t},0\right)=0$ and $\rho\left(\tilde{t},1\right)\,c^2\left(\tilde{t},1\right)=0$, it follows that

$$
S_{2,b}\left(\tilde{t}\right)=-\frac{1}{2\,\omega_{n,0}^2\,N_{n,0}}\int_0^1\frac{\partial\left[r^2\,(\delta r)_{n,0}\right]}{\partial m}
$$

$$
\left[\rho\,c^2\left(\frac{2}{\rho}\,\frac{\partial\rho}{\partial\tilde{t}}+\frac{1}{c^2}\,\frac{\partial c^2}{\partial\tilde{t}}+\frac{2}{r}\,\frac{\partial r}{\partial\tilde{t}}\right)\alpha_{n,0}\right.
$$

$$
\left. -2\,\frac{\rho\,c^2}{r}\left(\frac{1}{\rho}\,\frac{\partial\rho}{\partial\tilde{t}}+\frac{3}{r}\,\frac{\partial r}{\partial\tilde{t}}\right)(\delta r)_{n,0}\,\right]dm. \tag{19.60}
$$

Use of Eqs. (19.6) and (19.48) leads to

$$
S_{2,b}\left(\tilde{t}\right)=-\frac{1}{2\,\omega_{n,0}^2\,N_{n,0}}\int_0^1\left[\left(\frac{2}{\rho}\,\frac{\partial\rho}{\partial\tilde{t}}+\frac{1}{c^2}\,\frac{\partial c^2}{\partial\tilde{t}}+\frac{2}{r}\,\frac{\partial r}{\partial\tilde{t}}\right)c^2\,\alpha_{n,0}^2\right.
$$

$$
\left. -\left(\frac{2}{\rho}\,\frac{\partial\rho}{\partial\tilde{t}}+\frac{6}{r}\,\frac{\partial r}{\partial\tilde{t}}\right)\frac{c^2}{r}\,\alpha_{n,0}\,(\delta r)_{n,0}\,\right]dm. \tag{19.61}
$$

It results that

$$
S_2\left(\tilde{t}\right)\equiv-\frac{1}{2\,\omega_{n,0}^2\,N_{n,0}}\int_M\left(\delta q^1\right)_{n,0}\frac{\partial C_{11}}{\partial\tilde{t}}\left(\delta q^1\right)_{n,0}\,dm^*
$$

$$
=S_{2,a}\left(\tilde{t}\right)+S_{2,b}\left(\tilde{t}\right)
$$

$$
=-\frac{1}{2\,\omega_{n,0}^2\,N_{n,0}}\int_0^1\left[\left(\frac{2}{\rho}\,\frac{\partial\rho}{\partial\tilde{t}}+\frac{1}{c^2}\,\frac{\partial c^2}{\partial\tilde{t}}+\frac{4}{r}\,\frac{\partial r}{\partial\tilde{t}}\right)c^2\,\alpha_{n,0}^2\right.
$$

$$
\left. -\left(\frac{4}{\rho}\,\frac{\partial\rho}{\partial\tilde{t}}+\frac{12}{r}\,\frac{\partial r}{\partial\tilde{t}}\right)\frac{c^2}{r}\,\alpha_{n,0}\,(\delta r)_{n,0}\,\right]dm. \tag{19.62}
$$

19.6.3 The Third Part

From Eq. (5.39), it follows that

$$F_{1,1}\,(\delta r)_{n,0} \equiv \frac{1}{\dfrac{\partial r}{\partial m}}\,\frac{\partial \Phi'_{n,0}}{\partial m} = -\rho\,(\delta r)_{n,0}. \tag{19.63}$$

Partial differentiation of the operator $F_{1,1}$ with respect to the slow time variable then yields

$$\frac{\partial F_{1,1}}{\partial \tilde{t}}\,(\delta r)_{n,0} = -\frac{\partial \rho}{\partial \tilde{t}}\,(\delta r)_{n,0}, \tag{19.64}$$

so that

$$\begin{aligned} S_3\left(\tilde{t}\right) &\equiv -\frac{1}{2\,\omega_{n,0}^2\,N_{n,0}} \int_M \left(\delta q^1\right)_{n,0} \frac{\partial F_{11}}{\partial \tilde{t}} \left(\delta q^1\right)_{n,0} dm^* \\ &= \frac{1}{2\,\omega_{n,0}^2\,N_{n,0}} \int_0^1 \frac{\partial \rho}{\partial \tilde{t}}\,(\delta r)_{n,0}^2\,dm. \end{aligned} \tag{19.65}$$

Because of this result, it is convenient to define $S_4\left(\tilde{t}\right)$ as

$$\begin{aligned} S_4\left(\tilde{t}\right) &\equiv S_1\left(\tilde{t}\right) + S_3\left(\tilde{t}\right) \\ &= -\frac{1}{2\,\omega_{n,0}^2\,N_{n,0}} \int_0^1 12\,\frac{g}{r^2}\,\frac{\partial r}{\partial \tilde{t}}\,(\delta r)_{n,0}^2\,dm. \end{aligned} \tag{19.66}$$

19.6.4 Global Result

For conclusion, the relative rate of change of an isentropic period $\Pi_{n,0}$ in a radially pulsating star that evolves on its Helmholtz–Kelvin time-scale, is given by

$$\frac{1}{\Pi_{n,0}}\,\frac{\Pi_{n,0}}{d\tilde{t}} = S_2\left(\tilde{t}\right) + S_4\left(\tilde{t}\right). \tag{19.67}$$

The terms $S_2\left(\tilde{t}\right)$ and $S_4\left(\tilde{t}\right)$ consist of integrals over the total mass of the star. The integrals involve the rates of change of the radial distance, $\partial r/\partial \tilde{t}$, of the mass density, $\partial \rho/\partial \tilde{t}$, and of the square of the isentropic sound velocity, $\partial c^2/\partial \tilde{t}$, in the various layers of the evolving star. They also involve the eigenfunctions $(\delta r)_{n,0}\left(\tilde{t}, m\right)$ and $\alpha_{n,0}\left(\tilde{t}, m\right)$ related to the radial pulsation mode. All quantities are considered at the slow time $\tilde{t}$.

An explicit expression for the rate of evolutionary change of a non-radial pulsation period can be derived in a similar way from Eq. (19.40), but the derivation is much more complicated. In the development of the term

$$\frac{1}{\omega_{n,0}^2 N_{n,0}}\left[-\frac{1}{2}\int_M \overline{\left(\delta q^j\right)_{n,0}}\,\frac{\partial U_{jk}}{\partial \tilde{t}}\left(\delta q^k\right)_{n,0} dm^*\right],$$

the transverse components of the Lagrangian displacement must be included besides the radial component, and, for the partial derivatives $\partial \Phi'_{n,0}/\partial q^i, i = 1,2,3,$ Eqs. (7.4) and (7.5) must be used. In addition, the two terms

$$\frac{1}{\omega_{n,0}^2 N_{n,0}}\left[\frac{1}{4}\,\omega_{n,0}^2\int_M \frac{\partial g_{jk}}{\partial \tilde{t}}\,\overline{\left(\delta q^j\right)_{n,0}}\left(\delta q^k\right)_{n,0} dm^*\right]$$

and

$$\frac{1}{\omega_{n,0}^2 N_{n,0}}\left\{\frac{1}{4}\int_M \overline{\left(\delta q^j\right)_{n,0}}\,U_{jk}\left[g^{ik}\,\frac{\partial g_{ir}}{\partial \tilde{t}}\left(\delta q^r\right)_{n,0}\right] dm^*\right\}$$

must be developed, which involve the partial derivatives $\partial g_{22}/\partial \tilde{t}$ and $\partial g_{33}/\partial \tilde{t}$. The first term was developed by Bruggen & Smeyers (1987).

19.7 Rate of Change of a Period in a Radially Pulsating Star Subject to a Homologous Contraction or Expansion

A simple application of Eq. (19.67) is the determination of the rate of change of a period in a radially pulsating star that is subject to a homologous contraction or expansion. In a homologously contracting or expanding star, the relative rate of contraction or expansion is the same for all layers, i.e.,

$$\frac{\partial}{\partial m}\left(\frac{1}{r}\frac{\partial r}{\partial \tilde{t}}\right) = 0. \tag{19.68}$$

By use of Eq. (19.6), one derives that

$$\frac{\partial}{\partial m}\left(\frac{1}{r}\frac{\partial r}{\partial \tilde{t}}\right) = -\frac{1}{r}\frac{\partial r}{\partial m}\left(\frac{1}{\rho}\frac{\partial \rho}{\partial \tilde{t}} + \frac{3}{r}\frac{\partial r}{\partial \tilde{t}}\right), \tag{19.69}$$

so that the relative rate of change of the mass density is given by

$$\frac{1}{\rho}\frac{\partial \rho}{\partial \tilde{t}} = -\frac{3}{r}\frac{\partial r}{\partial \tilde{t}}. \tag{19.70}$$

The pressure in the layer characterised by the mass m is given by the integration of Eq. (19.3), which expresses the instantaneous hydrostatic equilibrium,

$$P = \int_m^1 \frac{m'}{r^4}\,dm'. \tag{19.71}$$

Differentiation with respect to $\tilde{t}$ yields

$$\frac{\partial P}{\partial \tilde{t}} = -4 \int_m^1 \frac{m'}{r^5} \frac{\partial r}{\partial \tilde{t}}\, dm'. \tag{19.72}$$

Because of Eq. (19.68), the equation can also be written as

$$\frac{\partial P}{\partial \tilde{t}} = -\frac{4}{r} \frac{\partial r}{\partial \tilde{t}} \int_m^1 \frac{m'}{r^4}\, dm', \tag{19.73}$$

so that the relative rate of change of the pressure is given by

$$\frac{1}{P} \frac{\partial P}{\partial \tilde{t}} = -\frac{4}{r} \frac{\partial r}{\partial \tilde{t}} \tag{19.74}$$

(Kippenhahn & Weigert 1990).

Suppose that the star is composed of a mixture of a classical ideal gas and radiation. The pressure in a stellar layer is then given by

$$P = \frac{\mathscr{R}}{\bar{\mu}} \rho T + \frac{1}{3} a T^4, \tag{19.75}$$

or, by

$$P = \frac{\mathscr{R}}{\bar{\mu}} \rho T \frac{1}{\beta} \tag{19.76}$$

with

$$1 - \beta = \frac{a T^4}{3 P}. \tag{19.77}$$

Here $\mathscr{R}$ is the gas constant, $\bar{\mu}$, the mean molecular weight, which is supposed to be constant during the contraction or expansion, a, the radiation density constant, and β, the ratio of the gas pressure to the total pressure.

By differentiation of Eqs. (19.76) and (19.77), one obtains

$$\left.\begin{aligned} \frac{1}{P} \frac{\partial P}{\partial \tilde{t}} &= \frac{1}{\rho} \frac{\partial \rho}{\partial \tilde{t}} + \frac{1}{T} \frac{\partial T}{\partial \tilde{t}} - \frac{1}{\beta} \frac{\partial \beta}{\partial \tilde{t}}, \\ \frac{1}{1-\beta} \frac{\partial (1-\beta)}{\partial \tilde{t}} &= \frac{4}{T} \frac{\partial T}{\partial \tilde{t}} - \frac{1}{P} \frac{\partial P}{\partial \tilde{t}}. \end{aligned}\right\} \tag{19.78}$$

Elimination of $\partial \beta / \partial \tilde{t}$ leads to

$$\frac{1}{T} \frac{\partial T}{\partial \tilde{t}} = -\frac{1}{r} \frac{\partial r}{\partial \tilde{t}}, \tag{19.79}$$

so that

$$\frac{1}{\beta}\,\frac{\partial \beta}{\partial \tilde{t}} = 0. \tag{19.80}$$

For a mixture of a classical ideal gas and radiation, the isentropic coefficient Γ_1 is given by

$$\Gamma_1 = \beta + \frac{2\,(4-3\beta)^2}{8-7\beta}. \tag{19.81}$$

It then follows that

$$\frac{1}{\Gamma_1}\,\frac{\partial \Gamma_1}{\partial \tilde{t}} = 0. \tag{19.82}$$

Hence, the relative rate of change of the square of the isentropic sound velocity is given by

$$\frac{1}{c^2}\,\frac{\partial c^2}{\partial \tilde{t}} = -\frac{1}{r}\,\frac{\partial r}{\partial \tilde{t}}. \tag{19.83}$$

In Eq. (19.67), the terms $S_2\left(\tilde{t}\,\right)$ and $S_4\left(\tilde{t}\,\right)$ can then be written as

$$\left.\begin{aligned} S_2\left(\tilde{t}\right) &= \frac{1}{2\,\omega_{n,0}^2\,N_{n,0}}\,\frac{3}{r}\,\frac{\partial r}{\partial \tilde{t}} \int_0^1 c^2\,\alpha_{n,0}^2\,dm, \\ S_4\left(\tilde{t}\right) &= -\frac{1}{2\,\omega_{n,0}^2\,N_{n,0}}\,\frac{3}{r}\,\frac{\partial r}{\partial \tilde{t}} \int_0^1 4\,\frac{g}{r}\,(\delta r)_{n,0}^2\,dm, \end{aligned}\right\} \tag{19.84}$$

so that

$$\frac{1}{\Pi_{n,0}}\,\frac{d\Pi_{n,0}}{d\tilde{t}} = \frac{1}{2\,\omega_{n,0}^2\,N_{n,0}}\,\frac{3}{r}\,\frac{\partial r}{\partial \tilde{t}} \int_0^1 \left[c^2\,\alpha_{n,0}^2 - 4\,\frac{g}{r}\,(\delta r)_{n,0}^2\right] dm. \tag{19.85}$$

Finally, by the use of Eq. (8.104) for the square of an eigenfrequency, in which the eigenfunctions are normalised such that $N_{n,0} = 1$, it results that

$$\frac{1}{\Pi_{n,0}}\,\frac{d\Pi_{n,0}}{d\tilde{t}} = \frac{3}{2}\,\frac{1}{R}\,\frac{dR}{d\tilde{t}}. \tag{19.86}$$

For any radial pulsation in a homologously contracting or expanding star, the relative rate of change of the isentropic period is thus equal to three halves of the relative rate of change of the star's radius at the time $\tilde{t}$ considered. This result agrees with that derived by Demaret (1974), as it appears from his Eqs. (59) and (78).

Appendix A
Green's Fundamental Formula of Potential Theory

In Green's fundamental formula of potential theory, the solution of Laplace's equation $\nabla^2 \Phi = 0$ in a domain bounded by a closed surface is expressed in terms of the distribution of Φ and $\partial\Phi/\partial n$ on that surface. The formula is obtained by means of Green's integral theorem, which is itself derived from Gauss' integral theorem.

Be **a** a vector field that is differentiable in the domain $V(t)$ bounded by the surface $S(t)$, so that the divergence of the vector field is determined at all points in the domain. According to Gauss' integral theorem,

$$\int_V \nabla \cdot \mathbf{a}\, dV = \int_S \mathbf{a} \cdot \mathbf{n}\, dS, \tag{A.1}$$

where **n** is the outward unit normal on the infinitesimal surface dS.

Suppose that the vector field **a** is defined in terms of two scalar functions U and W as

$$\mathbf{a} = U\, \nabla W - W\, \nabla U. \tag{A.2}$$

The two scalar functions are supposed to satisfy the conditions that their first derivatives are continuous and that their second derivatives exist. Since

$$\left.\begin{aligned} \nabla \cdot (U\, \nabla W) &= \nabla U \cdot \nabla W + U\, \nabla^2 W, \\ \nabla \cdot (W\, \nabla U) &= \nabla W \cdot \nabla U + W\, \nabla^2 U, \end{aligned}\right\} \tag{A.3}$$

the divergence of equality (A.2) leads to

$$\nabla \cdot \mathbf{a} = U\, \nabla^2 W - W\, \nabla^2 U. \tag{A.4}$$

Use of Gauss' integral theorem then yields

$$\int_V \left(U\, \nabla^2 W - W\, \nabla^2 U\right) dV = \int_S \left(U\, \frac{\partial W}{\partial n} - W\, \frac{\partial U}{\partial n}\right) dS. \tag{A.5}$$

This identity is known as Green's integral theorem or Greens's second identity.

P. Smeyers and T. Van Hoolst, *Linear Isentropic Oscillations of Stars: Theoretical Foundations*, Astrophysics and Space Science Library 371,
DOI 10.1007/978-3-642-13030-4, © Springer-Verlag Berlin Heidelberg 2010

In order to derive Green's fundamental formula of potential theory, one sets the function W equal to the reciprocal of the distance d between an arbitrary point of the integration domain V, with position vector $\mathbf{r}'$, and a given field point P, with position vector $\mathbf{r}$:

$$W = d^{-1} \equiv \left|\mathbf{r}' - \mathbf{r}\right|^{-1} \\ = \left[\left(x' - x\right)^2 + \left(y' - y\right)^2 + \left(z' - z\right)^2\right]^{-1/2}. \tag{A.6}$$

The function d^{-1} is harmonic, i.e. $\nabla^2 d^{-1} = 0$.

When the field point P is situated in the integration domain V, the point $\mathbf{r}' = \mathbf{r}$ is a singular point. One excludes this point from the integration domain by surrounding it with a small sphere with centre at P and radius r. Be S_1 the surface of this sphere. Use of Green's integral theorem on the domain V_1 that is contained between the spherical surface S_1 and the surface S yields

$$-\int_{V_1} \frac{1}{d} \nabla^2 U \, dV = \int_S \left(U \frac{\partial(1/d)}{\partial n} - \frac{1}{d} \frac{\partial U}{\partial n} \right) dS \\ + \int_{S_1} \left(U \frac{\partial(1/d)}{\partial n} - \frac{1}{d} \frac{\partial U}{\partial n} \right) dS. \tag{A.7}$$

In both surface integrals, the unit normal must be considered outward from the domain V_1.

By the introduction of a system of spherical coordinates with origin at the point P, one has, on the spherical surface S_1, that $d = r$, $\partial/\partial n = -\partial/\partial r$, and $dS = r^2 \sin\theta' \, d\theta' \, d\phi'$. In the second surface integral of the right-hand member, it results that

$$\int_{S_1} \frac{1}{d} \frac{\partial U}{\partial n} dS = -r \int_0^{2\pi} \int_0^{\pi} \frac{\partial U}{\partial r} \sin\theta' \, d\theta' \, d\phi'. \tag{A.8}$$

Since U is regular in the domain V, the integral tends to zero as $r \to 0$. Moreover,

$$\int_{S_1} U \frac{\partial(1/d)}{\partial n} dS = \int_0^{2\pi} \int_0^{\pi} U \sin\theta' \, d\theta' \, d\phi', \tag{A.9}$$

so that

$$\lim_{r \to 0} \int_{S_1} U \frac{\partial(1/d)}{\partial n} dS = 4\pi \, U(P). \tag{A.10}$$

In the left-hand-member of equality (A.7), one has

$$\lim_{r \to 0} \int_{V_1} \frac{1}{d} \nabla^2 U \, dV = \int_V \frac{1}{d} \nabla^2 U \, dV. \tag{A.11}$$

If the function U is harmonic in the domain V, it follows that

$$U(P) = \frac{1}{4\pi}\int_S \left(\frac{1}{d}\frac{\partial U}{\partial n} - U\frac{\partial(1/d)}{\partial n}\right) dS. \tag{A.12}$$

This identity is referred to as Green's fundamental formula of potential theory (Kellogg 1929, Ramsey 1940, Sigl 1973).

Appendix B
The Thermodynamic Isentropic Coefficients

B.1 Reversible Thermodynamic Processes

A definition of reversible thermodynamic processes given by Sommerfeld (1964b) is:

> Reversible processes are not, in fact, processes at all, they are sequences of states of equilibrium. The processes which we encounter in real life are always irreversible processes ... Instead of using the term "reversible processes" we can also speak of infinitely slow, quasi-static processes during which the system's capacity for performing work is fully utilized and no energy is dissipated. In spite of their not being real, reversible processes are most important in thermodynamics because definite equations can be obtained only by considering reversible changes; irreversible changes can only be described with the aid of inequalities when equilibrium thermodynamics is used.
>
> The actual criterion for a process to be reversible states that during its course there are no lasting changes of any sort in the surroundings if the process is allowed to go forward and then back to the original state.

B.2 Thermodynamic Relations

For small reversible thermodynamic transformations in a fluid in which the work is done by the pressure, Gibbs' equation is valid:

$$dU = T\,dS - P\,d\tau \tag{B.1}$$

(Batchelor 1967). U is the specific internal energy, S the specific entropy, P the pressure, and T the absolute temperature. The independent variables are S and τ, so that the internal energy, U, is regarded as a function of these two variables. P and T are two conjugate variables given by

$$T = \left(\frac{\partial U}{\partial S}\right)_\tau, \quad P = -\left(\frac{\partial U}{\partial \tau}\right)_S. \tag{B.2}$$

Since dU is a total differential, the double partial derivative $\partial^2 U/\partial\tau\,\partial S$ may be taken in two ways. The relation then holds

$$\left(\frac{\partial T}{\partial \tau}\right)_S = -\left(\frac{\partial P}{\partial S}\right)_\tau . \tag{B.3}$$

An independent variable can be replaced by its conjugate by means of a Legendre transformation. If one wants to replace, for example, the independent variable S by its conjugate T, it is necessary to subtract the product of the two conjugate variables, TS, from the dependent variable U. One then obtains the new function of state

$$F = U - T\,S, \tag{B.4}$$

known as the Helmholtz free energy. Its differential takes the form

$$dF = dU - T\,dS - S\,dT = -P\,d\tau - S\,dT. \tag{B.5}$$

The free energy, F, is thus function of the two independent variables τ and T. The two conjugate variables are S and P, and are given by

$$S = -\left(\frac{\partial F}{\partial T}\right)_\tau , \quad P = -\left(\frac{\partial F}{\partial \tau}\right)_T . \tag{B.6}$$

The relation then follows

$$\left(\frac{\partial S}{\partial \tau}\right)_T = \left(\frac{\partial P}{\partial T}\right)_\tau . \tag{B.7}$$

Differentiation of Eq. (B.1) with respect to τ while T is kept constant yields

$$\left(\frac{\partial U}{\partial \tau}\right)_T = T\left(\frac{\partial S}{\partial \tau}\right)_T - P. \tag{B.8}$$

Because of relation (B.7), it results that

$$\left(\frac{\partial U}{\partial \tau}\right)_T = T\left(\frac{\partial P}{\partial T}\right)_\tau - P. \tag{B.9}$$

B.3 Definitions of the Isentropic Coefficients

The isentropic coefficients measure the ratios of the differentials dP, $d\rho$, dT that are responses to reversible isentropic changes in a thermodynamic system. They are defined as

$$\Gamma_1 = \left(\frac{\partial \ln P}{\partial \ln \rho}\right)_S , \quad \frac{\Gamma_2 - 1}{\Gamma_2} = \left(\frac{\partial \ln T}{\partial \ln P}\right)_S , \quad \Gamma_3 - 1 = \left(\frac{\partial \ln T}{\partial \ln \rho}\right)_S . \tag{B.10}$$

The three isentropic coefficients are related to each other as

$$\Gamma_3 - 1 = \Gamma_1 \, \frac{\Gamma_2 - 1}{\Gamma_2}. \tag{B.11}$$

They reduce to the ratio of the specific heats at constant pressure and constant volume, $\gamma = C_P/C_V$, for a classic ideal gas. The specific heats are considered here per unit mass. The isentropic coefficients were introduced by Chandrasekhar (1939) for a mixture of gas and radiation.

The isentropic coefficient Γ_1 is related to the isentropic compressibility coefficient,

$$\kappa_S = -\frac{1}{\tau}\left(\frac{\partial \tau}{\partial P}\right)_S ,$$

as

$$\Gamma_1 = \frac{1}{P\,\kappa_S} \, . \tag{B.12}$$

By way of illustration, the variation of the isentropic coefficient Γ_1 in the surface layers of a solar model of Christensen-Dalsgaard et al. (1993) is represented in Fig. B.1 as a function of the relative radial distance from the centre of the model, $x = r/R_\odot$. The coefficient has nearly the value 1.67 in the largest part of the solar model but displays two minima near $x = 1$: the first minimum is situated in the layers of the first ionisation of the helium atoms, and the second, sharper, minimum, in the layers of the ionisation of the hydrogen atoms.

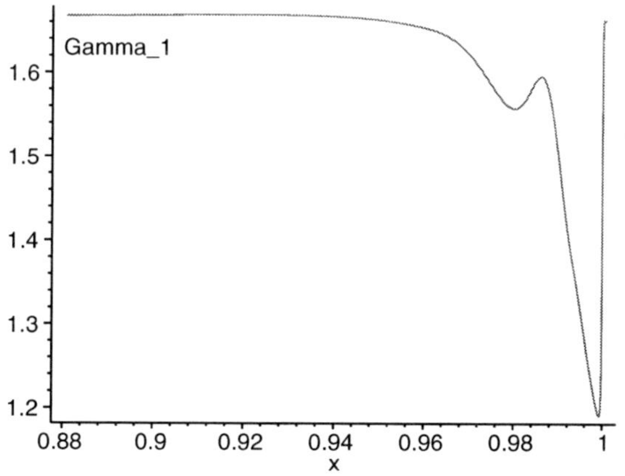

Fig. B.1 The isentropic coefficient Γ_1 in the surface layers of the solar model of Christensen-Dalsgaard et al. (1993), as a function of the relative radial distance from the centre, $x = r/R_\odot$

B.4 Equations for the Isentropic Coefficients

B.4.1 Equation for Γ_1

When the internal energy, U, is regarded as a function of τ and T, and T is regarded as a function of τ and P, it follows from Eq. (B.1) that, for a reversible isentropic process,

$$\left(\frac{\partial U}{\partial P}\right)_\tau dP + \left[\left(\frac{\partial U}{\partial \tau}\right)_P + P\right] d\tau = 0. \tag{B.13}$$

Since

$$\left(\frac{\partial U}{\partial P}\right)_\tau = \frac{\left(\frac{\partial U}{\partial T}\right)_\tau}{\left(\frac{\partial P}{\partial T}\right)_\tau}, \tag{B.14}$$

one also has

$$\left(\frac{\partial U}{\partial T}\right)_\tau dP + \left[\left(\frac{\partial U}{\partial \tau}\right)_P + P\right]\left(\frac{\partial P}{\partial T}\right)_\tau d\tau = 0. \tag{B.15}$$

On the other hand,

$$dP = \left(\frac{\partial P}{\partial T}\right)_\tau dT + \left(\frac{\partial P}{\partial \tau}\right)_T d\tau \tag{B.16}$$

leads to

$$\left(\frac{\partial P}{\partial T}\right)_\tau = -\left(\frac{\partial \tau}{\partial T}\right)_P \left(\frac{\partial P}{\partial \tau}\right)_T. \tag{B.17}$$

Elimination of $(\partial P/\partial T)_\tau$ from Eq. (B.15) yields

$$\left(\frac{\partial U}{\partial T}\right)_\tau dP - \left[\left(\frac{\partial U}{\partial T}\right)_P + P\left(\frac{\partial \tau}{\partial T}\right)_P\right]\left(\frac{\partial P}{\partial \tau}\right)_T d\tau = 0. \tag{B.18}$$

This equation can be written as

$$\frac{dP}{P} - \Gamma_1 \frac{d\tau}{\tau} = 0 \tag{B.19}$$

with

$$\Gamma_1 = -\frac{C_P}{C_V}\chi_\tau \tag{B.20}$$

and

$$\chi_\tau = \left(\frac{\partial \ln P}{\partial \ln \tau}\right)_T . \tag{B.21}$$

C_P and C_V are the specific heats, respectively, at constant pressure and at constant volume, for a unit mass.

Equation (B.19) is a generalisation of Poisson's equation for a reversible isentropic process in a perfect gas (Sommerfeld 1964b). Furthermore, χ_τ is related to the isothermal compressibility coefficient,

$$\kappa_T = -\frac{1}{\tau}\left(\frac{\partial \tau}{\partial P}\right)_T , \tag{B.22}$$

as

$$\chi_\tau = -\frac{1}{P\,\kappa_T} . \tag{B.23}$$

Because of this relation and relation (B.12), Eq. (B.20) also corresponds to the equation that relates the isentropic and the isothermal compressibility coefficients

$$\kappa_S = \frac{C_V}{C_P}\,\kappa_T . \tag{B.24}$$

Since

$$C_P - C_V = -\frac{P\tau}{T}\frac{\chi_T^2}{\chi_\tau} \tag{B.25}$$

with

$$\chi_T = \left(\frac{\partial \ln P}{\partial \ln T}\right)_\tau , \tag{B.26}$$

and χ_τ is known to be negative, $C_P > C_V$. Consequently, for any thermodynamic system, the isentropic coefficient Γ_1 is positif, and the isentropic compressibility coefficient is smaller than the isothermal compressibility coefficient.

B.4.2 Equation for Γ_2

Secondly, when the internal energy, U, is regarded as a function of T and τ, and τ is regarded as a function of P and T, it follows from Eq. (B.1) that, for a reversible isentropic process,

$$\left[\left(\frac{\partial U}{\partial T}\right)_P + P\left(\frac{\partial \tau}{\partial T}\right)_P\right] dT + \left[\left(\frac{\partial U}{\partial P}\right)_T + P\left(\frac{\partial \tau}{\partial P}\right)_T\right] dP = 0. \tag{B.27}$$

By use of the equality

$$\left(\frac{\partial U}{\partial P}\right)_T = \left(\frac{\partial U}{\partial \tau}\right)_T \left(\frac{\partial \tau}{\partial P}\right)_T \tag{B.28}$$

and the definition of the specific heat at constant pressure, C_P, the equation can also be written as

$$C_P\,dT + \left[\left(\frac{\partial U}{\partial \tau}\right)_T + P\right]\left(\frac{\partial \tau}{\partial P}\right)_T dP = 0. \tag{B.29}$$

In virtue of relation (B.9), it follows that

$$\frac{dT}{T} + \frac{\Gamma_2 - 1}{\Gamma_2}\frac{dP}{P} = 0 \tag{B.30}$$

with

$$\frac{\Gamma_2 - 1}{\Gamma_2} = -\frac{P\tau}{T}\frac{1}{C_P}\frac{\chi_T}{\chi_\tau} \tag{B.31}$$

or, after elimination of C_P by means of Eq. (B.20),

$$\frac{\Gamma_2 - 1}{\Gamma_2} = \frac{P\tau}{T}\frac{\chi_T}{\Gamma_1 C_V}. \tag{B.32}$$

B.4.3 Equation for Γ_3

Finally, by means of relation (B.11), it follows from Eq. (B.32) that

$$\Gamma_3 - 1 = \frac{P\tau}{T}\frac{\chi_T}{C_V}. \tag{B.33}$$

An additional useful thermodynamic relation is

$$\Gamma_1 = (\Gamma_3 - 1)\,\chi_T - \chi_\tau \tag{B.34}$$

(Cox & Giuli 1968, Kippenhahn & Weigert 1990, Hansen & Kawaler 1994).

Appendix C
Lagrange's Equations of Motion

Lagrange's equations of motion are useful for the description of motions of mass elements in terms of generalised coordinates. Be x^1, x^2, x^3 the three orthogonal Cartesian coordinates of a mass element that are defined in an inertial frame of reference, and be the time-dependent transformation formulae from the Cartesian coordinates to the generalised coordinates q^1, q^2, q^3 of the mass element of the form

$$x^k\left(t, q^1, q^2, q^3\right), \quad k = 1, 2, 3. \tag{C.1}$$

The generalised coordinates q^1, q^2, q^3 are considered to be functions of the time t and three parameters a^1, a^2, a^3 that characterise the mass element:

$$q^i\left(t, a^1, a^2, a^3\right), \quad i = 1, 2, 3. \tag{C.2}$$

One then also has

$$\dot{q}^{\,i}\left(t, a^1, a^2, a^3\right), \quad i = 1, 2, 3. \tag{C.3}$$

The kinetic energy, per unit mass, is given by

$$T_{\text{kin}} = \frac{1}{2}\delta_{ij}\dot{x}^{\,i}\dot{x}^{\,j} \tag{C.4}$$

or, more explicitly, by

$$T_{\text{kin}} = \frac{1}{2}\delta_{ij}\left(\frac{\partial x^i}{\partial t} + \frac{\partial x^i}{\partial q^k}\dot{q}^{\,k}\right)\left(\frac{\partial x^j}{\partial t} + \frac{\partial x^j}{\partial q^\ell}\dot{q}^{\,\ell}\right). \tag{C.5}$$

It can be expressed as a sum of terms of degrees 0, 1, 2 in the generalised velocity components:

$$T_{\text{kin}} = \frac{1}{2}\mu + \nu_k\,\dot{q}^{\,k} + \frac{1}{2}g_{k\ell}\,\dot{q}^{\,k}\dot{q}^{\,\ell} \tag{C.6}$$

with

$$\mu = \delta_{ij} \frac{\partial x^i}{\partial t} \frac{\partial x^j}{\partial t}, \quad \nu_k = \delta_{ij} \frac{\partial x^i}{\partial t} \frac{\partial x^j}{\partial q^k}, \quad k = 1, 2, 3 \tag{C.7}$$

(Brillouin 1960). As well as the covariant components of the metric tensor $g_{k\ell}$, μ and the ν_k are functions of the time t and the three generalised coordinates q^1, q^2, q^3.

If the gravitational force is the acting force, Lagrange's equations of motion in an ideal fluid are

$$\frac{d}{dt} \frac{\partial T_{\text{kin}}}{\partial \dot{q}^j} - \frac{\partial T_{\text{kin}}}{\partial q^j} = -\frac{\partial \Phi}{\partial q^j} - \frac{1}{\rho} \frac{\partial P}{\partial q^j}, \quad j = 1, 2, 3. \tag{C.8}$$

A derivation from Hamilton's variational principle that applies to continuous systems in the absence of pressure is given, for example, by Goldstein (1957).

The terms in the left-hand member can be developed as

$$\left.\begin{aligned} \frac{\partial T_{\text{kin}}}{\partial q^j} &= \frac{1}{2} \frac{\partial \mu}{\partial q^j} + \frac{\partial \nu_k}{\partial q^j} \dot{q}^k + \frac{1}{2} \frac{\partial g_{k\ell}}{\partial q^j} \dot{q}^k \dot{q}^\ell, \\ \frac{\partial T_{\text{kin}}}{\partial \dot{q}^j} &= \nu_j + g_{jk} \dot{q}^k. \end{aligned}\right\} \tag{C.9}$$

The total time derivative involves the implicit time dependence through the generalised coordinates and velocity components. One then has

$$\begin{aligned} \frac{d}{dt} \frac{\partial T_{\text{kin}}}{\partial \dot{q}^j} &= g_{jk} \ddot{q}^k + \frac{1}{2} \frac{\partial g_{jk}}{\partial q^\ell} \dot{q}^k \dot{q}^\ell \\ &\quad + \frac{1}{2} \frac{\partial g_{j\ell}}{\partial q^k} \dot{q}^k \dot{q}^\ell + \frac{\partial \nu_j}{\partial t} + \frac{\partial \nu_j}{\partial q^\ell} \dot{q}^\ell + \frac{\partial g_{jk}}{\partial t} \dot{q}^k. \end{aligned} \tag{C.10}$$

Substitution into Eqs. (C.8) yields

$$\begin{aligned} & g_{jk} \ddot{q}^k + \frac{1}{2} \left(\frac{\partial g_{jk}}{\partial q^\ell} + \frac{\partial g_{j\ell}}{\partial q^k} - \frac{\partial g_{k\ell}}{\partial q^j} \right) \dot{q}^k \dot{q}^\ell \\ & \quad + \frac{\partial g_{jk}}{\partial t} \dot{q}^k + \left(\frac{\partial \nu_j}{\partial t} - \frac{1}{2} \frac{\partial \mu}{\partial q^j} \right) + \left(\frac{\partial \nu_j}{\partial q^k} - \frac{\partial \nu_k}{\partial q^j} \right) \dot{q}^k \\ & = -\frac{\partial \Phi}{\partial q^j} - \frac{1}{\rho} \frac{\partial P}{\partial q^j}, \quad j = 1, 2, 3. \end{aligned} \tag{C.11}$$

For the transformation of $\ddot{q}^k$, it is appropriate to consider the inverses of the transformation formulae (C.2):

$$a^j \left(t, q^1, q^2, q^3\right), \quad j = 1, 2, 3. \tag{C.12}$$

Substitution into Eqs. (C.3) leads to equations for the generalised velocity components of the form

$$\dot{q}^k \left[t, q^1\left(t, a^1, a^2, a^3\right), q^2\left(t, a^1, a^2, a^3\right), q^3\left(t, a^1, a^2, a^3\right)\right], k = 1, 2, 3. \quad \text{(C.13)}$$

It then follows that

$$\ddot{q}^k \equiv \left(\frac{\partial \dot{q}^k}{\partial t}\right)_{a^1, a^2, a^3} = \frac{\partial \dot{q}^k}{\partial t} + \frac{\partial \dot{q}^k}{\partial q^\ell}\,\dot{q}^\ell, \tag{C.14}$$

so that Eqs. (C.10) become

$$g_{jk}\,\frac{\partial \dot{q}^k}{\partial t} + \left[g_{jk}\,\frac{\partial \dot{q}^k}{\partial q^\ell} + \frac{1}{2}\left(\frac{\partial g_{jk}}{\partial q^\ell} + \frac{\partial g_{j\ell}}{\partial q^k} - \frac{\partial g_{k\ell}}{\partial q^j}\right)\dot{q}^k\right]\dot{q}^\ell$$
$$+\frac{\partial g_{jk}}{\partial t}\,\dot{q}^k + \left(\frac{\partial \nu_j}{\partial t} - \frac{1}{2}\,\frac{\partial \mu}{\partial q^j}\right) + \left(\frac{\partial \nu_j}{\partial q^k} - \frac{\partial \nu_k}{\partial q^j}\right)\dot{q}^k$$
$$= -\frac{\partial \Phi}{\partial q^j} - \frac{1}{\rho}\,\frac{\partial P}{\partial q^j}, \quad j = 1, 2, 3. \tag{C.15}$$

Multiplication by g^{ij} and use of the definition of the Christoffel three-index symbols of the second kind yield the contravariant components of the equation of motion in Eulerian form

$$\frac{\partial \dot{q}^i}{\partial t} + \left(\frac{\partial \dot{q}^i}{\partial q^\ell} + \Gamma^i_{k\ell}\,\dot{q}^k\right)\dot{q}^\ell + g^{ij}\,\frac{\partial g_{ik}}{\partial t}\,\dot{q}^k$$
$$+g^{ij}\left(\frac{\partial \nu_j}{\partial t} - \frac{1}{2}\,\frac{\partial \mu}{\partial q^j}\right) + g^{ij}\left(\frac{\partial \nu_j}{\partial q^k} - \frac{\partial \nu_k}{\partial q^j}\right)\dot{q}^k$$
$$= -g^{ij}\,\frac{\partial \Phi}{\partial q^j} - g^{ij}\,\frac{1}{\rho}\,\frac{\partial P}{\partial q^j}, \quad j = 1, 2, 3, \tag{C.16}$$

or, equivalently,

$$\frac{\partial \dot{q}^i}{\partial t} + \dot{q}^\ell\,\nabla_\ell\,\dot{q}^i + g^{ij}\,\frac{\partial g_{jk}}{\partial t}\,\dot{q}^k + g^{ij}\left(\frac{\partial \nu_j}{\partial t} - \frac{1}{2}\,\frac{\partial \mu}{\partial q^j}\right)$$
$$+g^{ij}\left(\frac{\partial \nu_j}{\partial q^k} - \frac{\partial \nu_k}{\partial q^j}\right)\dot{q}^k = -g^{ij}\,\frac{\partial \Phi}{\partial q^j} - g^{ij}\,\frac{1}{\rho}\,\frac{\partial P}{\partial q^j}, \quad j = 1, 2, 3. \tag{C.17}$$

The sum of the first two terms in the left-hand member is the generalised acceleration component i of the mass element, which is defined as the rate of change of the generalised velocity component $\dot{q}^i$ of the mass element while the element is followed in its motion, i.e., as the total time derivative

$$\frac{d\dot{q}^i}{dt} = \frac{\partial \dot{q}^i}{\partial t} + \dot{q}^\ell \, \nabla_\ell \, \dot{q}^i \tag{C.18}$$

(McConnell 1957).

In the particular case in which the transformation of the Cartesian coordinates into generalised coordinates does not involve the time explicitly, one has

$$\frac{\partial g_{ik}}{\partial t} = 0, \quad \mu = 0, \quad \nu_j = 0, \quad j = 1, 2, 3, \tag{C.19}$$

so that the contravariant components of the equation of motion in Eulerian form reduce to

$$\frac{\partial \dot{q}^i}{\partial t} + \dot{q}^\ell \, \nabla_\ell \, \dot{q}^i = -g^{ij} \frac{\partial \Phi}{\partial q^j} - g^{ij} \frac{1}{\rho} \frac{\partial P}{\partial q^j}, \quad j = 1, 2, 3. \tag{C.20}$$

For the derivation of the covariant components of the equation of motion, it is convenient to return to Eq. (C.15). A recombination of terms yields

$$\begin{aligned}
&\left[g_{jk} \frac{\partial \dot{q}^k}{\partial t} + \frac{\partial g_{jk}}{\partial t} \dot{q}^k \right] + \left[\frac{\partial}{\partial q^\ell} \left(g_{jk} \, \dot{q}^k \right) - \dot{q}^k \frac{\partial g_{jk}}{\partial q^\ell} \right] \dot{q}^\ell \\
&+ \frac{1}{2} \left(\frac{\partial g_{jk}}{\partial q^\ell} + \frac{\partial g_{j\ell}}{\partial q^k} - \frac{\partial g_{k\ell}}{\partial q^j} \right) \dot{q}^k \, \dot{q}^\ell + \left(\frac{\partial \nu_j}{\partial t} - \frac{1}{2} \frac{\partial \mu}{\partial q^j} \right) \\
&+ \left(\frac{\partial \nu_j}{\partial q^k} - \frac{\partial \nu_k}{\partial q^j} \right) \dot{q}^k = -\frac{\partial \Phi}{\partial q^j} - \frac{1}{\rho} \frac{\partial P}{\partial q^j}, \quad j = 1, 2, 3.
\end{aligned} \tag{C.21}$$

The equation can be rewritten as

$$\begin{aligned}
&\frac{\partial}{\partial t} \left(g_{jk} \, \dot{q}^k \right) + \dot{q}^\ell \frac{\partial}{\partial q^\ell} \left(g_{jk} \, \dot{q}^k \right) - \frac{1}{2} \left(\frac{\partial g_{jk}}{\partial q^\ell} + \frac{\partial g_{k\ell}}{\partial q^j} - \frac{\partial g_{j\ell}}{\partial q^k} \right) \dot{q}^k \, \dot{q}^\ell \\
&+ \left(\frac{\partial \nu_j}{\partial t} - \frac{1}{2} \frac{\partial \mu}{\partial q^j} \right) + \left(\frac{\partial \nu_j}{\partial q^k} - \frac{\partial \nu_k}{\partial q^j} \right) \dot{q}^k = -\frac{\partial \Phi}{\partial q^j} - \frac{1}{\rho} \frac{\partial P}{\partial q^j}, \quad j = 1, 2, 3.
\end{aligned} \tag{C.22}$$

By setting in the third term of the left-hand member

$$\dot{q}^k = g^{km} \, g_{mn} \, \dot{q}^n, \tag{C.23}$$

one obtains

$$\frac{\partial}{\partial t}\left(g_{jk}\,\dot{q}^k\right)+\left[\frac{\partial}{\partial q^\ell}\left(g_{jk}\,\dot{q}^k\right)-\Gamma^m_{j\ell}\,g_{mn}\,\dot{q}^n\right]\dot{q}^\ell+\left(\frac{\partial \nu_j}{\partial t}-\frac{1}{2}\frac{\partial \mu}{\partial q^j}\right)$$
$$+\left(\frac{\partial \nu_j}{\partial q^k}-\frac{\partial \nu_k}{\partial q^j}\right)\dot{q}^k=-\frac{\partial \Phi}{\partial q^j}-\frac{1}{\rho}\frac{\partial P}{\partial q^j},\quad j=1,2,3. \tag{C.24}$$

The covariant components of the equation of motion in Eulerian form are then

$$\frac{\partial}{\partial t}\left(g_{jk}\,\dot{q}^k\right)+\dot{q}^\ell\,\nabla_\ell\left(g_{jk}\,\dot{q}^k\right)+\left(\frac{\partial \nu_j}{\partial t}-\frac{1}{2}\frac{\partial \mu}{\partial q^j}\right)$$
$$+\left(\frac{\partial \nu_j}{\partial q^k}-\frac{\partial \nu_k}{\partial q^j}\right)\dot{q}^k=-\frac{\partial \Phi}{\partial q^j}-\frac{1}{\rho}\frac{\partial P}{\partial q^j},\quad j=1,2,3, \tag{C.25}$$

where $\nabla_\ell\left(g_{jk}\,\dot{q}^k\right)$ is the covariant derivative of the covariant vector component $g_{jk}\,\dot{q}^k$ with respect to the generalised coordinate q^ℓ.

In the particular case in which the transformation of the Cartesian coordinates into the generalised coordinates does not involve the time explicitly, equalities (C.19) are valid, and Eq. (C.25) reduces to

$$\frac{\partial}{\partial t}\left(g_{jk}\,\dot{q}^k\right)+\dot{q}^\ell\,\nabla_\ell\left(g_{jk}\,\dot{q}^k\right)=-\frac{\partial \Phi}{\partial q^j}-\frac{1}{\rho}\frac{\partial P}{\partial q^j},\quad j=1,2,3. \tag{C.26}$$

Appendix D
Spherical Harmonics

The spherical harmonics are solutions of the partial differential equation

$$\mathscr{L}^2 Y = \lambda(\lambda+1)\, Y, \tag{D.1}$$

where $\mathscr{L}^2$ is the Legendrian defined by Eq. (2.51). On the solutions, conditions of regularity and uniqueness are imposed for $0 \leq \theta \leq \pi$ and $0 \leq \phi \leq 2\pi$ and the condition that $Y_\lambda(\theta, \phi + 2\pi) \equiv Y_\lambda(\theta, \phi)$. The equation then admits of solutions only for integer values of λ. No attention must be paid to negative values of λ, since $-\lambda - 1$ and $-\lambda$ lead to identical products $\lambda(\lambda+1)$.

The real eigenfunctions are

$$\begin{aligned} &P_\lambda^{\mu}(\cos\theta)\, \cos(\mu\phi), \quad P_\lambda^{\mu}(\cos\theta)\, \sin(\mu\phi), \\ &\qquad \lambda = 0, 1, 2, \ldots, \quad \mu = 0, 1, 2, \ldots, \lambda. \end{aligned} \tag{D.2}$$

They are called *tesseral* spherical harmonics of degree λ and order μ. In the particular cases $\mu = \lambda$ and $\mu = 0$, they are called, respectively, *sectorial* and *zonal* spherical harmonics.

Moreover, the complex eigenfunctions are

$$\begin{aligned} &Y_\lambda^{\mu}(\theta, \phi) = P_\lambda^{|\mu|}(\cos\theta)\, \exp(\mathrm{i}\mu\phi), \\ &\qquad \lambda = 0, 1, 2, \ldots, \quad \mu = 0, \pm 1, \pm 2, \ldots, \pm\lambda \end{aligned} \tag{D.3}$$

(Korn & Korn 1968).

The complex spherical harmonics obey the orthogonality relation

$$\int_0^{2\pi} \int_0^{\pi} \overline{Y_\ell^m}(\theta, \phi)\, Y_\lambda^{\mu}(\theta, \phi)\, \sin\theta\, d\theta\, d\phi = \delta_{\lambda,\ell}\, \delta_{\mu,m}\, N_{\ell m} \tag{D.4}$$

with

$$N_{\ell m} = \frac{4\pi}{2\ell+1} \frac{(\ell + |m|)!}{(\ell - |m|)!}. \tag{D.5}$$

From this relation, the useful property follows

$$\int_0^{2\pi}\int_0^{\pi}\sin\theta\,d\theta\,d\phi$$
$$\left[\frac{\partial\overline{Y_\ell^m}(\theta,\phi)}{\partial\theta}\,\frac{\partial Y_\lambda^\mu(\theta,\phi)}{\partial\theta}+\frac{1}{\sin^2\theta}\,\frac{\partial\overline{Y_\ell^m}(\theta,\phi)}{\partial\phi}\,\frac{\partial Y_\lambda^\mu(\theta,\phi)}{\partial\phi}\right]$$
$$=\delta_{\lambda,\ell}\,\delta_{\mu,m}\,\ell(\ell+1)\,N_{\ell m}. \tag{D.6}$$

One derives the property by using the transformations

$$\frac{\partial\overline{Y_\ell^m}}{\partial\theta}\,\frac{\partial Y_\lambda^\mu}{\partial\theta}=\frac{1}{\sin\theta}\,\frac{\partial}{\partial\theta}\left(\sin\theta\,\frac{\partial Y_\lambda^\mu}{\partial\theta}\,\overline{Y_\ell^m}\right)-\overline{Y_\ell^m}\,\frac{1}{\sin\theta}\,\frac{\partial}{\partial\theta}\left(\sin\theta\,\frac{\partial Y_\lambda^\mu}{\partial\theta}\right), \tag{D.7}$$

$$\frac{\partial\overline{Y_\ell^m}}{\partial\phi}\,\frac{\partial Y_\lambda^\mu}{\partial\phi}=\frac{\partial}{\partial\phi}\left(\frac{\partial Y_\lambda^\mu}{\partial\phi}\,\overline{Y_\ell^m}\right)-\overline{Y_\ell^m}\,\frac{\partial^2 Y_\lambda^\mu}{\partial\phi^2}. \tag{D.8}$$

Taking into account that

$$\int_0^{\pi}\left[\frac{\partial}{\partial\theta}\left(\sin\theta\,\frac{\partial Y_\lambda^\mu}{\partial\theta}\,\overline{Y_\ell^m}\right)\right]d\theta=0,\quad \int_0^{2\pi}\left[\frac{\partial}{\partial\phi}\left(\frac{\partial Y_\lambda^\mu}{\partial\phi}\,\overline{Y_\ell^m}\right)\right]d\phi=0, \tag{D.9}$$

one obtains

$$\int_0^{2\pi}\int_0^{\pi}\left(\frac{\partial\overline{Y_\ell^m}}{\partial\theta}\,\frac{\partial Y_\lambda^\mu}{\partial\theta}+\frac{1}{\sin^2\theta}\,\frac{\partial\overline{Y_\ell^m}}{\partial\phi}\,\frac{\partial Y_\lambda^\mu}{\partial\phi}\right)\sin\theta\,d\theta\,d\phi$$
$$=\int_0^{2\pi}\int_0^{\pi}\overline{Y_\ell^m}\,(\mathscr{L}^2 Y_\lambda^\mu)\,\sin\theta\,d\theta\,d\phi. \tag{D.10}$$

By use of partial differential equation (D.1) and orthogonality relation (D.4), the property follows.

The spherical harmonics form a complete set of functions on the surface of a sphere with unit radius (see, e.g., Jackson 1999).

Appendix E
Singular Perturbation Problems of the Boundary-Layer Type

A simple and instructive example of a singular perturbation problem that arises from a reduction of the order of a differential equation due to an approximation, is presented in Van Dyke (1964) but stems actually from Friedrichs (von Mises & Friedrichs 1942). The example is also given in von Mises & Friedrichs (1971).

Consider the inhomogeneous, second-order, linear differential equation

$$\varepsilon \frac{d^2 f}{dx^2} + \frac{df}{dx} = a, \tag{E.1}$$

where a is a constant, and ε a small parameter. The solution is determined in the interval $0 \le x \le 1$ and must satisfy the boundary conditions

$$f(0) = 0, \quad f(1) = 1. \tag{E.2}$$

An exact solution is known as

$$f(x;\ \varepsilon) = (1-a)\frac{1-\exp(-x/\varepsilon)}{1-\exp(-1/\varepsilon)} + ax \tag{E.3}$$

and is represented in Fig. E.1 for $a = 0.7$ and $\varepsilon = 0.02$.

In a regular perturbation problem, one searches for an approximation of order zero in the small parameter by neglecting the terms that contain the parameter. In this case, the differential equation reduces to the first-order equation

$$\frac{df}{dx} = a. \tag{E.4}$$

Its general solution

$$f(x;\ \varepsilon) = ax + C \tag{E.5}$$

contains the only arbitrary constant C. Therefore, it cannot satisfy the two boundary conditions, unless $a = 1$.

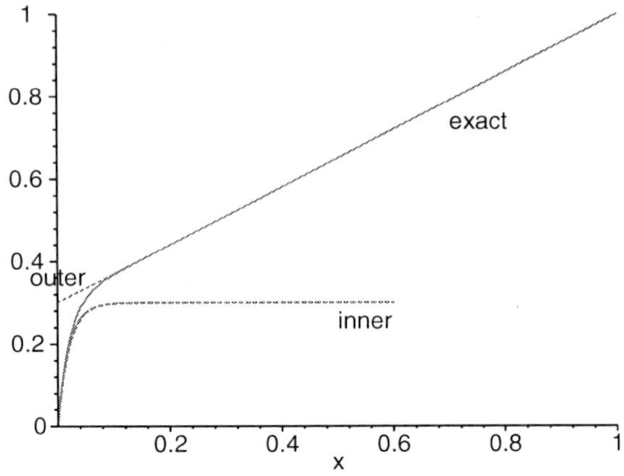

Fig. E.1 The exact function $f(x;\ \varepsilon)$ for $a = 0.7$ and $\varepsilon = 0.02$, and its outer and inner first asymptotic approximations

When one imposes the boundary condition at $x = 1$, the solution becomes

$$f(x;\ \varepsilon) = (1 - a) + a\,x. \tag{E.6}$$

This solution is a good approximation of the exact solution, except in a small *boundary layer* near $x = 0$, and is called the outer solution. It is given by

$$f^{(o)}(x) = (1 - a) + a\,x \tag{E.7}$$

and is represented in Fig. E.1.

The underlying reason why the outer solution is not a satisfactory approximation of the exact solution in the boundary layer is that the neglected term $\varepsilon\, d^2 f/dx^2$ there becomes of the same order of magnitude as the term df/dx, since the second derivative $d^2 f/dx^2$ is even much larger than the first derivative df/dx. An adequate approximation of order zero can be constructed only if the second derivative is incorporated into the differential equation that determines the approximation. For this purpose, boundary-layer theory is applied (see, e.g., Kevorkian & Cole 1981, 1996).

The boundary layer is stretched by the introduction of the coordinate

$$\zeta(x) = \frac{x}{\delta(\varepsilon)}, \tag{E.8}$$

where $\delta(\varepsilon) \to 0$ as $\varepsilon \to 0$. Setting $f(x) = f^{(i)}(\zeta)$, one transforms the differential equation into

$$\frac{\varepsilon}{[\delta(\varepsilon)]^2}\frac{d^2 f^{(i)}}{d\zeta^2} + \frac{1}{\delta(\varepsilon)}\frac{df^{(i)}}{d\zeta} = a. \tag{E.9}$$

The stretching is carried out to the limit where the term involving the second derivative of the function $f^{(\mathrm{i})}(\zeta)$ becomes of the same order in the small parameter as the term involving the first derivative. This limit is reached when

$$\delta(\varepsilon) = \varepsilon. \tag{E.10}$$

The differential equation then becomes

$$\frac{d^2 f^{(\mathrm{i})}}{d\zeta^2} + \frac{df^{(\mathrm{i})}}{d\zeta} = \varepsilon\, a. \tag{E.11}$$

At order zero in the small parameter, the differential equation reduces to

$$\frac{d^2 f^{(\mathrm{i})}}{d\zeta^2} + \frac{df^{(\mathrm{i})}}{d\zeta} = 0. \tag{E.12}$$

A general solution is

$$f^{(\mathrm{i})}(\zeta) = A\, \exp(-\zeta) + B, \tag{E.13}$$

where A and B are two arbitrary constants. This solution, which is called the inner solution, must satisfy the boundary condition at $x = 0$. Therefore,

$$B = -A, \tag{E.14}$$

so that the solution becomes

$$f^{(\mathrm{i})}(\zeta) = A\,[1 - \exp(-\zeta)]. \tag{E.15}$$

The inner solution $f^{(\mathrm{i})}(\zeta)$ is then matched with the outer solution $f^{(\mathrm{o})}(x)$ by the requirement that

$$\lim_{\zeta \to \infty} f^{(\mathrm{i})}(\zeta) = \lim_{x \to 0} f^{(\mathrm{o})}(x). \tag{E.16}$$

Observing that

$$\left.\begin{aligned} \lim_{x \to 0} f^{(\mathrm{o})}(x) &= 1 - a, \\ \lim_{\zeta \to \infty} f^{(\mathrm{i})}(\zeta) &= A, \end{aligned}\right\} \tag{E.17}$$

one has

$$A = 1 - a, \tag{E.18}$$

so that the inner solution becomes

$$f^{(\mathrm{i})}(\zeta) = (1 - a)\,[1 - \exp(-\zeta)]. \tag{E.19}$$

This form of the inner solution is represented in Fig. E.1.

Finally, one constructs a composite solution that is uniformly valid at order ε^0 in the interval $0 \leq x \leq 1$ by making the sum of the inner and the outer solution and subtracting the part common to both solutions. It results that

$$f^{(0)}(x;\varepsilon) = (1 - a)\,[1 - \exp(-x/\varepsilon)] + a\,x. \tag{E.20}$$

One readily verifies that the composite solution at order ε^0 corresponds to the first approximation of the exact solution in which the term $\exp(-1/\varepsilon)$ is neglected in comparison with 1. In the example chosen, the neglected term is equal to 1.9×10^{-22}.

Appendix F
Boundary Condition Relative to the Pressure on a Star's Surface

In the derivation of the condition relative to the pressure on the surface of an oscillating star, the surface is considered as a discontinuity surface. Discontinuity surfaces in flows of gases are surfaces on which distributions of commonly continuous quantities as velocity, pressure, mass density, …, display jumps. Generally, discontinuity surfaces in gases move, and their velocity at a point is different from that of the gas at the same point (Landau & Lifchitz 1971).

On a discontinuity surface, certain conditions must be satisfied because of the principles of the mechanics of continuous media. Here we derive the condition that follows from the principle for the rate of change of momentum.

Consider the motion of a continuous medium with respect to an inertial frame of reference, in which Cartesian coordinates x^1, x^2, x^3 are introduced. The velocity components of a mass element of the continuous medium are then given by $\dot{x}^1$, $\dot{x}^2$, $\dot{x}^3$. The points of a fictitious closed surface $S(t)$ have velocity components $\dot{x}^1_S$, $\dot{x}^2_S$, $\dot{x}^3_S$.

According to the principle for the rate of change of momentum, the momentum inside the volume $V(t)$ that is enclosed in the surface $S(t)$ changes because of the flux of momentum due to the motion of the continuous medium, and because of the acting forces. These forces are the resultant of the forces **F**, per unit mass, acting on the various mass elements inside the volume $V(t)$, and the tractions **t**, per unit surface, acting on the surface element $dS(t)$.

When **n** is the outward unit normal at a point of the surface $S(t)$, the i-component of the principle takes the form

$$\frac{d}{dt}\int_{V(t)} \rho\, \dot{x}^i\, dV = -\int_{S(t)} \rho\, \dot{x}^i \left(\dot{x}^j - \dot{x}^j_S\right) n_j\, dS + \int_{V(t)} \rho\, F^i\, dV + \int_{S(t)} t^i\, dS, \quad i = 1, 2, 3. \tag{F.1}$$

The three components t^i of the traction are related to the nine components T^{ji} of the stress tensor as

$$t^i = T^{ji}\, n_j. \tag{F.2}$$

In an ideal fluid,

$$T^{ji} = -\delta^{ji}\, P, \quad i, j = 1, 2, 3, \tag{F.3}$$

where δ^{ji} is a Kronecker delta. One then obtains

$$\frac{d}{dt}\int_{V(t)} \rho\,\dot{x}^i\, dV = -\int_{S(t)} \rho\,\dot{x}^i \left(\dot{x}^j - \dot{x}^j_S\right) n_j\, dS + \int_{V(t)} \rho\, F^i\, dV - \int_{S(t)} \delta^{ji}\, P\, n_j\, dS, \quad i = 1, 2, 3. \tag{F.4}$$

Imagine that the volume $V(t)$ is divided into two partial volumes $V_1(t)$ and $V_2(t)$ by a discontinuity surface $S^*(t)$. Be the volume $V_1(t)$ bounded by the surface $S_1(t)$, and the volume $V_2(t)$ by the surface $S_2(t)$. The discontinuity surface $S^*(t)$ is common to the two surfaces $S_1(t)$ and $S_2(t)$. The principle for the rate of change of momentum can now successively be applied to the total volume $V(t)$ and to the two partial volumes $V_1(t)$ and $V_2(t)$. By subtraction of the sum of the equalities that apply to the partial volumes from the equality that applies to the total volume, it results that

$$\int_{S^*(t)} \left\{ \left[\rho\,\dot{x}^i \left(\dot{x}^j - \dot{x}^j_S\right) n_j + \delta^{ji}\, P\, n_j\right]_1 + \left[\rho\,\dot{x}^i \left(\dot{x}^j - \dot{x}^j_S\right) n_j + \delta^{ji}\, P\, n_j\right]_2 \right\} dS = 0. \tag{F.5}$$

On the surface $S^*(t)$, one has that

$$\left(n_j\right)_2 = -\left(n_j\right)_1 \equiv -n_j, \tag{F.6}$$

so that the equality becomes

$$\int_{S^*(t)} \left\{ \left[\rho\,\dot{x}^i \left(\dot{x}^j - \dot{x}^j_S\right) + \delta^{ji}\, P\right]_1 - \left[\rho\,\dot{x}^i \left(\dot{x}^j - \dot{x}^j_S\right) + \delta^{ji}\, P\right]_2 \right\} n_j\, dS = 0. \tag{F.7}$$

Since the integral must be equal to zero for an arbitrary discontinuity surface $S^*(t)$ in any volume $V(t)$, the integrand must be identically zero. This condition can be expressed as

$$< \rho\,\dot{x}^i \left(\dot{x}^j - \dot{x}^j_S\right) + \delta^{ji}\, P >\, n_j = 0, \tag{F.8}$$

or, alternatively, as

$$< \rho \left[(\dot{x}^i - \dot{x}_S^i) + \dot{x}_S^i \right] \left(\dot{x}^j - \dot{x}_S^j \right) + \delta^{ji} P > n_j = 0. \tag{F.9}$$

The condition can be reduced by application of the principle of the mass conservation. This principle expresses that the amount of mass contained in the volume $V(t)$ changes only because of the mass flux through the surface $S(t)$, since mass is neither created nor destroyed inside the volume:

$$\frac{d}{dt} \int_{V(t)} \rho \, dV = - \int_{S(t)} \rho \left(\dot{x}^j - \dot{x}_S^j \right) n_j \, dS. \tag{F.10}$$

When the principle is applied in a way similar to that followed above for the principle for the rate of change of momentum, the condition leads to

$$< \rho \left(\dot{x}^j - \dot{x}_S^j \right) > n_j = 0, \tag{F.11}$$

so that condition (F.9) reduces to

$$< \rho \left(\dot{x}^i - \dot{x}_S^i \right) \left(\dot{x}^j - \dot{x}_S^j \right) + \delta^{ji} P > n_j = 0. \tag{F.12}$$

Among the discontinuity surfaces, one distinguishes between contact surfaces and shock waves. A contact surface separates two parts of a medium without gas streaming through it, and a shock wave, with gas streaming through it. Here we restrict ourselves to contact surfaces.

Since, at both sides of a contact surface,

$$\rho \left(\dot{x}^i - \dot{x}_S^i \right) = 0, \quad i = 1, 2, 3, \tag{F.13}$$

Condition (F.12) becomes simply

$$< P > n^i = 0, \quad i = 1, 2, 3, \tag{F.14}$$

so that

$$< P >= 0. \tag{F.15}$$

When the surface of an equilibrium star is regarded as a contact surface, and the pressure outside the star is identically zero, the condition implies that the pressure on the equilibrium surface of a star must be equal to zero:

$$P(R) = 0. \tag{F.16}$$

In a similar way, the pressure on the surface of an oscillating star must also be equal to zero:

$$P_{\mathrm{p}}(R + \delta R) = 0. \tag{F.17}$$

Because of the linear approximation,

$$P_{\mathrm{p}}(R + \delta R) = P(R) + \delta P(R), \tag{F.18}$$

and condition (F.16), condition (F.17) implies that the Lagrangian perturbation of the pressure must vanish on the star's equilibrium surface:

$$\delta P(R) = 0. \tag{F.19}$$

Appendix G
The Curl of a Vector Field

In tensor analysis, the operation of taking the curl of a vector field η leads to the creation of an antisymmetric tensor of rank $\left\{ {0 \atop 2} \right\}$:

$$T_{ik} = \nabla_k \, \eta_i - \nabla_i \, \eta_k . \tag{G.1}$$

Note that the operation of taking the covariant derivative applies to the covariant components of the vector field. Since

$$\left. \begin{aligned} \nabla_k \, \eta_i &= \frac{\partial \eta_i}{\partial q^k} - \Gamma^s_{ik} \, \eta_s , \\ \nabla_i \, \eta_k &= \frac{\partial \eta_k}{\partial q^i} - \Gamma^s_{ki} \, \eta_s , \end{aligned} \right\} \tag{G.2}$$

and the Christoffel three-index symbols of the second kind are symmetric in the lower indices, the definition becomes

$$T_{ik} = \frac{\partial \eta_i}{\partial q^k} - \frac{\partial \eta_k}{\partial q^i} \tag{G.3}$$

(Adler et al. 1965).

In the three-dimensional Euclidian space of classical mechanics, the curl of a vector field is introduced as the vector which has the contravariant components with respect to the local *coordinate* basis associated with an orthogonal coordinate system

$$(\nabla \times \eta \,)^n = g_{\mathrm{m}}^{-1/2} \, \epsilon^{nts} \, \nabla_t \, \eta_s , \tag{G.4}$$

where g_{m} is the determinant of the metric tensor, and ϵ^{nts} the antisymmetric symbol defined as

$$\epsilon^{nts} = \begin{cases} +1 \text{ when } nts \text{ is an even permutation of } 123, \\ -1 \text{ when } nts \text{ is an uneven permutation of } 123, \\ \;\;\, 0 \text{ when any two indices are equal} \end{cases} \tag{G.5}$$

(McConnell 1957, Misner et al. 1973, Hartle 2003).

For the system of spherical coordinates r, θ, ϕ introduced by means of transformation formulae (2.1), the contravariant components of the vector $\nabla \times \boldsymbol{\eta}$ with respect to the local coordinate basis are

$$\left.\begin{aligned}
(\nabla \times \boldsymbol{\eta})^r &= \frac{1}{r^2 \sin\theta}\left(\frac{\partial \eta_\phi}{\partial \theta} - \frac{\partial \eta_\theta}{\partial \phi}\right), \\
(\nabla \times \boldsymbol{\eta})^\theta &= \frac{1}{r^2 \sin\theta}\left(\frac{\partial \eta_r}{\partial \phi} - \frac{\partial \eta_\phi}{\partial r}\right), \\
(\nabla \times \boldsymbol{\eta})^\phi &= \frac{1}{r^2 \sin\theta}\left(\frac{\partial \eta_\theta}{\partial r} - \frac{\partial \eta_r}{\partial \theta}\right).
\end{aligned}\right\} \tag{G.6}$$

The covariant components η_r, η_θ, η_ϕ of the vector $\boldsymbol{\eta}$ are related to the contravariant components η^r, η^θ, η^ϕ as

$$\eta_r = \eta^r, \quad \eta_\theta = r^2\,\eta^\theta, \quad \eta_\phi = r^2 \sin^2\theta\,\eta^\phi. \tag{G.7}$$

Generally, the components of the vectors $\boldsymbol{\eta}$ and $\nabla \times \boldsymbol{\eta}$ are considered with respect to the local *orthonormal* basis

$$\mathbf{e}'_1 = \frac{\partial}{\partial r}, \quad \mathbf{e}'_2 = \frac{1}{r}\frac{\partial}{\partial \theta}, \quad \mathbf{e}'_3 = \frac{1}{r \sin\theta}\frac{\partial}{\partial \phi}. \tag{G.8}$$

If η'^r, η'^θ, η'^ϕ are the components of the vector $\boldsymbol{\eta}$ with respect to this basis, one has, in accordance with resolution (1.9) of a vector in terms of the basis vectors,

$$\begin{aligned}
\boldsymbol{\eta} &= \eta^r \frac{\partial}{\partial r} + \eta^\theta \frac{\partial}{\partial \theta} + \eta^\phi \frac{\partial}{\partial \phi} \\
&= \eta'^r \frac{\partial}{\partial r} + \eta'^\theta \frac{1}{r}\frac{\partial}{\partial \theta} + \eta'^\phi \frac{1}{r \sin\theta}\frac{\partial}{\partial \phi},
\end{aligned} \tag{G.9}$$

so that

$$\eta^r = \eta'^r, \quad \eta^\theta = \frac{1}{r}\,\eta'^\theta, \quad \eta^\phi = \frac{1}{r \sin\theta}\,\eta'^\phi. \tag{G.10}$$

The components of the vector $\nabla \times \boldsymbol{\eta}$ with respect to the local orthonormal basis then are

$$\left.\begin{aligned}
(\nabla \times \boldsymbol{\eta})'^r &= \frac{1}{r \sin\theta}\frac{\partial}{\partial \theta}\left(\sin\theta\,\eta'^\phi\right) - \frac{1}{r \sin\theta}\frac{\partial \eta'^\theta}{\partial \phi}, \\
(\nabla \times \boldsymbol{\eta})'^\theta &= \frac{1}{r \sin\theta}\frac{\partial \eta'^r}{\partial \phi} - \frac{1}{r}\frac{\partial}{\partial r}\left(r\,\eta'^\phi\right), \\
(\nabla \times \boldsymbol{\eta})'^\phi &= \frac{1}{r}\frac{\partial}{\partial r}\left(r\,\eta'^\theta\right) - \frac{1}{r}\frac{\partial \eta'^r}{\partial \theta}.
\end{aligned}\right\} \tag{G.11}$$

From the expression for the radial component $(\nabla \times \boldsymbol{\eta})'^r$, the property follows that the curl of a purely radial vector has no radial component. Furthermore, the components of the vector $\nabla \times \nabla \times \boldsymbol{\eta}$ with respect to the local orthonormal basis are given by

$$\left.\begin{aligned}
&(\nabla \times \nabla \times \boldsymbol{\eta})'^r \\
&\quad = \frac{1}{r^2 \sin\theta} \frac{\partial}{\partial r} \left[\frac{\partial}{\partial \theta} \left(r \sin\theta \, \eta'^\theta \right) + \frac{\partial}{\partial \phi} \left(r \, \eta'^\phi \right) \right] \\
&\quad\quad - \frac{1}{r^2} \left[\frac{1}{\sin\theta} \frac{\partial}{\partial \theta} \left(\sin\theta \, \frac{\partial \eta'^r}{\partial \theta} \right) + \frac{1}{\sin^2\theta} \frac{\partial^2 \eta'^r}{\partial \phi^2} \right], \\
&(\nabla \times \nabla \times \boldsymbol{\eta})'^\theta \\
&\quad = \frac{1}{r^2 \sin^2\theta} \frac{\partial^2}{\partial \theta \, \partial \phi} \left(\sin\theta \, \eta'^\phi \right) + \frac{1}{r} \frac{\partial^2 \eta'^r}{\partial r \, \partial \theta} \\
&\quad\quad - \frac{1}{r} \left[\frac{\partial^2}{\partial r^2} \left(r \, \eta'^\theta \right) + \frac{1}{r \sin^2\theta} \frac{\partial^2 \eta'^\theta}{\partial \phi^2} \right], \\
&(\nabla \times \nabla \times \boldsymbol{\eta})'^\phi \\
&\quad = \frac{1}{r \sin\theta} \frac{\partial^2 \eta'^r}{\partial r \, \partial \theta} + \frac{1}{r^2 \sin\theta} \frac{\partial^2 \eta'^\theta}{\partial \theta \, \partial \phi} \\
&\quad\quad - \frac{1}{r} \frac{\partial^2}{\partial r^2} \left(r \, \eta'^\phi \right) - \frac{1}{r^2} \frac{\partial}{\partial \theta} \left[\frac{1}{\sin\theta} \frac{\partial}{\partial \theta} \left(\sin\theta \, \eta'^\phi \right) \right].
\end{aligned}\right\} \tag{G.12}$$

Two types of vector fields of a particular interest are the toroidal and the poloidal vector fields. A vector field $\mathbf{u}$ is a toroidal vector field when

$$\mathbf{u} = \nabla \times (T \, \mathbf{1}_r) , \tag{G.13}$$

where $T(r, \theta, \phi)$ is a scalar function, and $\mathbf{1}_r$ the local unit vector in the radial direction. Moreover, a vector field $\mathbf{v}$ is a poloidal vector field when

$$\mathbf{v} = \nabla \times \nabla \times (S \, \mathbf{1}_r) , \tag{G.14}$$

where $S(r, \theta, \phi)$ is also a scalar function.

By use of equalities (G.11), one verifies that a toroidal vector field has the following components with respect to the local orthonormal basis:

$$u_r = 0, \quad u_\theta = \frac{1}{r \sin\theta} \frac{\partial T}{\partial \phi}, \quad u_\phi = -\frac{1}{r} \frac{\partial T}{\partial \theta}. \tag{G.15}$$

Similarly, by use of equalities (G.12), it follows that a poloidal vector field has the following components with respect to the local orthonormal basis:

$$v_r = \frac{1}{r^2}\mathscr{L}^2 S, \quad v_\theta = \frac{1}{r}\frac{\partial^2 S}{\partial r\,\partial\theta}, \quad v_\phi = \frac{1}{r\,\sin\theta}\frac{\partial^2 S}{\partial r\,\partial\phi}, \tag{G.16}$$

where the operator $\mathscr{L}^2$ is the Legendrian.

By definition, the curl of a toroidal vector field is a poloidal vector field. Furthermore, the curl of a poloidal vector field is a toroidal vector field. Indeed, by means of equalities (G.11) and (G.16), one derives for the components of the vector $\mathbf{w} = \nabla \times \nabla \times \nabla \times (S\,\mathbf{1}_r)$ with respect to the local orthonormal basis:

$$w_r = 0, \quad w_\theta = \frac{1}{r\,\sin\theta}\frac{\partial T^*}{\partial\phi}, \quad w_\phi = -\frac{1}{r}\frac{\partial T^*}{\partial\theta} \tag{G.17}$$

with

$$T^* = -\left(\frac{\partial^2 S}{\partial r^2} - \frac{1}{r^2}\mathscr{L}^2 S\right) \tag{G.18}$$

(Chandrasekhar 1961).

Appendix H
Eigenvalue Problem of the Vibrating String

Consider a vibrating string with length L whose end points are fixed. As usually, we let coincide the x-axis with the vibrating string at rest, and the origin of the x-coordinates with the initial point of the string. Be $u(x, t)$ the transverse displacement at a point of the string.

The eigenvalue problem of the vibrating string is determined by the partial differential equation

$$\frac{1}{c'^2}\frac{\partial^2 u}{\partial t^2} = \frac{\partial^2 u}{\partial x^2}, \tag{H.1}$$

where c' is the velocity of propagation of the wave along the string, and by the conditions

$$\left.\begin{aligned} &u(0,t) = 0, \quad u(L,t) = 0, \\ &\frac{\partial u(x,0)}{\partial t} = 0, \quad u(x,0) = f(x) \end{aligned}\right\} \tag{H.2}$$

(see, e.g., Mandl 1957).

One seeks solutions of the form

$$u(x,t) = [\exp(\mathrm{i}\lambda t)]\, X(x). \tag{H.3}$$

After the separation of the time, the eigenvalue problem is determined by the ordinary differential equation

$$\frac{d^2 X}{dx^2} + \lambda\, X = 0, \tag{H.4}$$

where $\lambda = \sigma^2/c'^2$, and the boundary conditions are

$$X(0) = 0, \quad X(L) = 0. \tag{H.5}$$

A general solution of the differential equation is

$$X(x) = A \sin\left(\sqrt{\lambda}\, x\right) + B \cos\left(\sqrt{\lambda}\, x\right). \tag{H.6}$$

The boundary condition at $x = 0$ requires that

$$B = 0. \tag{H.7}$$

In the usual procedure of resolution, one imposes that the solution also satisfies the boundary condition at $x = L$, so that

$$A \sin\left(\sqrt{\lambda}\, L\right) = 0. \tag{H.8}$$

When one excludes the trivial solution $X = 0$, the parameter λ must satisfy the eigenvalue equation

$$\lambda_n = \left(\frac{n\pi}{L}\right)^2, \quad n = 1, 2, 3, \ldots \tag{H.9}$$

With an eigenvalue λ_n, is associated the eigenfunction

$$X_n(x) = \sin\left(\frac{n\pi}{L}\, x\right). \tag{H.10}$$

The eigenvalue problem can also be solved in an alternative way, which is convenient for the resolution of the eigenvalue problem of the spheroidal modes in stars because of the singularities in the governing equations at $r = 0$ and $r = R$. Instead of imposing solution (H.6) to satisfy the boundary condition at $x = L$, one constructs a solution of the differential equation from that point. To this end, it is convenient to pass on to the independent variable

$$z = L - x, \tag{H.11}$$

and to introduce the function $Z(z) = X(x)$. Differential equation (H.4) then becomes

$$\frac{d^2 Z}{dz^2} + \lambda Z = 0. \tag{H.12}$$

A general solution is given by

$$Z(z) = C \sin\left(\sqrt{\lambda}\, z\right) + D \cos\left(\sqrt{\lambda}\, z\right), \tag{H.13}$$

and the boundary condition at $z = 0$ requires that

$$D = 0. \tag{H.14}$$

The solutions $X(x)$ and $Z(z)$, and their first derivative with respect to, say the independent variable x, must be continuous at all points. Therefore, one imposes, at an arbitrary point $x_f = L - z_f$,

$$\left.\begin{aligned} A \sin\left(\sqrt{\lambda}\, x_f\right) &= C \sin\left(\sqrt{\lambda}\, z_f\right), \\ A \cos\left(\sqrt{\lambda}\, x_f\right) &= -C \cos\left(\sqrt{\lambda}\, z_f\right). \end{aligned}\right\} \tag{H.15}$$

A necessary and sufficient condition for the existence of a non-trivial solution for the constants A en C is then

$$\sin\left(\sqrt{\lambda}\, L\right) = 0. \tag{H.16}$$

One then finds again eigenvalue equation (H.9).

Appendix I
The Euler–Lagrange Equations of Hamilton's Variational Principle for a Perturbed Star

Consider a field of Lagrangian displacements that are defined in a spherically symmetric equilibrium star and depend on time t and three spherical coordinates r, θ, ϕ. The spherical coordinates are the coordinates of the mass elements in the equilibrium star and can be regarded as Lagrangian parameters that characterise the mass elements during their motions. As elsewhere, they are treated as generalised coordinates q^1, q^2, q^3. The components of the Lagrangian displacement with respect to the local coordinate basis are then $\delta q^i\,(t, q^1, q^2, q^3)$, with $i = 1, 2, 3$.

The Lagrangian density considered depends on the components of the Lagrangian displacement, their first derivatives with respect to time, and their covariant derivatives with respect to the generalised coordinates:

$$\mathscr{L} = \mathscr{L}\left(\delta q^i, \frac{\partial\left(\delta q^i\right)}{\partial t}, \nabla_j\, \delta q^i\right), \quad i, j = 1, 2, 3. \tag{I.1}$$

Its first variation is given by

$$\delta\mathscr{L} = \frac{\partial\mathscr{L}}{\partial\left(\delta q^i\right)}\,\delta\left(\delta q^i\right) + \frac{\partial\mathscr{L}}{\partial\left[\partial\left(\delta q^i\right)/\partial t\right]}\,\delta\,\frac{\partial\left(\delta q^i\right)}{\partial t} + \frac{\partial\mathscr{L}}{\partial\left(\nabla_j\,\delta q^i\right)}\,\delta\left(\nabla_j\,\delta q^i\right), \tag{I.2}$$

so that Hamilton's variational principle (8.1) takes the form

$$\int_{t_1}^{t_2}\int_V \left[\frac{\partial\mathscr{L}}{\partial\left(\delta q^i\right)}\,\delta\left(\delta q^i\right) + \frac{\partial\mathscr{L}}{\partial\left[\partial\left(\delta q^i\right)/\partial t\right]}\,\delta\,\frac{\partial\left(\delta q^i\right)}{\partial t} + \frac{\partial\mathscr{L}}{\partial\left(\nabla_j\,\delta q^i\right)}\,\delta\left(\nabla_j\,\delta q^i\right)\right] dV\,dt = 0, \tag{I.3}$$

where V is the volume of the equilibrium star.

The second and the third term are transformed by partial integration:

$$\int_{t_1}^{t_2} \frac{\partial \mathscr{L}}{\partial \left[\partial \left(\delta q^i\right)/\partial t\right]} \delta \frac{\partial \left(\delta q^i\right)}{\partial t} dt = \int_{t_1}^{t_2} \frac{\partial}{\partial t} \left[\frac{\partial \mathscr{L}}{\partial \left[\partial \left(\delta q^i\right)/\partial t\right]} \delta \left(\delta q^i\right) \right] dt$$

$$- \int_{t_1}^{t_2} \left[\frac{\partial}{\partial t} \frac{\partial \mathscr{L}}{\partial \left[\partial \left(\delta q^i\right)/\partial t\right]} \right] \delta \left(\delta q^i\right) dt, \quad \text{(I.4)}$$

$$\int_V \frac{\partial \mathscr{L}}{\partial \left(\nabla_j \delta q^i\right)} \delta \left(\nabla_j \delta q^i\right) dV = \int_V \nabla_j \left[\frac{\partial \mathscr{L}}{\partial \left(\nabla_j \delta q^i\right)} \delta \left(\delta q^i\right) \right] dV$$

$$- \int_V \left[\nabla_j \frac{\partial \mathscr{L}}{\partial \left(\nabla_j \delta q^i\right)} \right] \delta \left(\delta q^i\right) dV. \quad \text{(I.5)}$$

Since the components of the Lagrangian displacement are kept constant at times t_1 and t_2, the first term in the right-hand member of equality (I.4) is equal to zero. After use of Gauss' integral theorem, Hamilton's variational principle then becomes

$$\int_{t_1}^{t_2} \int_V \left[\frac{\partial \mathscr{L}}{\partial \left(\delta q^i\right)} - \frac{\partial}{\partial t} \frac{\partial \mathscr{L}}{\partial \left[\partial \left(\delta q^i\right)/\partial t\right]} - \nabla_j \frac{\partial \mathscr{L}}{\partial \left(\nabla_j \delta q^i\right)} \right] \left[\delta \left(\delta q^i\right)\right] dV dt$$

$$+ \int_S \frac{\partial \mathscr{L}}{\partial \left(\nabla_j \delta q^i\right)} \left[\delta \left(\delta q^i\right)\right] n_j \, dS = 0, \quad \text{(I.6)}$$

where S is the surface of the equilibrium star.

The integrals are equal to zero for all variations $\delta \left(\delta q^i\right)$, with $i = 1, 2, 3$, if the displacement field satisfies the Euler–Lagrange equations

$$\frac{\partial}{\partial t} \frac{\partial \mathscr{L}}{\partial \left[\partial \left(\delta q^i\right)/\partial t\right]} + \nabla_j \frac{\partial \mathscr{L}}{\partial \left(\nabla_j \delta q^i\right)} - \frac{\partial \mathscr{L}}{\partial \left(\delta q^i\right)} = 0, \quad i = 1, 2, 3, \quad \text{(I.7)}$$

and the conditions on the surface of the equilibrium star

$$\frac{\partial \mathscr{L}}{\partial \left(\nabla_j \delta q^i\right)} n_j = 0, \quad i = 1, 2, 3 \quad \text{(I.8)}$$

(see, e.g., Schlögl 1956, Goldstein 1950).

Appendix J
Acoustic Waves

J.1 Acoustic Waves in a Uniform Gas

The velocity of sound in a uniform gaseous medium is derived from the equations that govern linear, isentropic perturbations in a compressible gas. In these equations, the gradients of pressure and mass density are considered to be identically zero (Landau & Lifchitz 1971). The perturbation of the gravitational force is neglected, so that the pressure force is the only acting force. The linearised equation of motion given by Eq. (2.31) can then be written in the vectorial form

$$\frac{\partial^2 \boldsymbol{\xi}}{\partial t^2} = -\frac{1}{\rho}\,\nabla P'. \tag{J.1}$$

Furthermore, Eqs. (2.33) and (4.1) reduce to

$$\frac{\rho'}{\rho} + \alpha = 0, \tag{J.2}$$

$$\frac{P'}{P} - \Gamma_1\,\frac{\rho'}{\rho} = 0. \tag{J.3}$$

By taking the divergence of both members of Eq. (J.1), one has

$$\frac{\partial^2 \alpha}{\partial t^2} = -\frac{1}{\rho}\,\nabla^2 P'. \tag{J.4}$$

Elimination of P' by means of Eqs. (J.3) and (J.2) leads to the wave equation

$$\frac{\partial^2 \alpha}{\partial t^2} = c^2\,\nabla^2 \alpha, \tag{J.5}$$

where $c^2 = \Gamma_1 P/\rho$.

Solutions are sought of the form

$$\alpha = \alpha_0 \exp\left[\mathrm{i}\left(\sigma t - k_x x - k_y y - k_z z\right)\right], \tag{J.6}$$

where α_0 is an arbitrary constant. Substitution into wave equation (J.5) yields the dispersion equation for plane acoustic waves

$$\sigma^2 = c^2\, k^2, \tag{J.7}$$

where $k^2 = k_x^2 + k_y^2 + k_z^2$.

The differences Δx, Δy, Δz between the coordinates of two points where a propagating wave with wave numbers k_x, k_y, k_z has the same amplitude, are related to the time interval needed for the propagation of the wave from one point to the other point as

$$\sigma\, \Delta t - k_x\, \Delta x - k_y\, \Delta y - k_z\, \Delta z = 0, \tag{J.8}$$

so that

$$\sigma = \mathbf{k} \cdot \frac{\Delta \mathbf{r}}{\Delta t}. \tag{J.9}$$

By use of equality (J.7), it follows

$$c = \frac{\mathbf{k}}{k} \cdot \frac{\Delta \mathbf{r}}{\Delta t}. \tag{J.10}$$

Hence, c is the fase or propagation velocity of the acoustic wave.

In virtue of Eqs. (J.2) and (J.3), ρ' and P' satisfy the same wave equation as α. From Eq. (J.1), it follows that

$$\frac{\partial^2 \xi}{\partial t^2} = c^2\, \nabla \alpha. \tag{J.11}$$

Hence, the motions of the mass elements have the same direction as the wave motion. Therefore, acoustic waves are said to be longitudinal waves (Sommerfeld 1964a). See also Sect. 5.4.

J.2 Vertical Propagation of Acoustic Waves in a Plane Isothermal Layer

The vertical propagation of acoustic waves in a plane isothermal gaseous layer in hydrostatic equilibrium was studied by Lamb (1908) for an unbounded layer.

Consider a Cartesian frame of reference whose z-axis is perpendicular to the plane gaseous layer and is directed in the sense opposite to that of the gravitational force. The gaseous layer satisfies condition (2.11) of the hydrostatic equilibrium, which now takes the form

$$\frac{dP}{dz} = -\rho\, g. \tag{J.12}$$

The gravity is regarded as constant in the gaseous layer.

Suppose that the isothermal gas layer obeys the equation of state

$$P = \frac{\mathscr{R}}{\bar{\mu}}\, \rho\, T, \tag{J.13}$$

where $\mathscr{R}$ is the gas constant, and $\bar{\mu}$ the mean molecular weight. The isentropic velocity of sound is then given by

$$c = \left(\Gamma_1 \frac{\mathscr{R}}{\bar{\mu}}\, T \right)^{1/2}, \tag{J.14}$$

so that it is constant in the gaseous layer when the isentropic coefficient Γ_1 is constant.

By dividing both members of condition (J.12) of hydrostatic equilibrium by P and introducing the pressure height scale,

$$H_P \equiv \left(-\frac{1}{P} \frac{dP}{dz} \right)^{-1}, \tag{J.15}$$

one has

$$\frac{1}{H_P} = \frac{\rho}{P}\, g. \tag{J.16}$$

The square of the isentropic velocity of sound can be related to the pressure height scale as

$$c^2 = \Gamma_1\, g\, H_P. \tag{J.17}$$

From Eq. (J.13), it follows that the density height scale,

$$H_\rho \equiv \left(-\frac{1}{\rho} \frac{d\rho}{dz} \right)^{-1}, \tag{J.18}$$

is equal to the pressure height scale.

For the description of purely vertical displacements of the mass elements in the isothermal gaseous layer, one can follow Lamb's approach by starting from equations expressed in terms of Lagrangian perturbations. When ζ is the Lagrangian displacement in the vertical direction, and the Eulerian perturbation of the gravitational force is neglected, the first Eq. (2.52) can be written as

$$\rho\,\frac{\partial^2\zeta}{\partial t^2} = -\rho\,g\,\frac{\partial\zeta}{\partial z} - g\,\delta\rho - \frac{\partial(\delta P)}{\partial z}. \tag{J.19}$$

The sum of the first two terms in the right-hand member of the equation is equal to zero, since Eq. (2.53) reduces to

$$\frac{\delta\rho}{\rho} = -\frac{\partial\zeta}{\partial z}. \tag{J.20}$$

Elimination of δP from Eq. (J.19) by means of Eq. (3.24) yields

$$\frac{\partial^2\zeta}{\partial t^2} = c^2\,\frac{\partial^2\zeta}{\partial z^2} - \frac{c^2}{H_\rho}\,\frac{\partial\zeta}{\partial z}. \tag{J.21}$$

Searching for solutions that depend on the time by a factor $\exp(\mathrm{i}\sigma t)$, one obtains the second-order differential equation

$$\frac{d^2\zeta}{dz^2} - \frac{1}{H_\rho}\,\frac{d\zeta}{dz} + \frac{\sigma^2}{c^2}\,\zeta = 0. \tag{J.22}$$

One passes on to a differential equation that does contain no term with the first derivative of the dependent variable, by introducing the new dependent variable

$$u = \zeta\,\exp[-z/(2H_\rho)]. \tag{J.23}$$

It results that

$$\frac{d^2u}{dz^2} + \left(\frac{\sigma^2}{c^2} - \frac{1}{4H_\rho^2}\right)u = 0. \tag{J.24}$$

Solutions that represent propagating waves are possible when

$$\sigma^2 > \frac{c^2}{4H_\rho^2} = \frac{c^2}{4}\left(\frac{1}{\rho}\,\frac{d\rho}{dr}\right)^2. \tag{J.25}$$

In the right-hand member, $c/(2H_\rho)$ is the cutoff angular frequency for acoustic waves that propagate purely vertically in a plane isothermal layer, as derived by Lamb (see also Sutmann et al. 1998).

List of Symbols

a	Radiation density constant
a^1, a^2, a^3	Lagrangian parameters characterising a mass element
C_P	Specific heat at constant pressure, per unit mass
C_V	Specific heat at constant volume, per unit mass
c	Isentropic sound velocity, $(\Gamma_1 P/\rho)^{1/2}$
c_c	Central value of the isentropic sound velocity
$D_{n,\ell}$	Frequency separations for 5 min-solar oscillations
dQ	Quantity of energy that is absorbed, positively or negatively, by a unit mass
$d\omega$	Infinitesimal solid angle
$\mathbf{F}$	Gravitational force, per unit mass
$\mathbf{F}_{\text{Rad}}$	Flux of radiative energy
$\mathscr{F}^i$	Component i of the local energy flux, $i = 1, 2, 3$
G	Gravitational constant
g	Gravity
g_s	Gravity on the star's surface
g_m	Determinant of the components of the metric tensor
g_{ij}	Covariant component of the metric tensor, $i, j = 1, 2, 3$
g^{ij}	Contravariant component of the metric tensor, $i, j = 1, 2, 3$
H	Hamiltonian
$\mathscr{H}$	Energy density
H_P	Pressure height scale
H_ρ	Density height scale
i	Imaginary unit
k_r	(generally) Complex local wavenumber of a wave propagating in the radial direction
L	Lagrangian
$\mathscr{L}$	Lagrangian density
$\mathscr{L}^2$	Legendrian
M	Total mass of a star
$m(r)$	Mass contained inside the sphere with radius r
$\mathbf{n}$	Unit normal
P	Pressure

P_c	Central pressure
$P_\ell(\cos\theta)$	Legendre's polynomial of the first kind, $\ell = 0, 1, 2, \ldots$
$P_\ell^m(\cos\theta)$	Associated Legendre polynomial of the first kind, $\ell = 0, 1, 2, \ldots$, $m = 0, \ldots, \ell$
q^1, q^2, q^3	Generalised coordinates
R	Total radius of a star
r	Radial coordinate of the spherical coordinates r, θ, ϕ
$\mathscr{R}$	Gas constant
S	Specific entropy
S_ℓ^2	Square of the Lamb frequency
T	Absolute temperature
T^{ij}	Components of the stress tensor, $i, j = 1, 2, 3$
T_{kin}	Kinetic energy of mass motion, per unit mass
T_{tot}	Total kinetic energy of mass motion in a star
t	Time
$\mathbf{t}$	Traction acting on an infinitesimal surface element, per unit surface
U	Specific internal energy
U_{tot}	Total internal energy of a star
$\mathbf{U}$	Tensorial integro-differential operator applying to linear, isentropic displacement fields
V	Volume of a star
V_{tot}	Total potential energy of a star
$X(\mathbf{r})$	Determinant of the displacement gradients
x^1, x^2, x^3	Orthonormal Cartesian coordinates
$Y_\ell^m(\theta, \phi)$	Real or complex spherical harmonic of degree ℓ and order m
α	Divergence of the Lagrangian displacement of a mass element
β	Ratio of the gas pressure to the total pressure
$\Gamma_1, \Gamma_2, \Gamma_3$	Isentropic coefficients
$\Gamma_{1,c}$	Central value of the isentropic coefficient Γ_1
$\Gamma_{1,R}$	Value of the isentropic coefficient Γ_1 on a star's surface
Γ^i_{jk}	Christoffel three-index symbols of the second kind, $i, j, k = 1, 2, 3$
γ	Ratio of the specific heat at constant pressure to the specific heat at constant volume
δ_{ij}	Kronecker's delta, $i, j = 1, 2, 3$
$\delta x^1, \delta x^2, \delta x^3$	Components of the Lagrangian displacement with respect to the orthonormal basis associated with the Cartesian coordinates x^1, x^2, x^3
$\delta q^1, \delta q^2, \delta q^3$	Generalised components of the Lagrangian displacement with respect to the local coordinate basis associated with generalised coordinates q^1, q^2, q^3
$\delta r, \delta\theta, \delta\phi$	Components of the Lagrangian displacement with respect to the local orthogonal coordinate basis associated with the spherical coordinates r, θ, ϕ

ϵ_n	Neuman factor, $n = 0, 1, 2, \ldots$
ϵ^{mki}	Antisymmetric symbol, $m, k, i = 1, 2, 3$
ε_1	Amount of energy that is liberated in a stellar medium by thermonuclear reactions, per unit mass and unit time
θ	Colatitude of the spherical coordinates r, θ, ϕ
κ_S	Isentropic compressibility coefficient
κ_T	Isothermal compressibility coefficient
$\bar{\mu}$	Mean molecular weight
ν	Cyclic frequency
$\boldsymbol{\xi}$	Lagrangian displacement of a mass element
ξ_r, ξ_θ, ξ_ϕ	Components of the Lagrangian displacement with respect to the local orthonormal basis associated with the spherical coordinates r, θ, ϕ
Π	Oscillation period
ρ	Mass density
$\rho_{\rm c}$	Central mass density
$\bar{\rho}$	Mean mass density
σ	Angular frequency
σ_0	Cutoff angular frequency for acoustic waves affected by a density stratification
τ	Specific volume
$\tau_{\rm dyn}$	Dynamic time scale of a star
$\tau_{\rm HK}$	Helmholtz–Kelvin time scale of a star
Φ	Gravitational potential, per unit mass
Φ_e	External gravitational potential, per unit mass
Φ_i	Internal gravitational potential, per unit mass
ϕ	Azimuthal angle of the spherical coordinates r, θ, ϕ
χ	$\Phi' + P'/\rho$
χ_T	Logarithmic partial derivative of the pressure with respect to the temperature, $(\partial \ln P/\partial \ln T)_{\tau,\bar{\mu}}$
χ_τ	Logarithmic partial derivative of the pressure with respect to the specific volume, $(\partial \ln P/\partial \ln \tau)_{T,\bar{\mu}}$
$\chi_{\bar{\mu}}$	Logarithmic partial derivative of the pressure with respect to the mean molecular weight, $(\partial \ln P/\partial \ln \bar{\mu})_{T,\tau}$
Ω	Gravitational potential energy of a star
ω	Dimensionless angular frequency, $\omega = \left[(GM)/R^{\,3}\right]^{1/2} \sigma$
∇^2	Laplacian
∇_j	Operator of partial differentiation with respect to the generalised coordinate q^j when applied to a scalar quantity, and operator of covariant differentiation with respect to the same coordinate when applied to a vector or tensor component
δ	When applied to a quantity Q, δQ, the operator yields the Lagrangian perturbation of that quantity
$'$	When applied to a quantity Q, Q', the operator yields the Eulerian perturbation of that quantity

$\dfrac{\partial}{\partial q^1}, \dfrac{\partial}{\partial q^2}, \dfrac{\partial}{\partial q^3}$	Local coordinate basis associated with the generalised coordinates q^1, q^2, q^3
$\dfrac{\partial}{\partial r}, \dfrac{\partial}{\partial \theta}, \dfrac{\partial}{\partial \phi}$	Local orthogonal coordinate basis associated with the spherical coordinates r, θ, ϕ
$\mathbf{1}_h$	Unit vector in the horizontal direction
$\mathbf{1}_r, \mathbf{1}_\theta, \mathbf{1}_\phi$	Local orthonormal basis associated with the spherical coordinates r, θ, ϕ
$\dfrac{\partial\left(x^1, x^2, x^3\right)}{\partial\left(a^1, a^2, a^3\right)}$	Jacobian or functional determinant
overbar	A horizontal line above a quantity denotes the complex conjugate of that quantity
$< x >$	Difference $x_1 - x_2$

References

M. Abramowitz, I.A. Stegun, *Handbook of Mathematical Functions – with Formulas, Graphs, and Mathematical Tables*. (Dover Publications, New York, 1965)

R. Adler, M. Bazin, M. Schiffer, *Introduction to General Relativity*. (McGraw-Hill, New York, 1965)

C. Aerts, ENEAS: the European Network of Excellence in AsteroSeismology. Commun. Asteroseismology **143**, 5–9 (2003)

C. Aerts, J. Christensen-Dalsgaard, D.W. Kurtz, *Asteroseismology*. (Springer, Dordrecht, 2010)

C. Aerts, A. Thoul, J. Daszyńska, R. Scuflaire, C. Waelkens, M.A. Dupret, E. Niemczura, A. Noels, Asteroseismology of HD 129929: core overshooting and nonrigid rotation. Science **300**, 1926–1928 (2003)

M.L. Aizenman, P. Smeyers, An analysis of the linear, adiabatic oscillations of a star in terms of potential fields. Astrophys. Space Sci. **48**, 123–136 (1977)

M. Aizenman, P. Smeyers, A. Weigert, Avoided crossing of modes of non-radial stellar oscillations. Astron. Astrophys. **58**, 41–46 (1977)

H. Ando, Y. Osaki, Nonadiabatic nonradial oscillations: an application to the five-minute oscillation of the Sun. Publ. Astron. Soc. Jpn. **27**, 581–603 (1975)

A.L. Andrew, Higher modes of non-radial oscillations of stars by variational methods. Aust. J. Phys. **20**, 363–368 (1967)

A.L. Andrew, Non-radial oscillations of massive stars by variational methods. Acad. R. Belg. Bull. Cl. Sci. 5e sér. **54**, 1046–1055 (1968)

V.V. Babikov, *Methods of Phase Functions in Quantum Mechanics*. (Nauka, Moscow, 1976)

J.N. Bahcall, R.K. Ulrich, Solar models, neutrino experiments, and helioseismology. Rev. Mod. Phys. **60**, 297–372 (1988)

N. Baker, R. Kippenhahn, The pulsations of models of δ Cephei stars. Z. Astrophys. **54**, 114–151 (1962)

N. Baker, R. Kippenhahn, The pulsations of models of delta Cephei stars. II. Astrophys. J. **142**, 868–889 (1965)

G.K. Batchelor, *An Introduction to Fluid Dynamics*. (Cambridge University Press, London, 1967)

T.R. Bedding, R.P. Butler, H. Kjeldsen, I.K. Baldry, S.J. O'Toole, C.G. Tinney, G.W. Marcy, F. Kienzle, F. Carrier, Evidence for Solar-like oscillations in β Hydri. Astrophys. J. **549**, L105–L108 (2001)

H.R. Beyer, On some vector analogues of Sturm-Liouville operators. MPA preprint (1994)

H.R. Beyer, The spectrum of adiabatic stellar oscillations. J. Math. Phys. **36**, 4792–4814 (1995a)

H.R. Beyer, The spectrum of radial adiabatic stellar oscillations. J. Math. Phys. **36**, 4815–4825 (1995b)

H.R. Beyer, B.G. Schmidt, Newtonian stellar oscillations. Astron. Astrophys. **296**, 722–726 (1995)

B.A. Bolt, J.S. Derr, Free bodily vibrations of the terrestrial planets. Vistas Astron. **11**, 69–102 (1969)

F. Bouchy, F. Carrier, p-mode observations on α Cen A. Astron. Astrophys. **374**, L5–L8 (2001)

P. Brassard, G. Fontaine, F. Wesemael, S.D. Kawaler, M. Tassoul, Adiabatic properties of pulsating DA white dwarfs. I. The treatment of the Brunt-Väisälä frequency and the region of period formation. Astrophys. J. **367**, 601–611 (1991)

M. Breger, G. Handler, R. Garrido, N. Audard, W. Zima, M. Paparo, F. Beichbuchner, L. Zhi-ping, J. Shi-yang, L. Zong-li, Z. Ai-ying, H. Pikall, A. Stankov, J.A. Guzik, M. Sperl, J. Krzesinski, W. Ogloza, G. Pajdosz, S. Zola, T. Thomassen, J.-E. Solheim, E. Serkowitsch, P. Reegen, T. Rumpf, A. Schmalwieser, M.H. Montgomery, 30+ frequencies for the δ Scuti variable 4 Canum Venaticorum: results of the 1996 multisite campaign. Astron. Astrophys. **349**, 225–235 (1999)

L. Brillouin, *Les Tenseurs en Mécanique et en Elasticité*. (Masson, Paris, 1960)

T.M. Brown, R.L. Gilliland, Asteroseismology. Annu. Rev. Astron. Astrophys. **32**, 37–82 (1994)

P. Bruggen, P. Smeyers, Theoretical expressions for evolutionary period changes in non-radially pulsating stars. Astron. Astrophys. **186**, 170–176 (1987)

D. Brunt, The period of simple vertical oscillations in the atmosphere. Q.J. Roy. Meteorol. Soc. **53**, 30–32 (1927)

D. Brunt, *Physical and Dynamical Meteorology*. (Cambridge University Press, New York, 1934)

J.R. Buchler, P.A. Yecko, Z. Kolláth, The nature of strange modes in classical variable stars. Astron. Astrophys. **326**, 669–681 (1997)

F. Carrier, F. Bouchy, P. Eggenberger, in *Asteroseismology across the HR diagram*, ed. by M. Thompson, M.S. Cunha, M.J.P.F.G. Monteiro. p-Mode observations on the K0 IV star δ Eridani. Reprinted from Astrophysics and Space Science, vol. 284, No. 1. (Kluwer Academic, Dordrecht, 2003), pp. 311–314

S. Chandrasekhar, *An Introduction to the Study of Stellar Structure*. (The University Press, Chicago, 1939)

S. Chandrasekhar, *Hydrodynamic and Hydromagnetic Stability*. (Clarendon, Oxford, 1961)

S. Chandrasekhar, A general variational principle governing the radial and nonradial oscillations of gaseous masses. Astrophys. J. **139**, 664–674 (1964)

S. Chandrasekhar, *Ellipsoidal Figures of Equilibrium*. (Yale University Press, New Haven, 1969)

S. Chandrasekhar, N.R. Lebovitz, On the oscillations and the stability of rotating gaseous masses. Astrophys. J. **135**, 248–260 (1962a)

S. Chandrasekhar, N.R. Lebovitz, On the oscillations and the stability of rotating gaseous masses. II. The homogeneous, compressible model. Astrophys. J. **136**, 1069–1081 (1962b)

S. Chandrasekhar, N.R. Lebovitz, The equilibrium and the stability of the Jeans spheroids. Astrophys. J. **137**, 1172–1184 (1963)

S. Chandrasekhar, N.R. Lebovitz, Non-radial oscillations of gaseous masses. Astrophys. J. **140**, 1517–1528 (1964)

G. Chanmugam, White dwarfs — Pulsation periods explained in terms of non-radial oscillations and gravity modes. Nat. Phys. Sci. **236**, 83 (1972)

J. Christensen-Dalsgaard, On adiabatic non-radial oscillations with moderate or large ℓ. Mon. Not. R. Astron. Soc. **190**, 765–791 (1980)

J. Christensen-Dalsgaard, The effect of non-adiabaticity on avoided crossings of non-radial stellar oscillations. Mon. Not. R. Astron. Soc. **194**, 229–250 (1981)

J. Christensen-Dalsgaard, in *Advances in Helio- and Asteroseismology*, ed. by J. Christensen-Dalsgaard, S. Frandsen. An overview of helio- and asteroseismology. International Astronomical Union Symposium 123, (International Astronomical Union, Reidel Publishing Company, Dordrecht, 1988), pp. 3–18

J. Christensen-Dalsgaard, Some aspects of the theory of solar oscillations. Geophys. Astrophys. Fluid Dynam. **62**, 123–152 (1991)

J. Christensen-Dalsgaard, D.O. Gough, On the dipolar f mode of stellar oscillation. Mon. Not. R. Astron. Soc. **326**, 1115–1121 (2001)

J. Christensen-Dalsgaard, D.J. Mullan, Accurate frequencies of polytropic models. Mon. Not. R. Astron. Soc. **270**, 921–935 (1994)

J. Christensen-Dalsgaard, A.J. Cooper, D.O. Gough, An upper bound to the periods of radial pulsation of the Sun. Mon. Not. R. Astron. Soc. **203**, 165–179 (1983)

J. Christensen-Dalsgaard, D. Gough, J. Toomre, Seismology of the Sun. Science **229**, 923–931 (1985)

J. Christensen-Dalsgaard, C.R. Proffitt, M.J. Thompson, Effects of diffusion on solar models and their oscillation frequencies. Astrophys. J. **403**, L75–L78 (1993)

A. Claverie, G.R. Isaak, C.P. McLeod, H.B. van der Raay, T. Roca Cortes, Solar structure from global studies of the 5-minute oscillation. Nature **282**, 591–594 (1979)

E.A. Coddington, N. Levinson, *Theory of Ordinary Differential Equations*. (McGraw-Hill, New York, 1955)

R. Courant, D. Hilbert, *Methoden der Mathematischen Physik*. (Springer, Berlin, 1968)

T.G. Cowling, On the motion of the apsidal line in close binary systems. Mon. Not. R. Astron. Soc. **98**, 734–743 (1938)

T.G. Cowling, The non-radial oscillations of polytropic stars. Mon. Not. R. Astron. Soc. **101**, 367–375 (1941)

J.P. Cox, Pulsating stars. Rep. Prog. Phys. **37**, 563–698 (1974)

J.P. Cox, *Theory of Stellar Pulsation*. (The University Press, Princeton, 1980)

J.P. Cox, R.Th. Giuli, *Principles of Stellar Structure*. (Gordon and Breach, New York, 1968)

J.P. Cox, C. Whitney, Stellar pulsation. IV. A semi-theoretical period-luminosity relation for classical Cepheids. Astrophys. J. **127**, 561–572 (1958)

J.P. Cox, D.S. King, A.N. Cox, J.C. Wheeler, C.J. Hansen, S.W. Hodson, Pulsations of the R Coronae Borealis stars. Space Sci. Rev. **27**, 529–535 (1980)

F.A. Dahlen, J. Tromp, *Theoretical Global Seismology*. (The University Press, Princeton, 1998)

W. Däppen, J. Perdang, Non-linear stellar oscillations. Non-radial mode interactions. Astron. Astrophys. **151**, 174–188 (1985)

I. De Boeck, T. Van Hoolst, P. Smeyers, The subseismic approximation for low-frequency modes of the Earth applied to low-degree, low frequency g-modes of non-rotating stars. Astron. Astrophys. **259**, 167–174 (1992)

P. Degroote, C. Aerts, A. Baglin, A. Miglio, M. Briquet, A. Noels, E. Niemczura, J. Montalban, S. Bloemen, R. Oreiro, M. Vučković, K. Smolders, M. Auvergne, F. Baudin, C. Catala, E. Michel, Deviations from a uniform period spacing of gravity modes in a massive star. Nature **464**, 259–261 (2010)

J. Demaret, Vibrational stability of stars in thermal imbalance: a solution in terms of asymptotic expansions. I: 'Isentropic oscillations'. Astrophys. Space Sci. **31**, 305–331 (1974)

J. De Ridder, C. Barban, F. Carrier, A. Mazumdar, P. Eggenberger, C. Aerts, S. Deruyter, J. Vanautgaerden, Discovery of solar-like oscillations in the red giant ε Ophiuchi. Astron. Astrophys. **448**, 689–695 (2006)

J. De Ridder, C. Barban, F. Baudin, F. Carrier, A.P. Hatzes, S. Hekker, T. Kallinger, W.W. Weiss, A. Baglin, M. Auvergne, R. Samadi, P. Barge, M. Deleuil, Non-radial oscillation modes with long lifetimes in giant stars. Nature **459**, 398–400 (2009)

S.L. Detweiler, J.R. Ipser, A variational principle and a stability criterion for the nonradial modes of pulsation of stellar models in general relativity. Astrophys. J. **185**, 685–707 (1973)

F.-L. Deubner, Observations of low wavenumber nonradial eigenmodes of the Sun. Astron. Astrophys. **44**, 371–375 (1975)

F.-L. Deubner, D. Gough, Helioseismology: oscillations as a diagnostic of the solar interior. Annu. Rev. Astron. Astrophys. **22**, 593–619 (1984)

G. Dogan, J. Christensen-Dalsgaard, M. Takata, in *Interpretation of Asteroseismic Data*, ed. by M. Breger, W. Dziembowski, M. Thompson. Dipole modes of stellar oscillations. Commun. Asteroseismology, **157**, 301–302 (2008)

N. Dunford, J.T. Schwartz, *Linear Operators, Part I General Theory*. (Interscience, New York, 1957)

J. Dyson, B.F. Schutz, Perturbations and stability of rotating stars – I. Completeness of normal modes. Proc. Roy. Soc. Lond. A **368**, 389–410 (1979)

W.A. Dziembowski, Nonradial oscillations of evolved stars. I. Quasiadiabatic approximation. Acta Astronom. **21**, 289–306 (1971)

W.A. Dziembowski, Oscillations of giants and supergiants. Acta Astronom. **27**, 1–32 (1977)

W.A. Dziembowski, in *A Half-Century of Stellar Pulsation Interpretations*, ed. by P.A. Bradley, J.A. Guzik. Helio- and asteroseismology. Astronomical Society of the Pacific Conference Series, vol. 135, 427–436 (1998)

C. Eckart, *Hydrodynamics of Oceans and Atmospheres*. (Pergamon, Oxford, 1960)

A.S. Eddington, *The Internal Constitution of the Stars*. (The University Press, Cambridge, 1926)

A.S. Eddington, *The Internal Constitution of the Stars*. (The University Press, Cambridge, reedition, 1988)

J. Eisenfeld, A completeness theorem for an integro-differential operator. J. Math. Anal. Appl. **26**, 357–375 (1969)

R. Emden, *Gaskugeln: Anwendungen der Mechanischen Wärmtetheorie auf Kosmologische und Meteorologische Probleme*. (Teubner, Leipzig, 1907)

I. Epstein, Pulsation properties of giant-star models. Astrophys. J. **112**, 6–21 (1950)

L. Eyer, C. Aerts, A search for new γ Doradus stars in the Geneva photometric database. Astron. Astrophys. **361**, 201–206 (2000)

E. Fossat, G. Ricort, Photospheric oscillations. I. Large scale observations by optical resonance method. Astron. Astrophys. **43**, 243–252 (1975)

S. Frandsen, F. Carrier, C. Aerts, D. Stello, T. Maas, M. Burnet, H. Bruntt, T.C. Teixeira, J.R. de Medeiros, F. Bouchy, H. Kjeldsen, F. Pijpers, J. Christensen-Dalsgaard, Detection of Solar-like oscillations in the G7 giant star ξ Hya. Astron. Astrophys. **394**, L5–L8 (2002)

J.L. Friedman, B.F. Schutz, Lagrangian perturbation theory of nonrelativistic fluids. Astrophys. J. **221**, 937–957 (1978)

K.O. Friedrichs, in *Studies and Essays presented to R. Courant on his 60th Birthday, Jan. 8, 1948*, ed. by K.O. Friedrichs, O.E. Neugebauer, J.J. Stoker. Criteria for the discrete character of the spectra of ordinary differential equations. (Interscience, New York, 1948), pp. 145–160

K.O. Friedrichs, *Spectral Theory of Operators in Hilbert Space*. (Springer, New York, 1973)

G. Frobenius, Ueber die Integration der linearen Differentialgleichungen durch Reihen. J. Math. **76**, 214–235 (1873)

M. Gabriel, On the multiplicity of the eigenvalues of nonradial stellar oscillations. Astron. Astrophys. **82**, 8–13 (1980)

M. Gabriel, The $D_{n,\ell}$ values and the structure of the solar core. Astron. Astrophys. **226**, 278–283 (1989)

M. Gabriel, R. Scuflaire, Properties of non-radial stellar oscillations. Acta Astronom. **29**, 135–149 (1979)

A. Gautschy, Pulsating post-asymptotic giant branch stars. Mon. Not. Roy. Astron. Soc. **265**, 340–346 (1993)

A. Gautschy, The becoming of the theory of stellar pulsations. Basel Preprint Series **86**, 1–23 (1996)

A. Gautschy, The development of the theory of stellar pulsations. Vistas Astron. **41**, 95–115 (1997)

A. Gautschy, W. Glatzel, The secular and thermal spectrum of hydrogen-deficient carbon stars. Mon. Not. Roy. Astron. Soc. **253**, 509–521 (1991)

A. Gautschy, H. Saio, Stellar pulsations across the HR diagram: Part 1. Annu. Rev. Astron. Astrophys. **33**, 75–113 (1995)

H. Goldstein, *Classical Mechanics*. (Addison-Wesley, Reading, 1957)

M. Goossens, P. Smeyers, Gravity modes in composite polytropic stars. Astrophys. Space Sci. **26**, 137–151 (1974)

D.O. Gough, Our first inferences from helioseismology. Phys. Bull. **34**, 502–507 (1983)

D.O. Gough, To the editors of 'The Observatory': Astereoseismology. Observatory **116**, 313–314 (1996)

D.O. Gough, The birth of asteroseismology. Science **291**, 2325–2327 (2001)

G. Grec, E. Fossat, Photospheric oscillations: III. Search for long period modes. Astron. Astrophys. **55**, 411–420 (1977)

G. Grec, E. Fossat, M. Pomerantz, Solar oscillations: full disk observations from the geographic South Pole. Nature **288**, 541–544 (1980)

D.B. Guenther, The p-mode oscillation spectra of an evolving 1 $M_\odot$ sun-like star. Astrophys. J. **375**, 352–365 (1991)

C.J. Hansen, Secular stability: applications to stellar structure and evolution. Annu. Rev. Astron. Astrophys. **16**, 15–32 (1978)

C.J. Hansen, S.D. Kawaler, *Stellar Interiors: Physical Principles, Structure, and Evolution.* (Springer, New York, 1994)

J.B. Hartle, *Gravity: An Introduction to Einstein's General Relativity.* (Addison Wesley, San Francisco, 2003)

D. Heynderickx, C. Waelkens, P. Smeyers, A photometric study of β Cephei stars. II. Determination of the degrees ℓ of pulsation modes. Astron. Astrophys. Suppl. Ser. **105**, 447–480 (1994)

C.O. Hines, Internal atmospheric gravity waves at ionospheric heights. Can. J. Phys. **38**, 1441–1481 (1960)

C. Hunter, On secular stability, secular instability, and points of bifurcation of rotating gaseous masses. Astrophys. J. **213**, 497–517 (1977)

M. Hurley, P.H. Roberts, K. Wright, The oscillations of gas spheres. Astrophys. J. **143**, 535–551 (1966)

E.L. Ince, *Ordinary Differential Equations.* (Dover Publications, New York, 1956)

T. Ishizuka, Non-radial oscillations: Lagrange's formalism. Publ. Astron. Soc. Jpn. **22**, 125–127 (1970)

P. Iweins, Mémoire de Licence Sciences Mathématiques, Université de Liège (1964)

P. Iweins, P. Smeyers, Sur le comportement asymptotique des oscillations linéaires du modèle homogène compressible. Acad. R. Belg. Bull. Cl. Sci. 5e sér. **54**, 164–176 (1968)

J.D. Jackson, *Classical Electrodynamics.* (Wiley, New York, 1999)

E. Jahnke, F. Emde, *Tables of Functions – with Formulae and Curves.* (Dover Publications, New York, 1945)

S.J. Jimenéz-Reyes, J. Quintero Nerkhorn, P.L. Pallé, HELAS Board: HELAS IT-platform: A new tool for the European helio- and asteroseismology community. Commun. Asteroseismology **153**, 108–115 (2008)

C.A. Jones, P.H. Roberts, The boundary layer method for pulsating stars. Geophys. Astrophys. Fluid Dynam. **14**, 61–101 (1979)

S. Kaniel, A. Kovetz, Schwarzschild's criterion for instability. Phys. Fluid. **10**, 1186–1193 (1967)

S. Kato, W. Unno, Pulsational instability of stars in quasi-equilibrium. Publ. Astron. Soc. Jpn. **19**, 1–19 (1967)

S.D. Kawaler, Taking the pulse of a massive star. Science **300**, 1885–1886 (2003)

O.D. Kellogg, *Foundations of Potential Theory.* (Springer, Berlin, 1929)

E.C. Kemble, *The Fundamental Principles of Quantum Mechanics.* (McGraw Hill, New York, 1937)

J. Kevorkian, J.D. Cole, *Perturbation Methods in Applied Mathematics.* (Springer, New York, 1981)

J. Kevorkian, J.D. Cole, *Multiple Scale and Singular Perturbation Methods.* (Springer, New York, 1996)

R. Kippenhahn, A. Weigert, *Stellar Structure and Evolution.* (Springer, Berlin, 1990)

M. Kiriakidis, K.J. Fricke, W. Glatzel, The stability of massive stars and its dependence on metallicity and opacity. Mon. Not. Roy. Astron. Soc. **264**, 50–62 (1993)

K. Kolenberg, E. Guggenberger, T. Medupe, Johnson photometry of southern Blazhko targets. Commun. Asteroseismology **153**, 67–83 (2008)

Z. Kopal, Nonradial oscillations of the standard model. Astrophys. J. **109**, 509–527 (1949)

Z. Kopal, *Close Binary Systems.* The International Astrophysics Series, vol. 5. (Chapman & Hall, London, 1959)

G.A. Korn, T.M. Korn, *Mathematical Handbook – for Scientists and Engineers.* (McGraw-Hill, New York, 1968)

H. Lamb, On the vibrations of an elastic sphere. Proc. Lond. Math. Soc. **13**, 189–212 (1882)

H. Lamb, On the theory of waves propagated vertically in the atmosphere. Proc. Lond. Math. Soc. Ser. 2 **7**, 122–141 (1908)

H. Lamb, *Hydrodynamics*. (The University Press, Cambridge, 1932)
L. Landau, E. Lifchitz, *Mécanique des Fluides*. (Mir, Moscou, 1971)
A.U. Landolt, A new short-period blue variable. Astrophys. J. **153**, 151–164 (1968)
R.E. Langer, On the asymptotic solutions of ordinary differential equations, with reference to the Stokes' phenomenon about a singular point. Trans. Am. Math. Soc. **37**, 397–416 (1935)
R.E. Langer, On the connection formulas and the solutions of the wave equation. Phys. Rev. **51**, 669–676 (1937)
G. Laval, C. Mercier, R. Pellat, Necessity of the energy principles for magnetostatic stability. Nucl. Fusion **5**, 156–158 (1965)
D.F. Lawden, *An Introduction to Tensor Calculus and Relativity*. (Methuen, London, 1962)
N.R. Lebovitz, On Schwarzschild's criterion for the stability of gaseous masses. Astrophys. J. **142**, 229–242 (1965a)
N.R. Lebovitz, On the onset of convective instability. Astrophys. J. **142**, 1257–1260 (1965b)
N.R. Lebovitz, On the necessity of Schwarzschild's criterion for stability. Astrophys. J. **146**, 946–949 (1966)
P. Ledoux, Sur la théorie des oscillations radiales d'une étoile. Astrophys. Norvegica **3**, 193–215 (1940)
P. Ledoux: Stellar models with convection and with discontinuity of the mean molecular weight. Astrophys. J. **105**, 305–321 (1947)
P. Ledoux: Contributions à l'étude de la structure interne des étoiles et de leur stabilité. Mémoires de la Société Royale des Sciences de Liège, Série 4, **9**, 7–294 (1949)
P. Ledoux, in *Astrophysics II: Stellar Structure*, ed. by S. Flügge. Stellar stability. Handbuch der Physik, vol. 51 (Springer, Berlin, 1958), pp. 605–688
P. Ledoux, Sur la forme asymptotique des pulsations radiales adiabatiques d'une étoile. I. Acad. R. Belg. Bull. Cl. Sci. 5e sér. **48**, 240–254 (1962)
P. Ledoux: Le problème de la stabilité des étoiles. Physique du XXe Siècle **8**, 79–110 (1963a)
P. Ledoux, in *Star Evolution*, ed. by L. Gratton. Stellar stability and stellar evolution. Proceedings of the International School of Physics "Enrico Fermi", Course XXVIII. (Academic, New York, 1963b), pp. 394–445
P. Ledoux, in *Stellar Instability and Evolution*, ed. by P. Ledoux, A. Noels, A.W. Rodgers. Non-radial oscillations. International Astronomical Union Symposium, vol. 59. (Reidel, Dordrecht, 1974), pp. 135–173
P. Ledoux, J. Perdang, Asymptotic approximation for nonradial stellar oscillations. Bulletin de la Société Mathématique de Belgique **32**, 133–159 (1980)
P. Ledoux, P. Smeyers, Sur le spectre des oscillations non radiales d'un modèle stellaire. Compte-Rendus de l'Académie des Sciences de Paris **262**, 841–844 (1966)
P. Ledoux, Th. Walraven, in *Astrophysics II: Stellar Structure*, ed. by S. Flügge. Variable stars. Handbuch der Physik, vol. 51 (Springer, Berlin, 1958), pp. 353–604
U. Lee, Stability of the delta Scuti stars against nonradial oscillations with low degrees ℓ. Publ. Astron. Soc. Jpn. **37**, 279–291 (1985)
U. Lee, H. Saio, Energy of oscillation: the role of the negative energy mode in overstable nonradial oscillations of rotating stars. Astrophys. J. **360**, 590–603 (1990)
J.W. Leibacher, R.F. Stein, A new description of the solar five-minute oscillation. Astrophys. Lett. **7**, 191–192 (1971)
J.W. Leibacher, R.F. Stein, in *The Sun as a Star*, ed. by S.D. Jordan. Oscillations and pulsations. (Centre National de la Recherche Scientifique, Paris. National Aeronautics and Space Administration, Washington, D.C., 1981), pp. 263–287
R.B. Leighton, in *Aerodynamic Phenomena in Stellar Atmospheres*, ed. by R.N. Thomas. Untitled comments on supergranulation and 5-min oscillations. International Astronomical Symposium 12, Il Nuovo Cimento Supplement Series 10, vol. 21, 321–327 (1961)
R.B. Leighton, R.W. Noyes, G.W. Simon, Velocity fields in the solar atmosphere. I. Preliminary report. Astrophys. J. **135**, 474–509 (1962)
K.G. Libbrecht, Solar and stellar seismology. Space Sci. Rev. **47**, 275–301 (1988)

K.G. Libbrecht, M.F. Woodard, J.M. Kaufman, Frequencies of solar oscillations. Astrophys. J. Suppl. Ser. **74**, 1129–1149 (1990)

I. Lopes, S. Turck-Chièze, The second order asymptotic theory for the solar and stellar low degree acoustic mode predictions. Astron. Astrophys. **290**, 845–860 (1994)

D. Lynden-Bell, J.P. Ostriker, On the stability of differentially rotating bodies. Mon. Not. Roy. Astron. Soc. **136**, 293–310 (1967)

F. Mandl, *Quantum Mechanics*. (Butterworths, London, 1957)

A.J. McConnell, *Applications of Tensor Analysis*. (Dover Publications, New York, 1957)

M. Martić, J. Schmitt, J.-C. Lebrun, C. Barban, P. Connes, F. Bouchy, E. Michel, A. Baglin, T. Appourchaux, J.-L. Bertaux, Evidence for global pressure oscillations on Procyon. Astron. Astrophys. **351**, 993–1002 (1999)

C.W. Misner, in *Astrophysics and General Relativity*, ed. by M. Chrétien, S. Deser, J. Goldstein. Gravitational collapse. Brandeis University, Summer Institute in Theoretical Physics, 1968, vol. 1. (Gordon and Breach, New York, 1969), pp. 113–215.

C.W. Misner, K.S. Thorne, J.A. Wheeler, *Gravitation*. (Freeman, San Francisco, 1973)

L. Mittag, M.J. Stephen, W. Yourgrau, in *Variational Principles in Dynamics and Quantum Theory*, ed. by W. Yourgrau, S. Mandelstam. Variational principles in hydrodynamics. (Pitman & Sons, London, 1968), pp. 142–161

P.M. Morse, H. Feshbach, *Methods of Theoretical Physics*. (McGraw-Hill, New York, 1953)

D.J. Mullan, g-mode pulsations in polytropes: high-precision eigenvalues and the approach to asymptotic behavior. Astrophys. J. **337**, 1017–1022 (1989)

D.J. Mullan, R.K. Ulrich, Radial and nonradial pulsations of polytropes: high-precision eigenvalues and the approach of p-modes to asymptotic behavior. Astrophys. J. **331**, 1013–1028 (1988)

A. Olech, J. Kaluzny, I.B. Thompson, W. Pych, W. Krzeminski, A. Schwarzenberg-Czerny, RR Lyrae variables in the globular cluster M55. The first evidence for nonradial pulsations in RR Lyrae stars. Astron. J. **118**, 442–452 (1999)

F.W.J. Olver, The asymptotic solution of linear differential equations of the second order in a domain containing one transition point. Philos. Trans. R. Soc. Lond. A **249**, 65–97 (1956)

F.W.J. Olver, *Asymptotic and Special Functions*. (Academic, New York, 1974)

Y. Osaki, Nonradial oscillations of a 10 solar mass star in the main-sequence stage. Publ. Astron. Soc. Jpn. **27**, 237–258 (1975)

J.W. Owen, The non-radial oscillations of centrally condensed stars. Mon. Not. R. Astron. Soc. **117**, 384–392 (1957)

C.L. Pekeris, Nonradial oscillations of stars. Astrophys. J. **88**, 189–199 (1938)

J. Perdang, On some group-theoretical aspects of the study of non-radial oscillations. Astrophys. Space Sci. **1**, 355–371 (1968)

J. Perdang, Local stellar stability. Astrophys. Space Sci. **44**, 177–215 (1976)

W.D. Pesnell, Weight functions in adiabatic stellar pulsations I: radially symmetric motion. Publ. Astron. Soc. Pac. **99**, 975–985 (1987)

W.D. Pesnell, Nonradial, nonadiabatic stellar pulsations. Astrophys. J. **363**, 227–233 (1990)

W.D. Pesnell, J.R. Buchler, The structure of the thermal modes in pulsating stars. Astrophys. J. **303**, 740–748 (1986)

J. Provost, G. Berthomieu, P. Morel, Low-frequency p- and g-mode solar oscillations. Astron. Astrophys. **353**, 775–785 (2000)

J. Quaegebeur, Private communication (2002)

K.-H. Rädler, Zur Dynamotheorie kosmischer Magnetfelder. II. Darstellung von Vektorfelden als Summe aus einem poloidalen und einem toroidalen Anteil. Astron. Nachr. **295**, 73–84 (1974)

A.S. Ramsey, *An Introduction to the Theory of Newtonian Attraction*. (The University Press, Cambridge, 1940)

L. Rayleigh, (Sir John William Strutt): *The Theory of Sound*. Macmillan, London (1877). Second revised version, Dover Publications, New York (1945)

A. Reisenegger, Multipole moments of stellar oscillation modes. Astrophys. J. **432**, 296–301 (1994)

T. Reye, *Die Wirbelstürme, Tornados und Wettersaülen*. (Genesius, Halle, 1872)

R.D. Richtmyer, *Principles of Advanced Mathematical Physics*, vol. 1. (Springer, New York, 1978)
F. Riesz, B. Sz.-Nagy, *Leçons d'Analyse Fonctionnelle*. (Gauthier-Villars, Paris, 1955)
A. Ritter, Wiedemann's Annalen **5–20** (1878–1883)
A. Ritter, Untersuchungen über die Höhe der Atmosphäre und die Constitution gasförmiger Weltkörper. Annalen der Physik und Chemie, Neue Folge **8**, 157–183 (1879)
W. Ritz, Über eine neue Methode zur Lösung gewisser Variationsprobleme der mathematischen Physik. J. reine angew. Math. **135**, 1–61 (1909)
H. Robe, Sur les oscillations non radiales d'une sphère. Acad. R. Belg. Bull. Cl. Sci. 5e sér. tome **51**, 598–603 (1965)
H. Robe, Les oscillations non radiales des polytropes. Ann. d'Astrophys. **31**, 475–482 (1968)
H. Robe, L. Brandt, Sur les oscillations non radiales des polytropes. Ann. d'Astrophys. **29**, 517–524 (1966)
P.H. Roberts, *An Introduction to Magnetohydrodynamics*. (Longmans, London, 1967)
L. Robin, *Fonctions Sphériques et Fonctions Sphéroïdales*. Tome I, (Gauthier-Villars, Paris, 1957)
E.L. Robinson, S.O. Kepler, R.E. Nather, Multicolor variations of the ZZ Ceti stars. Astrophys. J. **259**, 219–231 (1982)
S. Rosseland, The theory of oscillating fluid globes. Publications from Oslo University Observatory **1**, No. 2, 3–22 (1932)
S. Rosseland, *The Pulsation Theory of Variable Stars*. (The University Press, Oxford, 1949)
M.L. Roth, A. Weigert, More on avoided level crossing of non-radial stellar oscillations. Astron. Astrophys. **80**, 48–52 (1979)
I.W. Roxburgh, S.V. Vorontsov, in *Inside the Stars*, ed. by W.W. Weiss, A. Baglin. Asymptotic theory of low-degree stellar acoustic oscillations. International Astronomical Union Colloquium 137, Astronomical Society of the Pacific Conference Series, vol. 40, p. 535–537 (1993)
I.W. Roxburgh, S.V. Vorontsov, The seismology of stellar cores: a simple theoretical description of the 'small frequency separations'. Mon. Not. R. Astron. Soc. **267**, 297–302 (1994a)
I.W. Roxburgh, S.V. Vorontsov, The asymptotic theory of stellar acoustic oscillations: a fourth-order approximation for low-degree modes. Mon. Not. R. Astron. Soc. **268**, 143–158 (1994b)
I.W. Roxburgh, S.V. Vorontsov, An asymptotic description of solar acoustic oscillation of low and intermediate degree. Mon. Not. R. Astron. Soc. **278**, 940–946 (1996)
H. Saio, N.H. Baker, A. Gautschy, On the properties of strange modes. Mon. Not. R. Astron. Soc. **294**, 622–634 (1998)
E. Sauvenier-Goffin, Note sur les pulsations non-radiales d'une sphère homogène compressible. Bulletin de la Société Royale des Sciences de Liège **20**, 20–38 (1951)
A.K. Schenk, P. Arras, É.É. Flanagan, S.A. Teukolsky, I. Wasserman, Nonlinear mode coupling in rotating stars and the r-mode instability in neutron stars. Phys. Rev. D **65**, 1–43 (2001)
F. Schlögl, in *Mathematical Methods I*, ed. by S. Flügge. Randwerteprobleme. Handbuch der Physik, vol. 1. (Springer, Berlin, 1956), pp. 218–352
D.C. Schwank, Epstein weight functions for non-radial oscillations of polytropes. Astrophys. Space Sci. **43**, 459–468 (1976)
K. Schwarzschild, Ueber das Gleichgewicht der Sonnenatmosphäre. Göttingen Nachrichten **1**, 41–53 (1906)
R. Scuflaire, Sur la stabilité de configurations statiques vis-à-vis de perturbations adiabatiques. Acad. R. Belg. Bull. Cl. Sci. 5e sér. **57**, 1126–1136 (1971)
R. Scuflaire, The non radial oscillations of condensed polytropes. Astron. Astrophys. **36**, 107–111 (1974)
H. Shapley, On the nature and cause of Cepheid variation. Astrophys. J. **40**, 448–465 (1914)
H. Shibahashi, in *Progress of Seismology of the Sun and Stars*, ed. by Y. Osaki, H. Shibahashi. Introductory review of solar and stellar oscillations. Proceedings of the Oji International Seminar Hakone, Japan 1989. (Springer, Berlin, 1990), pp. 3–19
H. Shibahashi, Y. Osaki, Overstability of gravity modes in massive stars with the semiconvective zone. Publ. Astron. Soc. Jpn. **28**, 199–214 (1976)
R. Sigl, *Einführung in die Potentialtheorie*. (Wichmann, Karlsruhe, 1973)

P. Smeyers, Sur les oscillations non radiales adiabatiques d'étoiles massives. Acad. R. Belg. Bull. Cl. Sci. 5e sér. **49**, 128–141 (1963)

P. Smeyers, Les oscillations non radiales et la convection dans les étoiles. Ann. d'Astrophys. **29**, 539–548 (1966a)

P. Smeyers, Les oscillations linéaires et adiabatiques de la sphère homogène. Acad. R. Belg. Bull. Cl. Sci. 5e sér. **52**, 1126–1142 (1966b)

P. Smeyers, Les oscillations non radiales des étoiles massives. Bulletin de la Société Royale des Sciences de Liège, 36e année, n° 7-8, 357–392 (1967)

P. Smeyers, Note on variational principles governing small departures from equilibrium in gaseous spheres and in horizontal fluids. Ricerche Astronomiche **8**, 359–366 (1973)

P. Smeyers, in *Theoretical Problems in Stellar Stability and Oscillations*, ed. by A. Noels, M. Gabriel. Non-radial oscillations. Proceedings of the 25th Liège International Astrophysical Colloquium, pp. 68–92 (1984)

P. Smeyers, Asymptotic representation of low- and intermediate-degree p-modes in stars. Astron. Astrophys. **407**, 643–653 (2003)

P. Smeyers, The second-order asymptotic representation of higher-order non-radial p-modes in stars revisited. Astron. Astrophys. **451**, 237–249 (2006)

P. Smeyers, P. Bruggen, Linear radial pulsations in stars in thermal imbalance. Astron. Astrophys. **141**, 297–303 (1984). Erratum. Astron. Astrophys. **144**, 516 (1985)

P. Smeyers, A. Moya, The asymptotic representation of higher-order g^+-modes in stars with a convective core. Astron. Astrophys. **465**, 509–524 (2007)

P. Smeyers, Tassoul, M.: Asymptotic approximations for higher order nonradial oscillations of a spherically symmetric star. Astrophys. J. Suppl. Ser. **65**, 429–449 (1987)

P. Smeyers, P. Veugelen, A. Weigert, Variational principles for the linear and adiabatic oscillations of a spherical star, and related definitions of canonical energy density and of canonical energy flux. Geophys. Astrophys. Fluid Dynam. **16**, 55–71 (1980)

P. Smeyers, R. Briers, M. Tassoul, K. Degryse, R. Polfliet, T. Van Hoolst, in *Seismology of the Sun & Sun-Like Stars*, ed. by E.J. Rolfe. Asymptotic approximations of non-radial oscillation modes of the Sun. (European Space Agency SP-286, Paris, 1988), pp. 623–627

P. Smeyers, I. De Boeck, T. Van Hoolst, L. Decock, Asymptotic representation of linear, isentropic g-modes of stars. Astron. Astrophys. **301**, 105–122 (1995)

P. Smeyers, T. Vansimpsen, I. De Boeck, T. Van Hoolst, Asymptotic representation of high-frequency, low-degree p-modes in stars and in the Sun. Astron. Astrophys. **307**, 105–120 (1996)

R.C. Smith, K.J. Fricke, On the bubble method of stability analysis. Mon. Not. R. Astron. Soc. **172**, 577–584 (1975)

D.E. Smylie, M.G. Rochester, Compressibility, core dynamics and the subseismic wave equation. Phys. Earth Planet. Inter. **24**, 308–319 (1981)

D.E. Smylie, A.M.K. Szeto, M.G. Rochester, The dynamics of the Earth's inner and outer cores. Rep. Prog. Phys. **47**, 855–906 (1984)

Y. Sobouti, The potentials for the g-, p-, and the toroidal-modes of self-gravitating fluids. Astron. Astrophys. **100**, 319–322 (1981)

A. Sommerfeld, *Mechanics of Deformable Bodies*. Lectures on Theoretical Physics, vol. 2. (Academic, New York, 1964a)

A. Sommerfeld, *Thermodynamics and Statistical Mechanics*. Lectures on Theoretical Physics, vol. 5. (Academic, New York, 1964b)

R.F. Stein, J. Leibacher, Waves in the solar atmosphere. Annu. Rev. Astron. Astrophys. **12**, 407–435 (1974)

T.E. Sterne, Modes of radial oscillation. Mon. Not. R. Astron. Soc. **97**, 582–593 (1937)

R. Stothers, J.A. Frogel, The linear star model and dynamical stability. Astrophys. J. **148**, 305–309 (1967)

G. Sutmann, Z.E. Musielak, P. Ulmschneider, Acoustic wave propagation in the solar atmosphere III. Analytic solutions for adiabatic excitations. Astron. Astrophys. **340**, 556–568 (1998)

M. Takata, Momentum conservation and mode classification of the dipolar oscillations of stars. Publ. Astron. Soc. Jpn. **57**, 375–389 (2005)

M. Takata, First integrals of adiabatic stellar oscillations. Publ. Astron. Soc. Jpn. **58**, 759–775 (2006a)

M. Takata, Analysis of adiabatic dipolar oscillations of stars. Publ. Astron. Soc. Jpn. **58**, 893–908 (2006b)

M. Takata, Private communication (2009)

J.-L. Tassoul, Sur l'instabilité convective d'une masse gazeuse inhomogène. Ann. d'Astrophys. **30**, 363–370 (1967)

J.-L. Tassoul, *Theory of Rotating Stars*. (The University Press, Princeton, 1978)

M. Tassoul, Asymptotic approximations for stellar nonradial pulsations. Astrophys. J. Suppl. Ser. **43**, 469–490 (1980)

M. Tassoul, Second-order asymptotic approximations for stellar nonradial acoustic modes. Astrophys. J. **358**, 313–327 (1990)

M. Tassoul, J.-L. Tassoul, Asymptotic approximations for stellar pulsations. Astrophys. J. **153**, 127–134 (1968)

A.H. Taub, Stability of general relativistic gaseous masses and variational principles. Commun. Math. Phys. **15**, 235–254 (1969)

P.A. Thompson, *Compressible-Fluid Dynamics*. (McGraw-Hill, New York, 1972)

W. Thomson, Oscillations of a liquid sphere, Sects. 55–58 in Dynamical problems regarding elastic spheroidal shells and spheroids of incompressible liquid. Philos. Trans. R. Soc. Lond. **153**, 583–619 (1863)

R.C. Tolman, On the stability of spheres of simple mechanical fluid held together by Newtonian gravitation. Astrophys. J. **90**, 541–567 (1939)

I. Tolstoy, The theory of waves in stratified fluids including the effects of gravity and rotation. Rev. Mod. Phys. **35**, 207–230 (1963)

I. Tolstoy, *Wave Propagation*. (McGraw-Hill, New York, 1973)

R.K. Ulrich, The five-minute oscillations on the solar surface. Astrophys. J. **162**, 993–1002 (1970)

R.K. Ulrich, in *Solar Seismology from Space*, ed. by R.K. Ulrich, J. Harvey, E.J. Rhodes Jr., J. Toomre. Conference summary. A Conference at Snowmass, Colorado, pp. 3–5. JPL Publication 84–84 (1984)

W. Unno, On the stability of the solar core. Publ. Astron. Soc. Jpn. **27**, 81–99 (1975)

W. Unno, Y. Osaki, H. Ando, H. Saio, H. Shibahashi, *Nonradial Oscillations of Stars*. (The University Press, Tokyo, 1989)

V. Väisälä, Über die Wirkung der Windschwankungen auf die Pilotbeobachtungen. Societas Scientiarum Fennica. Commentationes Physico-Mathematicae II. **19**, 2–46 (1925)

Yu.V. Vandakurov, Astronomicheskii Zhurnal **44**, 786–797 (1967). English translation: The frequency distribution of stellar oscillations. Soviet Astronomy **11**, 630–638 (1968)

R. Van der Borght, Wan Fook Sun: Non-radial adiabatic oscillations of very massive stars. Acad. R. Belg. Bull. Cl. Sci. 5e sér. **51**, 978–985 (1965)

M. Van Dyke, *Perturbation Methods in Fluid Mechanics*. (Academic, New York, 1964)

T. Van Hoolst, Nonlinear, Nonradial Oscillations of Stars. Doctoral Dissertation, Katholieke Universiteit Leuven (1992)

T. Van Hoolst, Nonlinear, nonradial, isentropic oscillations of stars: Hamiltonian formalism. Astron. Astrophys. **286**, 879–889 (1994)

T. Van Hoolst, P. Smeyers, The quantities $D_{n,\ell}$ as measures of small frequency separations in the Sun and their origin. Astron. Astrophys. **248**, 647–655 (1991)

T. Van Hoolst, W.A. Dziembowski, S.D. Kawaler, Unstable non-radial modes in radial pulsators: theory and an example. Mon. Not. R. Astron. Soc. **297**, 536–544 (1998)

R. von Mises, K.O. Friedrichs, *Fluid Dynamics, Brown University, Summer Session for Advanced Instruction and Research in Mechanics, 1941*. (Providence, Rhode Island, 1942)

R. von Mises, K.O. Friedrichs, *Fluid Dynamics*. (Springer, New York, 1971)

J. von Neumann, E. Wigner, Über das Verhalten von Eigenwerten bei adiabatischen Prozessen. Phys. Z. **30**, 467–470 (1929)

S.V. Vorontsov, Astronomicheskii Zhurnal **68**, 808–824 (1991). English translation: Asymptotic theory of acoustic oscillations of the sun and stars. Soviet Astronomy **35**, 400–408 (1991)

S.V. Vorontsov, V.N. Zharkov, Helioseismology: theory and interpretation of experimental data. Sov. Sci. Rev. E, Astrophys. Space Phys. Rev. **7**, 1–103 (1989)

C.L. Waelkens, in *Stellar Pulsation*, ed. by A.N. Cox, W.M. Sparks, S.G. Starrfield. Periodic line profile and photometric variations in mid-B stars. Lecture Notes in Physics, vol. 274. (Springer, Berlin, 1987), pp. 75–78

C. Waelkens, Slowly pulsating B stars. Astron. Astrophys. **246**, 453–468 (1991)

B. Warner, E.L. Robinson, Non-radial pulsations in white dwarf stars. Nat. Phys. Sci. **239**, 2–7 (1972)

H. Weyl, Über gewöhnliche Differentialgleichungen mit Singularitäten und die zugehörigen Entwicklungen willkürlicher Funktionen. Mathematische Annalen **68**, 220–269 (1910a)

H. Weyl, Ueber gewöhnliche lineare Differentialgleichungen mit singulären Stellen und ihre Eigenfunktionen (2. Note.). Nachrichten von der Gesellschaft der Wissenschaften zu Göttingen, Mathematisch-Physikalische Klasse, Band **1910**, 442–467 (1910b)

H. Weyl, Ramifications, old and new, of the eigenvalue problem. Am. Math. Soc. **56**, 115–139 (1950)

W.A. Whitaker, Heating of the solar corona by gravity waves. Astrophys. J. **137**, 914–930 (1963)

B. Willems, T. Van Hoolst, P. Smeyers, Asymptotic representation of nonradial g^+-modes in stars with a convective core. Astron. Astrophys. **318**, 99–107 (1997)

C.L. Wolff, The five-minute oscillations as nonradial pulsations of the entire Sun. Astrophys. J. **177**, L87–L92 (1972)

C.L. Wolff, Some simple properties of stellar pulsation modes. Astrophys. J. **227**, 943–954 (1979)

L. Woltjer, The stability of force-free magnetic fields. Astrophys. J. **128**, 384–391 (1958)

J.H. Woodhouse, Asymptotic results for elastodynamic propagator matrices in plane-stratified and spherically-stratified earth models. Geophys. J. Int. **54**, 263–280 (1978)

K. Wright, Chebyshev collocation methods for ordinary differential equations. Comput. J. **6**, 358–365 (1964)

J. Zalewski, Strongly nonadiabatic pulsations and strange modes. Publ. Astron. Soc. Jpn. **44**, 27–43 (1992)

S.A. Zhevakin, Physical basis of the pulsation theory of variable stars. Annu. Rev. Astron. Astrophys. **1**, 367–400 (1963)

Index

Printed by Printforce, the Netherlands